Two Manuals Combined into One Book

AF882181

Military Police

And

Tactical Employment of Nonlethal Weapons

Department of Defense

Pentagon
Publishing

FM 3-19.4 (FM 19-4)

Field Manual
No. 3-19.4

Headquarters
Department of the Army
Washington, DC, 4 March 2002

Military Police Leaders' Handbook

Contents

Distribution Restriction: Approved for public release; distribution is unlimited.

*This publication supersedes FM 19-4, 7 May 1993.

Page

Page

Preface

This field manual (FM) addresses military police (MP) maneuver and mobility support (MMS), area security (AS), internment and resettlement (I/R), law and order (L&O), and police intelligence operations (PIO) across the full spectrum of Army operations. Although this manual includes a discussion of corps and division MP elements, it primarily focuses on the principles of platoon operations and the tactics, techniques, and procedures (TTP) the platoon uses to accomplish its mission.

This FM provides the capabilities and organization of the MP, demonstrates the flexibility and diversity of MP in adapting to any mission throughout the full spectrum of Army operations, and characterizes the MP as a combat-force multiplier. Additionally, this manual identifies the fact that the Army will not conduct operations alone and defines the role of the MP in support of joint, multinational, and interagency operations.

The MP TTP are organized by the MP functions of MMS, AS, I/R, LO, and PIO with supporting tasks, both individual and collective, to help illustrate the functions.

NOTE: United States (US) policy regarding the use and employment of antipersonnel land mines (APLs) outlined in this FM is subject to the convention on certain conventional weapons and executive orders (EOs). Current US policy limits the use of non-self-destructing APLs to (1) defending the US and its allies from armed aggression across the Korean demilitarized zone and (2) training personnel engaged in demining and countermine operations. The use of the M18A1 claymore in the command-detonation mode is not restricted under international law or EO.

Appendix A complies with current Army directives which state that the metric system will be incorporated into all new publications. *Appendix B* deals with media relations.

The proponent of this publication is Headquarters (HQ) United States Army Training and Doctrine Command (TRADOC). Send comments and recommendations on Department of the Army (DA) Form 2028 directly to Commandant, US Army Military Police School, ATTN: ATSJ-MP-TD, 401 MANSCEN Loop, Suite 2060, Fort Leonard Wood, Missouri 65473-8926.

Unless this publication states otherwise, masculine nouns and pronouns do not refer exclusively to men.

Chapter 1

Military Police Overview

This chapter provides information about how MP are organized, equipped, and trained to provide combat support (CS) across the full spectrum of Army operations.

INTRODUCTION

1-1. As a flexible economy-of-force organization, MP provide a wide range of diverse support because of their agility and versatility to adapt to any mission or environment. As a combat multiplier, they support the maneuver commander through the five MP functions. MP contribute to the commander's overall combat power by integrating efforts with those of other combat, CS, and combat service support (CSS) elements.

1-2. In addition to single-service operations, MP also support joint, multinational, and interagency activities. MP support air base defense in concert with Air Force Security Forces, operate joint and multinational checkpoints, conduct combined police patrols, and exchange police information and criminal intelligence with the host nation (HN), military, and civilian police agencies.

1-3. MP have the capability to expedite the movement of combat resources, provide critical asset security and protection, conduct I/R, contribute to force protection efforts through L&O operations, and gather and disseminate police information and intelligence.

MILITARY POLICE FUNCTIONAL AREAS

1-4. With the old battlefield missions, the term "operations" was used extensively and carried too broad of a meaning. To clarify the specific tasks of the MP, the battlefield missions have been redefined into the following five functional areas:

- MMS
- AS
- I/R
- L&O
- PIO

1-5. Each of these MP functions have task areas and tasks that support them. MP functions are the broadest areas for which tasks are placed. Some of these tasks will require groupings that might not be related to the entire function. Therefore, task areas were created to group specific tasks. Specific tasks consist of two types—collective and individual. Individual tasks are further divided into leader and soldier tasks *(Figure 1-1)*. The collective and individual tasks that support the MP task areas are found in the MP mission training plans (MTP) and MP soldier's manuals (SMs).

1-6. MP procedures are the lowest level of detail. They explain the "how to" at the task level. Procedures include the standing methods used by a unit to accomplish tasks, weapon and equipment operating steps, crew drills, and staff action and coordination. They are the building blocks of individual and collective task accomplishment and serve as the foundation of tactics and techniques. Procedures are explained in the unit standing operating procedures (SOPs), MTPs, SMs, and similar publications.

MANEUVER AND MOBILITY SUPPORT

1-7. The MMS function involves the measures necessary to enhance combat movement and the ability to conduct

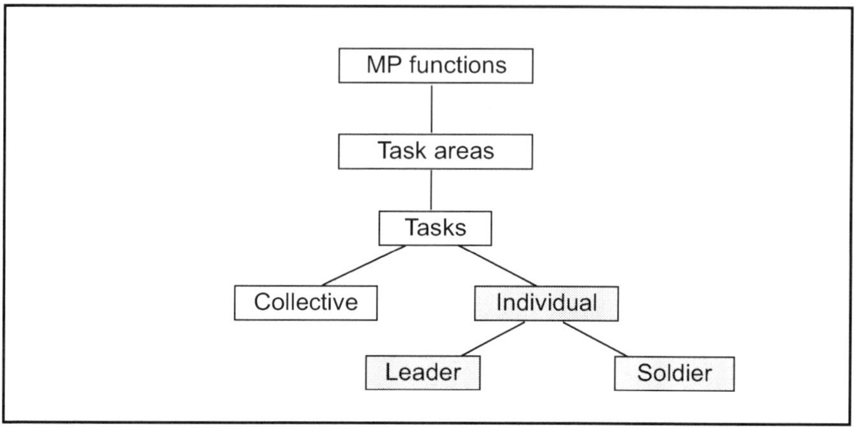

Figure 1-1. Relationship Between MP Functions and Tasks

movement of friendly resources in all environments. MP conduct MMS to ensure that the commanders receive personnel, equipment, and supplies when and where they are needed. The task areas that support the function of MMS include—

- MP support for river crossings, breaching, and passage-of-line operations.
- Straggler and dislocated civilian control.
- Route reconnaissance and surveillance.
- Main supply route (MSR) regulation enforcement.

1-8. The security and viability of the operational and tactical lines of communications (LOC) will be critical to continuous sustainment and recovery operations. MP ensure that logistics and supply operations are kept on time and arrive at the right place. Refer to *Chapter 5* for more information about MMS.

AREA SECURITY

1-9. The AS function consists of those security measures designed to give commanders freedom of maneuver and flexibility to conduct operations. The task areas that support AS include—

- Reconnaissance operations.
- Area damage control (ADC).
- Base and air base defense.
- Response force and tactical combat force (TCF) operations.
- Critical site, asset, and high-risk personnel security.
- Force protection and physical security.
- Antiterrorism.

1-10. MP performing AS contribute to securing and protecting the force and preserving combat power. Refer to *Chapter 6* for more information about AS.

INTERNMENT AND RESETTLEMENT

1-11. The I/R function consists of those measures necessary to provide shelter, sustain, guard, protect, and account for people (enemy prisoners of war [EPWs] and civilian internees [CIs], US military prisoners, and dislocated civilians [DC]). The task areas that support I/R include—

- EPW and CI handling.
- Populace and resource control.
- US military prisoner confinement.
- DCs control.

1-12. The international community, media, and public perceptions have increased sensitivity to the protection of human rights and the need for absolute accountability of interned, detained personnel, and refugees in military operations. Refer to *Chapter 7* for more information about I/R.

LAW AND ORDER

1-13. Task areas and tasks that minimize the effects of a criminal threat on friendly forces support the L&O function. MP conduct L&O to remove the conditions and opportunities that promote crime, thereby preventing diversion of military resources and maintaining military discipline. The task areas include—

- Law enforcement.
- Criminal investigations.
- US customs operations.
- Related L&O training.

1-14. Whether patrolling an installation's housing area, conducting counterdrug operations, enhancing security, or investigating war crimes, MP L&O capabilities are invaluable to the commander. Refer to *Chapter 8* for more information about L&O.

POLICE INTELLIGENCE OPERATIONS

1-15. The PIO function supports, enhances, and contributes to the commander's protection program, situational awareness, and battlefield visualization by portraying relevant threat information that may affect the operational and tactical environment. The task areas that support PIO include—

- Intelligence preparation of the battlefield (IPB).
- Active and passive roles.
- Police assessment process.

1-16. Whether in support of peacetime installation L&O or detecting threat forces in the battle space, the MP force employment provides the commander with substantial information and an intelligence source, particularly where the criminal element is the same as or closely aligned with the opposing forces (OPFOR) and government. Refer to *Chapter 9* for more information about PIO.

THREAT

1-17. Today, friendly forces encounter a broad range of traditional and nontraditional threats. No single dominating threat will be the undisputed focus of US security policy. Although overt attacks on the US and its strategic interests may be less common, stability operations and support operations will likely increase. Economic development and demographics, as well as the progression of social and cultural movements, will encompass an array of threat forces including nonnation forces (ethnic conflicts and terrorist activities) which challenge traditional nation and state environments. Additionally, nation and state forces (internal security forces and infantry-based and armor-mechanized-based armies) continue to present a global threat. These forces possess varying levels of military and advanced technology capabilities.

1-18. In recent military operations, a nontraditional criminal threat has emerged. The evolving criminal threat operates most often in the rear area, near ports, in built-up areas, and where troop populations are high. This threat is most likely to be detected at border crossings trying to disrupt the relocation efforts of DCs. They may commit crimes against particular ethnic groups or be at checkpoints and roadblocks trying to position weapons, explosives, or personnel in sustainment areas in order to disrupt military operations or kill friendly forces. Such a threat requires commanders to minimize its negative impact on friendly forces, resources, and operations. The MP continue to respond to nonmilitary threats including famine, health epidemics, illegal immigration, illegal drug traffic, and population dislocation.

MILITARY POLICE PLATOON ORGANIZATION AND LEADERSHIP

1-19. There are two basic MP platoon organizations, corps and division. Corps MP platoons are organized and equipped basically the same. Each division MP platoon supporting a different kind of division (such as heavy, light, airborne, or air assault) is designed under a different table(s) of organization and equipment (TOE).

CORPS MILITARY POLICE

1-20. Depending on the nature of the operation, corps MP are usually among the first forces deployed to support military operations around the world. They deploy early to areas devastated by natural or man-made disasters to assist disaster relief and damage assessment efforts. They provide security and force protection to friendly forces, critical facilities, and resources as units organize for military operations. In a developing theater, corps MP concentrate mission support to the main effort. Units whose assistance to the main effort is vital normally receive the highest priority for protection. Key facilities, such as traffic choke points, critical tunnels and bridges, and ammunition and fuel storage points may require special protection. As the theater matures, the focus may quickly change to other functions as MP adjust priorities to accommodate the change.

DIVISION MILITARY POLICE

1-21. Division MP are organized somewhat different depending on the type of division they are supporting. For example, a heavy division has one MP platoon providing direct support (DS) to each maneuver brigade and two MP platoons providing general support (GS) to the division's rear. Both airborne and air assault divisions have four MP platoons providing GS. A light infantry division is supported by three MP platoons.

1-22. In heavy divisions, where highly mobile forces are designed to move quickly over open ground, the overall need for MMS is significant. Division MP are likely to focus on expediting the forward movement of the critical combat resources into the division area. However, the priority could change quickly to removing EPWs from forward areas to freeing maneuver forces from guarding and caring for captives.

1-23. In airborne and air assault divisions, priority of MP support is most often needed for EPW operations and then for MMS to speed the movement of CS vehicles within the airhead.

1-24. For MP supporting any division, certain employment considerations remain constant. MP provide dedicated security for assets deemed critical by the division commander. This includes the division's main command post (CP) where MP operate outside the CP perimeter conducting screening missions designed to detect, disrupt, and delay enemy forces from disrupting the division's primary CP. Another consideration is MP accepting EPWs from capturing troops as far forward as possible.

SEPARATE BRIGADES

1-25. MP support to a separate brigade is normally provided by a four-squad MP platoon. The platoon is assigned to the brigade headquarters and headquarters company (HHC). The brigade has a separate provost marshal (PM) cell that serves as the command and control (C^2) element for the platoon. The brigade PM cell has operational control (OPCON) of all MP assets the same way the division PM has OPCON of the division MP assets. The brigade HHC provides sustainment support for both the PM cell and the MP platoon. The PM advises the separate brigade commander on matters pertaining to MP operations. The platoon leader directs the execution of the platoon's missions based on the priorities set forth by the PM and the supported commander.

1-26. The platoon can perform any of the five functions. The platoon leader may task organize the squads according to mission, enemy, troops, terrain, time available, and civilian considerations (METT-TC), and provide one squad to operate the EPW collecting point, one squad to provide a mobile security screen and occupy observation posts (OPs) around the brigade's CP, and two squads to conduct MMS and AS throughout the brigade's rear area.

INITIAL BRIGADE COMBAT TEAMS (IBCTs)

1-27. MP support to an IBCT may differ from that of other separate brigades. The IBCT is a preconfigured, ready-to-fight, combined-arms package. It is designed and optimized primarily for employment in small-scale contingencies operations (SSCO) in complex and urban terrain, confronting low-end and mid-range threats. The IBCT participates in war, with augmentation, as a subordinate maneuver component within a division or corps, in a variety of possible roles. It also participates in stability and/or support operations as an initial entry force. Civil unrest or complete turmoil normally characterizes these environments.

1-28. Organically, MP support to the IBCT is a two-person PM planning cell. The PM planning cell is located with the HHC's maneuver support cell and under the direct supervision of the brigade Operations and Training Officer (US Army) (S3). The role of the PM cell is significantly different from that of a traditional division PM or separate brigade PM. The main difference is the lack of organic or habitual MP assets in the IBCT. The absence of organic MP assets makes the job of the PM much more critical. The PM must—

- Understand the organization, capabilities, and limitations of the IBCT.

- Conduct effective liaison with higher HQ PM elements.
- Become an effective planner and anticipator of MP requirements.
- Task organize MP units effectively and efficiently.
- Assume C^2 of incoming MP forces or, if operating under a division, relinquish C^2 to a division or corps PM (if appropriate for effective employment of MP forces).

1-29. Depending on METT-TC, the IBCT may be augmented by MP elements ranging from a platoon to a battalion. Once the brigade receives MP augmentation, the PM then becomes a staff planner and coordinator for all MP activities.

1-30. Since the IBCT is a divisional brigade, the division PM and the IBCT PM play an important role in developing an optimum MP force package to support the brigade commander's concept of the operation. Despite the brigade's early time lines, the PM must consider and plan for MP augmentation forces as early as possible to free up valuable combat resources. SSCO that result in numerous EPWs, CIs, and refugees will hamper the maneuver force's freedom of movement.

1-31. Initially, MP priority of effort during the offense may be providing MMS for ground combat, CS, and CSS forces and taking control of EPWs and CIs. During the defense, the priority of MP support may shift to conducting AS and counterreconnaissance along the LOC, C^2 centers, and CSS bases. MP may also be required to conduct response force operations or become part of the TCF.

1-32. During stability and support operations, MP support may include—

- Order restoration.
- Crowd control.
- AS or force protection.
- DC operations.
- Noncombatant-evacuation operations.

PLATOON LEADER

1-33. The MP platoon leader is responsible to the company commander for the platoon's combat readiness, training, and discipline and the maintenance of its equipment. To be successful, the platoon leader must demonstrate an ability to lead soldiers and manage an organization, material, and time. He must be able to articulate the capabilities and limitations of the platoon to various non-MP staff sections. In combat, the platoon leader is responsible for accomplishing all the missions assigned to the platoon according to the commander's intent and preserving the platoon's fighting capability.

PLATOON SERGEANT (PSG)

1-34. The PSG leads elements of the platoon as directed by the platoon leader and assumes command of the platoon in the absence of the platoon leader. He directs the day-to-day activities of the platoon and ensures that the platoon has individual and team training and logistics needed to accomplish its mission. During tactical operations, he may assist in the control of the platoon.

TEAM AND SQUAD LEADER

1-35. The MP team leader is responsible to the squad leader for individual and team training and team discipline. He is responsible for the tactical employment and control of the team and the maintenance and operation of all vehicles and equipment organic to the team. During combat operations or anytime there is a threat, the team leader quickly assesses the situation, reports to his superiors, and takes appropriate action to protect the team according to the rules of engagement (ROE). A squad leader has the same responsibility for the squad as the team leader has for the team.

FORCE PROTECTION (FP) MEASURES

1-36. MP leaders at all levels must examine FP requirements and integrate FP measures throughout all the operations. Once higher HQ has established local FP policies, leaders set the example by complying with them. Leaders reduce the soldiers' exposure to hazards by strictly enforcing all the protective postures that may include—

- Traveling with at least two vehicles armed with at least one automatic weapon.
- Hardening of the vehicles.
- Wearing Kevlar® and body armor.
- Not driving off the road or cross-country.
- Placing off-post facilities off limits during nonduty hours.

1-37. MP leaders analyze and compensate for other threats such as disease, weather, crime, complacency, terrorism, morale, safety, and other considerations.

1-38. At the operational level, team and squad leaders establish a safety zone around their teams. The safety zone is the immediate area around the team where threat forces or events could harm the team or inflict casualties. In open terrain, the safety zone may extend out to the maximum effective range of the team's organic-weapon systems. However, when searching vehicles at a checkpoint or conducting crowd control, the safety zone may only be an arm's length from the team.

1-39. Team and squad leaders remain alert to threats that enter the team's safety zone. They must quickly assess any threat to the team and take appropriate action within the ROE to reduce the threat or move the team.

1-40. When required to operate in crowds, maintain eye contact with team members. Establish and maintain a safe distance between the team and the crowd. Never allow the team to become separated or surrounded.

1-41. When patrolling in built-up areas, the gunner scans the upper floors of the buildings and the streets to the vehicle's front, rear, and flanks and immediately reports any suspicious activity. The driver concentrates on the area directly in front of the vehicle looking for any unexploded munitions, scatterable mines, or other road hazards. All team members should stay awake, alert, and ready to react to danger.

MILITARY POLICE PLATOON MISSION, CAPABILITIES, AND LIMITATIONS

1-42. The platoon has one critical wartime mission which is to provide MP CS to an assigned area of operations (AO). MP CS consists of all five MP functions. The platoon performs its missions primarily mounted, taking full advantage of the high mobility multipurpose wheeled vehicle's (HMMWV's) versatility and the added protection and firepower of the armor security vehicle (ASV).

CAPABILITIES

1-43. The MP platoon is capable of operating day or night, in various terrain conditions, and under all weather and visibility conditions. Their mode of operation is possible through the deployment and employment of the three-person team throughout the battlefield. However, it is dependent on its parent unit for sustainment support. The platoon has self-protection capabilities such as nuclear, biological, chemical (NBC) detection equipment and a Platoon Early-Warning System (PEWS). The platoon's radio transmission range is increased with an OE-254 antenna. Because of extensive police training and law enforcement missions, the MP are highly skilled in the use of force and the employment of lethal and nonlethal technologies, information-collecting and dissemination, observation and surveillance, and crowd control. The MP platoon has a tremendous combat and noncombat information-collecting capability. This capability is the result of extensive area, zone, and route reconnaissance; daily contact with local nationals; conducting combined police patrols with HN military and civilian police agencies; and conducting field interviews. An MP platoon is capable of covering 500 square kilometers in rolling terrain; however, more severe terrain such as mountains, METT-TC, and mission objectives will affect this capability. For example, consider one mobile MP team per 10 kilometers of route coverage. For area coverage, begin with an estimate of one mobile MP team per 55 square kilometers.

1-44. Unlike most combat arms platoons, which maneuver together in formation, the MP platoon most often operates independently and dispersed over a large area. The platoon conducts combat operations, when required, through the employment of mobile combat systems containing three-man teams operating independently or in concert, and having vehicle crew-served and individual weapons capable of defeating a Level II threat and defending a position against dismounted threats.

1-45. Based on METT-TC, the platoon leader may task organize the platoon for certain missions. Normally, MP are employed as squads; however, individual teams may execute many MP tasks.

LIMITATIONS

1-46. During combat operations, the platoon is not organized and equipped to fight for extended periods unless it is augmented with indirect fire or close air support (CAS). Although the MP team is a lethal and highly mobile platform, it is not structured or equipped for prolonged autonomous missions. Leaders must use the MP team as a task organizational building block and avoid over tasking based solely on the number of teams available. The platoon has limited antiarmor capability and normally uses antiarmor weapons for self-protection and to break contact.

PEACETIME TRAINING

1-47. MP units train as they will fight. Peacetime training must replicate battlefield conditions and conform to Army doctrine. Leaders and soldiers must understand standardized doctrinal principles found in applicable manuals to ensure that training is conducted to standard. The following manuals provide the basic foundation for Army training:

- FMs.
- Training circulars (TCs).
- MTPs.
- Drill books.
- SMs.
- Army regulations (ARs)

1-48. *FM 25-100*, *FM 25-101*, and *TC 25-10* provide MP leaders with established training doctrine and assist the leaders in the development and execution of the training programs. These manuals introduce the concept of lane training and define it as a technique for training company, team, and smaller units on a series of selected soldier, leader, and collective tasks using specific terrain.

1-49. Lane training uses multiechelon techniques to maximize the efficient use of limited terrain and control conditions for formal or informal evaluations. Lane training is externally supported, resourced, and evaluated. It enables similar units to simultaneously or sequentially train on mission-related scenarios. Lane training is resource intensive, so commanders must maximize its benefit. Commanders narrow the focus and select only the most critical mission-essential task lists (METLs) items or collective tasks for training. Lane training is especially valuable for conducting specific METL tasks, situational training exercises (STXs), and other training events. It is often associated with training requiring movement over terrain; for example, movement to contact or conducting a route reconnaissance. Lane training can be modified to achieve benefits in L&O scenarios, such as special-reaction team (SRT) incidents, patrol incidents, traffic accidents, and so forth.

1-50. The lane training doctrine outlined in *FM 25-101* and *TC 25-10* can be tailored for small MP units by using the training execution model (TEM). The TEM follows the Army doctrine and training philosophy of hands-on METL training as outlined by *FMs 25-100* and *25-101*. Before the TEM can be implemented, the concept of the operation must be approved, evaluated, and directed from two levels up. For example, a squad leader must receive approval through the chain of command from his company commander to execute the training event; a platoon leader gets approval from the battalion commander and so forth.

1-51. The TEM incorporates the combined-arms training methodology and adjusts it to meet the MP training requirements. The TEM focuses the unit on the time available during the training cycle to train the most critical collective and individual tasks. The TEM consists of an eight-step training methodology that is based on leader certification of the lane expert and an observer/ controller (OC) as well as subordinate unit leaders. For more information about TEM refer to *Appendix C*.

Chapter 2

Battle Command

This chapter provides the techniques and procedures used by MP leaders at company and platoon level to C^2 their organizations.

OVERVIEW

2-1. Battle command is the art of battle decision making, leading, and motivating soldiers and organizations into action to achieve victory with the least cost to the organization. Commanders must visualize the current and future state of both friendly and enemy forces. The commander positions himself where he can guide and motivate the soldiers and influence the outcome of the missions.

2-2. The company commander is responsible for all that the unit does or fails to do. He cannot delegate this responsibility, and the final decision and responsibility rest with him. He discharges his responsibility through an established chain of command and holds each subordinate leader responsible for the actions of the platoon or the section.

2-3. The commander must be proficient in the tactical employment of the unit. He must know the capabilities and limitations of the soldiers and the equipment. A commander does this through a continuous cycle of planning, executing, and assessing training. Through this training, the commander gets to know the soldiers.

2-4. MP commanders prioritize, assign missions, and allocate resources where they can best support the higher echelon commander's intent. The company commander makes most of the tactical decisions. Technological advances in today's operational environments have reduced the time available for decision making while increasing the possibilities that must be considered.

2-5. Thorough and sound operational planning is the key to successful combat and CS operations. Commanders must identify the opportunities and anticipate and avoid problems. They must analyze their options before making the decisions on which subordinate leaders will base their actions. Commanders balance competing risks and then identify and develop the best course of action (COA).

MILITARY DECISION-MAKING PROCESS (MDMP)

2-6. The MDMP is a single, established, and proven analytical process used at all the echelons of the commands. This is a seven-step process used when adequate planning time and enough staff support are available *(Table 2-1)*. This process is a detailed, deliberate, sequential, and time-consuming process that helps the commander and his staff examine a battlefield situation and reach logical decisions. The commander uses the entire staff during the process to explore the full range of probable and likely enemy and friendly COAs and to analyze and compare his own organization's capabilities with the enemy's.

Table 2-1. MDMP

Step	Action	Step	Action
1	Receipt of the mission	5	COA comparison
2	Mission analysis	6	COA approval
3	COA development	7	Orders production
4	COA analysis		

2-7. At company level, the commander normally uses the MDMP in a time-constrained environment without enough staff. A unit can shorten the process if it fully understands the role of each step of the process and the requirements to produce the necessary products. The application of the MDMP at company level and below is called the troop-leading procedures (TLP). *Figure 2-1, page 2-4* shows the relationship between MDMP and TLP.

2-8. MP commanders plan successful operations by anticipating possible future events and planning contingencies. MP leaders enhance both planning and execution of the operations when they—

- Use the military planning and decision-making process.
- Develop short- and long-range goals.
- Identify goals and objectives with a recognizable end-state.
- Coordinate goals and actions internally and externally.
- Base their plans on objective planning factors.
- Review their plans, continuously, in light of the METT-TC and updated information.
- Assign responsibilities and express expectations.
- Identify the options that may develop during an operation.
- Stand ready to accommodate the changes.

MILITARY PLANNING

2-9. Commanders select and carry out the developed COAs using military planning. Military planning guidelines include—

- Forecasting requirements by analyzing and evaluating facts and trends to predict what may occur.

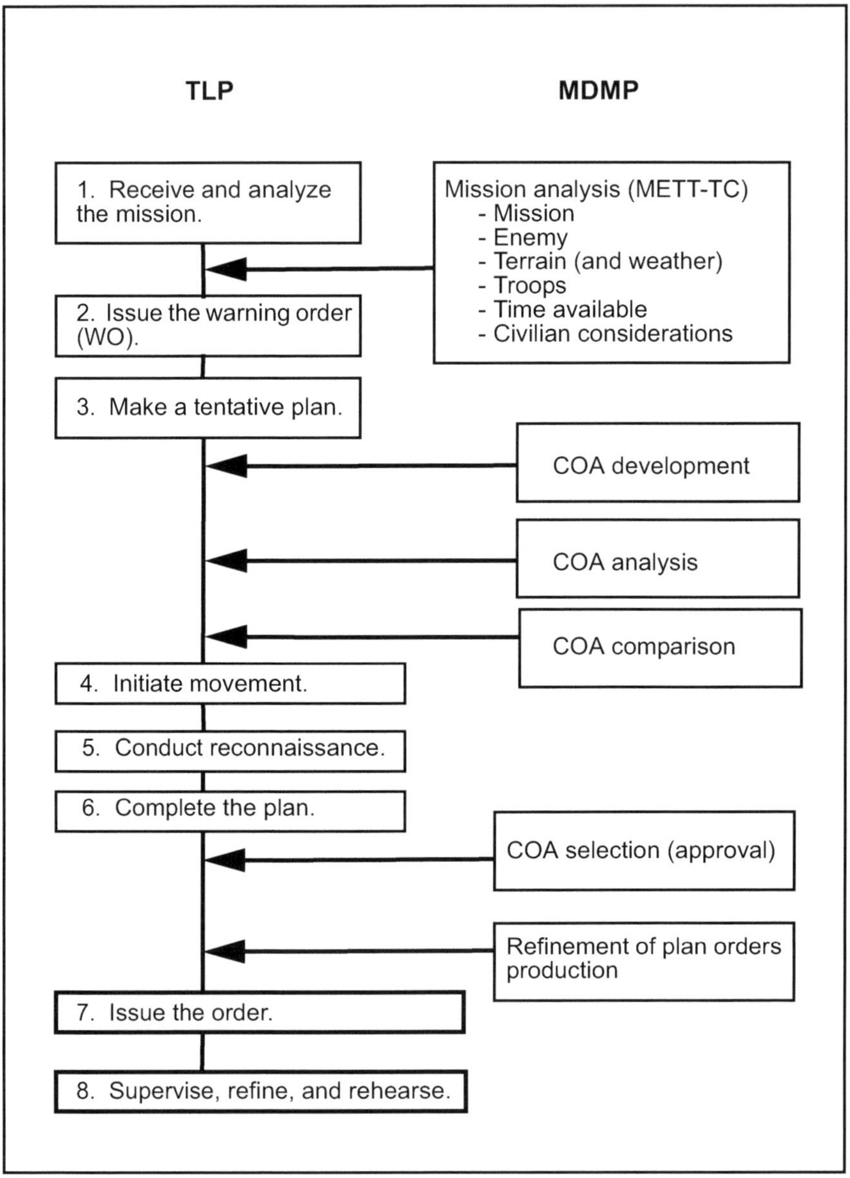

Figure 2-1. Relationship Between TLP and the MDMP

- Examining probable requirements and establishing priorities for further preparation.
- Studying implications and interrelationships of probable requirements.
- Analyzing the mission to determine tasks, their complexity, and their relative importance.
- Establishing guidance for further planning that will help keep all the elements focused on the commander's intent.
- Preparing studies and estimates to help formulate the COA and assess its feasibility.
- Selecting the COA, identifying the best course, and retaining other feasible courses for use in contingencies as alternate plans.
- Preparing the plan in detail and conducting rehearsals when time, resources, and security permit.

2-10. Use the following military planning guidelines to answer the three key questions of operational planning:

- What military condition must be produced to achieve the goal?
- What sequence of actions is most likely to produce that condition?
- How should resources be applied to accomplish that sequence of actions?

2-11. If the plan is not implemented immediately, retain it for later use. As events occur or new information becomes available, review and revise the plan accordingly.

ANALYSIS, FORECASTING, AND RISKS

2-12. Conducting a mission analysis is crucial to planning. The process begins by gathering facts and ascertaining current conditions, such as the—

- Higher-level mission and the commander's intent (one and two levels up).

- Current task organization (two levels down).
- Current unit status (locations, operation capabilities, and activities).
- Logistics situation (refer to *FM 101-5* for the logistic estimate format).

2-13. When facts are not available, the commander will need to develop assumptions. Assumptions must substitute for fact where information is not known. Keep in mind that as time passes between the receipt of a mission and the execution of a plan, facts are increasingly likely to have changed. Develop sound assumptions that can be used in place of facts.

2-14. Analyzing the higher-level mission and the commander's intent will help identify what tasks are required to accomplish the mission. As the mission is analyzed, identify both the *specified* and the *implied* tasks to be performed. *Specified tasks* are those stated in higher HQ orders and plans. *Implied tasks* (like crossing a river or passing through the lines of a unit lying between you and the objective) are not so stated, but must be accomplished to satisfy the overall operation. From among the specified and implied tasks, essential tasks that are crucial to the mission's success must be identified.

2-15. Integral to mission planning is the analysis of mission requirements in terms of time, space, and personnel. If MP are to balance the benefits of detailed planning against the need for immediate action, they must—

- Determine how much time there will be between receiving the mission and the deadline for having completed it.
- Know how long it will take to obtain and process information, make decisions, and issue orders.

- Know how long it will take subordinates to execute the orders, complete the mission, or carry out the operation.

2-16. Because each unit involved in an operation performs its planning based on the plans of the next higher level, allocation of adequate time for subordinate units to plan is a consideration at each level. Publishing SOPs reduces the number of details to be explained. It also promotes understanding and teamwork among commanders, staff, and troops.

2-17. When doing reverse planning, consider the classic allotment of one-third time for planning and two-thirds time for execution *(Table 2-2)*. Move backward from the time of execution to—

- Allocate time to accomplish each phase of an operation.
- Determine how much time is needed to rehearse.
- Determine how much time is needed for developing the plan.

Table 2-2. Reverse Planning Timetable

Time	Action
2230	Execute the mission.
2130-2215	Conduct inspection(s).
1845-2130	Conduct rehearsal(s).
1815-1845	Issue an operation order (OPORD).
1745-1815	Complete the plan.
1715-1745	Conduct the leaders' reconnaissance.
1630-1715	Issue a tentative plan.
1630	Issue a WO.
1600	Receive the mission.

2-18. When using terrain analysis, consider the layout of the battlefield. Appreciate the opportunities and limitations of the major terrain features, transportation networks, and built-up areas. Fit the operational concept and planning to that environment. Use the IPB to evaluate the area in terms of the military aspects of the terrain. Consider how to exploit the opportunities afforded by weather while minimizing its adverse effects.

2-19. Use current information on the threat to identify known enemy activities and threat capabilities that could affect this and future operations. Attempt to anticipate the enemy's objectives and intentions.

2-20. Consider available assets and determine acceptable levels of risk. At every echelon, MP disperse their assets and prioritize operations to meet the echelon commander's needs within the limits of the resources at hand. MP leaders must concentrate their efforts on key locations and accept risks elsewhere. When possible, recognize and moderate such risks in the choice of operations and in the contingency planning. The five steps to identify, analyze, and reduce risks are listed in *FM 100-14*.

COURSE OF ACTION IDENTIFICATION, DEVELOPMENT, AND SELECTION

2-21. The commander's selected COA will become the actual plan for accomplishing the mission. To ensure the best plan possible, identify several possible COAs, each significantly different from the others. In identifying COAs, do not overlook a check of the "basics" that include the—

- Commander's intent.
- Essential tasks.
- Effective use of C^2.
- Principles of war.
- Ethical considerations.
- Relative force ratio.

2-22. Consider preparing a COA statement (and sketch, if appropriate) for each option. The COA statement is the "how" of the operation. Ensure that it includes the following five elements:

- What – the type of action.
- When – the time the action will begin.
- Where – the assigned area.
- How – the use of available assets.
- Why – the purpose of the operation.

2-23. When developing the COA, analyze the relative combat power. Consider the initial array of the forces and develop the schemes of maneuver. Determine C^2 means and maneuver control measures.

2-24. Base doctrinal capabilities and planning on historical planning factors, and then relook the estimates in light of the available assets, the factors of METT-TC, the echelon commander's intent, and the mission's priorities. For example, when planning distribution of mobile assets for route coverage, begin with an estimate of one mobile MP team per 10 kilometers. For area coverage, begin with an estimate of one mobile MP team per 55 square kilometers.

2-25. When dispersing the assets into small combat elements, consider the classic ratios of friendly to enemy forces (3 to 1) to help ensure that the elements can concentrate enough combat power to accomplish the mission. Consider the speed and ease of reassembling the elements if dispersing them to distant sites.

2-26. Attempt to anticipate the enemy's likely moves. Consider ways to obstruct dangerous approaches to the area and avenues leading away from potential landing zones. Plan ways to combine the efforts of different resources, like enhancing the combat power for the base response and counterreconnaissance operations with fire support from field artillery or aviation. Attempt to determine the outcome of the operations by—

- Conserving unit strength through economy of force.
- Using terrain, weather, deception, and operations security (OPSEC) to your advantage.
- Focusing your efforts on enemy vulnerabilities.
- Ensuring unity of effort among subordinates and with your peers.

2-27. When planning for combat operations, whenever possible, develop a COA that avoids an enemy's strength and strikes at his weaknesses. Avoid head-on encounters with an enemy's forces. Seek to gain the element of surprise. When moving, plan to use indirect approaches and flank positions that do not attract immediate attention. Plan for fire support to increase MP combat power. Plan to operate on the enemy's flanks and rear, where direct fire is most effective, psychological shock is the greatest, and the enemy is least prepared to fight. Respond to and implement changes quickly and plan supplementary or alternative control measures to modify the plan as the situation dictates.

2-28. For a combat operation, the COA statement and sketches include the following:

- Allocated forces.
- Unit boundaries.
- Axes of advance.
- Routes for a forward or rearward passage of lines.
- Air axes for the maneuver of attack helicopters.
- Other control measures which may include—
  - Phase lines.
  - Assembly and holding areas.
  - Zones or sectors.
  - Battle positions.
  - Objectives.
  - Obstacles.
  - Routes.

2-29. Assess all of the feasible COAs after developing them. Consider constraints and restrictions on each COA. Weigh the available COA for the AO, for the level of responsibility, the commander's intent, and the mission's priorities. Decide on the best COA. For more information about the MDMP process, refer to *FM 101-5*.

2-30. Once the commander decides on a COA, he announces it in the form of orders that include his intent and concept of the operation. Based on these orders, the platoon leader uses TLP to organize his time during planning and preparation for the mission. Time management is the key. The platoon leader normally uses one-third of the available time to plan, prepare, and issue the order. The squad leaders have the remaining two-thirds of the time to prepare the squads for the mission. Whenever possible, TLP are integrated and accomplished concurrently rather than sequentially. Relationships between TLP and the MDMP are shown in *Figure 2-1, page 2-4*.

TROOP-LEADING PROCEDURES

2-31. TLP begin when the platoon leader is alerted for a mission and starts again when he receives a change or a new mission. Conducting TLP is an eight-step process *(Table 2-3, page 2-12)*. Steps 3 through 8 may not follow a rigid sequence. Several of the steps may be accomplished concurrently. In CS operations, platoon leaders rarely have enough time to go through each step in detail. However, the procedure must be followed, if only in abbreviated form. This ensures that nothing is left out of the planning and the preparation.

2-32. Risk management is an integral part of TLP and must be integrated into every aspect of mission

Table 2-3. The Eight Steps of TLP

Step	Action
1	Receive and analyze the mission.
2	Issue a WO.
3	Make a tentative plan.
4	Initiate movement.
5	Conduct a reconnaissance.
6	Complete the plan.
7	Issue the order.
8	Supervise, refine, and rehearse.

planning and execution. Identifying hazards and implementing control measures to mitigate those hazards will minimize operational accidents, thereby preserving combat power. Risk management is not a separate process, but rather infused into the entire TLP. For more information about risk management, refer to *FM 100-14*.

RECEIVE AND ANALYZE THE MISSION

2-33. TLP begin when the platoon leader is alerted for a mission and start again when he receives a change or a new mission. When the platoon leader receives the mission, he—

- Analyzes the mission.
- Considers the operations underway.
- Identifies the hazards associated with the mission, and considers the aspects of the current and future situations, environment, and known historical problems.
- Considers the time needed to plan and carry out the new mission. For example, will sleep plans be needed to ensure that all the teams are on a similar rest posture?

- Plans the use of available time. The most critical resource may be time, especially during daylight hours.
- Uses reverse planning to make a timetable. The timetable—
  - Identifies what must be done.
  - Works backwards from the time the soldiers have to be ready, allowing them time to do each task.
- Uses no more than one-third of the time for planning. The subordinate leaders need the remaining time to make preparations.

2-34. If time is too short to do the rest of the troop-leading steps in detail, at least do a fast mental review and—

- Make a quick map reconnaissance while sending for the subordinate leaders, depending on the level of the mission.
- Have the minimum control measures needed posted on their maps.
- Give an abbreviated order.
- Cite enemy and friendly situations.
- State the mission of the team, squad, or platoon and the concept of the operation.

2-35. If there is not enough time to do these actions, have the unit move out, then issue a fragmentary order (FRAGO) by radio or at the next scheduled halt. Continue the plan while moving.

ISSUE A WARNING ORDER

2-36. Issue an oral or written WO to the subordinate leaders as soon as possible. Give enough information for the unit to begin preparing for the mission. If needed, issue several WOs to keep subordinates informed. Refer to *Appendix D* for more information about WOs.

2-37. The unit SOP should detail what actions to take when a WO is received. Such actions may include drawing ammunition, rations, water, and communications gear and checking vehicles and equipment. Keep all the personnel informed of what they are to do and why they are to do it.

MAKE A TENTATIVE PLAN

2-38. Develop the plan based on the factors of METT-TC (using the OPORD format and the higher HQ order). The order may be specific about the tasks the unit is to do. The time available may be limited and the scheme of maneuver may be dictated. Nevertheless, the leader still must evaluate the mission in terms of METT-TC to see how MP elements can best carry out the commander's order. The leader must—

- Consider each factor and compare the COA to form a base for the plan.
- Include concepts for reconnaissance, coordination with adjacent and/or supporting units, and troop movement.
- Assess the identified hazards. Consider the impact of each hazard in terms of potential loss based on probability and severity.
- Identify control measures that will eliminate the hazards or mitigate them to an acceptable level. Make decisions on acceptable levels of risk based on potential benefits versus cost.
- Issue the plan, when firm, as an order. *Appendix D* contains the OPORD format.

INITIATE MOVEMENT

2-39. Instruct subordinate leaders to start moving to the assembly area (AA). Allow subordinate leaders enough time for their actions if the element has to move and reorganize for the mission.

CONDUCT A RECONNAISSANCE

2-40. Ensure that the terrain where the operation is to be conducted is reconnoitered. At a minimum, conduct a map reconnaissance. (A map reconnaissance is the easiest, but least reliable form of reconnaissance. It is a supplement to other types of reconnaissance.) Study the map for terrain features, natural barriers, and other characteristics. Have subordinate leaders help identify key terrain features. Refer to *FM 21-26* for map-reading skills. Follow up with a visual reconnaissance of the area to be used and the terrain over which you will operate. A visual reconnaissance can be done—

- On the ground. A ground reconnaissance is time consuming, but the most reliable type of reconnaissance. MP see terrain features up close and can note problems not easily seen using other reconnaissance methods.
- In the air. If available, air reconnaissance can cover terrain quickly. To do an air reconnaissance, show the pilot a map of the terrain to be reconnoitered. Specify the type of information to be gathered. Have one person in the plane track the patrol's route on a map. At critical points, if the aircraft can land, have part of the patrol dismount to make a ground reconnaissance while the rest of the patrol goes back into the air to provide overwatch security. If the aircraft cannot land, make a visual search for enemy activity or for the required information.

2-41. Use the information from the reconnaissance to verify or to change the plan and to modify the risk-assessment process. Adapt your tactics to the terrain and the abilities of the force. If the reconnaissance cannot be finished due to distance or enemy pressure, make the plan from what has been seen. Give instructions for later actions in general terms and confirm or change while moving over the terrain.

Provide a sketch that will aid with a sand table to help in the OPORD issue, if time permits.

COMPLETE THE PLAN

2-42. Add details or makes changes to the tentative plan (as a result of the reconnaissance and coordination with nearby and/or supporting agencies) and identify specific tasks for all the subordinate elements.

ISSUE THE ORDER

2-43. Issue an OPORD or FRAGO ensuring that—

- The soldiers know the plan.
- The instructions are stated clearly and concisely (use the OPORD format). Platoon and squad orders are usually issued orally. However, if time permits, they can be written. When the order is written, delete the service support and command and signal paragraphs if covered by the SOP.
- The soldiers are thoroughly briefed on the hazards associated with the mission and the control measures identified to mitigate the hazards.
- The subordinate leaders back brief the order and spot-check the soldiers.
- When possible, the order is given from the advantage point where soldiers can see the area in which they will operate. This lets the leader point out terrain features on the ground as well as on a map. If this cannot be done, use a terrain model (sand table) or a sketch to help explain the order.

SUPERVISE, REFINE, AND REHEARSE

2-44. Supervise, refine, and rehearse the preparation to ensure that the soldiers are ready for the mission. This includes—

- Using the feedback received from subordinates.

- Ensuring that every soldier knows the mission and understands the commander's intent for the operation. The unit must be able to carry out the mission in the absence of the leader that developed the plan.
- Continuing to prepare when the troop-leading steps have been completed.
- Having the soldiers rehearse their actions if there is enough time before an operation.

Rehearsals

2-45. Rehearsals build confidence and improve performance. They allow faults in a plan to surface. If possible, soldiers should rehearse on terrain and under conditions like those at the operation site. Give priority to actions to be taken in the objective area. A rehearsal is especially helpful when operating in reduced visibility.

2-46. Each type of rehearsal reflects an increase in mission realism and a corresponding increase in rehearsal benefit. Each technique increases the realism of the enemy, terrain, team and squad actions, and actual time and distance relationships. Rehearsal techniques fall into the following four categories:

- Back brief.
- Rock drill.
- Walk-through.
- Full-scale.

2-47. Regardless of the rehearsal technique, leaders must demand that soldiers demonstrate the known hazards associated with the mission and understand the control measures identified to counter them. Insist that rehearsals include prescribed control measures (either actions or equipment).

2-48. **Back Brief.** The back brief rehearsal is an event that occurs after an OPORD has been issued. The back brief is the quickest rehearsal technique. Subordinate leaders repeat back to the commander what he expects them to do and why, using a map or a sand table to explain their mission. The subordinate leader identifies all specified and implied tasks, determines their mission-essential tasks, and restates the mission. Items essential to the back brief are the—

- Commander's intent.
- Concept of the operation.
- Scheme of maneuver.
- Time to complete the tasks.

2-49. **Rock Drill.** A rock drill rehearsal is done by acting out the friendly and enemy actions based on the scheme of maneuver and the situation. Subordinate leaders rehearse their actions by moving objects, such as rocks, that represent them or the platoon. In acting out the plan, leaders can talk through their missions, critical tasks, actions, and decisions. All subordinate leaders act out their parts simultaneously so problems and disconnects in synchronization can be more clearly identified.

2-50. **Walk-Through.** A walk-through rehearsal is the acting out of actions that will occur during a mission using the actual vehicles and equipment that will be used to conduct the operation. Participants communicate with the same type of equipment they will use during the operation. During a walk-through, subordinate leaders rehearse—

- Movement techniques, both mounted and dismounted.
- Critical actions.
- Decision making.

2-51. Since the leaders are in a more realistic environment, they rehearse the finer aspects of synchronization, C^2, and squad and team actions. Aggressive portrayal of OPFOR is critical in walk-through rehearsals because it increases the rehearsal realism. This type of rehearsal is more difficult to orchestrate. It is the optimum balance between resource constraints and realism. Ensure that a walk-through rehearsal is the minimum goal for all the units.

2-52. **Full-Scale.** During a full-scale rehearsal, participants use real-time mounted and dismounted movements over the actual or similar terrain. It is normally conducted with all teams and squads to be used for the actual mission. At a minimum, one subunit must participate for a full-scale rehearsal. This type of rehearsal is the most resource intensive, but provides the most realistic training environment for the unit. It is often used to rehearse the operation plan (OPLAN) or OPORD when time is not an immediate constraint.

Inspection

2-53. The last action before an operation is inspecting. Allow ample time for subordinate leaders to correct problems. Inspect soldiers by checking their mental and physical readiness. Inspect their equipment and check for the following:

- Accountability.
- Serviceability.
- Weapons.
- Ammunition.
- Individual uniforms and equipment.
- Mission-essential equipment.
- Water and rations.
- Communications equipment.
- Vehicles.
- Camouflage.

2-54. The equipment used during a mission is based on the unit's SOP, a risk assessment, and special considerations. Ensure that the SOP specifies a combat load and a list of ammunition and equipment usually carried on missions. Base changes from the SOP combat load on the METT-TC.

2-55. *Appendix E* contains a sample precombat-inspection (PCI) checklist. Ensure that the soldiers have everything they need for the mission. Ensure that they—

- Know their duties.
- Have only the equipment needed.
- Are wearing their equipment correctly and securely.

ORDERS AND REPORTS

2-56. MP leaders translate their thoughts, evaluations, and decisions into understandable reports and orders. Battlefield communication requires standardized, streamlined procedures. Despite personal exhaustion or battle confusion, you must be able to rapidly report information or issue instructions that are simple, clear, and brief.

ORDERS

2-57. Combat orders are written or oral communications giving details of tactical operations and administration. The three most common types of combat orders at company level and below are—

- WO
- OPORD
- FRAGO

2-58. WOs and OPORDs generally have set formats. This helps ensure that the receiver understands the intent of the message and that all needed information is provided. Standardization helps save time in writing as well as interpreting orders. FRAGOs enhance what has been previously sent out in the OPORD, such as a change in the situation or mission. Refer to *Appendix D* for examples of orders.

Reports

2-59. MP report to higher HQ and provide information on which plans, decisions, and orders can be based. The information included in MP reports must be accurate and timely, and complete *negative* information ("There is no enemy at . . . . ") is often as important as positive information. Reports are the main record of operational events. The three broad categories of reports are—

- Administrative.
- Operational.
- Intelligence.

2-60. Commanders may specify report formats in their local tactical SOP. Treat friendly information, including administrative reports, as classified or sensitive in nature to keep information from falling into the enemy's hands. *FM 101-5-2* and *Appendix D* of this manual CONTAIN STANDARD REPORTING FORMATS.

SOPs

2-61. SOPs detail how forces will execute unit-specific techniques and procedures that commanders standardize to enhance effectiveness and flexibility. Commanders use the SOP to standardize routine or recurring actions not needing their personal involvement.

RULES OF ENGAGEMENT AND RULES OF INTERACTION (ROI)

2-62. MP usually are among the first CS forces deployed to troubled areas around the world. As a direct result of a peacetime L&O mission, MP continually train in the prudent use of force, crisis management, and operations requiring restrictive ROE and ROI.

2-63. ROE are the directives established by higher HQ that delineate the circumstances and limitations under which soldiers will initiate and/or continue engagement with belligerent forces. ROE may reflect the law of armed conflict and operational considerations, but are primarily concerned with the restraints on the use of force. ROE are the primary means by which commanders convey legal, political, diplomatic, and military guidance to soldiers. Leaders at every level must train their soldiers carefully and thoroughly concerning ROE and laws that govern armed conflict before deployment. During the conduct of the operation, leaders continue to train soldiers and stress firm, determined, and impartial execution of ROE to preclude inviting challenges from any of the belligerent parties.

2-64. ROI embody those human dimension skills needed to successfully interface with various categories of people. They spell out with whom, under what circumstances, and to what extent soldiers may interact with other forces and the civilian populace. ROI, when applied with good interpersonal communication (IPC) skills, improve the soldier's ability to accomplish the mission while reducing possible hostile confrontations. ROI and IPC, by enhancing the soldier's persuasion, negotiation, and communication skills, also improve his survivability. ROI founded on firm ROE provide the soldier with the tools to address nontraditional threats such as political friction, ideologies, cultural idiosyncrasies, and religious beliefs and rituals. ROI must be regionally and culturally specific. MP leaders must train soldiers on ROE and ROI using tactical vignettes or simulated events.

SITUATIONAL AWARENESS

2-65. Situational awareness is the ability to maintain a constant, clear mental picture of the tactical situation. This picture includes an understanding of both the friendly and enemy situations and of relevant terrain. It also includes relating events in time to form logical conclusions and make decisions that anticipate events. Since MP platoons normally operate dispersed, it is essential that all MP leaders maintain situational awareness so that they can make quick, sound tactical decisions. Situational awareness also permits MP leaders to anticipate events and relate separate pieces of information to form logical conclusions. One of the critical outcomes of situational awareness on the part of all MP is a reduction in fratricide incidents. Refer to *Appendix F* for fratricide avoidance.

BATTLEFIELD FRAMEWORK

2-66. The commander will structure the battlefield based on the conditions of METT-TC and his commander's intent. How he does this affects the MP platoon leader's mission planning and his ability to maintain situational awareness. Geographically, available assets, mission, and the AO influence the dispersion of MP assets. The framework of this AO can vary from an area that is dominated by several towns or large cities to an area that includes several bases and base clusters. MP can expect to operate in sustainment areas where there may be clear boundaries and closely tied adjacent units or in a decentralized structure with few secure areas and unit boundaries. Between these extremes are an unlimited number of possible variations. Maintaining situational awareness becomes more difficult as the battlefield becomes less structured. Modern, highly mobile operations with small forces lend themselves to a less rigid framework that challenges the MP ability to maintain an accurate *picture* of the battlefield. MP are the echelon commander's critical link to the battlefield, constantly gathering police, operational, and combat information.

BATTLEFIELD PICTURE

2-67. To have a clear picture of the battlefield, the MP leader must have virtually perfect knowledge of the friendly situation one level higher. This means the MP platoon leader must know the company situation and the location and mission of the adjacent company and platoons. It is also important that the platoon leader update his subordinate leaders periodically regarding the higher situation. The platoon leader must have a relatively complete knowledge of the terrain, and he must know as much as possible about the enemy. The requirement to maintain a real-time picture of the battlefield one level higher does not relieve the platoon leader of the requirement to understand the situation and the commander's intent two levels higher. The difference is that this understanding of the situation two levels higher does not have to be as specific or in real time.

2-68. Most of the information the platoon leader needs comes in the form of reports over communication channels. Subordinate leaders are required to periodically report their status. If an MP team is operating in an area that does not allow uninterrupted communications, the team leader coordinates with adjacent teams to relay his report. If an MP team does not report in a timely manner, the platoon leader must quickly determine the status of the overdue team.

2-69. If possible, the platoon leader monitors his platoon and company net. How effectively he can accomplish this is, to some degree, experience-dependent; however, there are techniques he can apply to relate the information he is receiving to the map and thereby track the tactical situation.

2-70. The platoon leader's map is the key to maintaining situational awareness. He plots all friendly position reports up to one level higher than his own and plots information from spot reports (SPOTREPs). He uses different colors for friendly and enemy elements to allow quick distinction. To avoid cluttering the map, he places a dot or symbol on the map where the element is located and labels the point with a number. The same number is then written in the map's margin (or beyond the AOs) with the complete SPOTREP or unit identification (ID) next to it. Include the time on this notation. As positions or reports are updated, the old symbol is crossed off and a new one with a corresponding notation is added; it is critical that updates to previous reports be clearly identified as such during transmission. This simple system can greatly increase the ability to track both friendly forces and enemy activity in a particular AO.

BATTLE SPACE

2-71. As mentioned earlier, an accurate picture of the battlefield provides the platoon leader with important tactical information, including friendly and enemy positions and relevant terrain. In turn, complete understanding of the military significance of this picture requires knowledge of the concept of battle space, the key element in the intellectual process of visualizing the battlefield.

2-72. At the most fundamental level, battle space is the three-dimensional *bubble* or area in which the platoon can acquire enemy forces and influence them with effective fires. This space is defined by the following numerous battlefield factors:

- The locations of the friendly forces, including the platoon's individual teams, OPs, and patrols.
- The effects of the terrain, weather, and movement.
- The ranges of all the available platoon weapons and sensing systems.

2-73. Each squad has its own battle space; the platoon battle space is the sum of the individual squads battle spaces. Platoon battle space is not restricted by boundaries; it can overlap with the battle space of adjacent units. For example, an enemy element that is spotted outside of the platoon's AO can still adversely affect the platoon's mission. Coordination is made with adjacent units to detect and destroy the threat.

2-74. Battle space has applications in all phases of mission planning and execution. During the planning process, it is a critical factor in selection of the routes and tentative positions. Once mission execution begins, the platoon leader's knowledge of the battle space is critical to issuing timely and effective orders as the situation changes.

2-75. The importance of battle space demands that the platoon leader direct most of his battle command effort toward managing and enhancing his space. He must be aware at every moment how battle space is changing as friendly and enemy forces move and terrain and visibility conditions change. He must evaluate how these changes affect his squads.

2-76. As the operation progresses, the platoon leader must take active measures to shape the battle space to his best advantage. One vital step in this process is to eliminate any gaps, or dead space, that exist within the *bubble*. The platoon leader can accomplish this in several ways, including maneuvering teams, repositioning OPs, and deploying patrols or remote sensors.

COMMAND POST OPERATIONS

2-77. Company- and larger-size elements have a tactical operations center (TOC) and platoons have CPs. The CP is wherever the platoon leader goes. It can be mobile or stationary. No matter the location, there must be communication with and command of the unit and a method for battle tracking. CP activities are a 24-hour operation. The leadership must ensure that there is a plan for continuous operations.

2-78. A mobile CP may be a HMMWV and an ASV. The platoon leader can make decisions on the move while having communication capabilities nearby. A leader's book can hold information on battle rosters, report formats, and a map of the battlefield.

2-79. A stationary CP may be the platoon leader's tent or office. Communications may be telephonic, by messenger, or by radio. Track the status of equipment and personnel by using charts. Use large maps to track the battle. Keep in mind that moving to another location requires taking charts and maps, so everything needs to be mobile.

2-80. Key personnel for CP operations are the platoon leader, the PSG, and the radio/telephone operator (RTO).

PLATOON LEADER

2-81. The platoon leader is responsible for C^2 of his organization and—

- Planning the missions according to the commander's guidance.
- Planning security to include the placement of crew-served weapons.
- Issuing orders and providing work priorities to subordinate leaders.
- Conducting PCI.
- Planning and conducting platoon rehearsals.
- Battle tracking (knowing) on a map the exact location of all the MP teams, to include the position of friendly platoons and known enemy, and a platoon-sector sketch.
- Adapting to new situations, making necessary adjustments, and issuing FRAGOs.

- Controlling the movement of the subordinate units.
- Ensuring that the communication net is established.
- Conducting risk assessment and continuously updating it.
- Reporting to higher HQ using the correct report format.

PLATOON SERGEANT

2-82. The PSG is responsible for the logistics of the operation and—

- Coordinating and providing all the needed class items.
- Ensuring equipment serviceability and accountability.
- Supervising the maintenance.
- Establishing and supervising a sleep plan.
- Supervising the security plan's execution.
- Supervising the maintenance of work priorities.
- Supervising PCIs.
- Supervising sanitation and hygiene.
- Adjusting, manning, and cross leveling the soldiers, weapons, and equipment.
- Coordinating and supervising morale services.
- Assisting the platoon leader in rehearsals.
- Supervising test firing.
- Assisting with battle tracking.
- Consolidating subordinate units' status reports for the platoon leader.
- Assisting with the reports.
- Providing technical and tactical advice to the platoon leader.
- Maintaining situational awareness.

RADIO/TELEPHONE OPERATOR

2-83. The RTO should be an experienced MP who is also the platoon leader's driver. The RTO—

- Communicates with higher HQ and separate units.
- Submits the required reports according to orders and SOPs.
- Maintains a record of communications.
- Maintains the radio and communication equipment.
- Assists with battle tracking.

Chapter 3

Shoot, Move, and Communicate

The ability of an MP unit to shoot, move, and communicate ensures its ability to detect, disrupt, and defend against the enemy and immeasurably adds to its survivability and maneuverability. MP are structured to be strategically, operationally, and tactically agile to respond to the increased range of worldwide MP requirements.

SHOOT

3-1. It is important that MP understand their shooting capabilities and limitations. Just as important is the understanding of firing techniques and associated fire distribution, reacting to air and armor attacks, calling for fire, and obtaining various fire support.

UNDERSTAND FIRE TECHNIQUES

3-2. Fire techniques include fire from or at a moving vehicle, fire distribution and control, and suppressive fire.

Fire From or at a Moving Vehicle

3-3. The key to forward maneuver is firing on the enemy. When maneuvering, the fire element—

- Attempts to destroy or suppress the enemy.
- Covers and protects the maneuver element as it advances.
- Moves, when possible, into its firing position undetected. Fire from an unexpected direction

has a greater effect than fire from a known position.

3-4. Firing on the move is less accurate than firing from a halt. However, to halt and fire takes more time and is more dangerous. A stationary vehicle is more likely to be hit than a moving vehicle. The team leader must decide whether to fire while moving or to fire from a short halt. He bases his decision on sound judgment and evaluation of the threat.

3-5. Crew-served weapons engage all targets on the move with free gunfire. To deliver this type of fire, the gunner removes the traversing and elevating (T&E) mechanism from the bottom of the receiver, allowing the gun to move freely in any direction. Accurate firing with crew-served weapons while moving is affected by—

- The terrain.
- The vehicle's speed.
- The team's proficiency.

3-6. When aiming from a moving vehicle or at a moving vehicle, or both, the gunner must lead the target. The speed of the firing vehicle, the time of flight, and the *angle of engagement* affect the amount of lead required. The time of flight is the required time it takes the projectile to move from the firing vehicle to the target. The angle of engagement is the angle found between the centerline of the vehicle and the gun when laid on the target. When a round is fired from the flank of a moving vehicle, the round drifts in the same direction and at the same speed as the vehicle. The longer the flight time and the larger the engagement angle, the greater the drift. Thus, the gunner must apply more lead to the shot. If a lead is required and the gunner is traversing left to keep on target, the gunner must lead left. If the gunner is traversing right to keep on the target, the gunner must lead right. This is true whether the firing vehicle is moving, the target is moving, or both are moving. *Table 3-1* shows the responsibilities of an MP team when firing on the move.

Table 3-1. Team Responsibilities When Firing While Moving

Position	Actions
Team leader	• Directs the driver. • Keeps the gunner oriented. • Senses the impact of the rounds–long, short, left, or right of the target. • Identifies additional targets. • Assists the gunner with reloading, if required. • Observes the surrounding terrain.
Gunner	• Develops a *feel* for the moving vehicle. • Tracks the position of the target with the MK19 grenade machine gun (GMG) despite the movement of the vehicle. • Remains alert to the sounds of the engine and transmission. These sounds indicate the type of terrain over which the vehicle is traveling and helps the gunner anticipate vehicle movements.
Driver	• Tries to maintain a steady gun platform while the gunner engages the targets. • Attempts to time the gear and direction changes so they occur immediately after firing and do not interfere with accuracy. • Informs the gunner of obstacles in the vehicle's path that might affect the gun's accuracy. • Announces "depression," "turn," and the like to warn the gunner of vehicle movements. • Announces, "steady" to let the gunner know when the vehicle is once again on a stable platform. The gunner assumes he has a stable platform unless the driver informs him otherwise.

Distribute Fire

3-7. MP leaders must distribute the fires of their organic weapons to destroy or suppress enemy positions. The following are the two methods to distribute fire on a target:

- Point fire. Point fire *(Figure 3-1, page 3-4)* is directed against one target (such as a machine gun position) with all the troops firing at the

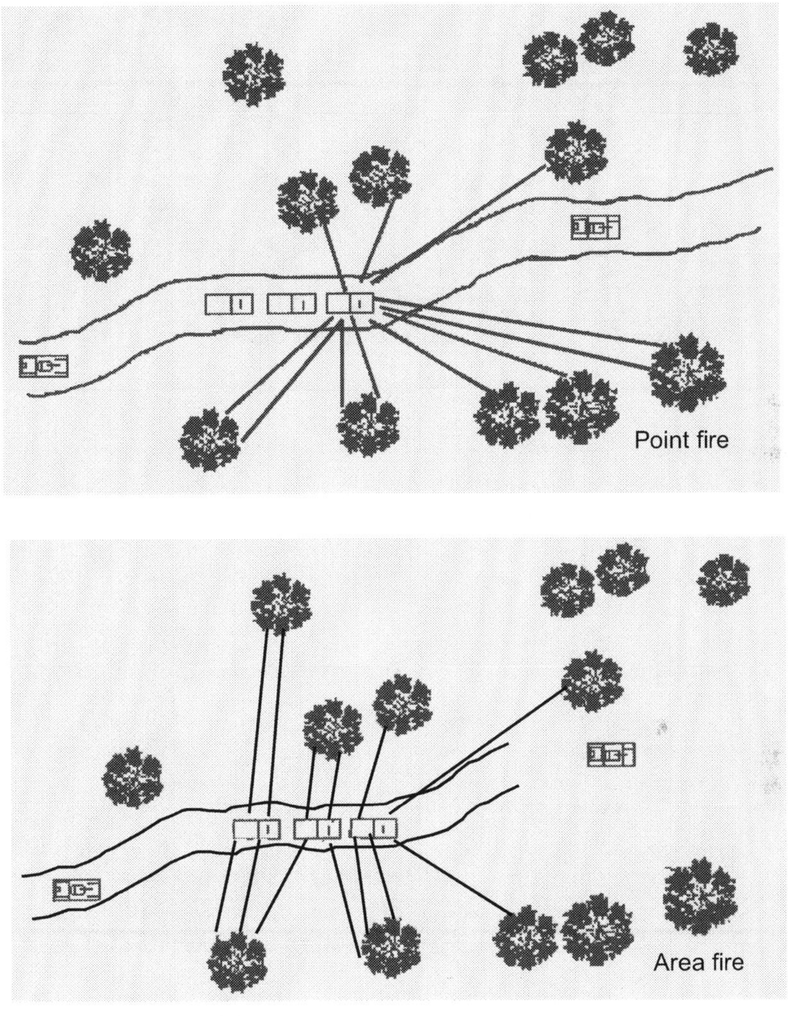

Figure 3-1. Fire Distribution

same target. Spreading out the base-of-fire element makes this type of fire particularly effective because the fire is directed from many sources.

- Area fire. Area fire *(Figure 3-1)* permits rapid cover of an entire area with fire from the left to the right and in depth, even if the enemy cannot be seen. This method is used without command and is the quickest and most effective way to bring all parts of a target under fire. Each member in the element is assigned a portion of the target. Fire is placed on likely locations for enemy positions rather than in a general area. If the leader wants fire on a wood line, he may shoot tracers to mark the center of the target. Soldiers to the left of the leader fire to the left of the tracers and soldiers to his right fire to the right of the tracers.

3-8. A rifleman fires his first shot on the part of the target that corresponds to his individual position. If he is left of the leader, he fires to the left of the leader's tracers. He then distributes his remaining shots over the part of the target extending a few meters right and left of his first shot. He covers the part of the target that he can hit without changing position.

3-9. A grenadier fires into the center of the target area of his team. He then distributes his shots over the remaining target area from the center to each side and from front to rear. A machine gunner covers part of the target depending on his position and how much of the target is in range. When possible, he covers the entire target of the team. When placing automatic suppressive fire on the enemy, the tendency is to shoot high. Therefore, he places the first bursts low and works up to the target. The squad leader tells the machine gunners where to shoot by assigning sectors of fire.

3-10. An MK19 gunner engages area targets with traversing and searching fire after the leader designates the width and depth of the target. If one MK19 GMG is being fired, the gunner engages the area target by adjusting his fire on the center of the mass, then traverses and searches to either flank. When he reaches the flank, he reverses direction and traverses and searches in the opposite direction. If two MK19 GMGs are being fired as a pair, the point of the initial lay and adjustment for both guns is on the midpoint of the target. After adjusting the fire on the center of the mass, fire is distributed by applying direction and elevation changes that give the most effective coverage of the target area. Usually, the right gun (number 1) fires on the right half, and the left gun (number 2) fires on the left half. *Appendix G* describes the MK19 qualification and familiarization tables and provides a sample scorecard.

Control Fire

3-11. Fire control is an essential component of fire distribution. A platoon leader must know what means he will direct the fire element to use when engaging the targets. He will communicate directly or use prearranged signals to identify the location of the target to the other units. He may use sound signals (such as voice, a horn, or a whistle), but must remember that they are only good for short distances and that their reliability and range are reduced by battle noise, weather, terrain, and vegetation. Use a radio to direct the base-of-fire element or adjust fires from reference points or landmarks, because a radio offers immediate voice communication. For example, he may say, "From the burning scout vehicle, northwest 50 meters, machine gun position." If portable radio equipment is not available, he uses prearranged visual signals, such as smoke or flares. A smoke round from a grenade launcher, unless it is being used for some other purpose, and a smoke canister can be used as a signal. Use these items during reduced visibility in addition to aiming stakes, illumination, night-vision devices, infrared chemical lights, and so forth.

Use Fire Commands

3-12. Leaders use fire commands to direct fire. A fire command has the following six parts:

- Alert. The leader alerts the soldiers to receive further instruction. He alerts the soldiers by name or unit designation, some type of visual or sound signal, personal contact, or any other practical way.
- Direction. The leader tells the soldiers the general direction to the target. In some cases, he pinpoints a target. The following are the three ways the leader can give the direction to the target:
  - Points with his armor rifle.
  - Fires tracer ammunition at a target.
  - Uses either target reference points (TRPs) or easily recognized man-made objects or terrain features. He gives the general direction just before giving the reference points.
- Description. The leader describes the target briefly but accurately and always gives the formation of the enemy soldiers.
- Range. The leader tells the soldiers the range to the target in meters.
- Method of fire. The leader tells the soldiers which weapons to fire, the type and amount of ammunition to fire, and the rate of fire.
- Command to fire. The leader tells the soldiers when to fire by using an oral command or visual signal. When he wants to control the exact moment of fire, he says, "At my command" (then pauses until ready to commence firing). When he wants to start firing on completion of the fire command, he just says, "Fire."

Use Subsequent Fire Commands

3-13. These commands adjust or change information given in the initial fire command. Only the elements that change are given.

Terminate Fire

3-14. Fire is terminated by the command or signal for *cease fire, end of mission*.

Suppress Fires

3-15. When the fire element is in position, it lays a heavy volume of fire on the enemy to suppress them. When the leader senses that the enemy is suppressed, he instructs the fire element to reduce its rate of fire as long as it keeps the enemy suppressed. As the movement element nears its objective, the fire element increases the rate of fire to keep the enemy down. This lets the movement element close enough to assault the enemy before the enemy can react. When the assault begins, or on a signal, the fire element stops firing, shifts its fire to another target, or walks its fire across the objective in front of the movement element, and then shifts or ceases fire.

3-16. Positions for fire elements are located so that movement of the maneuver element does not mask their fires. Fire element positions are often higher and usually to the flank of the maneuver element. The maneuver element neither masks the fire of the fire element nor moves outside the protective umbrella provided by the fire. A platoon or squad can point fire at one target or an area of several targets. In both cases, the leader must control the fire. He must ensure that the fire is directed on the enemy, not on the maneuver element.

Use Nonlethal Weapons (NLW)

3-17. The Department of Defense (DOD) defines NLW as weapons that are explicitly designed and primarily employed to incapacitate personnel or material while minimizing fatalities, permanent injury to personnel, and undesired damage to property and the environment. Unlike conventional weapons that destroy the targets principally through blast, penetration, and fragmentation, NLW employ means other than gross physical destruction to prevent the target from functioning.

3-18. NLW doctrine and concepts of operation are designed to reinforce deterrence and expand the range of options available to commanders. They enhance the capability of US forces to accomplish the following objectives:

- Discourage, delay, or prevent hostile actions.
- Limit escalation.
- Take military action in situations where use of lethal force is not the preferred option.
- Protect US forces.
- Disable equipment, facilities, and personnel temporarily.

NOTE: The zero probability of producing fatalities or permanent injuries is not a requirement of NLW. However, while complete avoidance of these effects is not guaranteed or expected, when properly employed, NLW significantly reduce them as compared with physically destroying the same target.

3-19. When drafting the ROE, it must be clearly articulated and understood that the role of NLW is an additional means of employing force for the particular purpose of limiting the probability of death or serious injury to noncombatants or belligerents. However, the use of deadly force must always remain an inherent right of individuals in instances when they, their fellow soldiers, or personnel in their charge are threatened with death or serious bodily harm. NLW add flexibility to the control of disturbances within the I/R facility and provide an environment where guard forces can permissively engage threatening targets *(Figure 3-2)* with limited risk of noncombatant casualties and collateral damage. Refer to *FM 90-40.*

3-20. The use of lethal force, employed under the standing ROE, will never be denied. At no time will forces be deployed without the ability to defend themselves against a lethal threat, nor will they forego normal training, arming, and equipping for combat. Nonlethal options are a complement to, not a replacement for, lethal force and seek to expand a proactive response across the range of military operations. Refer to *FM 90-40.*

3-21. The decision to use NLW against an adversary during a confrontation is delegated to the lowest possible level, preferably to the platoon or the squad. However, this requires that all personnel, not just the leaders, have a clear understanding of the ROE and the commander's intent. Refer to *FM 90-40.*

3-22. Commanders and public affairs officers must be prepared to address media questions and concerns regarding the use and role of NLW, and they must make it clear that the presence of NLW in no way indicates abandoning the option to employ deadly force in appropriate circumstances.

3-23. **Advantages of Employing Nonlethal Weapons**. NLW provide the commander with the flexibility to influence the situation favorably with reduced risk of noncombatant fatalities and collateral damage.

NLW	0	5	15	20	25	30	35	50	55	65	95	100	120
	Meters												
Modular crowd control munitions (MCCM)	Lethal zone			X	X	X	X	X	X	X	X	X	X
Stun hand grenade					X	X	X	X	X	X	X	X	X
12-gauge point		Lethal zone	**Nonlethal zone**				X	X	X	X	X	X	X
12-gauge area							X	X	X	X	X	X	X
40-millimeter point			**Nonlethal zone**						X	X	X	X	X
66-millimeter sting ball								**Nonlethal zone**					X
66-millimeter flash bang													X
66-millimeter CS grenade												X	X

Figure 3-2. Range of Munitions Contained in a Nonlethal Capability Set

3-24. NLW can be more humane, be consistent with the political and social implications implicit in humanitarian missions, be used during peacekeeping missions, and facilitate post-incident stabilization by reducing internee alienation and collateral damage.

3-25. The force that properly employs nonlethal options gains advantages over those who rely on lethal options alone, because the degree of provocation required to employ these options is substantially less. This advantage provides a more proactive posture and quicker response as well as a diminished likelihood of having a situation escalate to a point where deadly force is required to resolve a conflict within the I/R facility.

3-26. NLW options are less likely to provoke others and the use of NLW, in fact, may provoke a negative response. However, demonstrated restraint greatly diminishes feelings of anger and remorse when deadly force is required after nonlethal options fail.

3-27. **Military Police Nonlethal Weapons**. I/R facility commanders consider the use of force options discussed in *Chapter 2* and *AR 190-14* when dealing with disruptions within the compound. Facility commanders are encouraged by *AR 190-14* to substitute nonlethal devices for firearms when it is considered adequate for MP to safely perform their duties. Currently, MP have such nonlethal options as riot-control agents chlorobenzul-malononitrile (CS) and oleoresin capsicum (OC), military working dogs (MWD), an MP club, and a riot baton for crowd control. There are other nonlethal devices currently being tested and fielded that will be available to the I/R commander in the future.

3-28. **Nonlethal Training**. Soldiers and their leaders must be trained in the correct employment of NLW available to them. Soldiers and their leaders must understand the limited use of these systems in environments with restrictive ROE. Their training must be continuous at all levels to ensure that NLW are properly employed and that the leaders and the soldiers understand when and how to effectively employ them. They must understand that the incorrect application of NLW can have significant operational and political ramifications. Well-trained MP leaders, providing timely and clear guidance to MP soldiers using NLW, will ensure the mission's successful accomplishment.

3-29. Many NLW have both maximum effective and minimum safety ranges. Individuals struck short of the minimum safety range often suffer severe injuries or death while the effects of most nonlethal devices are greatly mitigated at longer ranges. In order to be effective, engage the threat within the *effective* zone, beyond the minimum safety range, and short of the maximum effective range.

3-30. When training with and planning for the use of NLW consider the following:

- Never apply NLW in a situation where deadly force is appropriate.
- Never apply NLW in a situation that will place troops in undue danger.
- Always cover NLW with deadly force.

3-31. **Nonlethal Tactics**. *FM 90-40* provides an in-depth discussion on the tactics associated with the employment of various NLW available to the commander, such as—

- Riot formations. Riot formations establish riot-control teams with a minimum response time. Because of the physical nature of riot control, individuals in riot control formations should not carry long rifles. Nonlethal attachments should follow closely behind the riot control formation. Lethal coverage should be provided for this entire formation. Refer to *FM 90-40*.

- Designated marksman. During a nonlethal engagement, the use of a designated marksman (DM) provides confidence and safety to those facing a riot. If a lethal threat is presented, the DM who is in an overwatch position and armed with a standard infantry rifle, mounted with a high-powered scope, can scan a crowd and identify agitators and riot leaders for apprehension as well as fire lethal rounds if warranted. Additionally, he is ideally suited for flank security and countersniper operations. Refer to *FM 90-40*.

React to An Air Attack

3-32. Passive and active air defense measures are used by MP to protect themselves from enemy air attacks during all missions.

3-33. **Passive Air Defense Measures.** MP employ passive air defense measures that include actions to avoid detection and air attack, and actions to limit damage if they are attacked. MP use active air defense measures to fight back against enemy aircraft. Enemy aircraft will attack and attempt to destroy any target seen. The passive air defense methods that limit enemy detection include—

- Concealment.
- Camouflage.
- Cover.

3-34. Dispersion is another passive air defense method. Its purpose varies from those of concealment, camouflage, and cover. While they are designed to hide MP and their vehicles and equipment from the view of the enemy, dispersion is employed to reduce the effects of an enemy air attack. Dispersion is important when MP are occupying a static position (such as a company TOC or platoon CP) or when they come under air attack. If MP come under air attack, they quickly disperse with their vehicles, move to a concealed position, if possible, and stop. Stationary vehicles are more difficult for the enemy to detect than moving vehicles.

3-35. Early warning (quick recognition of enemy aircraft) is a passive air defense method that affords MP an opportunity to take cover and one that may lead into active air defense measures. The warning may come through communication channels, OP and listening posts (LPs), or from convoy air guards. A whistle, a voice, a radio, or any other method can provide a warning.

3-36. All OP/LPs watch for enemy aircraft as a standard duty. When air sighting, the first person to see an enemy aircraft shouts, "Aircraft," then, "Front (Right, Left, or Rear)." In a convoy, air guards are given sectors of the sky to observe for enemy aircraft. When an enemy aircraft is spotted, the predetermined alarm (such as a horn or hand signal) is given until all vehicles are aware of the situation.

3-37. When an alarm is given, all dismounted troops take cover at once. They go below ground level, if possible. If the aircraft is not firing, MP withhold their fire to avoid disclosing their position and they allow the aircraft to pass. They stay concealed until the all clear is given. The MP leader initiates a size, activity, location, unit, time, and equipment (SALUTE) report (or refer to the SOP) for the sighting of hostile aircraft.

3-38. **Active Air Defense Measures**. Although passive measures are the first line of defense against air an attack, MP must be prepared to engage enemy aircraft. Low-flying hostile aircraft may appear suddenly from behind low hills, trees, or a haze. To gain surprise, they may attack with the sun behind them. Before MP fire at enemy aircraft, they must positively identify the aircraft as hostile. If the aircraft is making a firing run on the patrol, take cover and return fire. However, commanders may restrict active air defense when friendly aircraft are in the area.

3-39. In convoys, drivers alternately pull their vehicles off the road to the right and left, seeking concealment from air observation. Caution must be used if mines are a known threat. If the enemy aircraft is not attacking, the same actions stated earlier are taken. If the aircraft is attacking, MP dismount and seek cover away from the vehicle (the vehicle may be the aircraft's target) and return fire. All personnel remain under cover until the command is given to continue the mission.

3-40. Fire small arms at attacking aircraft during or after the first attack. MP direct fires to saturate the airspace through which the aircraft will fly without trapping the aircraft. For more detailed information refer to *FM 44-8*. When engaging hostile aircraft—

- Fire only on command unless under direct attack (being fired on by aircraft).
- Ensure that the direction of fire does not place rounds on friendly personnel, equipment, or positions.
- Deliver a large volume of fire.
- Lead a slow-moving aircraft and adjust fire by observing the flight of the rounds, especially if tracer rounds become available, using the free-gun technique of fire.
- Aim at the center mass of a grounded or hovering helicopter and a helicopter that is coming directly at your position. Cease-fire when the aircraft passes out of range.

3-41. Slow-moving rotary-winged aircraft that are on the ground, hovering, taking off, or landing are most successfully engaged by the MK19. The MK19 GMG's 40-millimeter ammunition is fired at a relatively slow speed and has a high trajectory at a long distance. The MK19 GMG rounds that do not hit the aircraft will detonate on impact with the ground. The location of friendly elements within range of the weapon must be considered at all times. *Figure 3-3* depicts the rules for selecting aiming points of various aircraft.

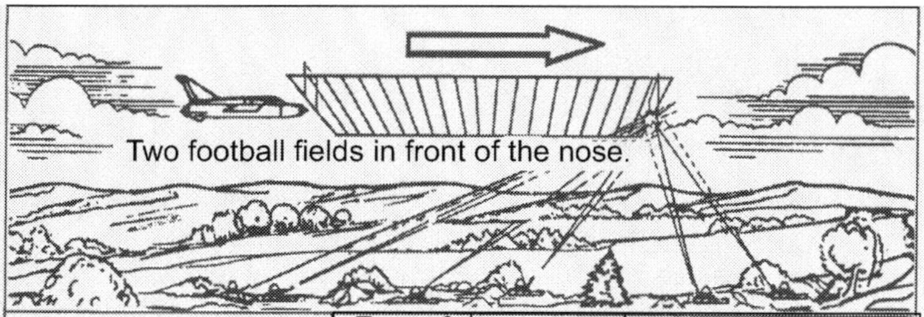

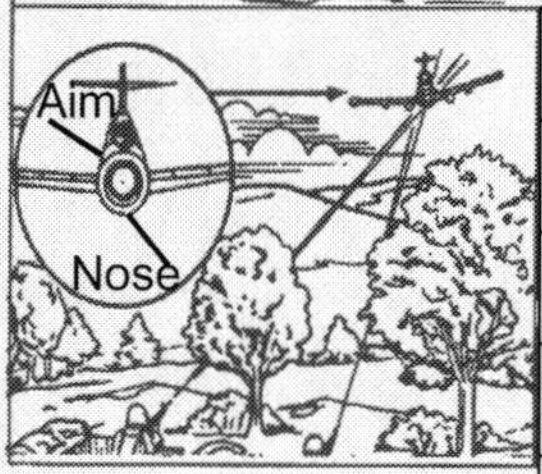

Type of Aircraft	Course	Aim Point
Jet	Crossing	Two football fields in front of the nose
Jet	Overhead	Two football fields in front of the nose
Jet	Directly at you	Slightly above the aircraft's nose
Helicopter	Crossing	One-half football field in front of the nose
Helicopter	Hovering	Slightly above the helicopter's body
Helicopter	Directly at you	Slightly above the helicopter's body

Figure 3-3. Rules for Selecting the Aim Point

REACT TO ARMOR

3-42. MP maneuver and operate in much of the battle space and can expect to encounter pockets of by-passed enemy armor within the rear area. MP engage enemy armor targets only for self-defense or when total surprise can be achieved. MP place antiarmor weapons on avenues of approach to defend against enemy armor. After initial contact, MP relocate immediately to avoid the fast-moving enemy armor and its firepower.

3-43. The (M136) antitank (AT) 4 provides antiarmor capability for MP teams. The AT4 is primarily employed against light armored vehicles, such as personnel carriers. It has a very limited capability against main battle tanks. The AT4 is issued as ammunition rather than as an individual weapon. It is carried and employed in addition to the basic weapon of MP.

3-44. The most stable firing positions for the AT4 are standing supported, prone, and prone supported. Whenever possible, use a supported position which provides more stability and aids in aiming.

Engage Armored Vehicles

3-45. The four methods to engage armored vehicles with the AT4 are—

- Single firing.
- Sequence firing.
- Pair firing.
- Volley firing.

3-46. Refer to *Table 3-2* for a description of the four methods. The best methods of engaging armored vehicles are pair and volley. Regardless of the method used, the closer the target, the better the chance for a first-round hit. Aim for the center mass of the target. The most vulnerable spots of armored vehicles are the top and the rear. The sides of the armored vehicles can also be penetrated.

3-47. An armored vehicle without the protection of dismounted infantry is vulnerable to a close attack by well-armed dismounted units. When an armored vehicle is buttoned up, visibility of the crew is restricted. This provides an opportunity for an armor-killer team to approach the vehicle with less risk of detection. The types of vehicles and the methods to engage enemy armored vehicles are—

- Stationary targets. Place the center post at the center of the visible mass for stationary targets. The firer does the same for vehicles that are moving toward or away from him.
- Slow-moving targets. Place the center post on the front leading edge of the vehicle (less than 10 miles per hour [mph]). This method is also applied to oblique moving targets.
- Fast-moving targets. These targets are moving more than 10 mph. If the target is moving to the left, place the right lead post at the center of the mass. If the target is moving toward the right, place the left lead post at the center of the mass.

Table 3-2. Methods of Engagement with the Light Antiarmor Weapon (LAW)/AT4

Method	Gunners	Firing Sequence	Probability of a Hit
Single firing	One	One gunner fires one AT4 at the target.	Low. Use only at ranges up to 200 meters for AT4s. Regardless of the method used, the closer the target the better the chances of a hit.
Sequence firing	One	One gunner fires two or more AT4s in turn. He prepares several weapons for firing before engaging the target. He gets the weapon, estimates the sight picture, and shoots the weapon in turn.	Good. If the first round misses, the gunner adjusts the range and the lead of the succeeding rounds until he gets a hit and fires until the target is destroyed.
Pair firing	Two or more	Each gunner fires one or more AT4s at a target, one at a time. They prepare several weapons for	Better. Two or more gunners track the target at one time, letting them get a target hit sooner. They can be ready to shoot as soon as an earlier

Table 3-2. Methods of Engagement with LAW/AT4 (Continued)

Method	Gunners	Firing Sequence	Probability of a Hit
		firing before engaging the target.	round hits. The first gunner sees a target, identifies it, and states the estimated range and lead to use. For example, the gunner, on spotting a fast-moving scout reconnaissance vehicle, says, "BMP 150 meters; fast target." He then fires at the target. If the first gunner misses, the second gunner gives a revised range and lead. This continues until one gets a hit. Once the correct range or lead has been found, all the gunners fire until the target is destroyed.
Volley firing	Two or more	Each gunner fires one or more AT4s on command or on signal until the target is destroyed. They prepare several weapons for firing before engaging the target.	This is the best method of engagement for an AT4 because the gunners shoot more rounds at a target at one time. This method is used only when the range and lead to the target have been determined. Range can be determined by using the map, pacing, or the results of pair firing after a target has been hit.

Estimate the Range of the Targets

3-48. A gunner has a better chance of hitting a target with the AT4 if he knows the range to the target. Determining the range is a learned skill and must be mastered by anyone who fires the AT4. Methods of range determination include—

- Using range finders.
- Measuring the distance on a map.

- Pacing.
- Firing the pair and sequence method.
- Using visual range estimation.

3-49. Visual range estimation is the least desirable method of range determination because of its inaccuracy. However, in the offense or in a hasty defense, it may be the only method available.

Disable Armored Vehicles

3-50. Armored vehicles are hard to destroy when firing at their front. Use the following three ways to disable them:

- Mobility kill. In this disabling method, the vehicle has stopped moving because a track or road wheel has been destroyed, or the vehicle has been hit in the engine compartment. The vehicle can no longer move but can return fire.
- Firepower kill. When a firepower kill has occurred, the main gun cannot return fire because of a hit in the turret, knocking out its capability to fire. The vehicle can still move, thus enabling it to get away.
- Catastrophic kill. In this kill, the vehicle is destroyed. To obtain a catastrophic kill, firers prepare to fire a second or third shot to destroy the vehicle.

CALL FOR FIRE

3-51. A call for fire is used to obtain fire support from other units. Fire support may be needed in the rear area if the enemy endangers key units or facilities. Fire support may come from mortars, artillery, Army aviation, and US Air Force aircraft. Before a mission, the commander will outline the TRPs and the priority of fires in the OPORD that will affect the call for fire response time. All MP must know how to call for and adjust fire. To call for fire the—

- Leader tells the RTO that a target has been seen.
- RTO initiates the call for fire while the target location is being determined.
- RTO sends the information as it is determined instead of waiting until a complete call for fire has been prepared.

3-52. MP may either go directly to the fire direction center (FDC) of the firing unit for artillery fire support or relay communications to MP leaders, the rear area CP fire support element, or when so directed, a TCF.

Use an Initial Call for Fire

3-53. Artillery fire support can provide the rear area with on-order fires to assist in countering threat incursions. A standard call-for-fire message is used to obtain artillery or other fire support. Regardless of the method of the target location used, the call for fire consists of six elements transmitted in three parts. There is a break and a read back after each part.

3-54. **First Transmission**. Send elements one and two during the first transmission.

- Element 1 includes the identification of the observer. This element tells the FDC who is calling and clears the net for the remainder of the call.
- Element 2 includes the WO. The type of fire support mission and the method of locating the target are identified in this element. See *Table 3-3* for the types of fire missions.

3-55. **Second Transmission**. Element 3 is sent during this transmission and includes the target's location. A target location may be provided by the grid coordinate

Table 3-3. Types of Artillery-Fire Missions

Mission	Description
Adjust fire	This mission is used when the observer is uncertain of the exact location of the target. The observer says, "Adjust fire."
Fire for effect	The observer uses this mission when he is certain of the location of the target. He is requesting a first-round fire for effect in this mission to validate the desired effect on the target with little or no adjustment. The observer says, "Fire for effect."
Suppress	This mission is used to quickly bring fire only on a preplanned target. A target description is not provided when requesting this mission. It is a simplified call for fire and is sent in one transmission. An example of what an observer may say is, "P25—this is P59—suppress AB2502—over."
Immediately suppress	This mission is similar to that of suppression. The difference is a planned target or a target of opportunity is firing at friendly soldiers or aircraft in this mission. The observer may say, "P25—this is P59—immediate suppression AB2503—over."

(normally six digits), a polar plot, or a shift from a known position method.

3-56. **Third Transmission**. Send elements 4, 5, and 6 during this transmission.

- Element 4 includes the target's description. A brief description of the target is given to the FDC using size, nature, activity, protection (SNAP). SNAP represents the size or shape, the nature or nomenclature, the activity, and the protection or posture of the target.
- Element 5 includes the method of engagement. This element consists of the type of adjustments, danger close, trajectory, ammunition, and distribution. The observer specifies how he wants to attack the target.

- Element 6 includes the methods of fire and control. The observer states who will give the command for fire to begin. If the observer wishes to control the time of firing, he will say, "At my command." If the observer does not say this, the FDC will fire as soon as the element is ready.

3-57. At a minimum, a call for fire must include the first four elements. Untrained observers need to send only the first four elements, and the FDC will decide the methods of engagement, fire, and control. Every MP must know that in order to put indirect fire on a target, the following information must be given to the FDC:

- Who he is.
- Where and what the target is.
- How close the target is to friendly troops.
- Where the target is in relation to his or other known positions.
- The direction from himself to the target. Determine the direction during—
  - Daylight with the mini-eye safe-laser infrared observation set (MELIOS), AN/PVS-6. The lightweight laser rangefinder is capable of determining ranges 50 through 9,995 meters. Refer to *TM 11-5860-202-10*.
  - Hours of darkness with the infrared (IR) illuminator, AN/PEQ-2A. The AN/PEQ-2A is for use with night vision devices (NVD) and can be used as either a handheld illuminator or pointer or can be weapon-mounted. In the weapon-mounted mode, the AN/PEQ-2A can be used to accurately direct fire as well as illuminate and designate targets. This item is fully waterproof and can be taken down to extended depths without risk of leakage.

3-58. Determining the direction to a target is an essential skill for the observer. Direction is an integral part of terrain map association, adjustment of fire, and target location. Use the following paragraphs to manually determine the direction to a target.

- Use a compress. Using an M2 or lensatic compress, the forward observer (FO) can measure the direction. The FO will add or subtract the grid of magnetic (GM) angle to determine the grid direction to send to the FDC.
- Scale from a map. Using a protractor or an observed fire (OF) fan, the FO can scale the direction from a map to an accuracy of 10 mils.
- Measure from a reference point. Using a reference point with a known distance, the FO can measure the angle between the reference point and the target and add or subtract the measured angle to or from the known direction to determine the direction to the target. The angle between the reference point and the target can be measured with binoculars or with the hand measurement technique depicted in *Figure 3-4.*
- Estimate. With a thorough terrain-map analysis, the FO can estimate the direction by visualizing the 8-cardinal directions (north [N], northeast [NE], east [E], southeast [SE], south [S], southwest [SW], west [W], and northwest [NW]).

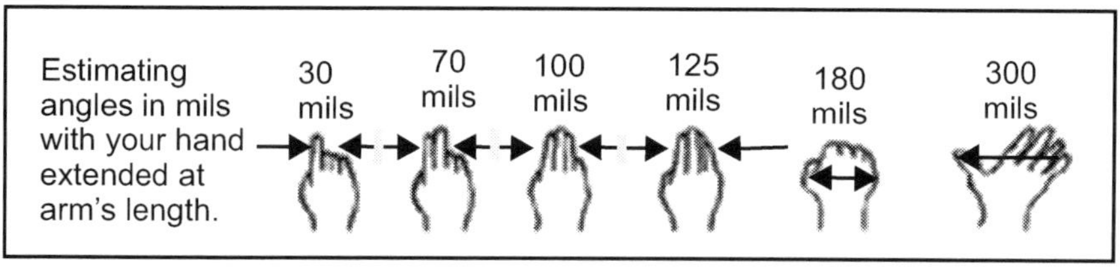

Figure 3-4. Determine the Direction to a Target

NOTE: The observer tries to be as accurate as possible, and the use of mils is preferred. All measured directions sent to the FDC will be rounded to the nearest 10 mils.

Identify the Elements of a Call for Fire

3-59. FDC personnel will help in the call for fire and subsequent adjustments by asking leading questions to obtain the information needed. Refer to *FM 6-30* for call for fire. The elements of a call for fire include the following:

- Observer identification. The observer identification tells who you are. Use the call signs from the signal operating instructions (SOI).
- Warning order. The WO alerts the firing units of the—
  - Type of mission. This includes adjusting fire, firing for effect, suppressing fire, and immediately suppressing fire.
  - Size of element to fire. Omission indicates a request for one field artillery (FA) battery. If the fire mission requires an element larger than a FA battery, state the size needed, such as 2 FA batteries (battalions).
  - Method of the target's location. The grid has no announcement. Announce the word "polar" for the polar plot. Shift from a known point by announcing the word "shift" followed immediately by the designation (target number) of the known point.
- Target location. The target location enables the FDC to plot the target.
  - Grid: Two-character, 6-digit grid, such as NA123456.
  - Polar: Direction (grid azimuth) and distance (meters) to the target from the observer's position. Give the difference in elevation if there is a vertical shift of over 35 meters between the observer and the target.
  - Shift from a known point or the direction to the target (grid azimuth). The three types of shifts are the lateral shift (left or right) in

meters, the flange shift (add or drop) in meters, and the vertical shift (up or down) over 35 meters from the known point and target.

- Target description. The target description helps the FDC to select the type and the amount of ammunition. A word picture of the target (for example, the number and type of vehicles or personnel observed).
- Method of engagement. The method of engagement tells the FDC how to attack the target, which includes the following:
  - Type of engagement. Area fire is standard without a request. Request precision fire only to destroy a point target.
  - Trajectory. A low-angle trajectory is standard without a request. An high-angle trajectory is at the request of the observer or when required due to masking terrain.
  - Danger close. Danger close is announced when applicable.
  - Ammunition. Ammunition is the type of projectile, the type of fuse action, and the volume of fire desired in the fire-for-effect phase stated in rounds per howitzer.
  - Distribution. Distribution is the type of sheaf desired and parallel is standard without request.
- Method of fire and control. The method of fire and control tells the FDC how you want to control the delivery and adjustment of the fire.
  - Method of fire. In area fire, the adjustment normally is conducted with one howitzer or with the center gun of a mortar platoon or section. If for any reason the observer

determines that *platoon right (left)* will be more appropriate, he may request it. Adjusting at extreme distances may be easier with two guns firing. The normal interval of time between rounds fired by a platoon or battery right (left) is 5 seconds. If the observer wants some other interval, he may so specify.

- Method of control. Use the following methods of control: *Fire when ready* is standard and no request is required; use *at my command* when weapons fire at the observer's command; use *cannot observe* when fire will not be observed; use *time on target* when rounds land at a specified time; use *continuous illumination* when the FDC will determine when to fire; use *coordinated illumination* when illumination rounds are fired only when the target is engaged; use *cease loading* when missions have two or more rounds in effect, causing the firing unit to stop loading rounds; and use *check firing* when there is a temporary halt in firing.

- Danger close. Include the term *danger close* in the method-of-engagement portion of the call for fire when the target is within 400 meters of any friendly troops or mortars and 600 meters for field artillery. When adjusting naval gunfire, announce "danger close" when the target is located within 750 meters when using naval guns that are 5 inches or smaller. For naval guns larger than 5 inches, announce "danger close" when the target is within 1,000 meters. The creeping method of adjustment will be used exclusively during danger close missions. The FO makes range changes by creeping the rounds to the target using corrections of less than 100 meters.

Plot a Target Location

3-60. Selecting a targeting method includes giving the directions in mils, degrees, or cardinal points of the compress (N, NE, S, SW, E, SE, W, and NW). Give a deviation of left or right and the distance in meters. Use the following paragraphs when plotting a target location:

3-61. Grid. Determine a 2-character, 6-digit grid for the target. Then, determine a grid direction to the target, and send it after the call for fire and before any subsequent corrections.

3-62. Polar. Determine the grid direction to the target. Determine the distance from the observer to the target. Determine if any significant vertical interval exists.

3-63. Shift. Refer to *Figure 3-5* and determine—

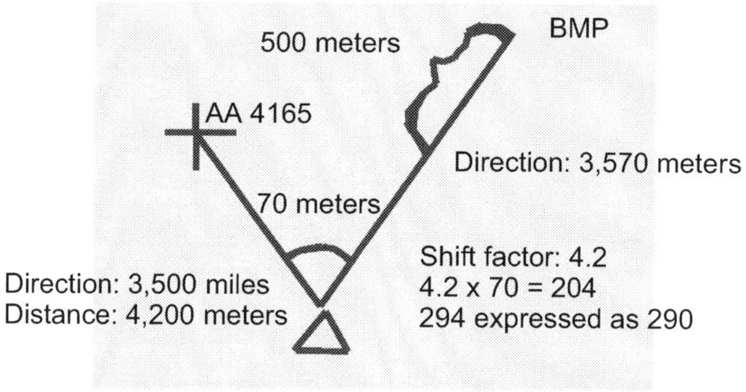

Figure 3-5. Call for Fire Elements

- The grid direction to the target.
- The lateral shift to the target from the known point. W = R times m (mil relation formula), when—

$W = R\ x\ m$

W = width of the lateral shift (the unknown)

R = shift factor, the distance to the known point divided by 1,000 and expressed to one decimal place

m = measured angle in mils from the known point to the target

- The range shift from the known point to the target.
- Any significant vertical interval that may exist.

Example of plotting a target location: Complete target location – direction 3,570, right 290, add 500

Adjust Fire

3-64. The objective in adjusting fire is to move the center of the impact to within 50 meters of the center of the target. The observer accomplishes this by sending the FDC subsequent corrections, which are deviation (lateral) and range corrections. The FDC can talk the observer through the adjustments if they are necessary. The burst is moved to, and kept on, the observer target line in order to get a positive range spotting. The observer target line is the line of sight (an imaginary line) between the observer and the target. When range spotting cannot be determined, the observer makes a request for a lateral correction to place the burst on the observer target line.

3-65. The observer makes range corrections to bracket the target between two successive rounds. The successive bracketing technique is used. After the first definite range spotting is determined, a correction is sent to the FDC to establish a bracket of known distance around the target (such as, one round over the target and one round short of the target). The observer then successively splits this bracket until he is within 50 meters of the target and calls for fire for effect.

3-66. Hasty bracketing is used if the nature of the target dictates that effective fires are needed faster than the above procedures can provide. Hasty bracketing depends on a thorough terrain analysis to give the observer an accurate initial-target location. A bracket is obtained on the first correction in a manner like that used for successive bracketing. Once the initial bracket is established, it is used as a yardstick to find subsequent corrections. The observer sends the FDC the correction to move the rounds to the target and calls for fire for effect. Hasty bracketing improves with observer experience and judgment. Fire for effect consists of one or more rounds from each gun of the unit firing at the target. Dispersion of the guns will cause the rounds to saturate the area with shell fragments. To end a fire mission, the observer states, "End of mission," and reports the results of the fire for effect. The observer may say, "End of mission, three T-62s neutralized, estimate two casualties, over."

Illuminate the Battlefield

3-67. Battlefield illumination can provide MP with enough light to aid in ground operations at night. Illumination can—

- Mark the targets for CAS.
- Increase visibility in areas of suspected enemy activity.
- Furnish the direction to the patrol activity.

3-68. Illumination is called for and adjusted like other indirect fires except the methods of engagement, fire, and control differ. The observer requests illuminating shells. The methods of fire and control differ in that the adjustment is based on how much visibility is needed in the target area. If the observer calls for—

- "Illumination," the observer gets one round from one gun.
- "Illumination, two guns," the observer gets one round each from two guns. The rounds will burst simultaneously.

- "Illumination, range and lateral spread," the observer gets one round each from four guns. The rounds will burst simultaneously in a diamond pattern.

3-69. The initial request for illumination must include the—

- Date when the illumination is needed, if illumination is preplanned.
- Purpose of the illumination.
- Requested time and duration of the illumination (for example, three minutes at 2150 hours or three minutes on call).
- Grid reference and, if needed, the height of the points or areas to be illuminated.
- Method of controls (any restrictions in the time and the place before and during the operation).

OBTAIN ARMY AVIATION FIRE SUPPORT

3-70. Army aviation provides the echelon commander with the ability to move combat resources across the battle space with little regard for the terrain's barriers. These units can provide surveillance or screen over a wide area in adverse weather and at night. Attack helicopter units provide the sustainment area with highly maneuverable antiarmor firepower. They are ideally suited for situations in which rapid reaction time is critical.

Control the Fire

3-71. While en route to a target area, the attack helicopter will contact the caller on the radio. For example, "1L22, this is 1X47, fire team arrives at estimated target area in 4 minutes, over." At this time, a call for fire is transmitted consisting of—

- The target's location and description.

- The proximity of the friendly unit to the target. The words *danger close* must be included when a friendly unit is 600 meters or less from the target. *Danger close* is required because some types of ordnance cannot be used in close proximity to friendly ground forces. When *danger close* is included, MP must mark the unit's location. The method of marking should be one that least reveals the position to the enemy, such as using panels or mirrors.
- The protection of the friendly units (such as good fighting positions, hasty positions, or exposed positions).
- The direction of the friendly unit from the target (cardinal direction).
- Other friendly fire support considerations, including artillery and mortars firing in the area and tactical aircraft (attack direction and altitude).
- Dangers to the flight. Report locations of known or suspected enemy antiaircraft weapons or other dangers to the flight (wires in the target's area, enemy artillery fire impacting in the target's area, or enemy aircraft).

3-72. When the helicopter arrives over the objective, the helicopter's fire team contacts the caller. The caller marks the target and states the method of adjustment. The caller can use the three following methods to mark the target:

- A reference either to a prominent terrain feature that can be identified from the air or to a known point.
- A direction to the target from a reference point, stated in mils or degrees.
- References to friendly fire (such as smoke grenades, tracers, smoke streamers, mortars, artillery, or marking rockets).

Use the Fire Adjustment Method

3-73. The three methods used to adjust the fire of a fire team are impact-observed, impact sound, and observer target. The preferred methods of adjustment for an attack helicopter in support of a ground force are impact observed and impact sound. Once established, the caller does not change the method of adjustment unless the situation dictates. If the method of adjustment is changed, the caller informs the fire team. When any adjustment is 50 meters or less, the observer transmits the adjustment and calls for fire for effect.

3-74. When using the impact-observed method of adjustment, the observer estimates the direction to the target by using a cardinal heading. He estimates the distance from the point of impact to the target in meters. When the observer cannot see the point of impact, he may use the impact sound method of adjustment. For this method of adjustment, the observer transmits, "Adjust fire. Impact sound. Over." The impact sound method differs from the impact-observed method in that the observer senses by sound, rather than sees, the direction of the impact and makes his corrections accordingly.

3-75. Although the impact-observed method is most preferred for adjusting the fire of the attack helicopters, the observer target method, which is less desirable, may be used. When using the observer target method, the observer must mark his location, possibly compromising his location.

3-76. To use the observer target method, the observer senses the direction, left or right, and the distance, in meters, from the impact to a point on the observer target line. Then, he senses the position of the point on the observer target line relative to the target, long or short, and the distance along the observer target line to the target. The sensing of the observer must be converted to corrections, such as right, left, add, and drop, and transmitted to the fire team. The chance of error for this method is greater than for the other methods. Unobserved rounds are handled the same as for the other methods. An example of an exchange of information between an observer and a fire team using the observer target method follows:

- Observer: "Left, five-zero. Add 100. Fire for effect. Over."
- Fire team: "Roger. Out." (The team commits against the target.)

3-77. When the target is suppressed or destroyed, the following transmission would occur:

- Observer: "End of mission. Target suppressed (destroyed). Over."
- Fire team: "End of mission. Out."

3-78. Reference points are used to visually locate the target. The eyes of the pilot are led to the reference point and from the reference point to the target, sometimes through a series of decreasingly obvious reference points. It is much harder for a pilot to find a target than to keep a target in sight. Any reference point must stand out or contrast with its surroundings.

Adjust Fire

3-79. Attack helicopter fire allows the pilot to observe the impact and effect of the ordnance on the target. This simplifies the adjustment procedure. However, the observer must still be prepared to adjust direct aerial fire. When adjusting aerial fire—

- Establish a reference point. The point of impact of the first round is the recommended reference point.
- Adjust for target strike. Do not try to bracket the target. The helicopter crew has direct visual contact with the target and needs only specific directions to fix the location.

- Transmit the corrections.

3-80. The following is a typical exchange of information between the observer and the fire team:

- Observer: "Adjust fire. Impact observed. Over."
- Fire team: "Impact observed. Out." (The team fires at the target.)
- Observer: "Northwest, seven-five. Over."
- Fire team: "Roger. Out." (The team fires at the target.)
- Observer: "North, three-zero. Fire for effect. Over."
- Fire team: "Roger. Out" (The team commits against the target.)
- Observer: "End of mission. Target suppressed (destroyed), over."
- Fire team: "End of mission. Over."

3-81. If the observer does not see the impact, the transmission would be—

- Observer: "Unobserved. Over."
- Fire team: "Unobserved. Over." (The team fires at the target.)

3-82. Adjustments continue until the mission is accomplished. *Table 3-4* shows how to direct a pilot to the target.

Table 3-4. Directing the Pilot to the Target

Ways of Directing the Pilot		How Used
Ammunition	Smoke rounds	Mortars, artillery, or grenade launchers. Phosphorous is usually the best because smoke clouds blossom quickly and are highly visible.
	Ordnance	Ordnance impacting the ground may be an adequate reference point.

Table 3-4. Directing the Pilot to the Target (Continued)

Ways of Directing the Pilot		How Used
	Illumination rounds	Good for guiding the strike aircraft to the target at night, but will not pinpoint small targets if the flares function at the usual height
	Trace fire	Used at night. The intersection of the two streams of traces or the impact point of one stream marks the target.
Fires	Grass or other	Sometimes used near the target as a night reference
Recogniz-able known points	Terrain features or landmarks	If clearly visible from the air, it can help when used with another location method (streams, roads, bridges, tree lines, cultivated areas, prominent hills).
	Friendly positions	When clearly recognizable from the air, it may be used day or night for locating close-in targets.

OBTAIN AIR FORCE TACTICAL-AIRCRAFT FIRE SUPPORT

3-83. During major enemy incursions in the rear area, fighter aircraft may be available to support ground operations by providing immediate CAS. CAS consists of air attacks against enemy targets that are close to the friendly forces. CAS requires detailed coordination with the maneuver of the ground forces to be effective. The coordination must be responsive, integrated, and controlled. Typical CAS targets are—

- Enemy troop concentrations.
- Fixed positions.
- Armored units of immediate concern to the ground forces.

3-84. CAS missions are flown at the request of the command level. They are planned, directed, and controlled by the Air Force through the Tactical Air Control System.

3-85. Air Force support is directed through a forward air controller who, in turn, talks to the pilots. The controller can be in an aircraft or operating on the ground. In most cases, the controller will come forward to a point where he can see the target. Once the target is in sight, he can adjust the aircraft to the target. If the controller cannot see the target, the observer will have to tell him how it can be identified. The observer must make sure that the controller knows where all the friendly elements close to the target are located.

3-86. If the observer is unable to talk to a forward air controller, he must contact a fire-support team operating in the maneuver area. Fire support teams have the equipment to talk directly to the pilots of the aircraft and are trained observers for CAS.

Mark Friendly Positions

3-87. Friendly positions are marked during close air strikes if there is no danger of compromise to enemy observers. This may require only a message (such as "All friendly positions are south of the target. Nearest are 500 meters."). As a rule, a mark is usually necessary when friendly troops are within 300 meters of the target. Marking of friendly positions may be overt or covert and include the following:

- Weapons fire. Weapons fire is useful as a signal if it is distinguishable from other types of fire. Tracers are especially useful.
- Smoke. Smoke grenades are the most commonly used overt marker, but they have limitations. Wind may move the smoke away from the source. Red and white smoke show up well with most backgrounds while some colors blend with their background.

- Mirrors. Signal mirrors and panels are probably the best covert ground-to-air devices for attracting attention. When the operator is proficient and the sun is shining, pilots can see a mirror flash for many miles away. Mirror signaling requires practiced training.
- Balloons. Balloons make a good covert marking system for use above a thick forest canopy.
- Flares. Rockets or 40-millimeter flares are good for attracting attention at night. If flares are fired in the direction of the aircraft, they can be mistaken as ground fire.
- Lights. Strobe lights produce a bright pulsing flash that is visible at night from 2 to 5 kilometers. Vehicle lights, such as unshielded red taillights, are visible to a pilot for several kilometers at night. Chemical glow lights may also be used.
- Ground commander's pointer. The pointer is a handheld device that is invisible to the naked eye, but its beam is visible through NVDs. It may be used by ground troops to clearly show air elements the location of friendly elements.

Select Attack Headings

3-88. A fighter aircraft is more likely to destroy its target if it attacks along the long axis of the target. Once the fighter aircraft knows where all the friendly units are and where the target is, the forward air controller tells the fighter pilot which attack heading to use. However, if the controller cannot see the target, the observer may have to recommend a direction of approach. The observer must remember that fighters should not attack across friendly positions.

3-89. An attack toward friendly units is undesirable because of ordnance dispersal patterns. An attack from behind and over friendly lines is also undesirable for several reasons. Some fighters dump empty cartridges overboard as they strafe. An empty 20-millimeter case weighs 114 grams and hits the ground at 167 kilometers per hour. An even greater hazard would be an inadvertent bomb release as the pilot repeatedly selects and arms his weapons systems while in the attack pattern.

MOVE

3-90. Movement by MP teams, squads, or platoons in combat is dependent on the mission, the terrain, and the likelihood of enemy contact. MP apply the fundamentals of movement which include—

- Moving on covered and concealed routes.
- Avoiding likely ambush sites.
- Enforcing camouflage, noise, and light discipline.
- Maintaining all-around security, to include air guards.
- Using formations and movement techniques based on METT-TC.

3-91. In addition to applying the fundamentals of movement, MP leaders ensure that they—

- Maximize the capabilities of HMMWVs and ASVs. This includes considering the speed, mobility, and firepower of the vehicles. Fire and move both vehicles as a weapon system.
- Make contact with the enemy using the smallest force possible. MP move with a small force in the lead with the rest of the unit ready to react. A team leads a squad and a squad leads a platoon. One team leads another when two vehicles are moving. This prevents the whole unit from being pinned down by enemy fire and provides the unit with the flexibility to maneuver.

- Use the terrain. The terrain offers natural cover against enemy fire and conceals them from enemy observation. MP leaders must devote constant attention to protect vehicles and prevent them from skylining. MP make use of all natural cover and concealment when moving or stopped. When MP do stop, they stagger their vehicles on the roadway.
- Control subordinate elements. MP leaders issue clear and complete orders to subordinate elements in order to maintain control. They issue OPORDs, which cite MP actions to be taken on contact and the immediate actions the teams should accomplish. The OPORDs also explains how the MP leader will direct subordinate elements through the use of hand and arm signals, pyrotechnics, and other visual signals.

USE MOVEMENT TECHNIQUES

3-92. Movement techniques are designed to minimize the exposure of the platoon to enemy fire and to place the platoon in a good position to react to enemy contact. They provide varying degrees of control, security, and flexibility. The selection of their use is based on METT-TC and the likelihood of enemy contact. Their effective use results in the platoon's contact with the enemy with the smallest platoon element.

3-93. MP employ the following three techniques of movement on the battlefield:

- Traveling.
- Traveling overwatch.
- Bounding overwatch.

3-94. While these techniques provide a standard method of movement, the leader must use common sense and sound judgment when employing them as he performs his missions and encounters different situations. The decision of which technique to use is based on terrain considerations and whether enemy contact is not likely, possible, or expected. The techniques are used in both the mounted and dismounted modes of movement.

Traveling

3-95. An MP leader selects the traveling method of movement when contact with the enemy is not likely and speed is required. This technique allows the lead and trail elements to move together as a unit. It is the fastest but least secure movement technique. Movement is continuous, and interval and dispersion are maintained between the squads as terrain and weather permit. The platoon does not intend to engage in combat, but it is dispersed to prevent destruction in case of unexpected air or ground attack. The distance between the vehicles is based on the factors of METT-TC. This method of movement, with MP mounted, is depicted in *Figure 3-6.*

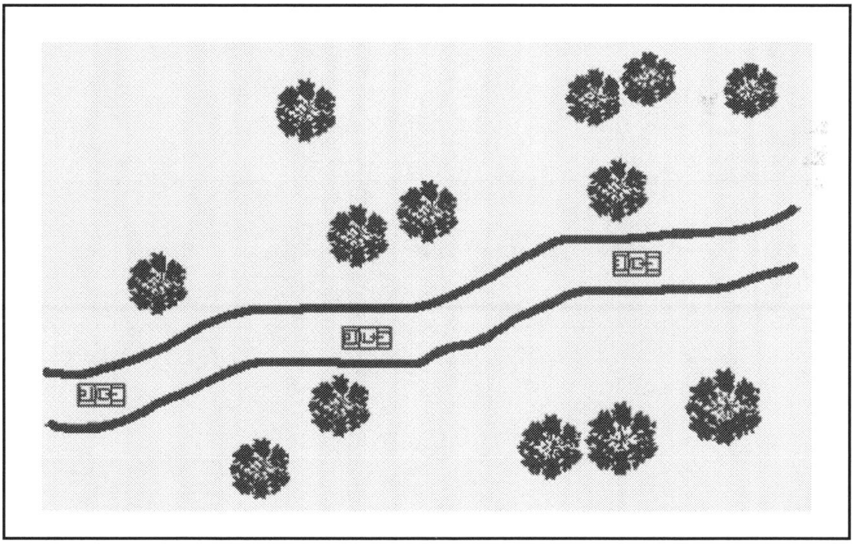

Figure 3-6. Traveling

Traveling Overwatch

3-96. Use the traveling overwatch method of movement when contact with the enemy is possible and speed is desirable. The lead element moves continuously along the best, covered and concealed routes for protection from possible enemy observation and direct fire. The trail element moves at variable speeds, continuously overwatching. It normally maintains contact with the lead element and may stop periodically for better observation. The trail element tries to stay one terrain feature behind the lead element, but close enough to provide immediate suppressive fire and maneuver for support. However, it remains far enough to the rear to avoid contact with the same enemy force that is engaging the lead element. This technique, with MP mounted, is depicted in *Figure 3-7*.

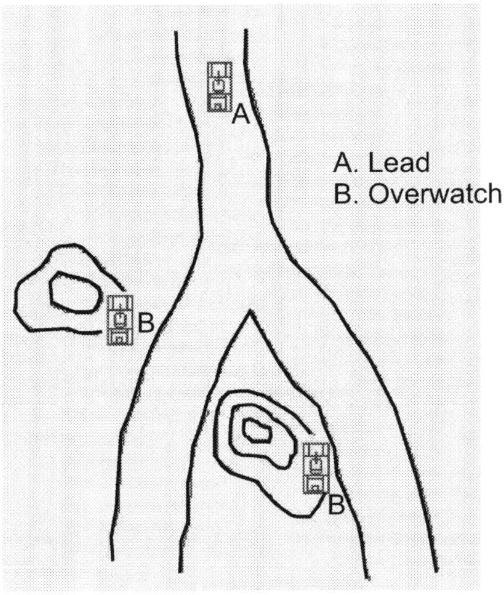

Figure 3-7. Traveling Overwatch

Bounding Overwatch

3-97. When MP expect to make contact with the enemy, they select the bounding overwatch method of movement. It is the slowest, but safest method of movement. In bounding overwatch, the trail element occupies a good, covered and concealed position to overwatch the lead elements. While one element moves, another is always stopped in position to overwatch the bounding element. Although, bounding overwatch is used when enemy contact is expected, use it when time is available regardless of the likelihood of enemy contact. It provides for immediate, direct suppressive fire on an enemy force that engages the bounding element with direct fire. A three-vehicle team uses the V-formation with bounding overwatch. The lead elements advance to a point (first move) where they can support the advance of the overwatch element. On signal, the overwatch element moves forward to a position abreast of the lead elements (second move) and halts. During its move, it is overwatched by both lead elements. The lead elements then move forward, secured by the overwatch. Maximum use is made of folds of the earth and concealment to mask movement from likely enemy positions. MP are mounted in this method of movement as shown in *Figure 3-8*.

PLATOON LEADER RESPONSIBILITIES

3-98. In the conduct of most tactical missions, the MP platoon moves as separate squads under the C^2 of the platoon leader. Regardless of which movement technique is directed, the platoon leader issues the squads an order explaining what each squad will do. This action becomes more critical as the likelihood of enemy contact increases. The platoon leader tells, and if possible, shows the squads—

- The enemy situation as he knows or suspects it to be.

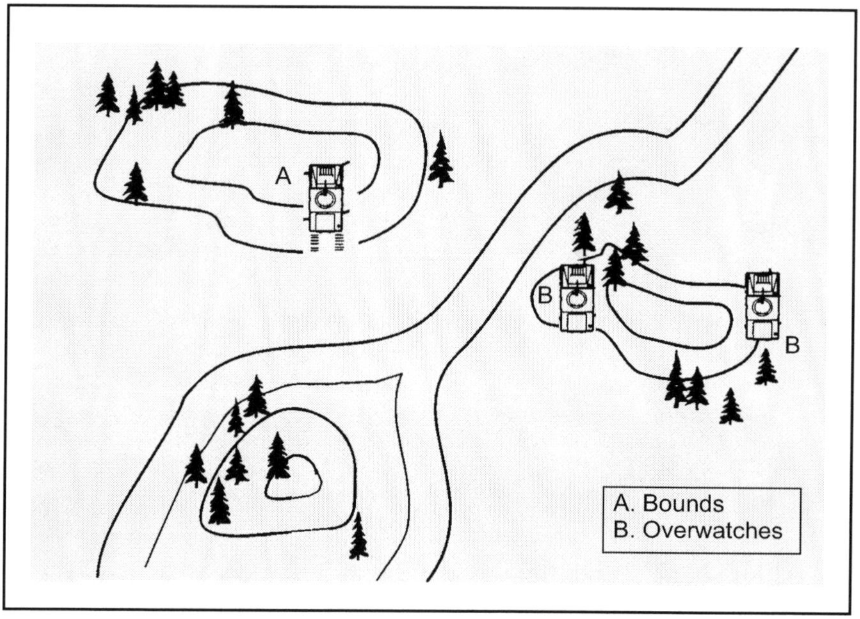

Figure 3-8. Bounding Overwatch

- The next overwatch position (objective for the bounding element).
- The route of the bounding element to that position.
- What he wants the squad to do after the bounding element gets to the next position.

USE EXTENDED AND TEMPORARY HALTS

3-99. When an MP platoon moves as an element, it uses the coil formation for extended halts. This formation provides the platoon with 360-degree observation and fields of fire. The coil is always executed from either the column or staggered column formation. The platoon uses the four-team organization. The lead team assumes the 12 o'clock position (the direction of travel). Teams occupy the 3, 6, and 9 o'clock positions in twos, facing in the appropriate direction. Ensure that there are 50 to 100 meters between the team's vehicles. The interval between the teams will be 100 to 200 meters (terrain dependent). The HQ element occupies the center of the formation. Each platoon must have its own SOP for the formation based on its METL, war plans, and most common organization; it should practice the SOP as a drill to ensure correct execution. *Figure 3-9* depicts an MP platoon in a coil formation.

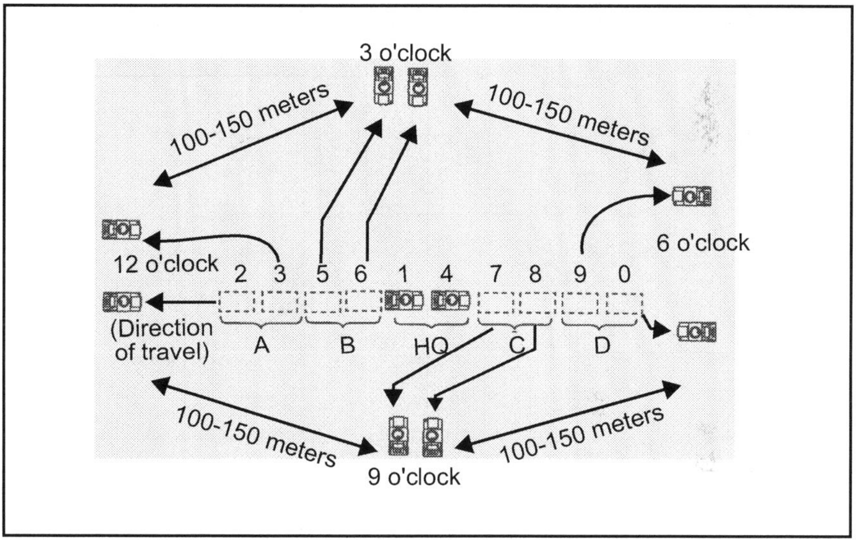

Figure 3-9. Platoon Coil Formation

3-100. MP use the herringbone formation for temporary halts from the march column. It provides them with a 360-degree observation and field of fire *(Figure 3-10).*

USE GRAPHIC CONTROL MEASURES

3-101. MP leaders use graphic control measures to regulate or direct the movement, positions, and fire of the platoon. Control measures—

Figure 3-10. Herringbone Formation

- Are not intended to restrict the exercise of initiative (the function of command). Leaders use control measures to clarify their intent, focus the platoon and squad effort, and ensure synchronization. Each control measure should have a specific purpose that contributes to mission accomplishment. If a control measure fails the purpose test, leaders should not use it.
- Can be drawn on a map, an overlay, a sketch, or a terrain model. Leaders should strive to keep control measures easily identifiable and simple. Graphic control measures may include the AA, the route, the release point (RP) and start point (SP), checkpoints, and so forth. *FM 101-5-1* discusses control measures and provides examples of their use.

CROSS A DANGER AREA

3-102. Danger areas are specific areas where there may be potential danger because of the increased risk of detection. Plan a way in which the patrol crosses danger areas to reduce the chances of a fight. MP make specific plans for crossing each known danger area and general plans for crossing unexpected danger areas. Patrols must be able to quickly modify these plans to fit the tactical situation. Typical danger areas include the following:

- Curves and blind spots on roads and trails.
- Streams.
- Open areas.
- Hill tops.

3-103. MP move cautiously at danger areas, using the bounding overwatch or variations of it to cross them. The MP leader decides how a patrol will cross danger areas based on the—

- Amount of time available.
- Size of the patrol.
- Size of the danger area.
- Fields of fire into the area.
- Amount of security available.

3-104. To cross a danger area, a patrol must designate nearside and farside rally points, secure the nearside and farside, and cross the danger area.

3-105. A small patrol may cross all at once, in pairs, or one element at a time. A large patrol normally crosses its elements one at a time. The leader positions security teams far enough out on both flanks and to the rear of the crossing point to give warning of the approaching enemy and to overwatch the crossing element. Once flank and rear security are positioned, the team crosses the danger area. The team crosses quickly, reconnoiters, and secures the far side of the danger area. The area on the farside must be large enough for a full patrol employment. When the team leader knows the farside is safe, he signals the rest of the patrol to cross. As each element crosses, it moves to an overwatch position or to the farside rally point until told to continue movement. When the patrol has crossed the danger area, the security teams cross and rejoin the patrol.

3-106. Open areas frequently afford the patrol the opportunity to observe the enemy from long ranges. Conversely, they often require that the patrol be exposed to possible enemy observation and fire for long periods of movement. The leader must make maximum use of the terrain and employ effective observation techniques to avoid exposing the patrol to a well-concealed and camouflaged enemy.

3-107. Before moving across a large open area, the patrol takes a thorough visual scan of the area. This should be done both dismounted and mounted, using all available optics. This scan focuses not only on finding potential enemy positions, but also on locating covered and concealed routes for bounding and a covered and concealed position to move to. If time and terrain permit, use dismounted troops to move to the far side of the open area and secure it. In very large open areas, use of dismounts may not be feasible because of the distances between covered and concealed positions.

3-108. Once the area has been cleared using visual means and/or dismounts, the squad or platoon moves across it. The patrol uses bounding overwatch because of the likelihood of enemy contact. If the open area is very large, the overwatch vehicle remains stationary until the bounding vehicle has moved a distance equal to half the effective range of the weapon system of the overwatching vehicle. When that point is reached, the overwatch vehicle must move out, even if the bounding vehicle has not yet reached a position of cover and concealment.

MOVE WHILE IN CONTACT

3-109. Maneuver is the technique used for moving while in contact with the enemy. Maneuver involves two actions that occur at the same time. One element moves to a position where it can engage the enemy while another element supports that movement with a base of fire. A patrol maneuvers to move forward, either to close with the enemy or to gain a better position for firing at the enemy. MP can also maneuver to get more information on the position and strength of the enemy. When the position of the enemy is unknown, it may result in an unexpected encounter known as a *chance contact*. MP use maneuver to move away and withdraw safely.

3-110. When maneuver begins, the MP leader most often goes with the base-of-fire element and controls its fire. The base-of-fire element covers the movement element by shooting at the enemy position. The movement element advances within the supporting range of the base-of-fire element, taking a position from which it can fire on the enemy. The movement element then becomes the base-of-fire element, and the former base-of-fire element begins moving. Depending on the distance to the enemy position and the amount of cover and concealment available, the base-of-fire element and the movement element alternate roles as needed to continue moving.

3-111. MP can maneuver mounted, dismounted, or in a combination of both. A fire element using the MK19 GMG will have difficulty moving dismounted. MP move mounted when the terrain protects them from enemy fire, and look for covered and concealed routes.

3-112. When receiving direct fire, the movement element uses maneuver while the base-of-fire element suppresses enemy fire. If the movement element is not receiving direct fire, it uses bounding overwatch or maneuvers internally.

3-113. MP of a dismounted movement element move based on the intensity of the enemy's fire. When facing intense enemy fire with little or no cover, MP may be forced to crawl. They will use the low or high crawl depending on the situation, the need for speed, and the example set by the leader. Although crawling is slow, it reduces exposure to enemy observation and fire. When MP are not moving forward, they place suppressive fires on the enemy. They may need to advance all the way into and through enemy positions by crawling.

3-114. Dismounted MP can use short rushes from one covered position to another when enemy fire allows brief exposure. To do this they should—

- Advance by short rushes to avoid the enemy's fire.
- Try to stay up no more than 3 to 5 seconds. This keeps the enemy from having time to track and engage them.
- Select the next covered position before beginning the rush.
- Rush from cover to cover.
- Not hit the ground just because 3 to 5 seconds are up.

3-115. MP of a mounted maneuver element move based on enemy fire and the terrain. When they move, they should—

- Use the terrain to mask their movement.
- Move quickly between protected positions so that the enemy cannot engage their vehicles.
- Dismount when the terrain no longer provides protection.

EXECUTE ACTIONS ON CONTACT

3-116. When MP encounter enemy forces, they must quickly execute actions on contact. Whether they remain undetected or are identified by the enemy, MP must first take measures to protect themselves, find out what they are up against, and then decide on a COA. To properly execute actions on contact, MP must take action consistent with the following fundamentals of reconnaissance:

- Remain focused on the reconnaissance objective.
- Report quickly and accurately.
- Maintain contact with the enemy.
- Retain the freedom to maneuver.
- Develop the situation rapidly.

3-117. The platoon leader specifies the actions on contact for the platoon. These specific instructions include the engagement criteria and the desired COA, based on the size and activity of the enemy force encountered. By knowing these details ahead of time, MP can develop the situation more rapidly and arrive at and execute the desired COA. The platoon strives to make contact with the smallest possible element. Visual contact in which the enemy is observed, but MP remain undetected, is the goal. This gives the platoon the greatest possible flexibility to maneuver and develop the situation.

3-118. The steps that make up the actions on contact must be thoroughly trained and rehearsed so that the platoon can react instinctively as a team whenever it encounters enemy forces. The four steps which are executed to allow the platoon to accomplish its mission according to the reconnaissance fundamentals are—

- Deploy and report.
- Evaluate and develop the situation.
- Choose a COA.
- Execute a COA.

Deploy and Report

3-119. The MP patrol that gains initial visual contact with the enemy deploys to covered terrain that has good observation and fields of fire. The MP patrol in visual contact sends a report using the SALUTE format. If the element in contact is unable to report or cannot report quickly, another squad or team must report. The elements not in contact temporarily halt in covered terrain, monitor the report, and plot the situation on their maps. The platoon or patrol leader immediately determines the COA.

Evaluate and Develop the Situation

3-120. The patrol concentrates on defining what they are up against. If they have not sent a SPOTREP at this point, they initially focus on getting enough information to send one. If the enemy has not detected them and time is available, the patrol attempts to confirm or determine in detail the enemy's size, composition, activity, orientation, and the locations of the enemy's weapon systems. They search for any additional information regarding the enemy that can help define the situation and update the SPOTREP.

Choose a Course of Action

3-121. Once the patrol has developed the situation and the platoon or patrol leader has enough information to make a decision, he selects a COA. The COA will be within the capabilities of the patrol and allow the patrol to continue the commander's concept of the operation. The platoon or patrol leader considers various possible COAs, including—

- Breaking contact and bypassing the enemy. This COA may be selected when the enemy sees the patrol before the patrol sees them and comes under—

- Sniper fire. In this situation, the patrol returns fire in the direction of the sniper and conducts the maneuver (fire and movement) to break contact with or destroy the sniper.
- Indirect fire. The patrol quickly gets out of the impact area and does not seek cover, as it may be pinned down by doing so. By continuing to move, the patrol is more difficult to hit. The patrol uses the clock system (described later) to break contact.
- Ambush. In an ambush, a patrol takes immediate action. Personnel in the kill zone return fire immediately and quickly move out of the kill zone. Elements not in the kill zone lay down a base of fire (and smoke if available) to support the withdrawal of the elements in the kill zone. The patrol breaks contact and reorganizes at the last rally point. After or while the elements in the kill zone are being extracted, the leader decides whether to destroy the ambush element or break contact based on the situation and the mission. If no guidance is given, the immediate action of the patrol is geared to breaking contact.

- Maintaining contact to support a hasty attack. This COA is appropriate when the MP patrol discovers enemy elements that the higher HQ wants destroyed. The patrol cannot destroy the enemy because it does not have the combat power (level III threat) or because it has other tasks to perform. In this situation, the patrol maintains contact and continues to develop the situation, focusing on supporting the hasty attack by a TCF. The patrol conducts additional reconnaissance and monitors any changes in the enemy's situation. It focuses on information to support a friendly hasty attack.

- Conducting a hasty attack. If contact is made and the MP patrol and the enemy element become aware of each other at the same time and at such a close range that maneuver is not feasible, the patrol may make an immediate assault. The elements nearest the enemy open fire and shout, "contact, front (right, left, or rear)." The patrol moves swiftly into the assault. It stops the assault if the enemy withdraws and breaks contact. If the enemy fights, the assault is continued until the patrol can break contact, the enemy is destroyed, or the enemy breaks contact. In most cases, the patrol cannot, or should not, mass its combat power to defeat an enemy force. If the patrol concentrates, it risks losing the capability to complete its mission and jeopardizes its ability to conduct subsequent missions. If the patrol attacks an enemy, it should only attack small dismounted formations or lightly-armored or unarmored reconnaissance vehicles. Except in self-defense, patrols should avoid attacking heavily armored vehicles or large formations.
- Establishing a hasty defense. The patrol will establish a hasty defense if it cannot bypass the enemy, all the teams are fixed or suppressed, and the patrol no longer has the ability to break contact by maneuver. Patrols should use a hasty defense when the enemy executes a hasty attack. The patrol maintains contact or fixes the enemy in place until additional combat power arrives or the patrol is ordered to move. If the patrol is required to conduct a hasty defense, the commander then becomes responsible for continuing to develop the situation.

Execute a Course of Action

3-122. The platoon leader updates his SPOTREP to the commander with any new information and then recommends a COA to the commander. The commander approves or disapproves the recommended COA based on how it will affect the parent unit's mission.

3-123. If the commander has anticipated the enemy situation the platoon is reporting, he will already have addressed the contingency in the OPORD and given guidance to his subordinates on what COA the platoon executes. In such a case, the platoon leader can evaluate the situation, choose a COA consistent with the higher commander's intent or concept, and execute it without further guidance. He keeps the commander informed of what he is doing as he executes the COA.

BREAK CONTACT

3-124. To break contact with the enemy without disorder, use the clock system. Use this system when the patrol and a larger enemy element see each other at the same time. The patrol must break contact or be destroyed. The direction the patrol moves is always 12 o'clock. When contact is made, the leader shouts a direction and distance to move (such as "4 o'clock, 300 meters" tells the patrol to move in the direction of 4 o'clock for 300 meters). If contact is broken, the patrol rallies at the designated distance and continues with the mission. If contact is not broken, then another direction and distance is given. This action continues until contact is broken. The leader can also use the clock system to shift or direct fire at a certain location.

CONSOLIDATE AND REORGANIZE

3-125. Once enemy resistance has ceased or the platoon or patrol has broken contact, leaders must quickly take steps to consolidate and prepare to defend against a counterattack. In consolidating on the objective, all-around security is critical because the enemy might counterattack from any direction. The leader must evaluate the terrain thoroughly. Platoons and patrols use the following two techniques when consolidating:

- Clock technique. In using this method, the leader designates either a compass direction or the direction of attack as 12 o'clock. He then uses clock positions to identify the left and right boundaries for squads. The leader positions key weapons along the most likely avenue of approach based on his assessment of the terrain *(Figure 3-11)*.

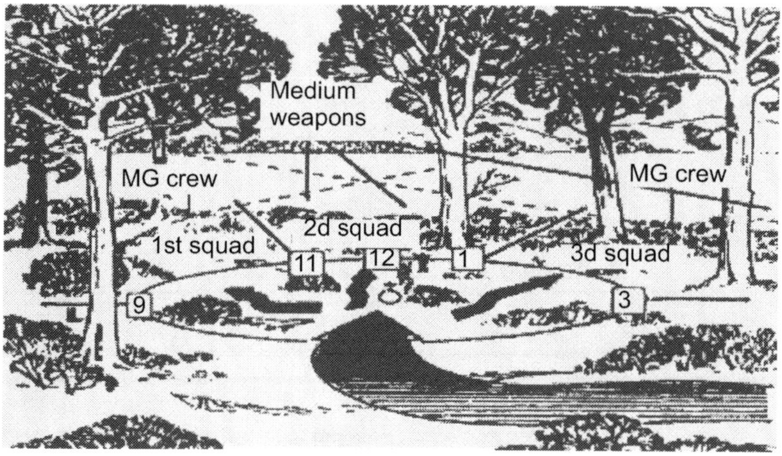

Figure 3-11. Clock Technique

- Terrain feature technique. In a similar manner, the leader identifies obvious terrain features *(Figure 3-12, page 3-58)* as the left and right limits for squads. In both techniques, he ensures that squad sectors of fire overlap each other and provide mutual support for adjacent units.

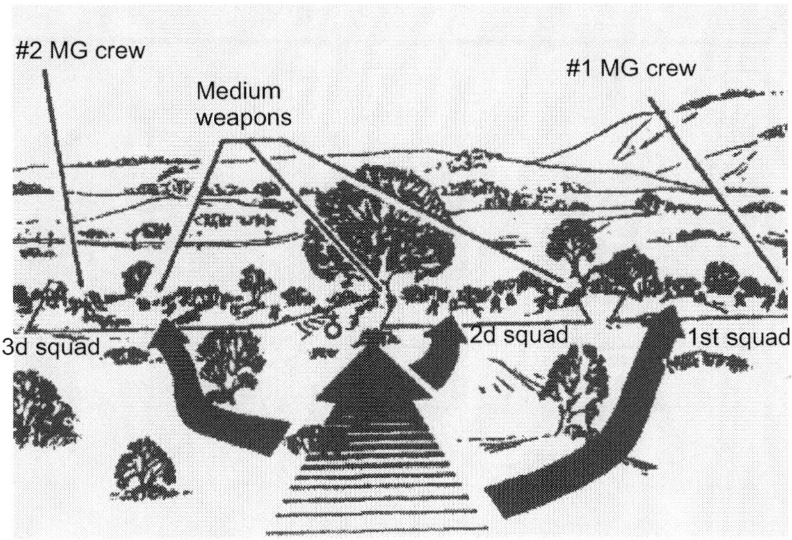

Figure 3-12. Terrain Feature Technique

3-126. Once platoons or patrols have consolidated, they begin to reorganize. Platoons reorganize to continue the mission. Reorganization involves the following:

- Reestablishing C^2.
- Remanning key weapons and redistributing ammunition and equipment.
- Clearing the objective of casualties and EPWs.
- Assessing and reporting the status of platoon personnel, ammunition, supplies, and essential equipment.

COMMUNICATE

3-127. On the battlefield, you must be able to communicate. Communication is the means through which battle command is exercised. MP on the battlefield must be able to communicate to maintain C^2 of their elements, call for fire or request other support, and respond to orders. The chain of command and succession of command must be known throughout the organization. There must be open lines of communication up, down, and laterally. Situations on the battlefield can change rapidly, and losing the ability to communicate for even a short duration can have a major impact on an operation. Commanders must provide for redundancy in means of communications. MP companies have the capability to place backup communication systems at key locations within an area of responsibility. METT-TC and the battlefield situation usually determine communication means. MP use any combination of the following systems:

- Sight and sound.
- Messenger.
- Wire and radio.

USE SIGHT-AND-SOUND SIGNALS

3-128. Visual signals are useful for sending prearranged messages over short distances, during radio silence, or when jamming interferes with radio transmissions. Arm-and-hand signals, flashlights, and pyrotechnics may send quick visual signals. Visual signals have some disadvantages, which include the following:

- They are less effective when visibility is limited.
- They may be seen and intercepted or imitated by the enemy.
- They may be masked by terrain features, reducing the chance of a message being received.
- They are easy to misunderstand.

NOTE: To overcome this last disadvantage, each man in the unit must be able to send, receive, and understand messages using visual signals.

3-129. Although arm-and-hand and light signals are standard throughout the Army, the meaning of pyrotechnic signals must be set in the C^2 portion of the OPORDs and in the SOI. To ensure that a pyrotechnic message was correctly received, confirm the message by some other means as soon as possible. For more information on visual signals, refer to *FM 21-60*.

3-130. Sound signals, like visual signals, work well only for short distances. Simple devices (such as whistles, horns, gongs, and explosives) may be used. Sound signals can be used to—

- Attract attention.
- Transmit prearranged messages.
- Spread alarms.

3-131. A well-known sound signal is the use of metal-on-metal to indicate an NBC hazard or attack. Battlefield noise may blend with or override sound signals, causing confusion and misunderstandings. Sound signals—

- Must be simple to understand.
- May be restricted for security reasons.
- Can be intercepted by the enemy.

USE MESSENGERS

3-132. Using messengers is the most secure way to communicate long messages and documents. However, it is also the slowest and messengers are vulnerable to enemy action. When using a messenger—

- Put the message in writing.
- Make the text clear, concise, and complete.
- Choose the most expedient transportation on hand.
- Encode the message (using the operational code in the SOI) if there is a risk that the messenger might be captured.

- Send a second messenger by a different route if a backup message is needed.

USE WIRE AND RADIO COMMUNICATION

3-133. Often, wire communication is more useful than radio. It is hard to jam, and unlike radio, more than one person can talk at one time. It is used most often for communicating with static posts. Wire communications cannot be secured and may be cut by the enemy. When a wire line has to be checked, MP are sent out in pairs. One MP looks for the cut and the other MP provides overwatch security. The enemy can take prisoners by cutting a line and capturing the soldiers who go to repair it.

3-134. Use a radio to communicate with mobile or distant elements. A secure voice radio is best. The enemy can intercept messages on an unsecured radio net. Regardless of the radio type, if the transmission is heard, the enemy may be able to detect the location of the radio or learn what the unit is doing. Keep the transmissions short. Know and use signal security and electronic counter-countermeasures. MP deny the enemy information from friendly telecommunications by following the SOI. To keep transmissions secure, MP—

- Authenticate all transmissions.
- Use only authorized codes.
- Use secure voice transmissions.
- Use encoded messages.

3-135. The enemy is kept from disrupting radio communications, and friendly emitters are protected from enemy detection, location, and identification by—

- Setting radios at low power.
- Placing antennas where terrain blocks the enemy's interception.
- Using remote radios and antennas.
- Using directional antennas.

- Using wire whenever possible.
- Observing listening silence.
- Using short transmissions.
- Using a random transmission schedule.
- Transmitting only when there is a need to do so.

Chapter 4

Combat Operations

This chapter provides the MP leader with the information needed to successfully complete a combat mission. When MP platoons conduct the tasks addressed in this chapter, they will mirror the actions of the company.

PREPARE FOR COMBAT

4-1. Units that are not directly engaged in combat often find it necessary to travel in order to position themselves for combat operations. During these movements, the battalion supports the company and the company supports and monitors the platoons with the movement plan. The move's success depends on the unit's discipline and ability to execute the plan. (Refer to *Appendix E* for precombat inspection checklists.)

MAINTAIN AND MOVE A COMBAT LOAD

4-2. Combat load is the quantity of supplies, in all classes, that the company must have on hand to sustain operations for a prescribed number of days. The company must be capable of moving the combat load, using organic transportation assets, into combat in a single delivery. To save time, the company combat loads vehicles while the quartering party readies the new site. The combat load ensures that a unit is ready for combat even when it is on the move.

4-3. The principles of combat loading are standard. All equipment, ammunition, and gear are loaded on the vehicles in a logical order and put in predesignated places. Knowing the location of each item allows for quick retrieval during the move. Combat loading also lends to a fast set up at the new site. Like the basic load, the company's combat load is mission-dependent. No single load plan can satisfy all the situations. MP leaders must consider the following:

- METT-TC.
- Vehicle and trailer capacities.
- Weight limits of the unit's vehicles and trailers, being careful not to overload them.
- Whether or not the equipment will fit (cube out). For equipment data, see the applicable technical manual (TM).

4-4. Unit SOPs has load plans tailored for various mission activities. Having a choice of load plans for various deployments reduces the load time. Load plans and diagrams are modified to suit METT-TC and vehicle and trailer capacities. The modifications are shown on the load diagram in the vehicle. HMMWVs may be loaded in many configurations, which include—

- Loading the basic equipment in the mounted standard brackets on the vehicle.
- Modifying and moving the brackets to meet mission or unit requirements.

PLAN A TACTICAL ROAD MARCH

4-5. The basic considerations in planning any road march are the METT-TC factors (especially the enemy situation and the mission); the march order; and the type, number, and characteristics of vehicles available for the movement. A company conducts a tactical road march for relocating and facilitating rapid movement of the vehicles with a prescribed speed and interval between the vehicles. When preparing for a tactical road march, the company uses the following planning sequence when time permits:

- Prepare and issue the WO as early as possible to allow maximum time for preparation.

- Prepare an estimate of the situation, specifying the organization of the march column.
- Organize and dispatch reconnaissance and quartering parties.
- Prepare detailed movement plans based on the organization of the march column and a review of the available reconnaissance information.
- Use the reconnaissance information to—
  - Choose sites for halts and RPs.
  - Spot problem areas along the route.
  - Select bypasses or alternate routes.
- Select fairly secure locations for halts.
- Choose areas that provide cover and concealment.
- Avoid choosing highly populated areas, curves in the road, or other hard-to-secure areas.
- Plan the timing so that the unit arrives at the SP just before it is scheduled to cross it. The time a unit must cross the SP is provided to the unit. As other units may be planning to use the route, each unit must cross the SP on time.
- Prepare and issue the march order.
- Prepare the overlays and issue them to the vehicle commanders and subordinate leaders. The road march overlay includes, at a minimum, the location of the SP, RP, scheduled halts, and checkpoints at critical points along the route.

COORDINATE AND DIRECT THE MARCH

4-6. The chain of command controls the column. The march leader—

- Coordinates the road march through the chain of command with the local movement control unit.
- Determines if the convoy needs a movement credit or a clearance to use the given route. If so, the march leader submits a *Department of*

Defense (DD) Form 1265 through the appropriate movement control center.

- Informs higher HQ and the supported units of the dates and times that the operations will stop at the old site and begin at the new site.
- Tasks subordinate leaders to attend a briefing to discuss unit readiness and load plans and forecast support requirements.
- Submits requests for support based on the forecast developed during the briefing. Requests may include fire support, refueling, vehicle recovery operations, and other support needed to complete the march.
- Issues an OPORD for the movement.
- Requests HQ personnel to prepare a movement table. Refer to *Appendix E* of *FM 55-10* for detailed information on the movement tables.
- Requires unit personnel to analyze the route reconnaissance information looking for likely enemy ambush sites.
- Ensures that a strip map, which may be included as an annex to the OPORD, is prepared. The strip map shows SPs; RPs; route numbers; place names; critical points; directional arrows; distances between the points; scheduled halt locations; and petroleum, oils, and lubricants (POL) refill points. Give copies to the unit drivers.
- Ensures that the drivers and assistant drivers are briefed.
- Ensures that radio communication is kept to a minimum during movement.
- Tasks subordinates to ensure that the road march plan is followed.
- Ensures that safety briefings are conducted and understood.

CONDUCT THE MARCH

4-7. The commander sets the conditions under which military traffic moves at night. The march leader ensures that personnel are aware of and abide by the set lighting conditions when the company moves at night. Conditions that are more restrictive may be imposed contingent on the threat environment (such as air raids). Lighting conditions may include normal lighting, reduced lighting, or blackout. If the situation warrants, travel by total blackout (use of night-vision goggles [NVGs]) may be prescribed. More often, travel is under partial blackout, using only enough light to see the road and to be seen by other road users. Minimal lighting reduces visibility from the air while it permits drivers to—

- Travel as quickly and safely as possible.
- Apply brakes in time.
- See the side of the road.

Movement Techniques

4-8. During a tactical road march, the unit leaders (such as the march leader and the platoon sergeant) travel in separate vehicles. This decreases the chance of a unit's primary leaders being lost in one enemy action. The convoy moves en route by close or open column march or by infiltration. In dusty conditions, space the vehicles so that the dust from one vehicle does not blind the driver of the next.

4-9. **Close Column**. In a close column, the elements are close together. Use a close column for marches during limited visibility, blackout conditions, and radio silence. Under these conditions, space the vehicles so that the driver can see the two lights in the blackout marker of the vehicle ahead. Visibility determines the set distance between the vehicles. A close column—

- Reduces the time it takes for the column to pass points on the route.

- Requires fewer guides, escorts, and markers for control than an open column does.
- Enhances movement through congested areas or over poorly marked routes.

4-10. **Open Column**. In an open column, elements are widely spaced as a passive defense measure. Generally, an open column is used during daylight, but may be used at night with infrared lights, blackout lights, or night-vision equipment. A distance of 50 to 100 meters or more may be designated between vehicles depending on METT-TC. Use an open column—

- When enemy contact is likely.
- To enhance security.
- Over dusty roads. Reducing dust is especially important when moving through areas contaminated by radioactive fallout.

4-11. **Infiltration.** Infiltration is the best passive defense against enemy observation and attack, although it may be difficult to control. To move by infiltration, vehicles are dispatched one at a time or in small groups at irregular intervals to keep traffic density low and to prevent undue massing of vehicles. Use infiltration—

- When time and road space allow.
- When maximum security, deception, and dispersion are needed.
- To maintain security during the march. When the unit approaches likely danger areas (such as bridges and tunnels), have one or more teams dismount. The teams check both sides of the road before the vehicles pass. This is critical when only a map reconnaissance was conducted before the move.

Movement Considerations

4-12. Consider events and control measures that are used during the move in order for the tactical road march to be successful. These measures include the SP and RP, checkpoints, restrictions, speed control, halts, disabled vehicles, and mined areas.

4-13. **Start Point**. A SP gives the vehicles of a march column a common point for starting their movement. When vehicles use more than one route, each route has an SP. The SP is a recognizable place along the route of march (such as a road intersection). It should not be in a defile, on a hill, or at a sharp curve in the road that could cause movement to slow. Ensure that the SP is far enough away from the AAs to allow the vehicles to be organized and move at the prescribed speed when they reach it. Elements of the company reconnoiters the route to the SP to determine the times needed for the serial to arrive at and clear the SP before starting the march.

4-14. **Release Point**. A RP provides all the vehicles of the march column with a common point for reverting to the commander's control. It is a point on the march route that is easy to recognize on the map and on the ground. Guides meet the vehicles as they arrive at the RP and lead them to their new positions, usually in an AA. Multiple routes and cross-country movement from the RP to the assembly areas allow vehicles to disperse rapidly. When leaders select a RP, avoid hills, defiles, and sharp curves that may cause elements to slow or stop on the route. Ensure that vehicles are not required to countermarch or pass through another element to reach their new position.

4-15. **Checkpoints**. Use checkpoints on a route for reference when providing instructions and identifying places where interference with movement might occur or timing may be critical.

4-16. **Restrictions**. Restrictions are points along the march route where the movement may be limited or obstructed during certain time periods (such as bridges, intersections, ferries, or bypasses). The march planner—

- Starts the move early enough to pass such a point before a restriction begins.
- Delays the start of the move to pass a restriction after it has ended.
- Plans to halt the column along the route until the restriction is lifted.

4-17. **Speed Control**. Vehicles in a column of any length may simultaneously encounter different types of routes and obstacles. This causes sections of the column to move at different speeds at the same time, producing an undesirable accordion or whip effect. The movement order specifies the march speed, march rate, and the maximum safe catch-up speed to reduce *column whipping*. The lead vehicle must not exceed the authorized maximum speed of the slowest vehicle in the column. To minimize vehicle congestion on the nearside of an obstacle, vehicle commanders and drivers must be alert and maintain the prescribed minimum following distance. Vehicles should make only gradual speed changes while maintaining their prescribed interval. Vehicle commanders must constantly be aware of the vehicle interval to their front and rear and adjust their speed accordingly.

4-18. **Halts**. Halts are conducted for various reasons. They—

- Allow following traffic to pass.
- Provide time for rest, mess activities, and personal comfort and relief.
- Permit refueling and maintenance of vehicles.
- Provide for maintenance and inspection of equipment.
- Allot time for adjustments in the schedule.

4-19. The time and duration of the halts are usually specified in the movement order or prescribed in the unit SOP. The SOP prescribes actions to take during halts. A short rest halt of 15 minutes is usually taken after the first hour of marching. A 10-minute short halt is taken every 2 hours thereafter. The prescribed march rate includes the time required for short halts. When possible, march elements using the same route should stop at the same time. Route characteristics may make it necessary to halt at a particular point on the route rather than simultaneously at a fixed time.

4-20. Long halts are planned in advance. The length of the halt is added to the total travel time. Locations for long halts are normally selected to allow all the vehicles to clear the road and permit proper dispersion. The unit commander schedules halts for refueling in advance.

4-21. The herringbone formation is used to provide security for the march column during temporary halts. During temporary halts, the MP teams move their vehicles to alternate sides on or off the road in a herringbone pattern that lets vehicles pass down the center of the column. Movement commanders give permission for execution of unscheduled halts. The first priority at any halt is local security. OPs are established and sectors of fire are assigned to each vehicle. These actions should be automatic and part of the unit SOP.

4-22. **Disabled Vehicles**. Disabled vehicles must not obstruct traffic. Their crews must move them off the road and report their status immediately to the PSG. Crews must immediately signal the follow-on vehicles to bypass and continue movement. They then establish security and post guides to direct traffic. If possible, crews repair their vehicles and rejoin the rear of the column just ahead of the trail element. Vehicles that have dropped from the column return to their positions only when the column has halted. The trail party recovers vehicles that cannot be repaired by their crews.

4-23. **Mined Areas**. When a company encounters mined areas, it must remember that the safety of the unit is the most important factor. It bypasses mined areas whenever possible, but considers how the delay will affect the outcome of the mission. Remember to—

- Be cautious. Mines may be used to force an element to take an alternate route into an ambush site.
- Screen the bypass route, if possible, before diverting an element.

NOTE: Refer to *Appendix H* for countermine operations.

4-24. Ensure that all efforts are made to bypass mined areas; however, if the element must cross a mined area when engineer assets are not available to breach the minefield, act quickly and cautiously. Mined areas, like other obstacles, are often covered by enemy fire. Before crossing—

- Detonate mines from a protected position.
- Detonate mine trip wires by rigging an object near the trip wire to fall on the wire.
- Use a hand grenade or direct fire to detonate mines.
- Detonate pressure-sensitive mines by rigging an A-frame over the mine and placing a heavy object, attached to a rope, over the mine. Take cover and allow the object to fall on the mine.
- Devise other methods to detonate the detected mines.
- Send a report to the next higher command when mines have been neutralized. Refer *FM 20-32* and *FM 21-75* for information on mines and countering mines.

ESTABLISH A NEW OPERATIONAL SITE

4-25. MP elements most often will collocate as part of an established base or base cluster. On occasion, MP may need to set up a base of their own. To set up at a new location, whether as part of an established base or base cluster or separately as a company or a platoon base, you must—

- Reconnoiter new sites.
- Pick the most favorable site and its alternate. Choose a site that—
  - Is easily accessible.
  - Can accommodate all the unit's vehicles and equipment.
  - Has a firm, well-drained surface.
  - Has some natural cover and concealment.
  - Is relatively easy to defend.
- Prepare and secure the site.
- Complete the move and establish communication.
- Establish local security to sustain survivability.

USE A QUARTERING PARTY

4-26. A quartering party is needed whenever a unit relocates. The quartering party's mission is to—

- Reconnoiter the area for enemy presence, booby traps, NBC contamination, and other hazards.
- Establish the dismount point and perimeter security.
- Establish communication.
- Identify the company CP.
- Designate vehicle locations.
- Position guides at the RP to meet the main party.
- Prepare the area for occupation and assist the units with occupation.

4-27. The platoon leader or PSG designates vehicles and personnel from the platoon that will be part of a battalion or company quartering party. The entire platoon may serve as the battalion or company quartering party.

4-28. While the unit loads for deployment, the quartering party moves to and readies the new site. The quartering party's job ends when the last vehicle in the main body arrives at the new site. The size of a quartering party is based on the tactical situation and the amount of work required to prepare the site for occupancy. A quartering party for platoon relocation would be smaller than that of a company. A company quartering party is likely to have personnel from the unit's HQ, each platoon, the maintenance and dining sections, and communications.

4-29. The quartering party leader ensures that the equipment and supplies are available to clear, secure, and set up the new site. A quartering party may need—

- NBC detecting and monitoring equipment.
- Mine detectors.
- Saws or axes to clear the wooded areas.
- White engineer tape.
- Portable route signing material.

4-30. The quartering party leader assigns tasks to the teams based on the size of the quartering party, the work to be done, and METT-TC. He ensures that each team has the equipment needed to complete its tasks (refer to *FM 7-10*) and that they are at the proper mission-oriented protection posture (MOPP) level if they are operating in an NBC environment.

March Halts

4-31. At march halts, quartering party teams set up local security. If the vehicles can leave the road, the teams form a 360-degree perimeter around the convoy. If the vehicles cannot leave the road, they are parked at an angle so alternate vehicles face opposite sides of the road. Each team is assigned a sector to observe, with sectors overlapping between the vehicles. Each team member has a specific area of responsibility. The troops remain alert and ready to take action on contact with the enemy. All personnel watch for enemy aircraft.

Site Arrival

4-32. When the quartering party arrives at the site, it clears and then secures the site. One or more teams, after dismounting their vehicles, search the area for mines, booby traps, items of intelligence value, or other signs of enemy presence.

4-33. If nuclear weapons have been used, at least one team using radiation detection, indication, and computation (RADIAC) meters monitors the site for radioactive contaminants. Monitoring for chemical and biological agents must be continuous because it is difficult to detect their first dispersal.

4-34. In urban areas, team members clear buildings that will be used by the unit. Team members clear the structures outside the perimeter if there is a possibility of enemy presence. The priority of buildings to be cleared and the number of teams needed are based on METT-TC. Refer to *FM 90-10-1*.

4-35. When the area is cleared, one or more teams perform the following functions:

- Set up the OPs and the LPs.
- Set up defensive positions on likely enemy avenues of approach. These positions provide early warning and limited protection during the occupation of the new site.
- Prepare the new site for the main body's arrival.

Company Move

4-36. When setting up a company site, the quartering party—

- Chooses a tentative location for the company CP.

- Sets up the company CP where it can best control the company, be well defended, and have LOC to the subelements.
- Uses buildings (in an urban area) to conceal the CP.
- Considers cover and concealment when choosing the CP location.
- Makes use of natural cover and concealment when possible.
- Uses camouflage screens and man-made cover and concealment where needed.
- Sets up the wire communication net. Marks those areas where other unit elements will be positioned, using signs or materials that cannot be easily seen by the enemy.
- Picks roads and trails that permit an easy flow of traffic.
- Chooses alternate exits and marks them for use as emergency exits.
- Designates parking areas for the heaviest, most awkward vehicles, such as 5-ton trucks.
- Selects a troop area and—
  - Marks the areas where latrines, garbage dumps, and tents will go.
  - Uses ground guides for vehicle movement in areas where troops are sleeping. (For safety, unit personnel should sleep only in the troop area).
  - Chooses a structure (in an urban area) that protects the troops from natural elements and has adequate latrine facilities.
- Locates the following:
  - The food service section inside the perimeter, well away from interior roads to keep dust from contaminating the food. Locate the serving line to take advantage of

cover and concealment. In urban areas, select a building for service and meal activities.

- The latrines away from the bivouac area. Place latrines at least 30 meters down slope from wells or other water sources, and at least 100 meters from the dining facility, downwind and down slope, if possible. In urban areas, use existing latrines if they can serve at least 8 percent of the unit at one time.
- The maintenance section where vehicles can arrive easily from the main road through the site. Ensure that vehicles are able to enter the maintenance tent at one end and exit at the other. Use existing garages for maintenance operations in urban areas.
- The supply section to meet space, roadway access, and drainage needs. In urban areas, use warehouse-type buildings for supply operations.
- The tactical communication section where it has space enough to support the whole operation. Usually it collocates with the maintenance section or the operations section.

4-37. When the main body arrives, the quartering party—

- Maintains security as the main body moves into the site.
- Maintains noise and light discipline.
- Ensures that the vehicles rapidly clear the approach route while maintaining vehicle intervals.
- Dismounts all personnel except drivers at the dismount point.

- Briefs the leader of the main body on the situation and the operational status.

4-38. The leader of the main body informs the higher HQ that the move has been completed. He reports the location coordinates for both the CP and the alternate CP by messenger or other secure means. The leader ensures that the entire party prepares fighting positions and other defensive measures.

Platoon Relocation

4-39. A quartering party, in advance of the platoon's relocation, has the same considerations scaled to size and need, as one in advance of a company. An MP platoon may collocate with a company HQ or an existing base. However, more often, a platoon's base must be set up where a platoon HQ can best—

- C^2 its squads.
- Communicate easily with its squads and higher HQ.
- Link squads, the company CP, and the supported unit.

4-40. The platoon HQ can operate from a static base or vehicles. If the platoon elements are going to operate in one location (as they would for an EPW holding area), the platoon leader sets up a static platoon HQ base. If the platoon elements must operate dispersed over a large area, the platoon leader must remain mobile. In such cases, a platoon leader could elect to set up a temporary platoon base as a rally point to report, resupply, and reorganize the platoon's resources.

4-41. The platoon bases are set up the same. The PSG selects a site that offers good cover and concealment. The site must be defendable and allow the HQ vehicle to be located near the tent. A small tent houses the platoon's HQ. Use a radio set control group to relay remote communication into the tent. An antenna increases the transmission distance and is located based on OPSEC principles. Wire communication is limited to the platoons that can hook into an existing wire net.

CONDUCT MILITARY POLICE BASE SELF-DEFENSE

4-42. When collocated with a base or a base cluster, the platoon is integrated into that base's or base cluster's self-defense planning and operations. Although bases and base clusters are more prevalent at corps and echelon above corps (EAC), the same principle applies to MP located at the division or brigade support areas. When an MP base is set up on its own, the base is responsible for its own security and protection.

Collocated

4-43. An MP platoon collocates with a base or a base cluster for logistical support and a means to conduct operations. When collocated, the platoon leader coordinates with the base defense operations center (BDOC) and the base cluster operations center (BCOC) to integrate the platoon's efforts with the base's and base cluster's efforts. The platoon's portion of the base's or base cluster's defense is to help provide early warning of the threat by area security and/or maneuver and mobility support in the area near the base or the base cluster. Because MP resources are austere, the platoon only shares sector efforts on the base's perimeter. MP are used as static posts (such as gates) only under extreme conditions. An MP platoon may be tasked to serve or augment the base cluster's response force. Before the platoon leader accepts this tasking, he consults with the company commander to—

- Ensure that the company commander knows that he may have one less platoon if he is tasked for the TCF or response force for the battalion.
- Allow the company commander to advise the base commander that the MP platoon could

serve as a response force to the base if it is not committed.

4-44. Each base has a BDOC that plans, coordinates, and supervises base defense operations. The BDOC initiates contingency planning that enables the base to—

- Increase the manning posture of the base contingent on the threat.
- Detect and defeat the threat within their capabilities.
- Hold against heavier enemy forces until response forces arrive.
- Maintain control of the fight within the base.
- Support the fire and movement of the response force operating outside the base.

4-45. Each base cluster has a BCOC to monitor base defense plans and establish the base cluster reaction force. The BCOC—

- Provides C^2 of the resources for planning, coordinating, and supervising the base cluster's defense.
- Coordinates base defense operations.
- Maintains communication with bases within the cluster as well as MP, BDOCs, and the sustainment area operations center. A great deal of intelligence is provided to a BDOC and BCOC through the rear operations net, which helps in planning the defense.

4-46. The platoon's plans for the interface of MP support into the base's self-defense plans address—

- Cover and concealment of personnel and equipment.
- Signal security.
- Reliable and redundant communication systems at all guard locations (such as land lines, radio

links to the BDOC, and telephone hookups to the center switch).

- Deception.
- Contingency planning.
- Improvement of base defense positions.
- Assistance of the area MP.
- Coordination with the BCOC or rear-area operation center (RAOC), as required.
- OPs/ LPs.
- Noise and light discipline.
- Immediate reaction to enemy threat or attack.
- Rehearsals of defense measures.

4-47. All plans and overlays depicting MP support are forwarded to the BCOC. There they are consolidated and forwarded to the RAOC. (If a base is not part of a base cluster, the base forwards all the plans and overlays directly to the RAOC.)

Separate Setup

4-48. Set up an MP platoon separately only when there is no other alternative. This is the least desirable means for a platoon to set up. When the platoon sets up as a base separately, it must be able to defend against a wide range of enemy activity. It integrates the defense of its base (including indirect-fire systems, air defense artillery, and tactical aircraft) with the defense efforts of other bases in the sustainment area. Engineers, dismounted troops, armored vehicles, and helicopters contribute to the overall security of the bases. Bases coordinate and synchronize their defense efforts to enhance their strengths and reduce their vulnerabilities. A base's defense priorities include—

- Establishing initial base security.
- Positioning crew-served weapons and troops on assigned sectors of fire.
- Clearing fields of fire and preparing range cards.

- Preparing fighting positions.
- Installing communication.
- Emplacing obstacles and mines.
- Improving primary fighting positions to include overhead cover.
- Preparing alternate and supplementary positions.
- Stockpiling ammunition, food, and water.
- Preparing routes and trenches between positions.
- Developing a counterattack plan.

4-49. Using the IPB process can help the platoon predict threats to base security. Platoons must be aware of the enemy's location, organization, direction of movement, and strength. In the same regard, platoons must have effective OPSEC to deny similar friendly information to the enemy. Platoons can continually improve base defenses by considering what avenues of approach and methods of attack the enemy could use, given the vulnerabilities of the platoon's base. The base defense plan must have overlays depicting weapons positions, sectors of fire, final protective fires, and reaction force contingencies. Plans are updated as often as possible.

4-50. The base's reaction force efforts are coordinated with the designated-area response force. Platoon leaders develop detailed employment plans and exchange as much information as possible with the response force and TCF commander before they are needed. Although the base's reaction force usually would not fight beyond the perimeter of the base, the reaction force must be ready to assist the response force or TCF when it arrives. The following should be considered:

- Command relationships before, during, and after linkup.
- Coordination of fire support before, during, and after linkup.

- Recognition signals and communication procedures to be employed.
- Follow-on operations required.
- Area damage control.

Setup in a Hide Position (Temporary Defensive Position)

4-51. If the platoon steps down from sustained continuous operations and cannot return to its base or base cluster, it may need to operate briefly from a temporary defensive position. When used properly, a temporary defensive position can enable the platoon to rest, recover, repair damaged equipment, and plan for future operations. It offers concealment with little chance of detection by the enemy. Platoons want to get the best security they can while tasking a minimum of soldiers to provide security. When the platoon leader decides to operate from a temporary defensive position, he notifies the company HQ.

4-52. Locate the position in or near the area of normal operations so that sustained operations can be resumed immediately, on order. METT-TC should be a primary concern. Easily defendable positions are preferred over those that are more difficult to defend. Ensure that the position has more than one exit route and provides communication with the next higher HQ that are enhanced or at least not interfered with by terrain. While built-up, urban areas afford suitable concealment for temporary defensive positions, it is essential that the capability to communicate be assessed before the selection of such a site. Other considerations of a temporary defensive position include—

- Keeping vehicles secure and available in a nearby location.
- Positioning vehicles so that key equipment can be moved or removed without displacement of the entire unit.

- Concealing equipment from the sides and overhead. This prevents detection from aerial observers and some side-looking airborne radar.
- Covering and concealing to reduce security and/or defense requirements.
- Providing enough space between vehicles to allow a vehicle to bypass another vehicle that is inoperable.
- Ensuring that the platoon follows signal security and uses noise and light discipline.
- Setting up fighting positions if the situation calls for them.

SET UP LOCAL SECURITY

4-53. Self-defense planning and coordination is done as soon as the base is set up. Prior planning and mission analysis are essential elements of a base defense. MP must be able to defend the site even before occupation is complete. When an MP element is located as part of an established base, it helps defend a portion of the larger unit's perimeter. Elements that set up separately usually defend their sites by deploying in a 360-degree perimeter. The techniques and principles of defense are the same for defending a separate squad, platoon, company, or base. To plan a perimeter defense, evaluate the situation. Analyze the terrain in terms of observation and fields of fire, cover and concealment, obstacles, key terrain, and avenues of approach (OCOKA). Defenses are placed where the threat is greatest.

4-54. The platoon leader establishes the CP and the OP. He locates the CP and the OP where he can best see and control the platoon. If this is not possible, he locates it where it can cover the most likely avenue of enemy approach. An alternate CP and OP, operated by the PSG, is placed where it can control the portion of the perimeter that cannot be seen or controlled by the main

CP and OP. The platoon leader then decides what other security measures and means of communication to use.

4-55. Platoon leaders must plan more than cover and concealment to counteract threat infrared, radar, thermal, and other sensors. The platoon leader uses the principles of camouflage and counters the recognition factors that make an object stand out from its background by—

- Locating soldiers, equipment, or structures where they are least discernible. This alone can reduce or eliminate many recognition factors.
- Using any mix of hiding, blending, disrupting, or disguising that conceals visibility.
- Maintaining camouflage discipline continuously.

4-56. When the number of soldiers that will defend a 360-degree perimeter is small, the platoon leader must—

- Vary the size of the defensive sectors.
- Identify alternate fighting positions.
- Retain flexibility of thinking.
- Decide what equipment—
  - Is needed to set up a perimeter defense.
  - Should stay in the vehicles.
  - Must be requisitioned or picked up later.

4-57. The equipment to improve defensive positions includes concertina wire, sandbags and tape (for cover and concealment), trip flares, pyrotechnic devices, mines, or PEWS, and other lethal and nonlethal technologies.

4-58. Platoons must be able to defend during day or night, in reduced visibility, and in a variety of weather conditions. The platoon leader or sergeant ensures that the platoon has the equipment it needs to defend under these conditions. The platoon must know how and when to use the equipment. During reduced visibility—

- Take steps to keep the enemy from observing or surprising the platoon.
- Require OPs and LPs. There should be at least one OP and one LP per squad. OPs and LPs report the enemy's advance and call for illumination and supporting fire. As in a daylight defense, MP manning OPs and LPs withdraw before they become engaged in close combat.
- Use patrols, illumination, PEWSs, and NVDs to help detect the enemy's advance.
- Use trip flares to provide warning and give some illumination. As a rule, do not fire until the targets are visible.
- Use camouflage, movement control, and light and noise discipline.
- Limit radio traffic to essential information.
- Ensure strict fire control to keep from disclosing the fighting positions.
- Ensure that gunners with crew-served and antiarmor weapons use NVDs.
- Provide illumination by using handheld flares or grenade launchers with illuminating rounds. Added light may be provided by fire support.
- Ensure that platoon leaders plan the use of messengers, visual signals, personal contact, or whistles to communicate with the squad leaders. Squad leaders plan to communicate with their team leaders and teams using personal contact or sound and visual signals.

CONSTRUCT FIGHTING AND SURVIVABILITY POSITIONS

4-59. Fighting positions help protect soldiers and their equipment from the enemy and from the enemy's small-arms fire and fragmentation weapons, while allowing soldiers full weapon system engagement. A fighting position provides cover and concealment from which to engage or defend against the enemy. See *FM 7-8* for individual fighting position construction.

4-60. Fighting positions do not protect against the destructiveness of artillery and other area weapons. Nevertheless, a dug-in fighting position may be the key to survivability. Digging in cannot remove a soldier's vulnerability by itself. It does reduce exposure to the enemy's acquisition, targeting, and engagement systems. Platoons must be able to construct their survivability positions, often without engineer assistance.

4-61. Locate fighting positions for crew-served weapons where gunners can stop dismounted attacks. The sectors of fire must cover infantry avenues of approach and provide the most grazing fire across the platoon or squad front. Overlap the sectors of fire with each other and those of adjacent squads. Prepare fighting positions so that their primary sectors of fire have the guns firing across the front of the unit. Prepare secondary sectors of fire so that the guns fire to the front.

4-62. Usually, one MP team occupies an M249 or MK19 fighting position. One member is the gunner, one is the assistant gunner, and one is the ammunition bearer or rifleman. Each gunner has a primary and a secondary sector of fire. The gunner fires in the secondary sector only on order or when there are no targets in the primary sector. Each gunner uses aiming stakes to set his weapon for a final protective line (FPL) or a principal direction of fire (PDF) within the primary sector. The FPL and PDF are control measures to help defend a position. In an attack, the gunner knows the primary areas. He engages the greatest threat and, on the order of the platoon leader or PSG, fires the FPL.

PREPARE SECTOR SKETCHES

4-63. After the crew-served weapons are in position, the squad leader positions the remaining MP to protect the gunners and to cover areas not covered by the gunner's. Using the range cards, the squad leader makes a squad sector sketch. He includes a rough sketch of the terrain around the weapon *(Figure 4-1)*. The squad sector sketches are used to plan defense and to control fire. Squad sector sketches show the following:

- The main terrain features in each sector of fire and the ranges to the features.
- Each primary fighting position.
- The primary and secondary sectors of fire for each position.
- MK19 and M249 FPL or PDF.
- The type of weapon at each position.
- The OP and LPs and squad leaders' positions.
- Dead space.
- Mines and obstacles.

4-64. The squad leader checks the range cards and the squad sector sketch for gaps or other flaws in the fire plan. He adjusts the weapons or the sectors as necessary. If the squad leader finds dead space, he takes steps to cover it with mines, grenade-launcher fire, or indirect fire. He then prepares two copies of the squad sector sketch. He keeps one copy and forwards the other copy to the platoon leader who makes a platoon sector sketch *(Figure 4-2, page 4-28)*. The platoon sector sketch shows the following:

- Squad sectors of fire.
- The crew-served and antiarmor weapons positions and sectors of fire, including FPL or PDF for the crew-served weapons and target reference points for the antiarmor weapons.
- Positions of the mines and the obstacles.

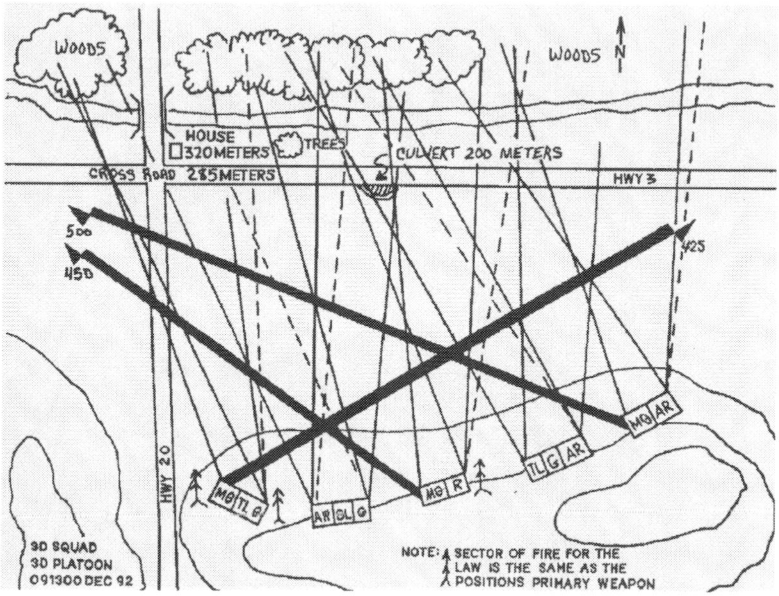

Figure 4-1. Squad Sector Sketch

- Indirect fire planned in the platoon's sector of fire.
- The OP and LPs and patrol routes (if any).
- The platoon CP and OP.

4-65. The platoon leader coordinates with the nearby units. He usually coordinates from left to right and from front to rear. The fires of units within the perimeter must be closely coordinated with the platoon's defensive fire plan. Squad leaders coordinate their fire plans with adjacent squads. All positions and units near the platoon are mutually supporting. The platoon leader makes sure gaps between the units are covered by fire, observation, patrols, OPs and LPs, or sensors. The units exchange information on—

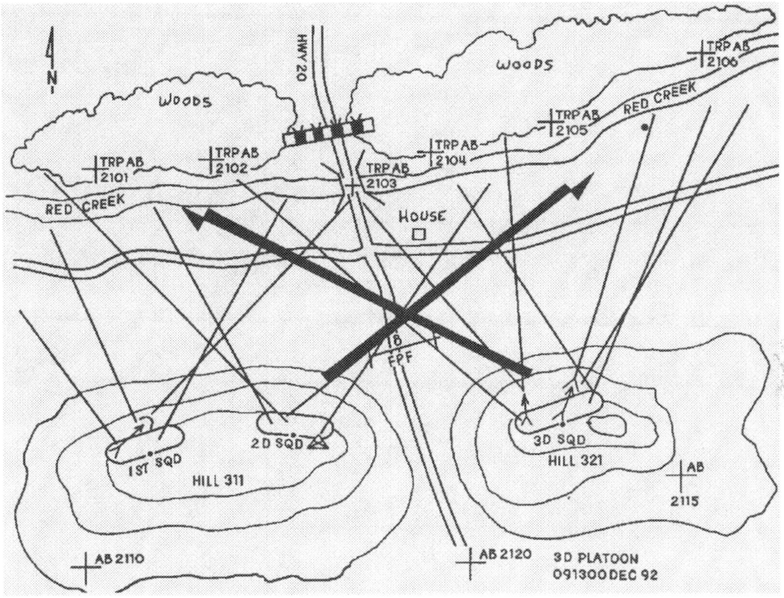

Figure 4-2. Platoon Sector Sketch

- The location of dead space between the elements and how to cover it.
- The locations of primary, alternate, and supplementary positions and sectors of fire for automatic weapons, antiarmor weapons, and subordinate elements.
- The locations of the OPs and LPs.
- The locations and types of obstacles and how they are covered by fire.
- Any patrols to be conducted, giving their size, type, times of departure and return, and routes.

PREPARE RANGE CARDS

4-66. The FPL for the M249 is the line where an enemy assault is to be checked by interlocking fire from all weapons. Use the M249 on the FPL for grazing fire no more than 1 meter above the ground, about hip high, across the front of the element. Use the MK19 or M203 to cover the dead space. To figure the dead space on the FPL, the gunner watches a person walking down the FPL and marks spaces that cannot be grazed. The gunner records all the dead space data on the range card *(Figure 4-3, page 4-30).* He prepares at least two copies of the range card, keeping one card at the position and giving one copy to the squad leader. Fire on a gunner's FPL is its final protective fire (FPF). FPF is usually used as a last resort to stop an enemy assault. All weapons fire on command, continuously, until the call to stop FPF is given.

4-67. When terrain prevents the use of an FPL, the gunner uses a PDF instead. He directs his fire toward the most threatening avenue of approach that leads to his position. His weapon is positioned to fire directly on this approach rather than across the squad's front.

4-68. Construct fighting positions for the MK19 like M249 fighting positions. However, be aware that it takes added effort to keep the M3 tripod from moving because of the recoil of the MK19. If gunners are using the M249 machine gun, they should use the tripod when firing at an angle and the bipod when firing to the front. When gunners change their fires from the oblique to the front, they must move the machine gun, but leave the tripod in place. If gunners are using the MK19, they position the tripod toward the primary sector of fire. However, because there is no bipod for the MK19, gunners must be prepared to adjust both the weapon and the tripod to the secondary sector, if required. After a crew is positioned and is assigned an FPL or a PDF, the team—

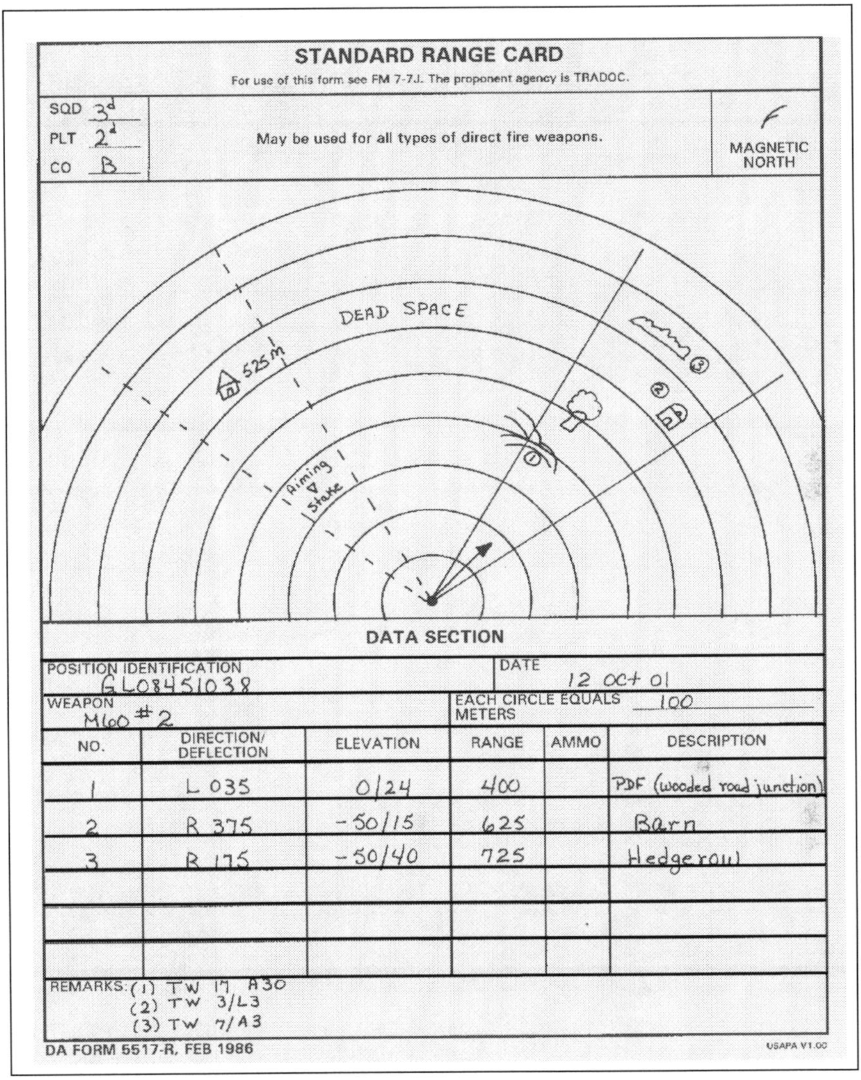
STANDARD RANGE CARD

For use of this form see FM 7-7J. The proponent agency is TRADOC.

SQD 3d
PLT 2d
CO B

May be used for all types of direct fire weapons.

MAGNETIC NORTH

DATA SECTION

POSITION IDENTIFICATION GL08451038

DATE 12 Oct 01

WEAPON M60 #2

EACH CIRCLE EQUALS 100 METERS

NO.	DIRECTION/ DEFLECTION	ELEVATION	RANGE	AMMO	DESCRIPTION
1	L 035	0/24	400		PDF (wooded road junction)
2	R 375	-50/15	625		Barn
3	R 175	-50/40	725		Hedgerow

REMARKS: (1) TW 17 A30
(2) TW 3/L3
(3) TW 7/A3

DA FORM 5517-R, FEB 1986 USAPA V1.00

Figure 4-3. Standard Range Card

- Marks the tripod's position and the limits of their sectors of fire with aiming stakes.
- Outlines the hole.

- Digs the firing platform first. This lessens their exposure if they have to shoot before the position is completely constructed. Dig the firing platform at a level that allows the gun to traverse the sectors of fire.
- Lowers the gun to reduce the gunner's profile. This also reduces the height of the frontal cover needed.
- Digs the hole deep enough to protect itself and still allow the gunner to shoot in comfort (usually about armpit deep).
- Places the dirt where frontal cover is needed.
- Uses the rest of the dirt to build the flank and rear cover when the frontal cover is high enough and thick enough. Sandbags, wire, hatchets, or saws can be useful for building overhead cover or improving the fighting positions.

4-69. The ammunition bearer digs a one-man fighting position to the flank. He positions himself where he can see and shoot to the front and the oblique. Usually the ammunition bearer is on the same side as the FPL or the PDF. From there he can see and shoot into the machine gun's secondary sector. He can also see the gunner and the assistant gunner. The ammunition bearer connects his position to the machine gun position by a crawl trench. This allows him to provide ammunition or replace one of the gunners.

SELECT FIGHTING POSITIONS IN BUILT-UP AREAS

4-70. Planning a defense of a platoon on urban terrain is similar to planning a defense in the countryside. Defensive positions must cover likely enemy avenues of approach, be mutually supporting, and provide cover and concealment. Use AT weapons on mounted avenues of approach. Machine guns cover dismounted approaches. AT4s and M203 grenade launchers work well in built-up areas. They are likely to hit enemy armored vehicles on the top or the side where armor is thin.

4-71. The method of defense (such as in-depth or linear) in the two areas is based on the same considerations. Use obstacles to canalize the enemy into kill zones or to deny key terrain. Orders must be very specific. Due to limited resources, use obstacles to channel, divert, or impede movement.

Position Locations

4-72. Select defensive positions in urban areas based on METT-TC. Often a squad occupies a building, but larger buildings may be defended by a platoon. Select buildings that—

- Are well built. Concrete and steel construction is preferred.
- Have strong floors to keep the structure from collapsing under the weight of debris.
- Have thick walls and floors so that the enemy cannot shoot through roofs and walls to kill defenders.
- Are constructed of nonflammable material. Avoid wood. Strong, fireproof construction provides protection from a nuclear attack as well as conventional firepower.
- Have few glass windows (or break and remove the glass).
- Provide good fields of fire. Buildings located next to vacant lots, alleys, and parks allow better fields of fire than buildings located next to other buildings.
- Allow mutual support between the buildings. No building should be subject to attack without troops in another building being able to provide supporting fire.

4-73. Locate positions so as not to establish a pattern. Avoid obvious firing locations like church steeples (remember the elements of OCOKA) by—

- Placing MK19s in the building where they can cover assigned sectors of fire and FPL.
- Having the squad automatic riflemen and grenadiers cover enemy approach routes to the building.
- Placing most rifle positions at or near ground level to have overhead protection and provide grazing fire on approaches.
- Positioning some MK19 gunners higher to get a longer range. In addition, they can fire into areas that would be dead space for ground-level weapons.
- Positioning AT4s (remember the back blast) so that they can fire down on tracked infantry fighting vehicles and wheeled scout reconnaissance vehicles.

Building Improvement

4-74. Change the outside of the building as little as possible, but inside the building—

- Improve the fighting positions to provide overhead and frontal cover. Use firing ports to avoid enemy observation.
- Cut or blow holes between rooms and floors so the soldiers can move quickly by a covered and concealed route to other firing positions in the building.
- Seal off unused basements to prevent enemy entry.
- Barricade doors, halls, and stairs and take down fire escapes to keep the enemy out of the building.
- Reinforce positions with sandbags, solid debris, beds, furniture, and so forth.

- Screen or block windows and other openings. This keeps the enemy from seeing which windows are manned and throwing hand grenades into the building. When firing from the windows or holes in the walls, be sure the muzzle of your weapon does not protrude beyond the wall. This conceals the muzzle flash.
- Remove combustible materials to limit the danger of fire.
- Turn off electricity and gas.
- Stockpile water and dirt to fight fires.
- Wear armored vests, earplugs, and goggles for protection from dust and debris.

Other Considerations

4-75. Operating in urban terrain can be challenging, so consider the following:

- The employment of weapons is different (shorter ranges).
- The position locations are different, such as LAWs and AT4s go on the upper floors of the buildings (refer to *FM 23-25* for the safety considerations).
- The target acquisition is more difficult (such as, aiming stakes and layered fires are used extensively).
- The CSS will center more on stockpiling materials in positions rather than on traditional resupply methods.
- Controlling indirect fire is more difficult.
- Primary communication must be by messenger, wire, or visual signs rather than radio.
- Avenues of approach are more canalized.
- The three dimensions of the enemy (aboveground, ground level, or below ground)

may be in use. The enemy can easily isolate subordinate units.

- Civilians and fire hazards may be present.

ESTABLISH AND OPERATE AN OBSERVATION POST/LISTENING POST

4-76. OPs/LPs are selected locations from which to look and listen for enemy activity within an assigned area of observation. The OPs/LPs, the primary means of maintaining surveillance of an assigned avenue or a named area of interest (NAI), are positions from which MP observe the enemy and direct and adjust indirect fires against him. From the OPs/LPs, MP send SALUTE reports *(Figure 4-4, page 4-36)* to the commander when observing enemy activity. Use OPs/LPs for the following:

- On key terrain when the surveillance of a specific area is required.
- To prevent the enemy from a surprise attack on other friendly forces.
- As an early warning security measure in a defensive perimeter.
- For the monitoring of likely enemy avenues of approach, drop zones (DZ), and landing zones (LZ).

4-77. OPs/LPs can be performed either mounted or dismounted. A dismounted OP provides maximum stealth and has the greatest likelihood of remaining undetected by the enemy. The disadvantage of the dismounted OP is the time it takes to remount and move if necessary. If rapid movement or displacement is anticipated, the OP mounts or remains mounted.

4-78. A mounted OP/LP offers the advantages of rapid movement and protection because the enemy can easily detect them; however, it is potentially much less effective than a dismounted OP/LP.

SIZE: Give the size of the enemy unit as the number of troops or vehicles seen. Report 10 enemy infantrymen (not an infantry squad). Report three enemy tanks (not an enemy tank platoon).

ACTIVITY: Report what the enemy was doing. "They are emplacing antipersonnel mines in the road."

LOCATION: Report where the enemy was seen. Report the grid coordinates and the direction the enemy was heading. If a map is not available, relate the location to key terrain, such as the enemy's location is "on the Hahn Road, 300 meters south of the Kell River Bridge."

UNIT: An enemy soldier's unit may be hard to determine. Report markings or other distinctive features seen on the vehicles. Some countries have special uniforms and headgear. Some have colored tabs on the uniforms to show the type of unit, or the unit's actions may show its type. The kind of equipment it has may be peculiar to a certain type of unit. For example, a scout reconnaissance vehicle may indicate a reconnaissance unit; an amphibious tracked infantry-fighting vehicle may indicate an airborne unit.

TIME: Report the time the enemy was seen, not the time you are reporting.

EQUIPMENT: Report all the equipment the enemy is wearing or using. If you do not recognize an item of equipment or a type of vehicle, sketch it. Submit the sketch with the report.

Figure 4-4. Salute Report

SELECT OBSERVATION POST/LISTENING POST SITES

4-79. The platoon leader selects the general location for the platoon's OP/LP after analyzing METT-TC factors. From his analysis, he determines how many OPs and LPs to establish. He decides where they must be positioned to allow long-range observation along the avenues of approach assigned by his commander and to provide depth through the sector. Section and squad leaders select the exact positions for each OP/LP on the ground. The OP/LP must have the following characteristics:

- Covered and concealed routes to and from the OP/LP. Ensure that MP can enter and leave their OP/LP without being seen by the enemy.
- Unobstructed observation of the assigned area or sector. Ideally, the fields of observation of adjacent OPs/LPs overlap to ensure full coverage of the sector.
- Covered and concealed positions that are effective. MP select positions with cover and concealment to reduce their vulnerability on the battlefield. MP may need to pass up a position with favorable observation capability, but with no cover and concealment, to select a position that affords better survivability.
- Located where they will not attract attention. Do not locate OPs/LPs in such locations as a water tower, an isolated grove of trees, or a lone building or tree; these positions draw enemy attention and may be used as enemy artillery TRPs.
- Located where they are not silhouetted. Avoid hilltops and position OPs/LPs further down the slope of the hill or on the side, provided there are covered and concealed routes into and out of the position.

MAN THE OBSERVATION POST/LISTENING POST

4-80. Ideally, an MP team should man an OP/LP to maintain team integrity. Position OPs/LPs down the slope or on a flank of a hill, if there are covered withdrawal routes. Each of the OP's/LP's fields of observation overlap those of adjacent OPs/LPs. MP may have to selectively clear fields of observation. Ensure that MP are not seen when entering and leaving an OP/LP. Equip OP/LP teams to observe the area, report information, protect themselves, and call for and adjust indirect fire. OPs/LPs on a defensive perimeter need secure communications. Use of field phones or secured radios are usually best. However, messengers can be used. OPs/LPs may use portable radios to supplement wire communication. One MP observes the area while a second MP provides local security and records and reports information. The third MP rests or provides backup security. The team members switch jobs every 20 to 30 minutes because the observer's effectiveness decreases quickly after that time. The observer needs—

- A map of the area.
- A compass.
- Communication equipment (wire and radio).
- Observation devices, such as binoculars, observation telescope, and NVDs.
- An SOI extract.
- The report formats contained in the SOP.
- A radio (this may be the only means of communication from a remote site like a DZ or an LZ).

POSITION OBSERVATION AND LISTENING POSTS

4-81. Place OPs/LPs either in a linear configuration or in-depth. Linear placement *(Figure 4-5)* allows the platoon to observe the assigned sector from several OP/LP sites, reducing the chance of the enemy entering the sector without being observed. This method works well when the platoon has been assigned a large sector with few avenues of approach or is in desert-type terrain. In-depth OP/LP placement *(Figure 4-6, page 4-40)* allows the platoon to observe the entire sector by placing OP/LP sites where the platoon can observe the most likely avenues of approach in the sector as well as along the sector flanks. This method works well when the platoon is assigned a sector with several avenues of approach or is in heavily wooded terrain. In-depth placement allows for redundancy in observation and better sector coverage.

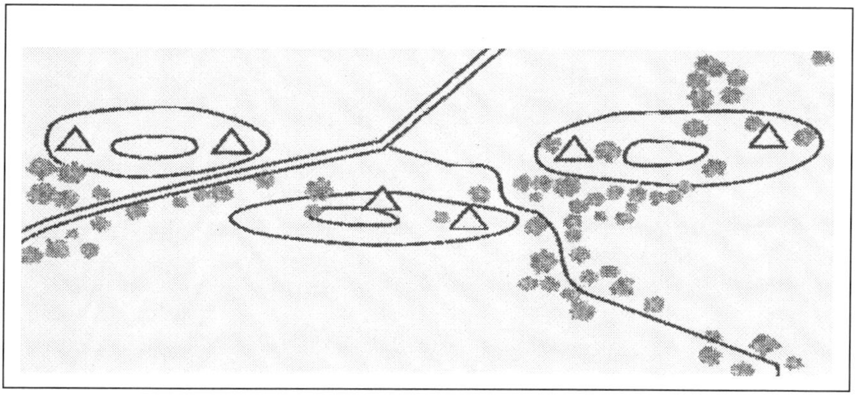

Figure 4-5. Linear OP/LP Placement

4-82. OP/LP team emplacement at night depends on METT-TC factors. The platoon leader places OPs/LPs close to the perimeter and within direct fire range of the defensive perimeter for protection. The team leader designates a specific location and primary direction of fire for the crew-served weapon. The OP/LP team builds a hasty-fighting position or a prepared-fighting position depending on METT-TC. The team leader designates a covered and concealed location behind OPs/LPs for the vehicle. The OP/LP team has a covered and concealed withdrawal route to the vehicle from the fighting position. The team camouflages the OP/LP and their vehicle while the gunner clears a field of fire and prepares a range card. The squad leader establishes communication with higher HQ and tells the team when and how to report. He tells them—

- If and when they should fire at the enemy.

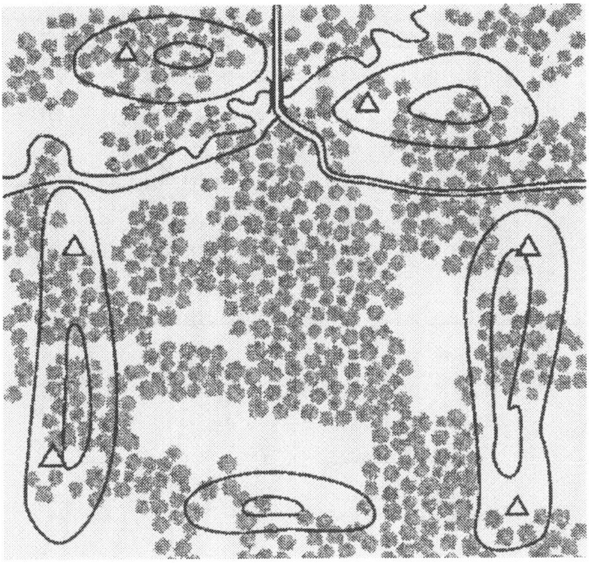

Figure 4-6. In-Depth OP/LP Placement

- How to get back to the squad if they must withdraw.
- What reentry signals to use.
- When they will be replaced, if known.
- To fight or withdraw according to his instructions.
- To be careful not to be drawn away by a small enemy element while the main element attempts to penetrate the perimeter.
- When to pull back or under what conditions they can withdraw without his order.

4-83. The frequency of relief for the OP/LP team depends on the team's physical condition and morale, the weather, the number of troops available, and the next operation. The squad leader carefully plans how each soldier receives rest. When OP/LP team is part of a defensive perimeter, it—

- Ensures that it has rearward cover.
- Builds fighting positions for protection and concealment.
- Uses trip flares, noisemaking devices, and NVDs to detect the enemy.
- Emplaces claymore mines for added protection.
- Coordinates with the perimeter on the reentry procedures to the perimeter from the withdrawal route.

DEFEND A SITE

4-84. Vigilance is the watchword for local security. When the OPs/LPs detect enemy elements, they notify their superior who calls for indirect fire, if it is available. When the enemy's advance threatens the OP/LP, order the OP/LP to withdraw. As the enemy approaches platoon positions, have the platoon increase its volume of fire. The platoon leader determines if the platoon can destroy the enemy from its assigned positions. If the platoon leader determines that the platoon can destroy the enemy, the platoon continues to fight with the following actions:

- The platoon leader or the FO continues to call for indirect fire as the enemy approaches. The platoon normally begins engaging the enemy at the maximum effective range. It attempts to mass fire and initiate them at the same time to achieve surprise. Long-range fire should disrupt enemy formations, channelize the enemy toward engagement areas, prevent or severely limit the enemy's ability to observe the location of friendly positions, and destroy the enemy as it attempts to breach tactical obstacles.

- The leaders control fire using standard commands, pyrotechnics, and other prearranged signals. The platoon increases the intensity of fire as the enemy closes within range of additional weapons. Squad leaders work to achieve a sustained rate of fire from their positions by having buddy teams fire their weapons so that both are not reloading them at the same time.
- The platoon and squad leaders consider the following when controlling and distributing fires:
  - The enemy's range.
  - The priority of the targets (what to fire at, when to fire, and why).
  - The nearest or most dangerous targets.
  - Shifts to concentrate fires on their own or as directed by higher HQ.
  - The ability of the platoon to engage dismounted enemy with grazing fires, and flank shots against enemy vehicles.
- The platoon leader initiates FPF as the enemy closes on the platoon's perimeter. The following actions occur at the same time:
  - The automatic weapons fire along interlocking PDF or FPLs. Other weapons fire at designated PDF. The M203 grenade launchers engage enemy in dead space or against enemy attempts to breach the protective wire.
  - The platoon continues the fight with claymore mines and hand grenades.
  - The platoon leader requests indirect FPF in support of his positions, if applicable.
  - The platoon continues to defend until the enemy is repelled or until the platoon is ordered to disengage.

4-85. If the platoon leader determines that the platoon can not destroy the enemy, he—

- Reports the situation to the company commander.
- Repositions the platoon to—
  - Continue fires into the platoon sector (engagement area).
  - Shift to alternate or supplementary positions.
  - Reinforce other parts of the company.
  - Counterattack locally to retake lost fighting positions.
  - Withdraw from an untenable position using fire and movement to break contact. (The platoon leader should not move his platoon out of position if it will destroy the integrity of the company's defense.)

NOTE: In any movement out of a defensive position, the platoon must employ all direct and indirect fire means available to suppress the enemy long enough for the platoon to move.

4-86. The platoon reorganizes after it has completed the initial battle with the enemy or relocated. The platoon positions key weapons, reestablishes security, provides first aid and prepares wounded soldiers for evacuation, and redistributes ammunition and supplies. The platoon relocates selected weapons to alternate positions if the leaders believe that the enemy may have pinpointed them during the attack and adjusts other positions to maintain mutual support. The platoon also reestablishes communication. It reoccupies and repairs positions and prepares for renewed enemy attack. The platoon repairs damaged obstacles and replaces mines and booby traps. When the platoon reorganizes, it performs the following actions:

- The squad and section leaders provide ammunition, casualty, and equipment (ACE) reports to the PSG. Team leaders provide fuel status. The PSG consolidates the ACE reports, reviews the consolidated ACE report with the platoon leader, and forwards it to the company commander.
- The platoon leader reestablishes the platoon's chain of command.
- The PSG coordinates for resupply and supervises the execution of the casualty and EPW evacuation plan.
- The platoon continues to improve positions. The platoon quickly reestablishes the OP/LP resumes patrolling as directed.

4-87. If the enemy gets through the FPF, repel it by close combat. If the perimeter is penetrated, move teams to block the penetration and cover friendly troops moving to alternate or supplementary positions. Even though the counterattack capability is limited, try to restore the perimeter. When the enemy is repelled—

- Reestablish security.
- Send patrols forward to maintain contact.
- Call for indirect fire on areas where the enemy is likely to regroup.
- Reorganize squads.
- Evacuate seriously wounded MP.
- Redistribute and resupply ammunition.
- Repair positions and continue to improve them.
- Keep the next higher commander informed throughout the conduct of the defense.

LAY HASTY PROTECTIVE MINEFIELDS

4-88. When possible, lay a hasty protective minefield as part of the unit's defensive perimeter. It can stop, delay, or restrict movement. MP often lay mines to restrict enemy movement near a defensive perimeter or at ambush sites. In the defense, platoons and squads lay hasty protective minefields to supplement weapons, prevent surprise, and give early warning of enemy advance. Hasty minefields must be covered by fire. Ensure that adjacent units are informed of the mine locations.

4-89. Platoons and squads must have permission from higher HQ to install hasty protective minefields. Higher HQ may, however, delegate approval authority to the company commander for emplacement of a hasty protective minefield. Requests for permission go through the normal chain of command.

4-90. If the company is not authorized mines in its basic loads, a special request may be needed. The enemy threat to the rear area requires commanders to issue mines as an additional protective measure. The M18A1 antipersonnel mine (claymore) and the M21 AT mine are the two mines most likely to be available to rear-area units for a hasty protective minefield. Refer to *FM 20-32* and *FM 21-75*.

4-91. MP generally will have claymores available to them, which is mainly a defensive weapon. However, the ways in which the claymore is used is limited only by the imagination. Plan the use of claymore mines to suit METT-TC. Emplace the mines—

- On likely dismounted avenues of approach.
- To cover dead space not covered by FPF of crew-served weapons.
- Outside the hand grenade range, but within the range of small-arms weapons.
- Where they are covered by observation and fire.
- Where back blast will not injure friendly forces.
- Beside buildings or other sturdy structures in urban terrain.

- Strapped to boards (for detonation from around corners).

4-92. MP record the exact location of the mines on *DA Form 1355-1-R (Figure 4-7)* when emplacing the minefield. This enables anyone to recover them. If possible, the unit should recover the mines before the unit relocates, and the same persons who emplaced them should recover them.

PATROLS

4-93. MP are organized and equipped to conduct mounted operations. The primary offensive weapon system is the MK19. It is designed as a mounted or static position weapon. However, an MP platoon may conduct limited dismounted operations with its other organic weapons. Refer to *FM 7-8* for more information about patrol operations, including organization, planning, and execution. In general, an MP platoon may be required to conduct reconnaissance and combat patrols.

RECONNAISSANCE PATROLS

4-94. Dismounted reconnaissance patrols are directed by higher HQ and conducted to gather detailed information on the enemy, terrain, specific NAIs, or avenues of approach. When executed as part of a screen or other security mission, a reconnaissance patrol can ensure the security of an OP/LP or the platoon's defensive perimeter. This is also referred to as a security patrol. Refer to Chapter 6 for area and zone reconnaissance.

COMBAT PATROLS

4-95. An MP platoon may conduct a combat patrol to establish an ambush on a dismounted enemy avenue of approach.

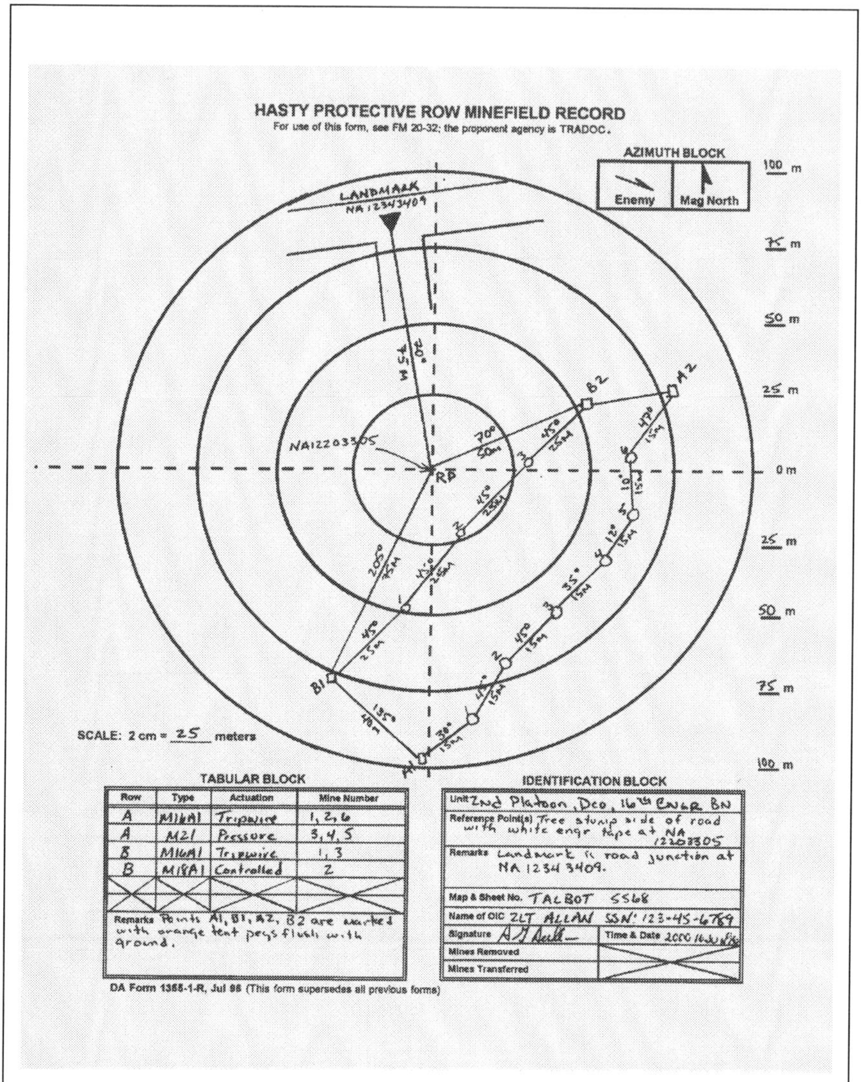

HASTY PROTECTIVE ROW MINEFIELD RECORD

For use of this form, see FM 20-32; the proponent agency is TRADOC.

AZIMUTH BLOCK: Enemy | Mag North

SCALE: 2 cm = 25 meters

TABULAR BLOCK

Row	Type	Actuation	Mine Number
A	M16A1	Tripwire	1, 2, 6
A	M21	Pressure	3, 4, 5
B	M16A1	Tripwire	1, 3
B	M18A1	Controlled	2

Remarks: Points A1, B1, A2, B2 are marked with orange tent pegs flush with ground.

IDENTIFICATION BLOCK

Unit	2nd Platoon, D co, 16th Engr BN
Reference Point(s)	Tree stump side of road with white engr tape at NA 12203305
Remarks	Landmark is road junction at NA 1234 3409.
Map & Sheet No.	TALBOT 5568
Name of OIC	2LT ALLAN SSN: 123-45-6789
Signature	A J Allan
Time & Date	2000 16 Ju…
Mines Removed	
Mines Transferred	

DA Form 1355-1-R, Jul 98 (This form supersedes all previous forms)

Figure 4-7. Sample DA Form 1355-1-R

Ambush

4-96. MP elements, normally no smaller than a platoon, use an ambush along suspected enemy routes and elsewhere against Level II threats in the rear area. An ambush enables a small unit with light weapons to harass or destroy a larger, better-equipped unit. An ambush may be a surprise attack from a concealed position on a moving or temporarily halted target or an attack by fire only. A successful ambush requires the following:

- Surprise. The platoon must seize and maintain control of the situation.
- Coordinated fire. The platoon must deliver a large volume of fire into the kill zone, using individual and crew-served weapons, mines, demolitions, and indirect fire (if available), and isolate the kill zone to keep the enemy from escaping or being reinforced.
- Control. Before, during, and after the ambush all elements must be able to communicate effectively with the platoon leader, primarily by using hand and arm signals.

4-97. The platoon leader organizes the platoon into assault, support, and security elements. An ambush is laid on an enemy's expected approach route. The platoon leader selects the site and members are positioned to provide—

- Good visibility of the avenues of approach and the kill zone.
- Good fields of fire into the kill zone.
- Cover and concealment.
- Obstacles between the teams and the kill zone.
- Covered and concealed withdrawal routes.

4-98. A good ambush site restricts the enemy's movement to one flank by natural or man-made obstacles. Natural obstacles include cliffs, steep embankments, swamps, steep grades, sharp curves in the road, narrow trails, streams, and heavily wooded areas. Man-made obstacles can include mines, booby traps, and roadblocks. The ambush is configured to suit the—

- Type of ambush.
- Terrain.
- Troops available.
- Weapons.
- Equipment.
- Ease of control.
- Overall combat situation.

4-99. To conduct an ambush, the platoon leader is positioned so he can best control the ambush elements, normally with crew-served weapons or the AT4, especially if the enemy has armor. The platoon leader—

- Positions the flank security elements.
- Emplaces obstacles and mines.
- Improves fighting positions, if time permits.
- Places a crew-served weapon to cover the left and right limits of the kill zone. These weapons must ensure that once an element is in the kill zone, it cannot leave it laterally.
- Reports to higher HQ when the ambush is in place.

4-100. The platoon leader initiates the ambush with a casualty-producing weapon, such as a claymore mine or a crew-served weapon. He ensures that there is a back-up method in case the primary means fails. The remainder of the platoon opens fire once the ambush has begun.

4-101. Most often, platoons will deploy a squad-size element for an attack on a single kill zone (a point ambush). If the company is deploying a platoon-size force to conduct a number of coordinated, related ambushes (an area ambush), the principles are the same. An area ambush works best where close terrain keeps enemy movement largely limited to trails or roads. For an area ambush—

- Choose one central ambush site around which you can control and organize the outlying ambushes.
- Select outlying ambush sites on the enemy's possible avenues of approach and escape from the central site.
- Set up and maintain communication with all the outlying sites.
- Assign the general locations of the outlying sites to the squad leaders. They will each set and conduct a point ambush.
- Direct the squad leaders to let the enemy pass through the kill zones until the central ambush begins.
- Provide specific instructions to the squad leaders in case the enemy detects an outlying site before the central ambush begins.

Ambush Types

4-102. There are three types of ambushes. They are the line and L-shaped ambush formations and the antiarmor ambush (refer to *FM 7-8* for more information).

4-103. **Line Ambush Formation**. A line formation (refer to *FM 7-8* for more information) is easy to control and is useful in all levels of visibility. The assault and support elements parallel the long axis of the kill zone to engage the enemy with flanking fire. The target may be so dispersed that it extends beyond the kill zone. Leaders must—

- Position the assault and support elements parallel to the enemy's movement route (such as on a road or trail or at a stream).

- Limit the kill zone to the size area that the ambush can cover with a great volume of fire.
- Place obstacles (such as claymore mines or existing natural obstacles) between the kill zone and the ambush element to prevent counter-ambush actions.
- Leave access lanes through the obstacles so the kill zone can be assaulted (if directed).

4-104. **L-Shaped Ambush Formation**. An L-shaped formation is useful on a straight stretch of a trail, road, or stream. It also works well at a sharp bend in a trail, road, or stream. The assault element is the long leg of an "L," paralleling the kill zone to provide flanking fire. The support element is the short leg, capping the end of the kill zone at a right angle to the assault element. The support element provides enfilade fire to interlock with fire from the other leg.

4-105. **Antiarmor Ambush**. The principles for an antiarmor ambush are the same as for an area or point ambush. MP are likely to encounter bypassed enemy armor in the rear area. The primary antiarmor weapon for MP is an AT4. This is a light antiarmor weapon with limited capability against medium and heavy armor vehicles. However, MP may be required to set up a hasty antiarmor ambush to destroy one or two light enemy armor vehicles. An antiarmor ambush is best performed by a platoon. The platoon leader positions the antiarmor weapons where they can engage the target from the rear, flank, or top. Multiple AT4s are used to ensure destruction. The platoon provides support and security.

CLEARING TECHNIQUES

HIGH-INTENSITY VERSUS PRECISION CLEARING TECHNIQUES

4-106. Precision clearing techniques do not replace other techniques currently being used to conduct building and room clearing during high-intensity combat. Specifically, they do not replace the clearing technique in which a fragmentation or concussion grenade is thrown into a room before US forces enter. Use precision room clearing techniques when the tactical situation calls for room-by-room clearing of a relatively intact building in which enemy combatants and noncombatants may be intermixed. They involve increased risk in order to clear a building methodically, rather than using overwhelming firepower to eliminate or neutralize all its inhabitants.

4-107. From a conceptual standpoint, standard high-intensity room-clearing drills can be thought of as a deliberate attack. The task is to seize control of the room, with neutralization of the enemy in the room the purpose. The fragmentation or concussion grenade can be thought of as the preparatory fire used before the assault. As in a deliberate attack against any objective, the assaulting element moves into position using covered and concealed routes. The preparatory fire is initiated when the assaulting element is as close to the objective as it can get without being injured by the enemy. The assault element follows the preparatory fire (fragmentation or concussion grenade) onto the objective as closely as possible. A rapid, violent assault overwhelms and destroys the enemy force and seizes the objective.

4-108. Compared to the deliberate attack represented by high-intensity room-clearing techniques, precision room-clearing techniques are more conceptually akin to a reconnaissance in force or perhaps an infiltration attack. During a reconnaissance in force, the friendly unit seeks to determine the enemy's locations, dispositions, strength, and intentions. Once the enemy is located, the friendly force is fully prepared to engage and destroy it, especially if surprise is achieved. The friendly force retains the options of not employing preparatory fire (fragmentation or concussion grenades) if it is not called for (the enemy is not in the room) or if it is inappropriate (there are noncombatants present also). The attacking unit may choose to create a diversion (use a stun-hand grenade) to momentarily distract the defender while it enters and achieves domination of the objective.

4-109. The determination of which techniques to employ is up to the leader on the scene and is based on his analysis of the existing set of METT-TC conditions. The deliberate attack (high-intensity techniques), with its devastating suppressive and preparatory fire neutralizes everyone in the room and is less dangerous to the assaulting troops. The reconnaissance in force (precision techniques) conserves ammunition, speeds up the clearing process, reduces damage, and minimizes the chance of noncombatant casualties. Unfortunately, even when well executed, it is very stressful and hazardous for friendly troops.

4-110. Certain precision room-clearing techniques, such as methods of squad and fire team movement, the various firing stances, weapon positioning, and reflexive shooting, are useful for all combat in confined areas. Other techniques, such as entering a room without first neutralizing the known enemy occupants by fire or explosives, are appropriate in only some tactical situations.

4-111. Generally, if an alerted enemy force that is determined to resist occupies a room or building, and if most or all noncombatants are clear, employ overwhelming firepower to avoid friendly casualties. In such a situation, use supporting fires, demolitions, and fragmentation grenades to neutralize a space before friendly troops enter.

4-112. In some combat situations, the use of heavy supporting fire and demolitions would cause unacceptable collateral damage or unnecessarily slow the unit's movement. In other situations, often during stability and support operations, enemy combatants are so intermixed with noncombatants that US forces can not in good conscience use all available supporting fire. At such times, room-by-room clearing may be necessary and precision room-clearing techniques are most appropriate.

PRINCIPLES OF PRECISION ROOM-CLEARING

4-113. Battles that occur at close quarters, such as within a room or hallway, must be planned and executed with care. Units must train, practice, and rehearse precision room-clearing techniques until each fire team and squad operates smoothly. Each unit member must understand the principles of precision room-clearing, such as surprise, speed, and controlled violence of action.

Surprise

4-114. This is the key to a successful assault at close quarters. The fire team or squad clearing the room must achieve surprise, if only for seconds, by deceiving, distracting, or startling the enemy. Sometimes stun grenades are used to achieve surprise. These are more effective against a nonalert, poorly trained enemy than against alert, well-trained soldiers.

Speed

4-115. This provides a measure of security to the clearing unit. Speed allows soldiers to use the first few vital seconds provided by surprise to their maximum advantage. In precision room clearing, speed does not mean incautious haste. It can best be described as a careful hurry.

Controlled Violence of Action

4-116. This eliminates or neutralizes the enemy while giving him the least chance of inflicting friendly casualties. Controlled violence of action is not limited to the application of firepower only. It involves a soldier's mind-set of complete domination. Each of the principles of precision room-clearing has a synergistic relationship to the others. Controlled violence coupled with speed increases surprise. Hence, successful surprise allows increased speed.

FUNDAMENTALS OF PRECISION ROOM-CLEARING

4-117. The ten fundamentals of precision room-clearing address actions soldiers take while moving along confined corridors to the room to be cleared, while preparing to enter the room, during room entry and target engagement, and after contact. Team members should—

- Move tactically and silently while securing the corridors to the room to be cleared. Carry only the minimum amount of equipment.

NOTE: Rucksacks and loose items carried by soldiers tire them, slow their pace, and cause noise.

- Arrive undetected at the entry to the room in the correct order of entrance and be prepared to enter on a single command.
- Enter quickly and dominate the room. Move immediately to positions that allow complete control of the room and provide unobstructed fields of fire.
- Eliminate the entire enemy in the room by fast, accurate, and discriminating fires.

- Gain and maintain immediate control of the situation and all personnel in the room.
- Confirm whether enemy casualties are wounded or dead. Disarm and segregate the wounded. Search all enemy casualties.
- Perform a cursory search of the room. Determine if a detailed search is required.
- Evacuate all wounded and any friendly dead.
- Mark the room as cleared using a simple, clearly identifiable marking according to the unit SOP.
- Maintain security and be prepared to react to more enemy contact at any moment. Do not neglect rear security.

COMPOSITION OF THE CLEARING TEAM

4-118. Execute precision room-clearing techniques by the standard four-man fire team. Because of the confined spaces typical of building- and room-clearing operations, units larger than squads quickly become unwieldy. When shortages of personnel demand it, conduct room-clearing with two- or three-man teams; four-man teams are preferred. Using fewer personnel greatly increases the combat strain and risks.

BREACHING

4-119. An integral part of precision room-clearing is the ability to gain access quickly to the rooms to be cleared. Breaching techniques vary based on the type of construction encountered and the types of munitions available to the breaching element. Techniques range from simple mechanical breaching to complex, specialized demolitions.

Shotgun Ballistic

4-120. A useful method of breaching is the shotgun ballistic breach for forced entry of standard doors. Use a

12-gauge shotgun loaded with buckshot or slugs to breach most standard doors quickly. When done properly, the shotgun breach requires only a few seconds. The two standard techniques of shotgun breaching are the doorknob breach and the hinge breach. When attempting either technique, the gunner approaches the door from an angle, avoiding standing in the area directly in front of the door. While holding the stock of the shotgun in the pocket of his shoulder, the gunner places the muzzle tightly against the door, and aims down at a 45-degree angle.

4-121. **Doorknob Breach**. For the doorknob breach, *(Figure 4-8, page 4-58)* the aim point is a spot halfway between the doorknob and the frame, not at the doorknob itself. The gunner fires two quick shots in the same location, ensuring that the second shot is aimed as carefully as the first. Weak locks may fly apart with the first shot, but the gunner should always fire twice. Some locks that appear to be blown apart have parts still connected that will delay entry. If the lock is not defeated by the second shot, the gunner repeats the procedure.

4-122. **Hinge Breach**. The hinge breach technique is performed much the same as the doorknob breach, except the gunner aims at the hinges. He fires three shots per hinge—the first at the middle, then at the top and bottom *(Figure 4-8)*. He fires all shots from less than an inch away from the hinge. Because the hinges are often hidden from view, the hinge breach is more difficult. Regardless of which technique the gunner uses, immediately after he fires, he kicks the door in or pulls it out. He then pulls the shotgun barrel sharply upward and quickly turns away from the doorway to signal that the breach point has been cleared. This rapid clearing of the doorway allows the following man in the fire team a

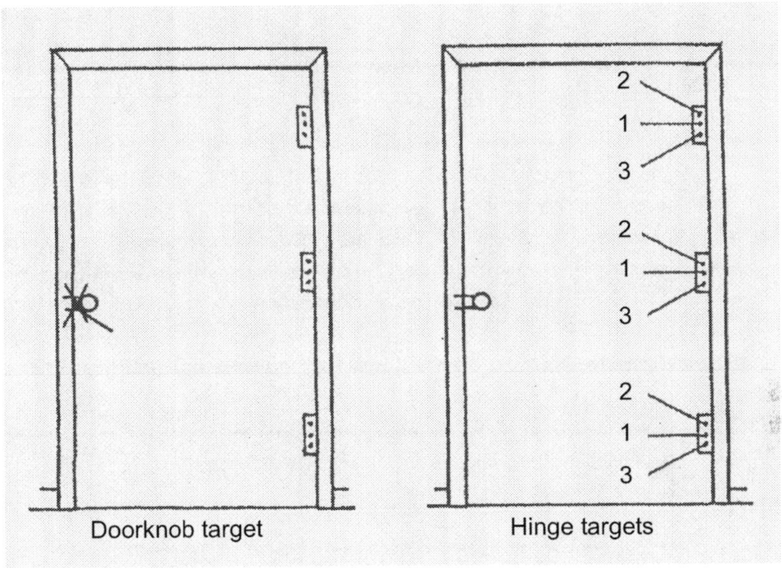

Figure 4-8. Aim Points for a Shotgun Breach of a Standard Door

clear shot at any enemy who may be blocking the immediate breach site.

> **WARNING**
> **Do not use small arms (5.56 or 7.62 millimeters) as a ballistic breach on doorknobs and hinges except as a last resort. It is unsafe and could result in death.**

4-123. Demolitions are often needed to defeat more elaborate barriers or to produce a desired effect to aid the initial entry.

Mechanical Breaching

4-124. Mechanical breaching is planned as a backup to a ballistic or explosive breach. Mechanical breaching is an assumed capability within all units. Taking the time to defeat weak barriers such as doors or windows by means of crowbars, saws, sledgehammers, battering rams, or axes is a decision that must be made based on the conditions of METT-TC.

Order of Movement

4-125. Clearing-team members must approach the breach point quickly, quietly, and in standard order. This approach preserves the element of surprise and allows for quick entry and domination of the room. The order of movement to the breach point is determined by the method of breach and intended actions at the breach point. The members of the fire team are assigned numbers 1 through 4, with the team leader normally designated number 3. If one member of the clearing team is armed with the squad automatic weapon (SAW) rather than an M16 rifle or carbine, he should be designated as number 4.

4-126. **Ballistic (Shot Gun) Breach**. The order of movement for a shotgun breach has the gunner up front, followed by the number 1 man, number 2 man, and then the number 3 man (team leader). After the door is breached, the gunner moves to the rear of the lineup and assumes the position of the number 4 man.

4-127. **Explosive (Demolition) Breach**. The order of movement for an explosive breach without engineer support is number 1, number 3 (team leader), number 2, and then number 4. The number 1 man provides security at the doorway. The number 3 man (team leader) carries the demolition charge and places it. The number 2 man provides security overhead, and the number 4 man provides rear security. After the demolition charge is placed, the team moves to covered positions and prepares to enter in the standard 1, 2, 3, 4 order.

4-128. **Mechanical Breach**. A suggested order of movement for a mechanical breach is the initial assault team in order, followed by the breach man/element. At the breach point the assault team leader will bring the breach team forward while the assault team provides local security. After the breach is conducted, the breach team moves aside and provides local security as the assault team enters the breach.

ENTRY CONSIDERATIONS

4-129. The entire team enters the room as quickly and smoothly as possible and clears the doorway immediately. If possible the team moves from a covered or concealed position already in their entry order. Ideally, the team arrives and passes through the entry point without having to stop.

4-130. The door is the focal point of anyone in the room. It is known as the *fatal funnel*, because it focuses attention at the precise point where the individual team members are the most vulnerable. Moving into the room quickly reduces the chance anyone will be hit by enemy fire directed at the doorway.

4-131. On the signal to go, the clearing team moves from covered or concealed positions through the door quickly and takes up positions inside the room that allow it to completely dominate the room and eliminate the threat. Team members stop movement only after they have cleared the door and reached their designated point of domination. The first man's position is deep into the far corner of the room. The depth of his movement is determined by the size of the room, any obstacles in the room such as furniture, and by the number and location of enemy and noncombatants in the room.

4-132. To make precision room-clearing techniques work, each member of the team must know his sector of fire and how his sector overlaps and links with the sectors of the other team members. Team members do not move to the point of domination and then engage their targets. They engage targets as they move to their designated point. However, engagements must not slow movement to their points of domination. Team members may shoot from as short a range as 1 to 2 inches. They engage the most immediate enemy threats first. Examples of immediate threats are enemy personnel who are—

- Armed and prepared to return fire immediately.
- Blocking movement to the position of domination.
- Within arm's reach of a clearing team member.
- Within 3 to 5 feet of the breach point.

CLEAR A SINGLE ROOM

4-133. Precision military operations on urbanized terrain (MOUT) room clearing is generally accomplished using a four-person team, consisting of three riflemen and one SAW gunner or breacher. Entry does not always require four personnel. If a specific room is to small or there are less than four clearing members due to attrition, the room may be cleared with less. However, never clear a room with less than two soldiers. If the leadership determines that the use of four-soldier teams is impractical for the situation, use the standard three-soldier team filling the numbers 1, 2, and 3 entry positions. The term *point of domination* refers to the two corners of the room assigned to the numbers 1 and 2 soldiers. These points allow the team to gain complete control of the objective. Each team member is assigned a different, but interlocking field of fire. This ensures mutual supporting fires. This type of entry is dynamic in nature. When completed with precision, this technique overwhelms the enemy and allows the team to move on to the next objective very quickly. When a room is cleared, the exiting procedure for leaving the room varies depending on the location of the entry point (either in the middle or at the corner of the room).

> **WARNING**
> **The muzzle of the clearing team's weapons should always be pointing wherever that soldier is looking. The clearing team should remain one foot away from all the walls to avoid injury due to bullets traveling down the wall (the path of least resistance).**

Stack Position

4-134. Once the approach to the room or building is completed, the team stacks *(Figure 4-9)* parallel to the outside wall. The number 1 soldier provides security on the entry point. This point is normally a door in a precision MOUT environment. The number 2 soldier provides security to the front of the team. The number 3 soldier (normally the team leader) is in the high-ready position. The high-ready position consists of the soldier placing the buttstock of the weapon in his shoulder pocket and raising the muzzle of the weapon from a 90-degree angle to a 45-degree angle. The number 4 soldier is in the low-ready position. The low-ready position consists of the soldier placing the buttstock of the weapon in his shoulder pocket and lowering the muzzle of the weapon to a 45-degree angle. The high- and low-ready positions aid in ensuring that the soldiers do not flag their fellow team members with their weapons during entry or clearing. A support element covers the teams from a distance, concentrating on the team's blind spots.

Initial Entry

4-135. When the situation allows (time, concealment, and so forth), it is important to ensure that each team member is prepared to enter the initial-entry point to the building before the assault. This is achieved while in

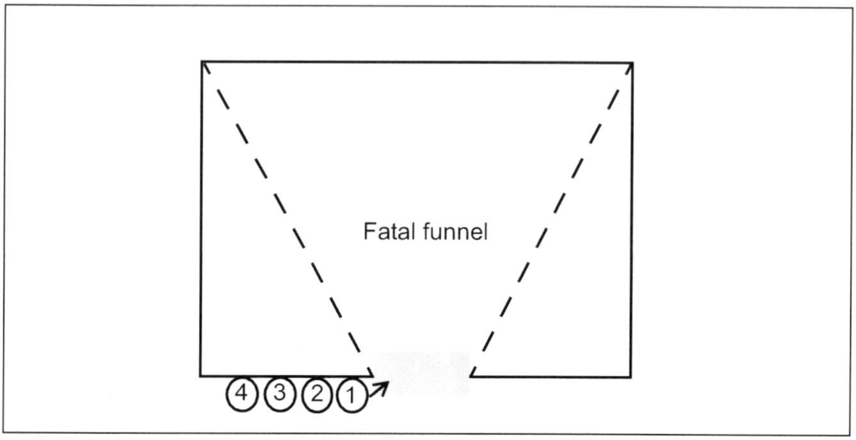

Figure 4-9. Stack

the stack position. When ready, the number 1 soldier nods his head while maintaining security on the door. When soldier number 4 is ready, he squeezes the shoulder of soldier number 3 with the nonfiring hand. This action continues up the line. When the signal reaches soldier number 1, he silently counts to three and then enters the building with the rest of the team following. If someone inside opens the door before this sequence is completed, the team will immediately enter the structure.

Initial-Entry Breach

4-136. *Figure 4-10, page 4-64,* depicts the process of a nonexplosive breach from the standard stack. Soldier number 4 moves up to the side of the door that contains the locking mechanisms. When soldier number 3 gives the signal, soldier number 4 breaches the door (by shotgun, battering-ram, and so forth) and kicks it open. Soldier number 4 then moves back from the door to allow the rest of the team to enter. Soldier number 4 follows them into the room. Refer to *FM 90-10-1* for shotgun breaching techniques.

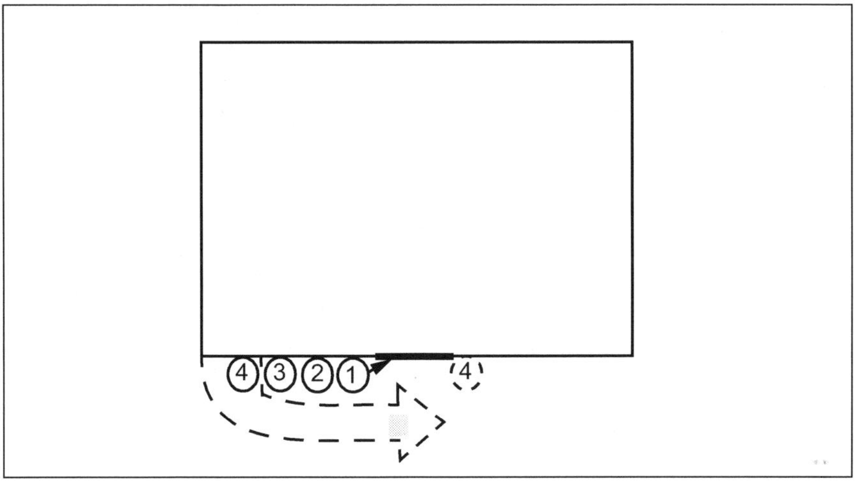

Figure 4-10. Nonexplosive Breach

Room Entry

4-137. Clear a room in the following manner when the entry point is located in the middle of the room:

- Soldier number 1 *(Figure 4-11)* enters the room using the path of least resistance, (moves across the doorway, into the room, and does not make a turn around the doorframe) clearing the fatal funnel as fast as possible. The term fatal funnel is derived from the natural reaction of the enemy soldiers inside a room (the large end of the funnel) to focus on and fire at the doorway (the spout of the funnel). The rounds are funneled into the doorway from almost any position in the room. An attacker should never move slowly through (or stop in) the fatal funnel. Besides providing a silhouetted target to the enemy, these actions will slow the entry of the rest of the team (reducing supportive fires). The number 1 soldier's first responsibility (in this scenario) is to clear the bottom right corner. This soldier will not be concerned with the middle of the room unless a threat is located to his direct front as he

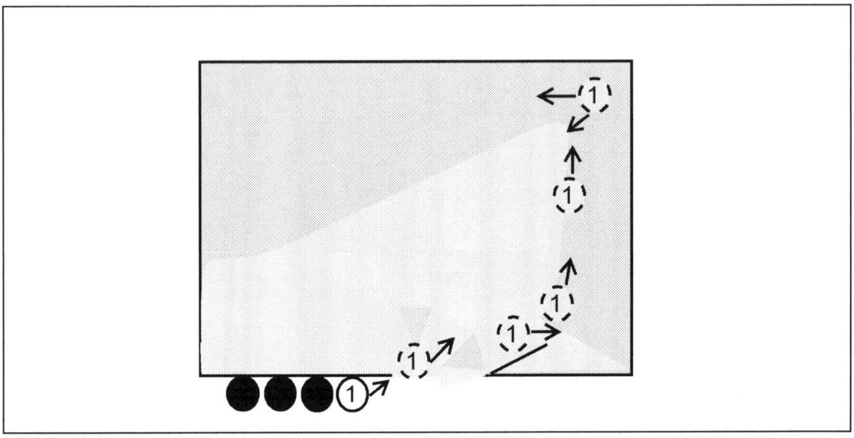

Figure 4-11. Number 1 Man, Middle-Door Entry

enters the room. If the number 1 soldier assesses this threat as being capable of hindering his movement to the top right corner of the room (his point of domination), he may engage that threat in a manner that avoids losing momentum. If the door obstructs or partially obstructs the route, soldier number 1 pushes hard through it. This helps to dissipate any loss of momentum and knocks an enemy soldier off balance if he is hiding behind the door. Once the number 1 soldier visually clears the corner, he moves up the wall to his front, clearing it as he goes. The number 1 soldier stops at his point of domination. He then turns and begins a scan from the top left corner to 1 meter off the number 2 soldier's weapon, located in the bottom left corner in this scenario (number 1 has farther to go, soldier number 2 should be in position once number 1 arrives).

- Soldier number 2 *(Figure 4-12, page 4-66)* enters the room immediately after the number 1 soldier enters. Soldier number 2, in this case, executes a hard left turn and moves in the opposite direction of soldier number 1. Soldier number 2's first concern is to clear the bottom left corner. Once this corner is visually cleared, he continues to clear up the left wall. Once soldier number 2 reaches this corner, he has reached the second point of domination. Soldier number 2 stops and scans from the top left corner to 1 meter off soldier number 1's weapon.

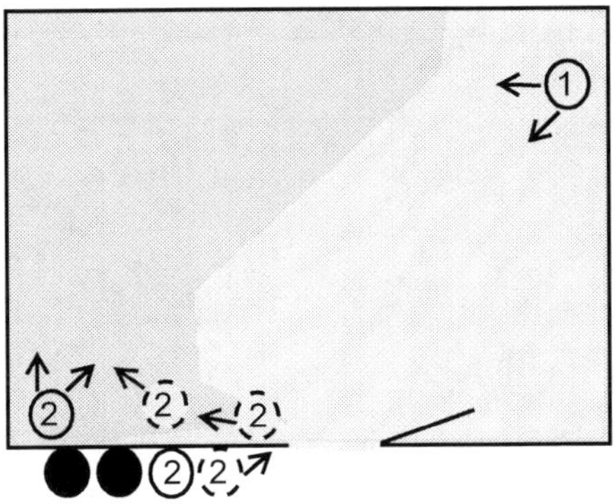

Figure 4-12. Number 2 Man, Middle-Door Entry

- Soldier number 3 enters the room at the same time as the number 1 and number 2 soldiers *(Figure 4-13)*. He moves his weapon down from the high-ready position, between soldiers number 1 and number 2 (using his height advantage, if it exists). Soldier number 3's first and primary concern is the middle of the room. If soldier number 3 enters the room quickly, effective supporting fires will be provided for the number 1 and number 2 soldiers as they clear the corners and move to their points of domination. Soldier number 3 begins with a

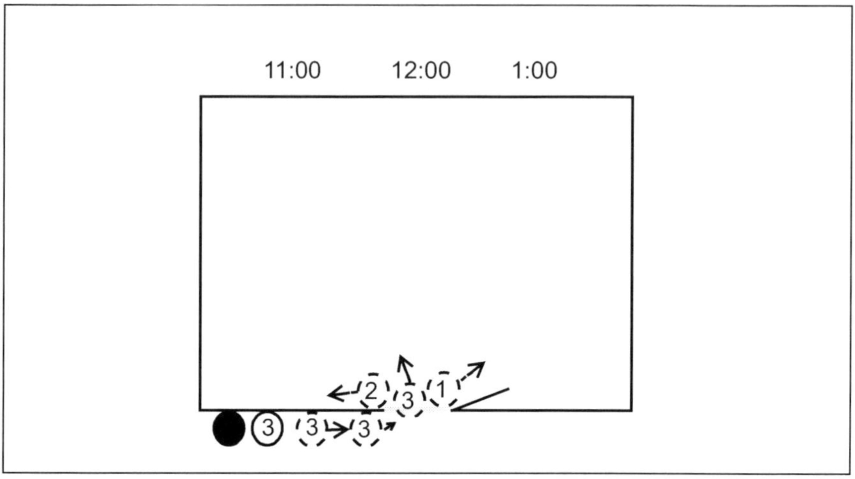

Figure 4-13. Number 3 Man, Middle-Door Entry

point of aim (in this scenario) at the 11 o'clock position. He indexes his weapon down the opposite wall, while moving laterally (cross step) out of the fatal funnel, in the opposite direction of soldier number 2 *(Figure 4-14)*. Soldier

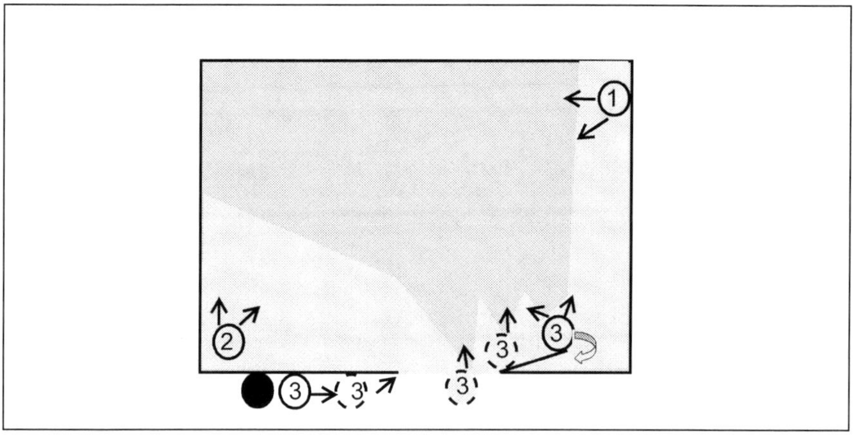

Figure 4-14. Number 3 Man, Middle-Door Entry

number 3 stops when out of the fatal funnel and begins to scan from 1 meter off soldier number 1's weapon to 1 meter off soldier number 2's weapon. Once one full scan is completed, soldier number 3 clears behind the door and then returns to scanning the room.

- Soldier number 4 clears in the manner depicted in *Figure 4-15* if the room is large enough for the SAW gunner or breacher to enter. He enters the room directly behind soldier number 3 and clears the fatal funnel as quickly as possible, moving in the opposite direction of soldier number 3. Soldier number 4's primary concern is the middle of the room. He begins with an overlapping point of aim at the 1 o'clock position and indexes his weapon down the far wall as he moves laterally towards the number 2 soldier. Soldier number 4 stops, once clear of the fatal funnel, and continues to scan from 1 meter off the number 2 soldier to 1 meter off the number 1 soldier.

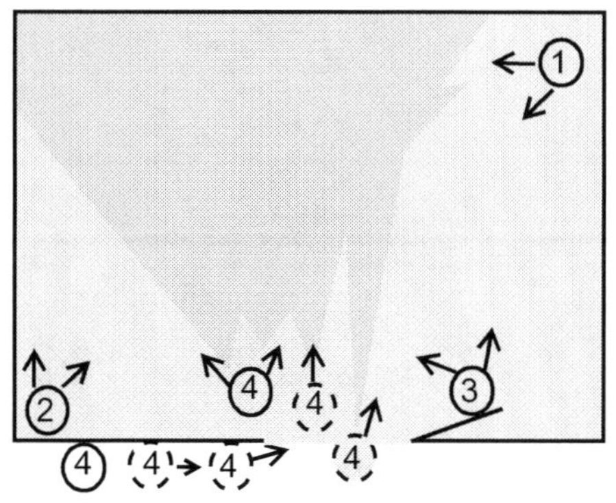

Figure 4-15. Initial Coverage

NOTE: The M249 or shotgun are not precision weapons. The number 4 soldier only engages when a clear shot at a confirmed combatant presents itself. The M249 is also very useful when the situation has degraded and heavier support is required. Leaders should train SAW gunners or breachers to be prudent in the use of their weapons in a precision MOUT environment.

- *Figure 4-16, page 4-70,* shows the coverage area of each team member once they have reached their positions. The room can be visually cleared very quickly because it has been dominated in depth. This is especially important considering the fact that most rooms contain furniture (hiding positions) and many are irregular shaped. A majority of the room is covered at all times, should an undiscovered or wounded enemy soldier attempt to engage the clearing team. The bottom right corner is the only area not effectively covered by fire at this time. Both the number 1 and number 3 soldiers cleared this area during the initial entry. If the number 1 soldier or number 3 soldier determines that a threat might exist in this corner, it is soldier number 3's responsibility to provide security in this area.

Communication

4-138. When the room has been dominated, communication is required between the team members in order to clear the room quickly and move on to the next objective. Whether scanning the room or providing security on possible threats (open doors, unidentified persons, or dead or wounded enemy), the team members cannot shift their gaze around the room to ascertain the current situation. To solve this problem, the team leader calls for a status when he determines that no immediate threat exists. This status or situation report (SITREP)

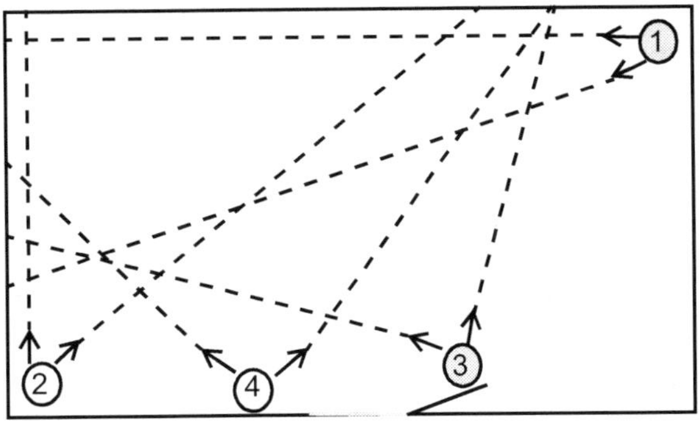

Figure 4-16. Number 1 Man, Center-Door Entry

consists of each clearing team member's current condition and any possible threat in his sector. The number 1 soldier states his situation to the rest of the team. The number 2 soldier follows with his report and so on down the line. The team leader absorbs all this information and then gives directions based on the information. The conversation should be at a level that each team member may hear, but no louder. If soldier number 1 fails to sound off, soldier number 2 states his status. If soldier number 1 is down, the team will know it. If soldier number 1 fails to sound off due to a mental lapse, he may give a SITREP once the other team members have given theirs. An example of this type of communication is as follows:

- Team leader (TL) (soldier number 3 in this scenario): "Status."
- Soldier number 1: "One up!"
- Soldier number 2: "Two, one down to my front!"
- Soldier number 3: "Three up!"
- Soldier number 4: "Four up!"
- TL: "Soldier number 1 secure the subject."

4-139. This scenario is based on a one-room dwelling objective for simplicity. Once the subjects are secured, they are extracted them from the dwelling by a support team and taken to a secure area where their status can be determined and first aid may be rendered, if needed. Meanwhile, the clearing team continues its mission. This would include (dependant on the mission) searching the room for any hidden threats, contraband, or enemy information or equipment. If there are multiple rooms to clear, the TL either directs one of the clearing team members to remain in the room as security (such as "soldier number 1 security") or (if available) calls for a reserve, who is positioned outside of the room, to enter the room and secure the subjects inside it until the building is cleared. The team would then be free to move on to the next room.

Corner-Door Position

4-140. When an entry point is located in the corner of the room, slight modifications are made to the middle-door entry method. The number 1 soldier *(Figure 4-17, page 4-72)* enters through the path of least resistance and clears the path to the corner to his front and stops. This is soldier number 1's point of domination. Soldier number 1 does not continue on to the next corner, as in the middle-door method, because it would take to long for him to reach this location. The number 2 soldier makes a hard turn around the doorframe and clears the path to the corner diagonally opposite of soldier number 1. This is soldier number 2's point of domination. The number 3 and number 4 soldiers use the far corner of the room as the 12 o'clock reference point *(Figure 4-18, page 4-72)*. All other actions in the room are the same as the middle-door method.

Rules

4-141. No matter what the room configuration, there are a few rules that should always be adhered to. These rules include the following:

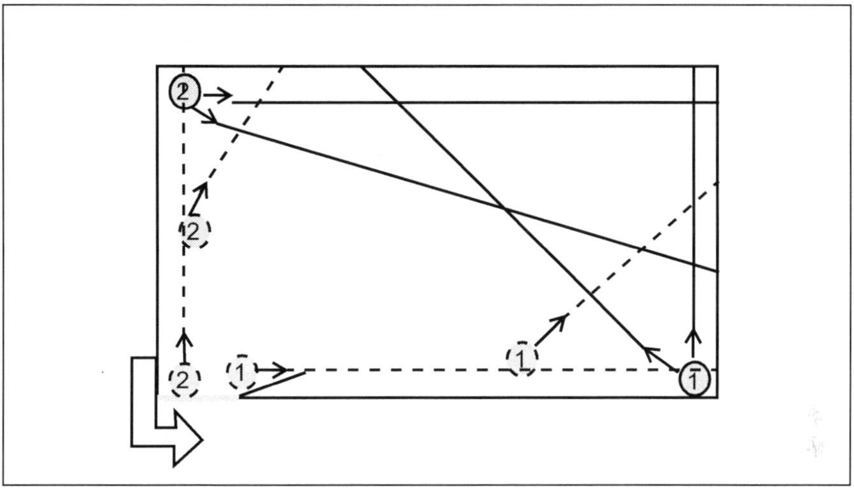

Figure 4-17. Number 1 and 2 Man, Corner-Door Entry

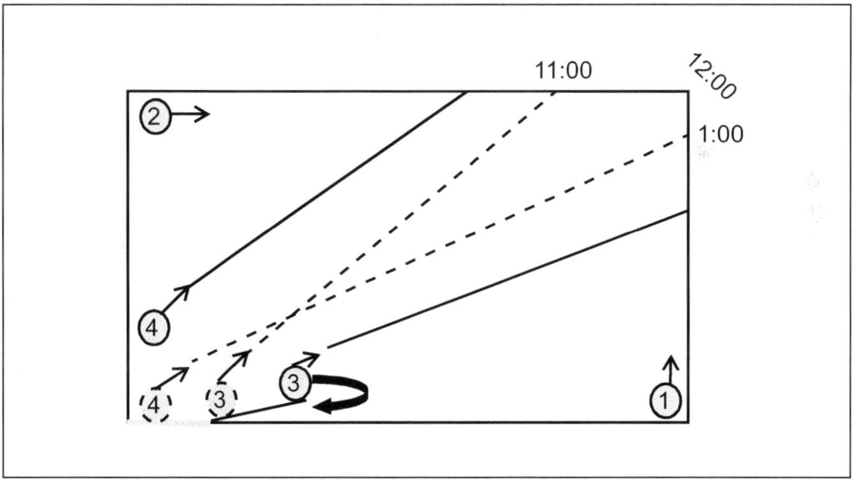

Figure 4-18. Number 3 and 4 Man, Corner-Door Entry

- Know that it does not matter if the person in front of you goes the wrong way when entering the room. Just go the opposite way as the soldier in front of you and it will work out.

- Enter the room as quickly and smoothly as possible and do not waste movements. Remember smooth is fast. The faster each team member picks up its initial point of aim, the more difficult it becomes for the defender. Even a prepared defender can be caught off guard.
- Clear (do not stop) the fatal funnel. The number 3 and number 4 soldiers are especially prone to stopping in the funnel.

NOTE: Ensure that the doorway is completely cleared before assuming a final position in the room.

- Stay focused. Never stop scanning your sector for targets unless—
  - You identify a threat in your sector. This threat could be an open door leading to an uncleared room, a person in the room other than one of the team members, an obstacle that cannot be cleared visually from your position, or anything else that you may determine as a threat. If such a threat exists, provide security for the team by covering it with your weapon.
  - You are ordered by the team leader to perform another task.
- Ensure that if you fall down while entering the room, you stay down and do not move. Do not get up until a team member places his hands on you and lifts you up. If a team member arbitrarily gets up, he may become an impediment to the rest of the team's movements. Another possibility is that he would rise into the path of a team member's bullet.
- Rehearse communication. As stated above, speed and momentum will make or break this type of

action. Poor communication techniques will slow your building clearing to a crawl.

- Ensure that each team member knows the procedures for each position. After the first room is cleared in a multiroom objective, a team member may find that he is in a different position in the stack than when the assault started. The team leader should always attempt to avoid being the number 1 soldier when entering a room. This can normally be accomplished, but in some situations it will be unavoidable.
- Ensure that you never enter a room alone. Two personnel are the minimum room-clearing team. If two soldiers are entering a room (either due to the size of the room or attrition), they should assume the number 1 and number 2 soldiers' positions.
- Ensure that you never flag a fellow team member. Flagging is defined as pointing your weapon at or sweeping your weapon across another team member. Muzzle awareness, at all times, is imperative.

CLEAR A CONNECTED ROOM

4-142. Clearing multiple rooms using connecting doors is preferable to using hallways. This action reduces exposure and compliments speed and momentum.

4-143. *Figure 4-19* shows a team that has just cleared a room using the corner-door technique. The door to the adjacent room is open, providing the number 2 and number 4 soldiers the opportunity to partially clear the second room. If a threat is identified in the adjoining room, it may be engaged from the cleared room. Once this firing has stopped, the SITREP will be restarted if necessary.

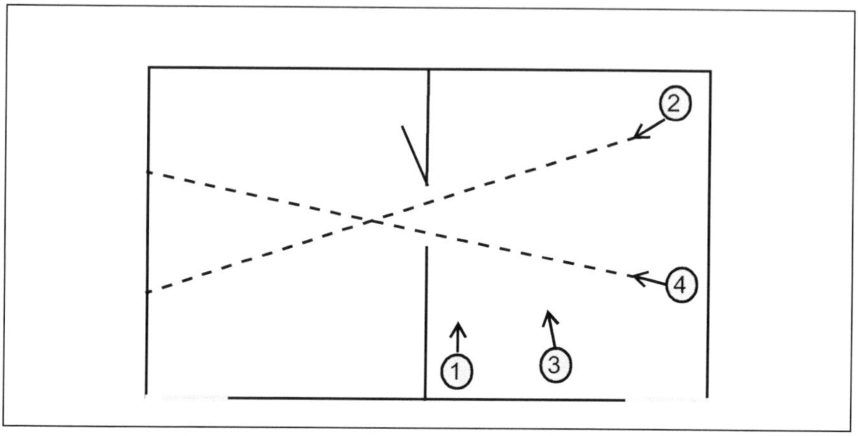

Figure 4-19. Partial Clearing of a Connected Room

4-144. *Figure 4-20, page 4-76,* shows how the team stacks in preparation for entry into the next room. When the team leader declares the first room clear, the team will automatically stack on the entrance to the next room. In this example, crossing the doorway would not only place a soldier in the fatal funnel unnecessarily, it would also mask the cover fire of the number 4 soldier. Therefore, soldiers number 1 and 3 stack to the left of the door and soldier number 2 stacks to the right. Soldier number 4 remains in a position to cover a portion of the adjacent room. In this example, soldier number 1 remains the first soldier to enter the room. He awaits a squeeze from soldier number 3. This squeeze acts as a signal to soldier number 1 that at least one other team member is behind him and is ready to enter the room. This form of communication reduces the time spent discussing the order of entry and eliminates the need for a verbal queue (an early warning for enemy soldiers inside the room) to enter the room. If the door had been positioned elsewhere in the room, it could have just as easily been the number 2 or number 3 soldier entering the adjacent room first.

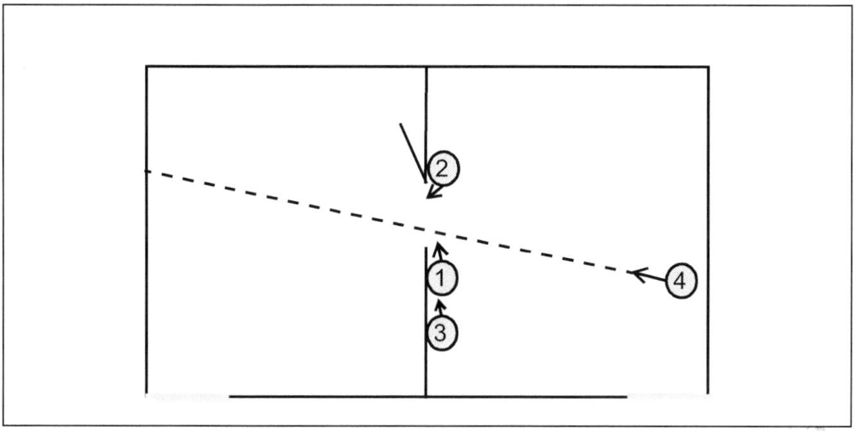

Figure 4-20. Team Stack

4-145. *Figure 4-21* depicts the position of the team members while clearing the second room. The soldiers' original numbers are left in place to demonstrate the difference in the positions from the first room. As the team enters the room, soldier number 4 lowers his weapon and moves quickly to the entry point to prevent flagging any team member. He then moves up to the entrance and follows the rest of the team into the next room.

4-146. *Figure 4-22* depicts a closed door to the adjacent room. A closed door is considered locked in all cases when a breacher is present (this is assuming the breacher has the means available to defeat the lock). Once the other team members are stacked, the breacher moves to the door and breaches it on the hand signal of soldier number 3. Soldier number 4 then steps back in order to clear the path for soldier number 2 (the third soldier to enter the room). If no breaching equipment is available, a quick attempt to open the door, from the side may be made. If the door is locked, kick it open. However, this should be the last choice for defeating a locked door. Kicking generally requires more than one

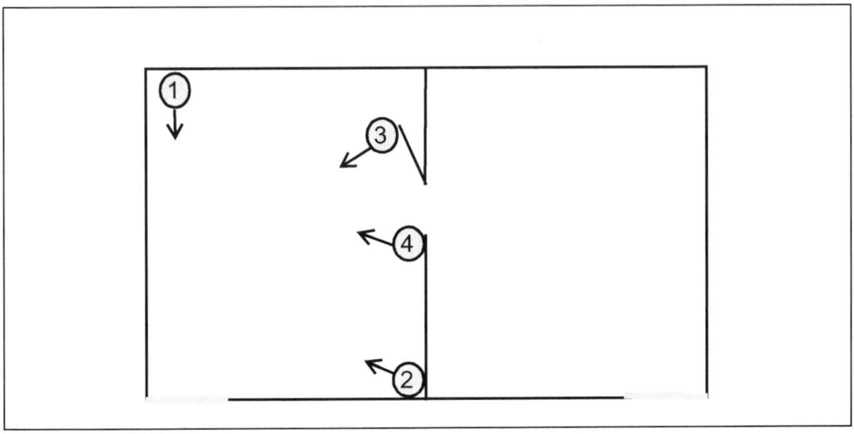

Figure 4-21. Team Members' Positions

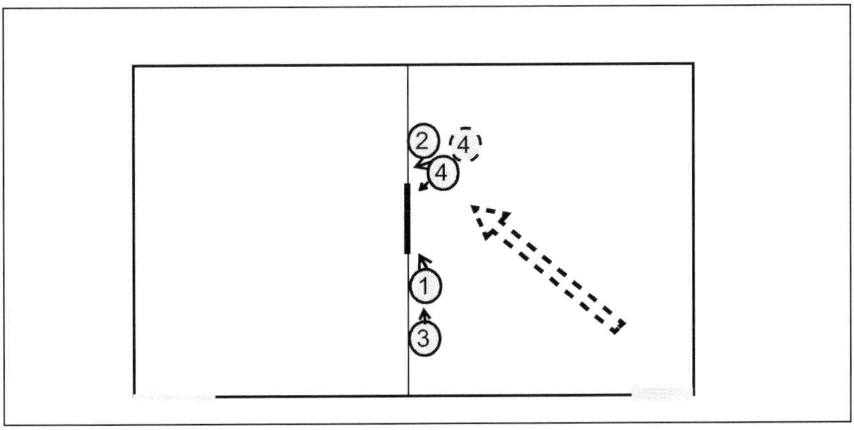

Figure 4-22. Team Positions When the Door to the Adjacent Room is Closed

attempt to defeat the lock. After the first attempt, the breacher's chances of drawing fire through the door are significantly increased.

NOTE: Do not attempt to open the door by hand. If the door is locked, this action will alert any enemy soldiers inside of the room of your intentions and

possibly expose the breacher to fire through the doorway.

4-147. *Figure 4-23* depicts the team preparing to enter the next room or a hallway. Soldier number 1 moves to a position just inside the doorway. He would have had security on the doorway from his original point of domination. The other team members stack to the left, remaining just inside the doorway. Soldier number 1 is now in the position to partially clear the next room or hallway. The team enters the room the same way as described in the initial coverage.

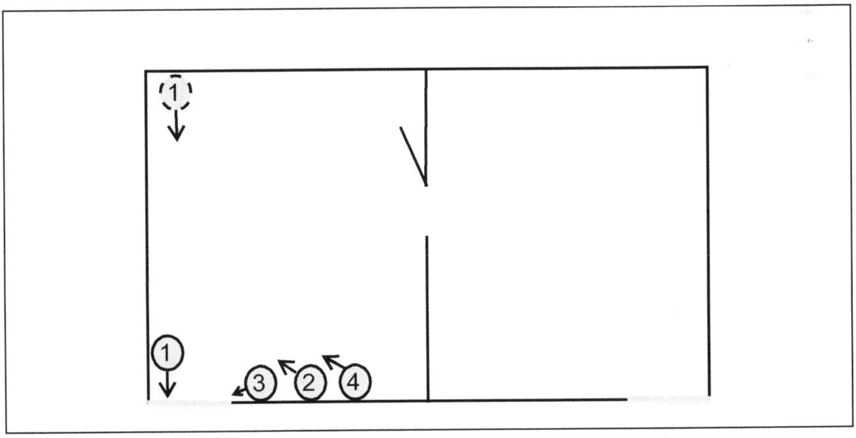

Figure 4-23. Stack Variation

CLEAR A FLOOR

4-148. When clearing a single-floor complex, the principles discussed in single- and connecting-room clearing are used. The only added features are the hallways and an additional squad, when available, to combat attrition. *Figures 4-24* through *4-28* depict the sequence of events in a floor-clearing mission.

4-149. *Figure 4-24* gives an example of a two-squad (four teams, plus HQ element) assault force, preparing

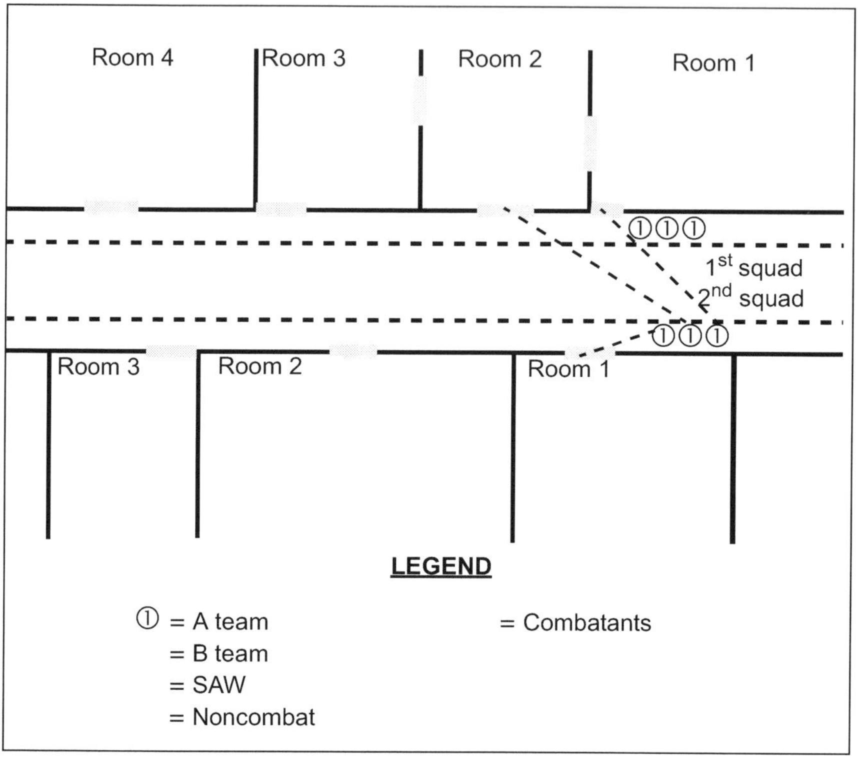

Figure 4-24. Clearing a Floor

to clear a floor. The first and second squads are stacked on adjacent opposite walls in the hallway. The first squad prepares to enter or clear the first room on their side. Soldier number 1 (first squad) is continually providing security on the door to his direct front. The second squad halts before exposing itself to the open door in front of the first squad. The second squad's number 1 soldier provides security on the first doorway to his direct front. If no entry point to his direct front exists, soldier number 1 provides security down the hallway. Soldier number 2 (second squad) provides security on the second doorway to the front of the first squad. Soldier number 3 (second squad) provides security on the door to the direct front of the first squad. He lifts that security when the first squad begins to enter that room. The SAW gunners in both teams provide security down the hallway. The HQ RTO (not depicted) provides rear security.

4-150. In *Figure 4-25,* the first squad clears the first room and moves into the second adjoining room. As the second room is cleared, soldier number 3 calls shot and engages an enemy threat in the next room. The B team of the first squad moves into the first room after the A

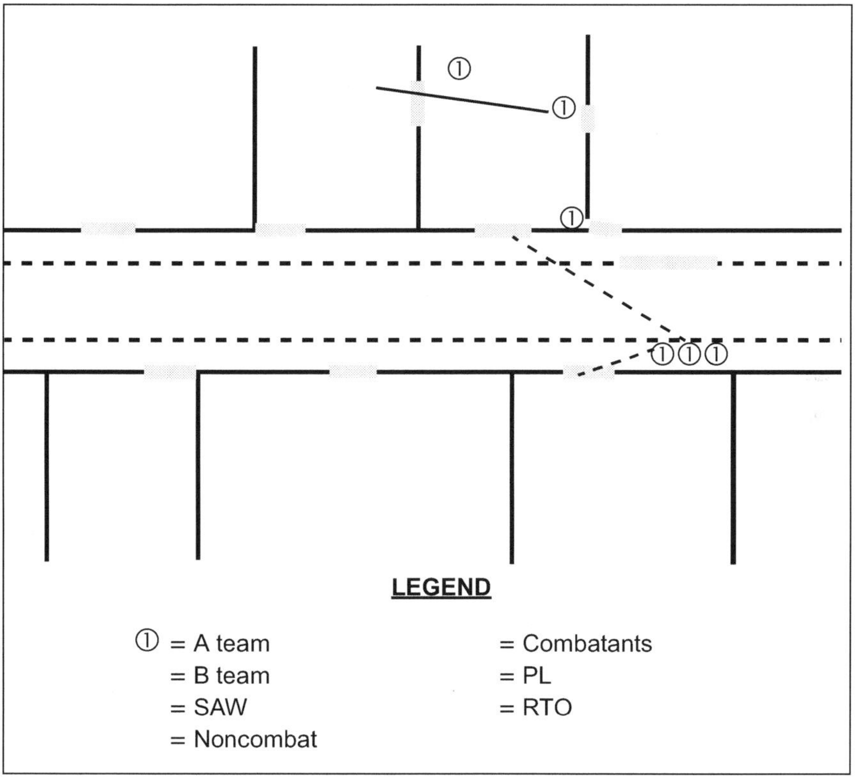

Figure 4-25. Clearing Adjoining Rooms with Hallway Security

team begins to clear the second room. The B team awaits permission from the A team leader to enter the second room. The HQ SAW gunner picks up the hallway security. The second squad has not moved from their original position. A team should never move past an uncleared room. In this case, the second squad will not receive a room clear signal from the first squad until the first squad reaches and clears the third room. This is because entry points interconnect the rooms on this side of the hallway. These entry points should be used in lieu of the hallway, whenever possible.

4-151. In *Figure 4-26, page 4-82,* the first squad has cleared the first three rooms. In this example, one soldier from the B team has been detached to secure the first room, a second has been left in the second room to secure the noncombatant, and a third is securing a noncombatant and a downed enemy soldier in the third room. Soldier number 1 (first squad) remains inside the doorway and provides security on the entry point of the second room to the second squad's front. Soldier number 2 provides security on the third room to second squad's front. The second squad receives the all-clear sign from the first squad (radio, thumbs-up, or hand and arm signals) and moves to their first-room entry point.

NOTE: If there are not enough personnel available to leave security in cleared rooms (especially in multilevel clearing), disarm and secure all combatants and noncombatants using handcuffs or flex cuffs.

4-152. If more than one floor is involved, leave a two-person security team on each floor while the building is cleared. In this instance, each room must be systematically recleared once the building is secured. A marking system (such as chalk or chemical lights) must be implemented in order to identify the rooms that have been cleared.

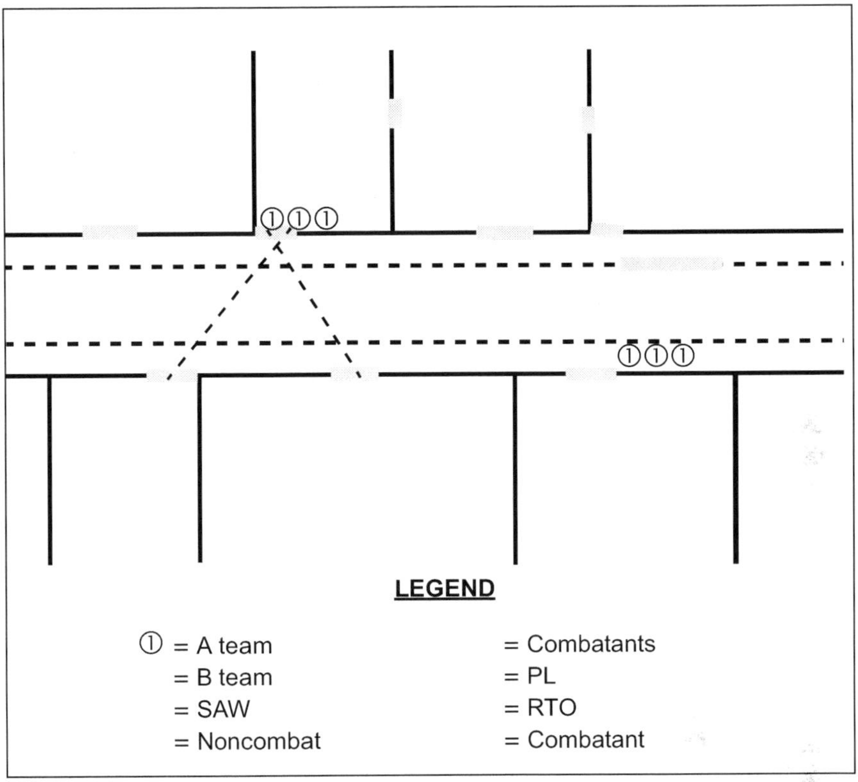

Figure 4-26. First Squad Clearing a Room with Hallway Security

4-153. In *Figure 4-27* the second squad clears all three rooms on their side of the hallway. The second squad was forced to clear each room by using the hallway as an access. In this example, the second squad's B team would not follow A team into the room. When the room is cleared, the first member in the B team stacks and (when instructed) enters in order to relieve the A team of security responsibilities. Once this takes place, A team is free to exit the room and move down the hallway to the next room.

NOTE: Before the team exits the room into the hallway, the lead team member announces that the team is about to exit the room. They should wait for permission to come out into the hallway.

4-154. Once the second squad has finished clearing their side, the number 1 soldier provides security on the next uncleared room on the right (first squad) side. The first squad receives the all clear signal and then moves up to the next room. Both B team SAW gunners continue to provide security down the hallway.

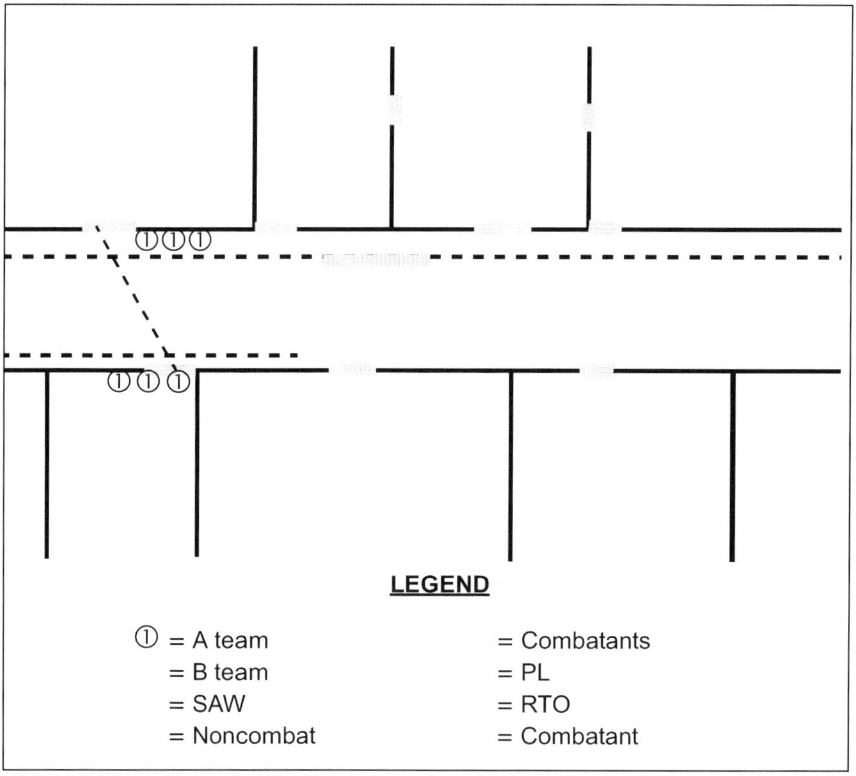

Figure 4-27. Second Squad Clearing Rooms

4-155. *Figure 4-28* depicts the events that take place once the building has been cleared. Two SAW gunners are dispatched to each end of the hallway for security. The rooms containing noncombatants or enemy personnel are searched first (mark them to indicate that they have been searched). Once this is completed, it is usually best to remove all of the combatants and noncombatants encountered during the clearing operation (using a small detachment) along an extraction route that is covered by a support element. The support team escorts the noncombatants to a designated holding area to determine their status, and then searches the remaining rooms.

CLEAR A STAIRWAY

4-156. When multilevel structures are encountered, stairs become an added obstacle that will require maneuver *(Figure 4-29, page 4-86)*. One of the more dangerous stair situations that a team is likely to encounter is a stairway with a turn between floors. Besides the blind spot at the turn, these stairways often have a loft that overlooks the bottom portion of the stairway. If the team is ascending, the number 1 soldier (as always) provides security to his direct front. Soldier number 2 secures the top row of stairs. Soldier number 3 secures the loft area. If a loft area does not exist, soldier number 3 secures the top of the second row of stairs. If the team is descending, each soldier has the same area of responsibility, except that the bottom of the stairway is secured, as opposed to the top.

CLEAR BLIND SPOTS

4-157. When a blind spot is encountered (such as stairs or a sharp corner), it can be cleared with a technique known as cutting the pie *(Figure 4-30, page 4-87)*. The soldier clears the dead space by slicing it up into pieces and then clearing one piece at a time. By doing this, the soldier exposes only a very small portion of his frame to

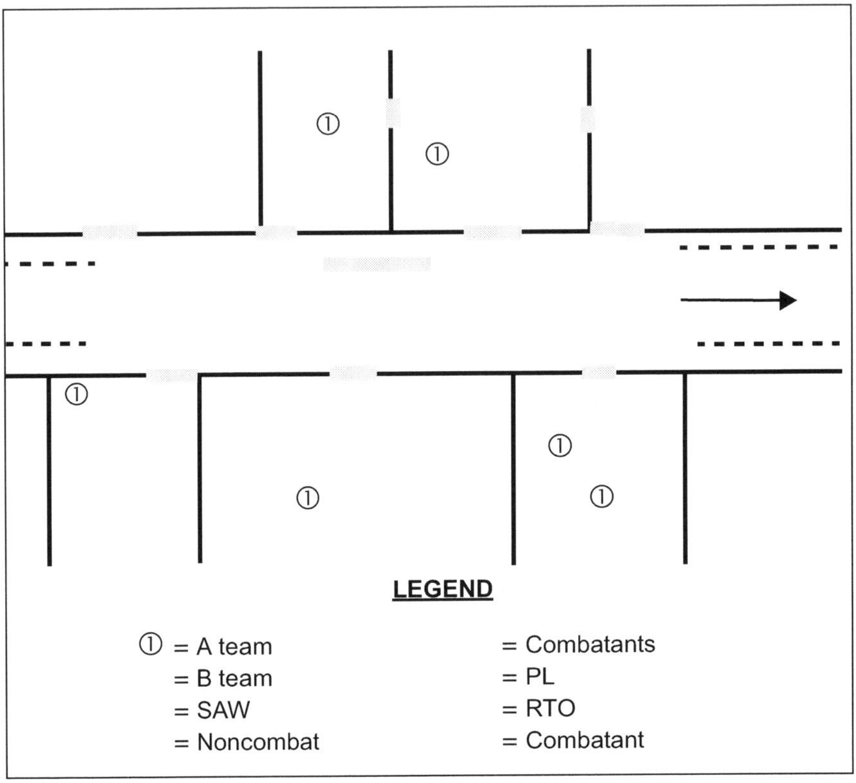

Figure 4-28. Events After Building is Closed

any enemy combatants, while placing himself in a position to methodically clear the dead space.

SECURE PERSONNEL

4-158. The clearing team secures and identifies all personnel found in a room or building that is being cleared or searched. Until identified, (which will not normally occur during the clearance of the building) they approach noncombatants in the same manner as combatants. They do not assume that the frail old man in civilian clothes, cowering in the corner, presents no

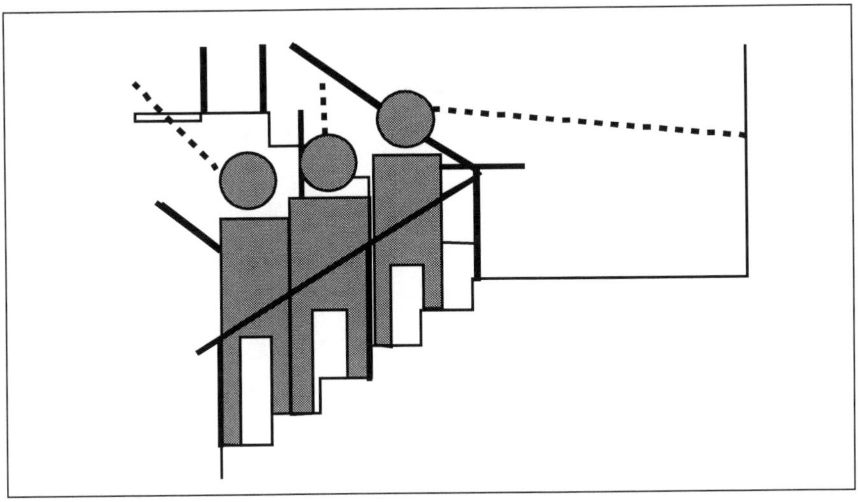

Figure 4-29. Clearing Stairs

threat. On contact, the soldier that first indexes his weapon onto the individual orders the individual to his knees, with his hands on his head, and his fingers interlocked. This soldier stays out of arms reach at all times. Do not allow the unidentified individual near any exits. The individuals will remain in that position, with security present, until the building or room has been cleared. When it is time to secure the individual, the soldier that has been providing security does all the talking. In a slow, loud (but calm) voice, the soldier gives short, easy to understand, instructions to the individual, such as "look at me," "sit up straight," and so forth. This talk has a dual purpose. It—

- Gives directions that helps in securing the individual and calms him down.
- Keeps the individuals attention oriented on him and not on the second securing soldier.

4-159. As the first soldier talks to the individual, a second team member unholsters his pistol for close-in work. The second soldier approaches the individual from

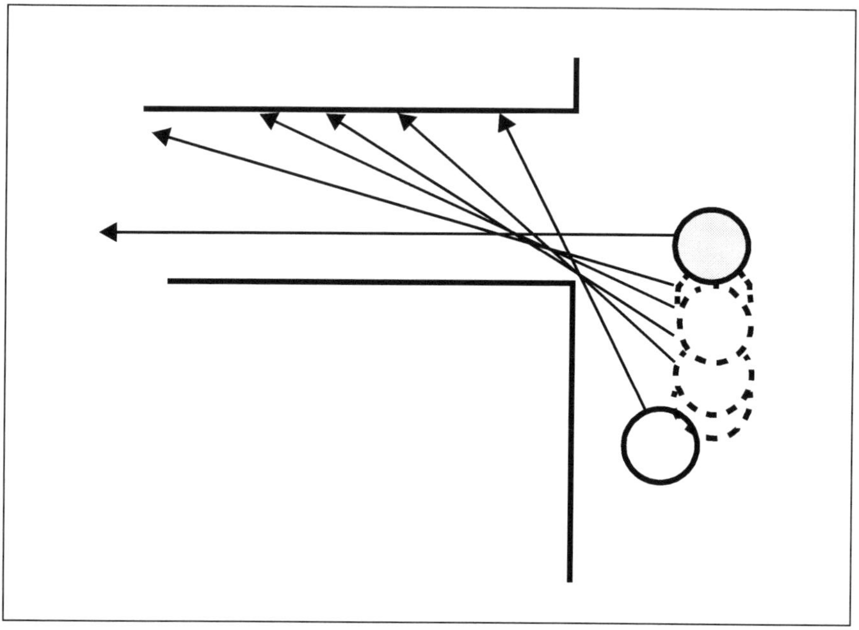

Figure 4-30. Cutting the Pie

a blind spot, grasps the middle three interlocked fingers, and places a knee in the middle of the individual's back. Once the individual is under control, the second soldier holsters the pistol and secures the hands of the individual behind his back. If there are numerous individuals to secure and there will be several minutes before they can be exfiltrated from the building, these individuals can be placed on their knees and tilted forward so that their foreheads are against a wall. Their bodies should be at such an angle that their heads support most of the weight. If the prisoner attempts to get up, he will fall to the floor. This position enables a few soldiers to watch several prisoners. If an enemy soldier becomes engaged and the combatant does not appear to be conscious, it must be determined whether the individual is deceased or not. The body should be approached in the same manner as described above. If a kick does not gain a response, a simple eye gouge or sternal rub should identify whether the combatant is indeed dead. It is always best to secure the hands of the individual before applying an eye gouge or sternal rub. Beware of the enemy soldiers *playing opossum*.

TRAINING FOR MASTERY

4-160. It is obvious that the techniques described above will work best in an environment that has quick-thinking and intelligent soldiers. If the above techniques are practiced using the crawl, walk, and then run method, squads will find themselves catching on very quickly. When the one-room technique is mastered, everything else will fall into place. As teams work together, they will come up with their own systems of signaling, security, and change up in positions. The clearing of a building will be one smooth flow. The platoon leader starts training with single rooms outlined in engineer tape. This system allows for ease in critiquing, and all personnel will be able to observe and learn from the positive and negative performances of each team. The only building layout that cannot be reproduced by tape is a multilevel building (due to the presence of stairwells). Once the platoon is proficient in the one-room technique, the platoon leader moves to connecting rooms outlined in engineer tape. He continues this process until a complete floor in a real building can be cleared smoothly. Once multilevels are covered, he adds simulated breaching, furniture, and combatant or noncombatant scenarios. Once this type of training is completed, an assault team should have the base knowledge required to overcome specific situations that are not covered in this chapter. The platoon leader remembers that this is a perishable skill. As time passes, new platoon members must be integrated into this training and older members must be retrained.

Chapter 5

Maneuver and Mobility Support

MMS, formerly known as battlefield circulation control, consists of those measures necessary to enhance combat movement and the ability to conduct movement of friendly resources in all environments. These measures ensure that commanders receive personnel, equipment, and supplies as needed. MMS is conducted across the full spectrum of military operations. The primary focus of MP during MMS is to ensure swift and uninterrupted movement of combat power and logistical support.

MANEUVER SUPPORT

5-1. Maneuver is the employment of forces on the battlefield in combination with fire (direct or indirect) or fire potential. It is the movement of combat forces to gain a positional advantage, usually to deliver or threaten delivery of direct and indirect fires. MP tasks that support maneuver include—

- MP support to river crossings.
- MP support to breaching operations.
- MP support to a passage of lines.
- Straggler control.
- DC control (refer to *Chapter 7* for more information about DC operations).

SUPPORT FOR RIVER CROSSINGS

5-2. A river is a significant obstacle that may slow, stop, or impede a unit's ability to maneuver. Units are restricted to moving in column formations along limited routes that come together at crossing sites. Friendly forces are vulnerable while crossing water obstacles. The challenge is to minimize the river's impact on the commander's ability to maneuver. The three types of river crossings include—

- Hasty.
- Deliberate.
- Retrograde.

5-3. MP traffic control is essential to help reduce exposure time and speed units across any obstacle. In addition, effective traffic control contributes to the flexibility of the crossing plan by enabling commanders to change the sequence, the timing, or the site of the crossing units. MP can switch units over different routes or hold them in waiting areas as directed by the tactical commander. This support is vital in reducing congestion, speeding the crossing of any obstacle (not just water), and enabling the maneuver forces to maintain momentum.

Hasty River Crossing

5-4. A hasty river crossing is a decentralized operation using organic, existing, or expedient crossing means. It is the preferred river crossing method. Conduct a hasty river crossing as a continuation of an attack to ensure little or no loss of momentum by the attacking force. The MP platoon in direct support of a crossing maneuver brigade, may be required to support the crossing without additional support.

Deliberate River Crossing

5-5. A deliberate river crossing requires planned and augmented MP support. Conduct a deliberate river crossing when a hasty crossing cannot be made successfully, normally when offensive operations must be renewed at the river, and when enemy forces must be cleared from the area. A buildup of firepower and equipment is needed on both entry and exit banks. Normally, MP support from corps is required to augment the division MP company.

Retrograde Crossing

5-6. Closely plan and control a retrograde crossing. Massed crossing forces could slow momentum or exceed bridge classification limits. Forces moving to the rear may retrograde to defensive positions beyond the water obstacle and may be slowed as they set up to defend the exit bank. MP support retrograde crossings the same as they do deliberate crossings.

River Crossing Planning

5-7. The crossing force commander plans the river crossing operation. He prepares an OPORD and specifies what support is required. The PM, based on the OPORD, plans MP support for the river crossing. The plan includes how MP assets will be used and what additional resources are needed. The MP commander supporting the operation plans and supervises the mission based on the OPORD and guidance from the PM. The OPORD normally gives OPCON of all units entering the crossing area to the crossing commander.

5-8. The MP leader supporting the crossing site develops a traffic control plan to support the circulation control plan. He must plan for—

- Traffic control posts (TCPs) and temporary route signs at—
  - Major crossroads on the MSR and near crossing sites and lateral boundaries to control traffic from adjacent unit areas that could interfere with division surface movements.
  - Staging areas and engineer regulating points (ERPs) to provide directions and

information, control movement to and from staging areas according to planned times, and relay messages between traffic HQ and the moving unit.
  - Holding areas on the entrance bank to direct traffic to crossing sites; on the exit bank, inside the traffic regulating line (TRL), to control movement; and on the exit bank, outside the TRL, to temporarily hold sections of a convoy or a unit until it can reassemble and continue its movement.
- Mobile patrols to operate along primary routes to control traffic, spot problems, guide and escort vehicles, and reroute traffic when necessary.
- Temporary EPW collecting points. Set up the collecting points outside the TRL. Evacuate EPWs through the crossing areas as quickly as possible so their transit does not impede the movement of friendly forces.

5-9. For brigade crossings, the MP leader may collocate with the brigade staff to form a small, temporary traffic control cell located at the brigade main CP or the brigade TOC. The brigade main CP controls the maneuver support force that consists of corps engineers, bridge companies, MP, and chemical units.

Control Measures

5-10. To ease control of large, fast-moving forces, the river crossing plan usually allots one crossing area for each maneuver brigade. The commander uses control measures to delineate areas of responsibility for subordinates and to ease traffic control. *Figure 5-1* shows the following control measures.

5-11. **Release Line (RL)**. As used in river crossing operations, RLs are used to delineate the crossing area. RLs are located on both the far shore and nearshore and

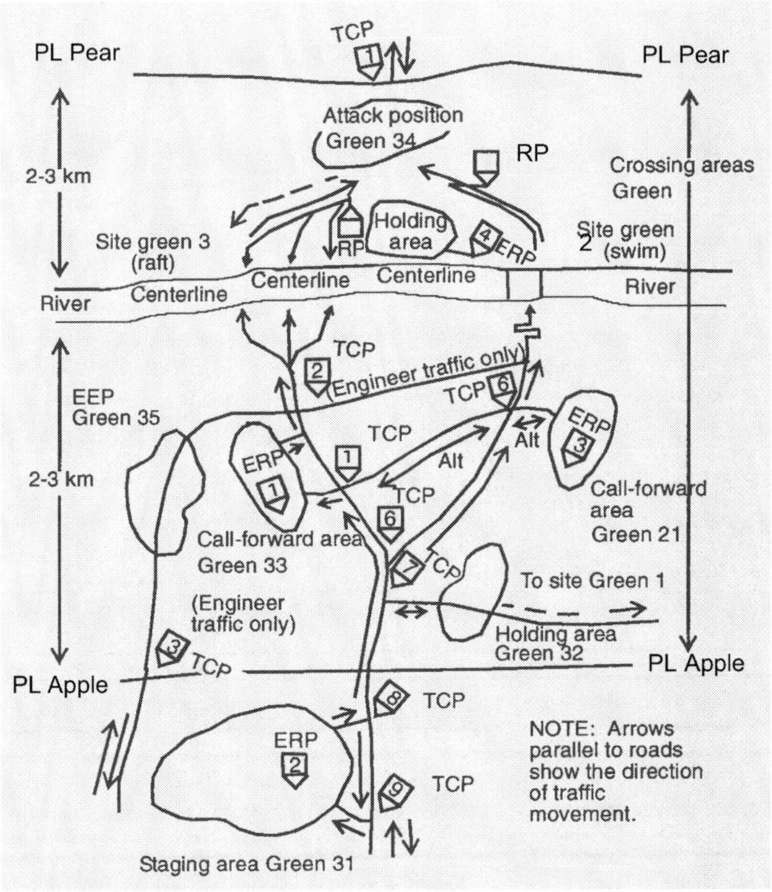

Figure 5-1. River Crossing Control Measures

indicate a change in the HQ that is controlling the movement. RLs are normally located within 3 to 4 kilometers of the river and on easily identifiable terrain features, if possible.

5-12. **Crossing Areas**. Crossing areas are controlled-access areas that decrease congestion at the river. This permits swift movement of the forces. Each lead brigade has a crossing area on both sides of the river that is defined by brigade boundaries and the RL. Crossing areas normally extend 3 to 4 kilometers on each side of the river, depending on the terrain and the anticipated battle.

5-13. **Waiting Areas**. Waiting areas are located adjacent to the routes or axes of advance. Commanders use the following waiting areas to conceal vehicles, troops, and equipment while waiting to resume movement or make final crossing preparations:

- Staging areas. These are battalion-size waiting areas outside the crossing area where forces wait to enter the crossing area. The brigade traffic control cell handles the units' movement into the staging areas. The crossing area commander (CAC) controls movement from the staging areas into the crossing areas. MP operate TCPs at the staging areas according to the crossing and traffic circulation plans. They emplace temporary signs along the route from the staging area through the crossing area to guide the convoys. Units make crossing preparations and receive briefings on vehicle speed and spacing in the staging areas. Staging areas—
  - Are located to support the crossing concept.
  - Are far enough back to permit the rerouting of the battalion along other roads or to alternate crossing sites.
  - Are easily accessible from major routes.
  - Have enough area for dispersing a battalion-size unit.
  - Provide concealment.
- Call-forward areas. These areas are company-size waiting areas located within the crossing

area. Engineers use them to organize units into raft loads; crews use them to make final vehicle crossing preparations. The CAC controls movement from the staging area to the call-forward area. The crossing site commander (CSC) directs movement from the call-forward area to the crossing site and on to the far-shore attack position. As a minimum, each CSC operates his own call-forward area. Call-forward are—

- Located to support the crossing plan.
- Company size within the crossing area.
- Easily accessible from routes.
- Planned with a minimum of one per crossing site.
- Collocated with ERPs.
- Used to organize units into raft loads.
- The final preparation areas before going to the crossing site.
- Normally operated by engineers.

- Holding areas. These areas are waiting areas that forces use during traffic interruptions. Units move into these areas when directed by TCP personnel and disperse rather than stay on the roads. Holding areas are battalion size outside of the crossing area and company size within it. Far-shore holding areas are used to organize return traffic. MP operate holding areas according to the crossing and traffic circulation plans and—
  - Are used as call-forward areas for return traffic from the far shore.
  - Are located to support the crossing plan.
  - Are easily accessible from routes.
  - Have enough area for dispersion.
  - Provide cover and concealment.

- Are defensible.
- Maximize traffic flow with minimum control.

- Attack positions. The attack positions are the last positions occupied or passed through by the assault echelon or the attacking force before crossing the line of departure. Within the bridgehead, the attack position is the last position before leaving the crossing area or bridgehead line.
  - Assembly areas. These are the areas where forces prepare or regroup for further action.

5-14. **Engineer Equipment Parks (EEPs)**. These are areas located a convenient distance from bridging and rafting sites for assembling, preparing, and storing bridge equipment and material. They are at least 1 kilometer from the river and hold spare equipment and empty bridge trucks that are not required at the crossing sites. EEPs should be located where they do not interfere with the traffic to the crossing sites and where equipment can be concealed and dispersed. Ideally, routes leading from the EEPs to the crossing sites are not the same routes used by units crossing the river.

5-15. **Traffic Control Posts**. In river crossings, TCP personnel assist the crossing-area HQ in traffic control by reporting and regulating the movement of units and convoys. TCP personnel relay messages between the crossing-area HQ and the moving units. The PM identifies locations that need or require TCPs. MP operate TCPs on both banks of the river to control traffic moving toward or away from it. TCPs are operated at major or critical crossroads and road junctions, staging areas, holding areas, and ERP.

5-16. **Engineer Regulating Point**. ERPs are technical checkpoints used to ensure that vehicles do not exceed the capacity of the crossing means. They help maintain traffic flow. Vehicles not allowed to cross are removed so that they do not cause a traffic backup at the actual crossing site. Engineers man the ERPs and report to the CSC. TCPs are collocated with the ERP to ensure that all vehicles clear the call-forward areas. An additional duty of ERP personnel is to give the drivers final instructions on site-specific procedures and other information, such as speed and vehicle intervals. As a minimum, each crossing site requires an ERP at its own call-forward area. If enough engineer assets are available, an ERP may be established at far-shore holding areas to regulate rearward traffic.

Route Execution

5-17. MP must be prepared to establish holding areas along movement routes on order. If the road network sustains damage, vehicles will need to be routed into the holding areas until traffic can be restored or rerouted. Refer to *paragraph 5-104* for more information about holding areas.

5-18. MP mobile patrols operate along primary routes, monitoring traffic, spotting problems, and rerouting traffic as necessary and conducting AS around the crossing area. They make frequent checks of temporary signs to prevent the enemy from tampering with them.

5-19. MP may be directed to screen the crossing unit's flanks and rear. The size of such an element is determined by METT-TC. In most environments this mission requires at least a squad. MP conduct screening missions to provide early warning of enemy approach and to provide real-time information, reaction time, and maneuver space for the crossing unit. The squad fights only for self-protection and remains within its capabilities. Refer to *Chapter 6* for more information about screening missions.

5-20. Include at each crossing site a temporary EPW collection point. Initially the collection point will be on the entry bank. Once MP cross as part of the support force, a temporary collection point is established on the exit bank. A division central collection point is established outside of the crossing area. Refer to *Chapter 7* for more information about division forward collection points.

5-21. Rigid control of civilian movement is necessary to preclude congestion on movement routes. The PM coordinates for HN police support to ensure that the civilians who live in the crossing area are kept in place or, if necessary, quickly moved to designated areas away from the river. Normally, civilians are not allowed to cross the river or move along the edge of the river during the river crossing operation. Refer to *Chapter 7* for more information about DC resettlement.

MILITARY POLICE SUPPORT TO BREACHING OPERATIONS

5-22. Breaching operations are conducted to allow maneuver despite the presence of obstacles. Obstacle breaching is the employment of a combination of tactics and techniques to advance an attacking force to the far side of an obstacle that is covered by fire. Breaching operations begin when friendly forces detect an obstacle and begin to apply the breaching fundamentals. Breaching operations end when the battle handover has occurred between the follow-on forces and a unit conducting the breaching operation.

Support Planning

5-23. MP support to breaching operations is similar to MP support to river crossing operations. The employment of MP is based on METT-TC, available resources, and the commander's priorities. MP support to breaching operations includes—

- Operating TCPs at the breaching site and along routes leading to or departing from the breaching site.
- Operating holding areas.

- Providing mobile guides to escort the units.

5-24. The platoon leader coordinates with higher HQ and the engineer forces conducting the breach for essential information that includes the—

- Azimuth and distance to the final-approach marker or the 8-digit grid coordinate of the final-approach marker that is entered into the teams Global Positioning System (GPS) receiver.
- Lane marking pattern currently emplaced.
- Type of final-approach marker used.
- Traffic control plan and march order.

5-25. A combined-arms breach is a complex operation and requires precise synchronization. Breaching operations normally require the maximum use of TCPs to assist support, breach, and assault forces to move along various lanes. Refer to *paragraph 5-88* for more information about TCPs. Lanes are marked to safely pass units through the obstacle. The three levels of lane marking are—

- Initial.
- Intermediate.
- Full.

5-26. MP may provide TCPs and guide support to lanes at any level of marking. However, the main effort of MP support may come in later phases of the operation, when larger units (battalion and above) are passed to subsequent objectives, and time permits marking improvements to be made. The increase in traffic and the more diverse forces with different levels of driver experience will increase the need for MP traffic control operations. MP guides are simply mobile MP teams that escort units from one control measure or point to another. Guides and TCPs are essential when there are multiple lanes. *Figure 5-2, page 5-12* shows the flexibility that the combinations of multiple lanes and guides or TCPs provide the commander.

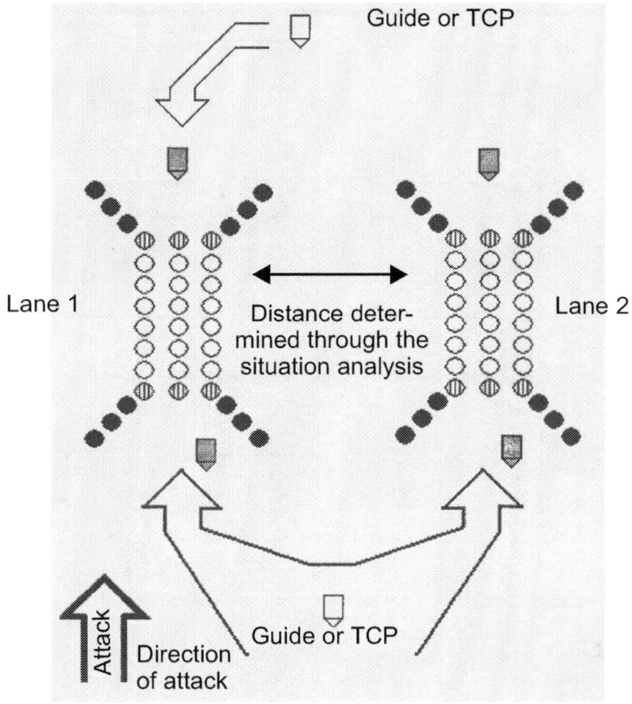

Figure 5-2. Multiple Lanes (Two-Way Traffic)

Movement Execution

5-27. The commander sets the priority of movement based on the situation. MP may concentrate their efforts on assisting the immediate passage of larger combat forces. Or their priority may quickly shift to ground evacuation of casualties or vehicle recovery operations. MP traffic control operations give the commander the ability to make last-minute changes in the traffic flow or lane usage

5-28. MP may be required to establish unit holding areas (battalion and company size) in the event that traffic is disrupted on the lanes due to enemy activity or the need to do maintenance or upgrade a lane. Refer to *paragraph 5-104* for more information about holding areas.

5-29. The commander collocates guides or TCPs at the far recognition marker when he feels the situation requires more positive control.

5-30. Guides and TCPs are briefed on this information and are kept up to date on changes to the traffic control plan and enemy activity in the AO.

5-31. The platoon leader plans for the possible need to establish a forward EPW collection point near the breaching operation. Refer to *Chapter 7* for more information about division forward collection point. He must also plan for an increase in the number of TCPs needed during limited visibility or in restrictive terrain. Refer to *FM 3-34.2* for more information about breaching operations.

PASSAGE OF LINES SUPPORT

5-32. This area describes how an MP leader is to plan and conduct MP support to the passage of lines. The MP elements described in the following paragraphs are supporting the passing and stationary units. MP conducting a battle handover or passage of lines to a TCF is discussed in *Chapter 6*.

5-33. A passage of lines is a tactical event normally associated with a battle handover. A passage may be designated as a forward or rearward passage of lines. Moving a maneuver unit through the positions of an emplaced unit that is in contact with the enemy is a critical action. It requires detailed coordination; planning; and close, continuous supervision of the movement.

5-34. The main focus of MP support to a passage of lines is normally employing special traffic control measures that include—

- TCPs.
- Temporary route signing.
- Checkpoints and roadblocks.
- Defiles.

5-35. MP may also provide guides to escort the passing unit en route to a release point or AA. Similar to MP support to breaching operations, guides provide the commander a means to change the sequence, timing, or lanes of the passing units.

Passage of Lines Planning

5-36. MP support the passage of lines operation to assist a maneuver unit in contact with the enemy to maintain movement. Depending on the scope of the operation, a division MP company may not be enough to support a passage of lines operation. METT-TC may necessitate the need for additional corps MP support.

Control Measures

5-37. When planning control measures for a passage of lines, MP leaders must consider the placement of the following:

- AAs where units prepare for further action.
- The battle handover line (BHL) where the stationary force assumes responsibility for the sector from the covering force.
- The forward edge of the battle area.
- Passage lanes along which the passing units move to avoid stationary units and obstacles.
- Passage points where units will pass through one another. They are located where the commanders want the units to execute the

passage of lines. Designate multiple passage points to help eliminate congestion.

- Contact points (designate an easily identifiable terrain feature) where the units will physically meet.
- SPs where unit elements come under the control of the commander responsible for the movement.
- Phase lines, used in controlling the timing of the operation, are usually recognizable terrain features extending across the zone of action.
- RPs where unit elements revert to their respective commanders and continue moving to their destinations.
- Travel routes from the point of origin to the destination.
- Checkpoints used to coordinate friendly movement. (Checkpoints are not used as reference points for reporting enemy locations.)

Passage of Lines Execution

5-38. MP support for a passage of lines is conducted at the platoon level. The company monitors the platoon and coordinates with higher HQ. The company operations section is required to conduct detailed coordination with the passing and stationary units that includes—

- Communication requirements.
- Recognition signals.
- Route SP.
- Time of passage.
- Passing lanes.
- Control measures to include TCPs, escort and guide vehicles, temporary route signs, or a combination of these.
- EPW and DC evacuation.

5-39. The platoon leader continuously coordinates with the company operations section to confirm the following:

- The size of the passing unit.
- Locations of AAs.
- Recognition signals.
- The actual time that the passage of lines will commence.

5-40. MP support for a passage of lines begins at the route SP, which serves as a contact point. Recognition signals are displayed at the SP. MP monitor the passing unit's command net during the entire passage. Radio silence is maintained during this time.

5-41. Depending on the situation, MP can support the passing unit with the following:

- TCPs.
- Escort and guide vehicles.
- Temporary route signs.
- A mix of these measures.

5-42. The platoon leader selects the method that best supports the moving unit's passage through the stationary unit. Temporary route signing will decrease the number of TCPs needed, but if routes are not well defined or they cross congested areas, expect to provide TCPs or escort vehicles.

5-43. MP must be prepared to initiate vehicle holding areas at designated locations along movement routes. If the road network sustains damage, vehicles will be routed into a holding area until traffic can be restored or rerouted. (Refer to *paragraph 5-104* for more information about holding areas.)

5-44. The platoon leader plans for the need to establish a temporary EPW collection point, and establishes it near the passage area but out of the view of the operation.

5-45. Strict control of the movement of civilians is necessary to preclude congestion on routes used for the operation. When necessary, MP establish evacuation routes to move DCs to a designated area rear of the staging areas. Local nationals who live in the immediate area will remain in place or be evacuated primarily by HN police from the area.

STRAGGLER CONTROL

5-46. MP conduct straggler control operations to assist commanders in maintaining combat strength by locating and returning stragglers to their units. MP identify stragglers at the TCPs, checkpoints, roadblocks, defiles, or while patrolling the MSR. For large numbers of stragglers, special collecting points are set up along the MSR.

Operate Straggler Posts

5-47. When operating a straggler CP, MP teams need to know what units are located or operating in their AO. Most stragglers are soldiers who have become accidentally separated from their command. Stragglers are identified by checking the following:

- Uniforms.
- Unit insignia.
- Bumper markings on the vehicles.
- Identification cards or tags.
- Passes or other authorization documents.

5-48. For each straggler identified, as a minimum, MP record—

- The straggler's name, rank, social security number (SSN), and nationality.
- The straggler's unit.
- The straggler's category ("injured" or "uninjured").
- Whether the straggler is armed or not.
- Where the straggler was coming from and his destination.
- Why and when the straggler left the unit.

- The location where the straggler was sent.

5-49. MP administer first aid to the injured, wounded, or ill stragglers. Seriously ill or injured soldiers are evacuated. Stragglers who have information of immediate tactical value are reported to higher HQ. Soldiers fit for duty who mistakenly became separated are returned to their units or a HQ within their chain of command. The soldiers unit is responsible for any transportation requirements.

5-50. MP treat deliberate stragglers, those who have deserted or are attempting to desert or are absent without leave (AWOL), with caution. These stragglers may resort to violence to avoid military control. MP search, disarm, and detain them. They hold these stragglers until transport and escort can be arranged to take them to their unit, the straggler collecting point, or another facility set by the SOP or the straggler control plan. MP safeguard confiscated property and documents, and dispose of them according to the straggler control plan.

5-51. MP handle stragglers from the HN or other allied forces the same as US stragglers. If the PM has coordinated with other national forces to set up joint straggler posts, allow MP from other national units to handle stragglers from their own forces.

Operate Straggler Collecting Points

5-52. When large numbers of stragglers exist and TCPs, mounted patrols, and straggler control posts are not able to handle the straggler flow, MP may be tasked to operate a straggler collecting point. MP temporarily hold stragglers at collecting points while they process them for return to their units, placement in medical channels, or placement in other military channels.

5-53. The PM operations section plans the location of a straggler collecting point, placed along a key MSR or at an intersection of the MSR. This allows quicker access to the straggler collecting point to aid in moving stragglers to their appropriate destination.

5-54. Often it is collocated where elements of medical, transportation, and MP units can share efforts to ease the disposition of stragglers. At a straggler collecting point, MP may need food, water, clothing, and shelter for stragglers. If a medical facility is not close by, request extra medical supplies and be prepared to administer first aid.

5-55. Guards separate the injured stragglers from the uninjured. They process each soldier at the collecting point. Guards record the key information on each soldier for a straggler report. They search, segregate, and guard stragglers who refuse to return to their unit. Guards assemble and forward the report to wherever the straggler control plan directs. They hold stragglers at the collecting point until transport arrives.

5-56. Detain stragglers who refuse to return to their unit until their unit provides an escort or until they are transported to a detainment facility.

MOBILITY SUPPORT

5-57. Mobility is the capability of military forces to move from place to place while retaining the ability to fulfill their primary mission. It includes those activities that enable a force to move personnel and equipment on the battlefield without delays due to terrain or obstacles. MP activities that support mobility include the following:

- Route reconnaissance and surveillance.
- MSR regulation enforcement.
- Special circulation control measures.

ROUTE RECONNAISSANCE AND SURVEILLANCE

5-58. MP conduct route reconnaissance and surveillance operations to gain detailed information on specific routes to be used as deployment routes, MSRs, or movement corridors. Mobile MP teams record and report the condition of the MSR and other critical roadways, identifying effects of weather on road surfaces, damage to routes, NBC contamination, and the presence of enemy activity. Platoon leaders use this information to develop a road reconnaissance report and a route reconnaissance overlay.

5-59. The platoon leader provides the MP reconnaissance reports and overlays are used to assist the division PM with the development of the division traffic control plan. Movement planners use information gathered from the MP route reconnaissance to update the highway traffic section's (HTS's) traffic circulation control plan and to formulate the highway traffic regulation plan. Refer to *FM 55-10* for more information about traffic circulation control plans and highway traffic regulation plans.

Plan

5-60. MP leaders plan route reconnaissance operations by examining intelligence reports and maps of the area surrounding the route to be reconnoitered. When time is critical, MP conduct a hasty route reconnaissance to obtain specific information only. The MP leader must clearly understand the following critical tasks to be accomplished:

- Find and report all enemy forces that can influence movement along the route.
- Determine the trafficability of the route.
- Reconnoiter any special areas that could influence movement on the route. These areas may consist of highly populated areas or key

terrain features. Additional teams may be needed to cover these areas based on METT-TC.

- Inspect all bridges on the route.
- Locate fords or crossing sites near all bridges on the route.
- Inspect all overpasses, underpasses, and culverts.
- Locate areas suitable for short halts and holding areas.
- Locate mines, obstacles, and barriers along the route.
- Locate a bypass around built-up areas, obstacles, and contaminated areas.
- Report route information.

5-61. A more detailed route reconnaissance would include additional information concerning the terrain, potential hazards, or obstacles and would include key terrain and built-up areas 2 to 3 kilometers on either side of all MSRs. A route reconnaissance this detailed requires considerably more time and personnel.

Execute

5-62. One MP squad can conduct a hasty route reconnaissance of only one route, (about 30 kilometers). MP platoons reconnoiter three routes within the boundaries of the platoon AO, if route reconnaissance is their primary focus.

5-63. The size of the reconnaissance patrol is determined by using METT-TC factors. In most environments route reconnaissance operations are not conducted with less than a squad. The squad leader organizes the squad into security teams and a reconnaissance team. The reconnaissance team records the information, completes a *DA Form 1248*, and prepares a reconnaissance overlay. The other teams provide security.

5-64. MP do not engage the enemy when conducting route reconnaissance except in self-defense or when ordered to do so. They report any visual contact with the enemy and maintain surveillance while gathering as much information as possible. They break visual contact only on order from proper authority.

5-65. Often the main purpose of a reconnaissance is to confirm information already known about a route. Additional information can be obtained from the division PM, the division transportation officer, and from HN police. MP teams talk to convoy commanders, vehicle drivers, local nationals, and highway control regulating teams to gain or verify information about well-traveled roads. Although this information is not as reliable as information gathered from driving a route, its reliability increases when several road users report the same condition.

5-66. Mounted MP patrols continuously collect data at the level of detail required by the PM or commander and report it by the fastest secure means available. The patrols travel the routes within the AO to—

- Identify and locate the recommended route.
- Check the driving time and distances between easily recognized points.
- Look for obstructions and restrictions (bridges, tunnels, steep grades, sharp curves, ferries, snow blockage, defiles, flooding, rock falls, and slides).
- Note the location and type of possible ambush sites on the route.
- Look for terrain where direct fire from the enemy could stop movement on the route.
- Identify natural defense, counterambush, and assembly locations.

- Identify areas where terrain restricts communications.
- Watch for enemy situations that could affect route security or conditions, such as—
  - Enemy elements positioned on key terrain.
  - The enemy emplacing mines and other obstacles.
  - Frequency changes or type of enemy fire in the area.
  - Enemy aerial interdiction.

5-67. To keep from overlooking critical terrain data, the squad leader prepares a checklist of items that may be included on the reconnaissance overlay, such as—

- The route classification formula.
- The identification and location of the reconnoitered route.
- The road distances between the points that are easily recognized both on the ground and on the map.
- The presence and lengths of steep grades (having a slope of 7 percent or greater).
- Curves having radii of less than 45 meters.
- Military load classifications (MLC) and limiting dimensions of bridges. Include suitable bypasses, classifying them as easy, difficult, or impassable.
- Locations and limiting data of fords, ferries, and tunnels. Include suitable classification of bypasses.
- Route restrictions (like underpasses) below minimum standards and, if appropriate, the distances such restrictions extend.
- Areas suitable for short halts, holding areas, or bivouacs that offer easy access to the roadway and adequate dispersion, cover, and

concealment. Include information on the shoulders.

- Rock fall and slide areas that may present a traffic hazard.
- Overhead clearance of less than 4.3 meters.
- Civil or military road numbers or other designations.
- Obstructions to traffic.

5-68. Roads that bisect heavily wooded areas are likely obstacle and ambush sites. Heavily loaded vehicles are vulnerable to ambushes and unable to circumvent obstacles easily. Steep grades and numerous S-turns, where logistical vehicles that are heavily loaded with supplies slow to a crawl, make good ambush points. For more information on route classification, refer to *FM 5-170* and *Appendix I* of this manual.

5-69. If enemy activity is suspected along a route, the squad should—

- Use caution when approaching critical locations.
- Deploy using traveling overwatch or bounding overwatch. Choose movement techniques according to the latest information on suspected enemy activity.
- Avoid danger areas.

5-70. Use caution when approaching a sharp bend or a defile in the road. Such areas are often mined and are ideal sites for an ambush. When necessary, the squad leader has the reconnaissance element conduct a dismounted reconnaissance of these areas while the security element provides overwatch.

5-71. MP check bridges for mines and booby traps. Before crossing a bridge, MP have the security element move to an overwatch position. They have the reconnaissance element dismount and check the bridge and its approaches for mines, booby traps, and demolition charges. If any are found, MP move to a covered and concealed area, report the information, and request engineer support to clear the mines. MP maintain surveillance of the bridge until the mines are cleared. They stop friendly forces and civilians from using the bridge until the engineers have cleared it. When the bridge is cleared, MP have the reconnaissance element gather critical data on the bridge's characteristics and continue the reconnaissance.

5-72. MP reconnoiter key terrain and built-up areas near the route. They move on and off the road to identify enemy activity. The type of terrain dictates whether a reconnaissance is conducted mounted or dismounted. Reconnoitering terrain can be time-consuming. The mission order and the time available for the reconnaissance determine how many and which terrain features are reconnoitered.

MAIN SUPPLY ROUTE REGULATION ENFORCEMENT

5-73. MP traffic control activities support movement control by enforcing highway regulation plans. Traffic enforcement measures, such as speed control and safety inspection checkpoints, help protect the force and ensure that only authorized traffic uses controlled routes. MP employ special circulation control measures, such as temporary route signing, TCPs, holding areas, defiles, and checkpoint and roadblock operations, to support combat and sustainment operations. Refer to *Chapter 6* for more information about checkpoints and roadblocks.

5-74. Highway regulations are set by the agency having jurisdiction over the road network. MSR regulation measures are stated in the command's highway regulation plan. They also may be in the unit SOPs and command directives. The HTS sets the route classification.

5-75. Mounted MP teams patrol MSRs to monitor traffic and road conditions. They gather information on friendly and enemy activity and assist stranded vehicles and crews. Road condition changes and enemy activity are reported immediately through MP channels.

SPECIAL CIRCULATION CONTROL MEASURES

5-76. MP limit, control, block, or direct mounted or dismounted forces traveling on the MSR, by employing special circulation control measures. Special circulation control measures include the following:

- Temporary route signing.
- TCPs.
- Holding areas.
- Defiles.

5-77. Many of these measures are used in MP support to breaching operations, river crossings, and passage of lines.

Temporary Route Signing

5-78. A signed military route system, like the signed US highway system, can enable road users to reach their destinations by following route signs and road markings displayed along the roadside. MP patrols monitor signs on a routine basis, checking specific signs before critical moves. Engineers erect permanent signs, but signs can be damaged, destroyed, or moved by weather, saboteurs, and battle.

NOTE: For more information about temporary route signing refer to *Appendix I*.

Traffic Control Post

5-79. TCPs are used to support MMS only when needed. They are used to preclude the interruption of traffic or unit movement along designated routes. TCPs are communication links to units using the MSR. Show on the traffic control plan and the traffic circulation plan the placement of TCPs. MP activities at a TCP include the following:

- Monitoring and assisting traffic authorized to use the MSR.
- Redirecting unauthorized vehicles to the road network they need.
- Providing route security for the MSR at critical locations or intersections.
- Monitoring for NBC contamination.
- Rerouting traffic as needed.
- Gathering information and reporting it.
- Providing information to passing units.
- Assisting stragglers and DCs.

5-80. **Plan**. An MP squad can operate one TCP for an extended period or three TCPs for a short duration. The platoon leader uses METT-TC to analyze the mission and estimate the situation. He decides the appropriate weapons, materials, and equipment needed and considers such factors as the movement of traffic and the degree of control required. If HN police support is needed, the company operations section or the PM arranges for it. The platoon leader uses overlays and the traffic control plan to determine the location for the TCP. The squad leader plans for emergency destruction of documents and equipment in case the TCP is attacked. The squad leader plans for continuous operations by—

- Developing an adequate sleep plan.
- Arranging for maintenance and refueling.
- Arranging for additional rations.
- Constructing fighting positions.
- Camouflaging all vehicles and equipment.

5-81. **Execute**. TCPs are manned at points where two or more MSRs converge or where confusion could affect vehicle movement. They are used to help protect the force at critical locations where civilian or military traffic can cause an accident. Operations in which TCP will be maximized include the following:

- Deliberate river crossings.
- Deliberate breach operations.
- Defile operations.

5-82. When METT-TC requires a TCP to be manned by one MP squad, the squad leader—

- Analyze the terrain location.
- Positions the teams.
- Directs the squad's vehicles to a covered and concealed position near the squad's fighting position. Use camouflage nets, if needed.
- Selects a fighting position from which the squad can cover and secure the TCP.
- Maintains communication.

5-83. Once the squad has occupied the actual TCP site, the squad leader establishes security and provides a grid coordinate to higher HQ. One team in the overwatch covers the TCP while another team watches the flow of traffic from a covered and concealed position near the road. When needed, a member of this team moves to the center of the road to direct the flow of traffic while the other members provide security. If necessary, the squad leader has the third team resting in a covered and concealed position. The squad leader ensures that all three teams communicate by wire (the preferred method) or radio. If neither is available, they use arm and hand signals.

5-84. At a TCP, the main purpose is to ensure smooth and efficient use of the road network according to the traffic circulation plan. The plan contains—

- Military route numbers and directions of travel.
- Light lines and blackout areas.
- Highway regulation points and MP TCP.
- Route control classification.

5-85. Vehicles too wide or heavy for a road will be denied access. MP reroute them to alternate MSRs. No authorization is needed for travel on an open route, but use of a classified route may be restricted to certain units, operations, or types of vehicles.

5-86. All vehicles on the dispatch route will have a current movement credit issued by the HTS. On a supervised route, normally a column of 10 or more vehicles or an individual vehicle of exceptional size or weight will have movement credit from HTS.

5-87. MP stop vehicles or convoys that are not following MSR regulations. They tell the convoy commander why the vehicles are halted. The convoy commander makes immediate corrections. When immediate corrections cannot be made, MP record the key information about the incident and report it through MP channels.

5-88. TCP is used to disseminate information about the AO. MP provide information to authorized personnel only. They support the commander's force protection program by providing friendly forces with current information on route conditions and enemy activity. At the TCP, MP disseminate information on the locations of contaminated areas, supply points, medical facilities, and other units on a need-to-know basis and should not volunteer more information than is needed to avoid creating security and/or intelligence issues.

5-89. MP actively seek information from road users by asking drivers what they have seen of suspected or actual enemy activity along the MSR. When language is not a barrier, MP talk to local civilians and to the HN civilian and military police to gain information about the road network in an AO. MP pass this information up the chain of command so that it can be verified as reliable.

5-90. MP are constantly on watch for enemy aircraft and suspicious activity by the local populace. When such activity is observed they relay SPOTREPs through MP channels. MP use the SALUTE format to report this information.

5-91. When the movement control agency requests it, MP keep track of military movements that pass through a TCP. This helps movement planners keep track of the progress of convoys. This information is compiled into a passing report that includes the—

- TCP location.
- Date.
- Convoy identification (unit or serial number).
- Time the first vehicle passed the TCP.
- Time the last vehicle passed the TCP.
- Number of vehicles in the convoy.

5-92. Usually, TCP passing reports are picked up at the TCP or transmitted by secure radio. The platoon leader compiles the TCP passing reports into one report. He forwards the report through MP channels or as directed by the commander. The report may be written or transmitted. If transmitted, a report is encoded according to the unit SOP. In some instances, the platoon leader may permit a squad leader to bypass the usual report channels and submit a passing report directly to the movement control agency.

Holding Areas

5-93. MP operate vehicle-holding areas to help regulate the traffic flow. Holding areas can be used as independent measures or with other measures like defiles or checkpoints to support large operations like river crossings or passage of lines. When MP operate holding areas, they direct vehicles, convoys, and troops into and out of the holding areas.

5-94. **Plan**. METT-TC and the size of the holding area determine the number of teams needed to operate it. When one MP squad operates a holding area, the squad leader designates one team to control the entrance to the holding area, another team to control the exit from the holding area, and one team to provide security. He also assigns each squad member a fighting position.

- The general location for a holding area may be designated by the echelon movement control unit, PM, or MP company commander. The exact location is selected by the MP leader with the mission. The holding area's location is noted on the traffic control plan and passed to the echelon movement control unit to keep the traffic circulation plan current. Select a site where—
- Parked vehicles can face the exit so that they can be driven from the area quickly.
- Vehicles can be dispersed.
- There is easy access to and from the roadway.
- The surface of the area is firm enough to hold the weight of the vehicles.
- The area is large enough to allow vehicles to be covered and concealed from air and ground observation.
- The area can be defended.

5-95. **Execute**. MP teams at the entrance and exit to the holding area are positioned in concealed locations. When vehicles approach the holding area, one MP moves to the center of the road and directs the vehicles in. He tells the vehicle driver or convoy commander where to park. He moves back to the concealed location when all vehicles have entered. Vehicles moving in the holding area should be ground-guided by members of the convoy.

5-96. The MP team at the exit operates in a similar manner. MP move from their cover and concealment only when necessary. Depending on the tactical situation, MP teams may use red lens flashlights to direct traffic at night. Colored chemical light sticks are posted to help drivers identify their locations within a holding area if the situation permits.

5-97. Ideally, the MP leader controls the holding area from a position overlooking the entrance and exit. The MP leader receives instructions on when to allow vehicles to pass from higher HQ. When the holding area is to support a river crossing site, the MP leader has a movement schedule to follow. He receives his movement information from the echelon movement control office or the crossing area commander.

5-98. When operating a large holding area, the MP leader may need additional personnel inside the holding area to direct traffic and parking and ensure that the units comply with the flow plan. Large holding areas will have a simple control plan, such as a subdivision system.

5-99. Take the following steps when using a subdivision system:

- Make a map or a sketch of the area showing the road net, trails, and major obstacles.
- Outline the holding area on the map or sketch.
- Divide the area into equal subdivisions and assign a letter or a name to each subdivision. This helps direct units to their section of the holding area.
- Erect signs showing the outline of each area.
- Develop a traffic flow plan and erect directional signs to help users.

5-100. Keep a count of vehicles in the subdivisions by number, size, and unit designation for each vehicle. At night, use chemical light sticks to identify the sections within the holding area and the exit.

5-101. Vehicle holding areas are controlled sites. MP instruct convoy commanders to keep drivers with their vehicles or assign them fighting positions. They do not allow convoy personnel to openly move around within the holding area. They use signs to help control traffic and maintain communications between positions by wire or hand and arm signals.

Defiles

5-102. Defiles keep traffic moving smoothly despite narrowed passageways. Controls at defiles ensure that traffic moves through the passage, one direction at a time, first from one end and then the other. A defile may be considered a critical site and its security a priority.

5-103. **Plan**. METT-TC and the estimated length of the operation are used to determine the actual size of the element needed to operate a defile. In most environments, a defile will not be operated with less than a squad. An MP platoon may be needed to conduct a large defile operation for extended periods. At a defile *(Figure 5-3)*, MP do the following:

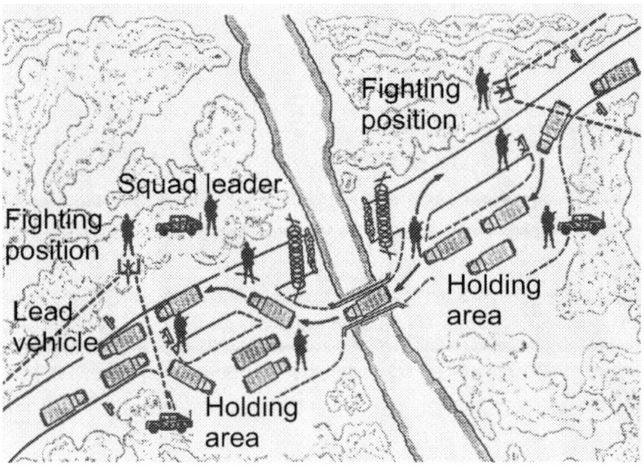

Figure 5-3. Defile

- Secure and defend the site.
- Brief drivers about obstructions.

- Control access so vehicles move through quickly.
- Ensure that vehicles enter one at a time.
- Reroute traffic when necessary.

5-104. Terrain or traffic needs may dictate a need for vehicle holding areas and signs or a TCP. Larger defiles require a holding area at each end. The placement of holding areas depends on the site available and the ease of communicating between the sites and the defile.

5-105. **Execute**. Because defiles involve restricted movement, they are an ideal target for the enemy. Security of a defile is a priority. Before a defile is put into operation, an area reconnaissance is conducted to detect the presence of enemy activity in and around the location of the defile. Once the area has been reconnoitered, the squad leader establishes security and reports the exact location of the defile, if not previously known, to higher HQ. He selects the crew-served weapons fighting position, picking key terrain that overlooks the defile. He ensures that the squad's vehicles are covered and concealed. As in most stationary MP operations, communications within the squad is primarily wire and arm and hand signals.

5-106. MP use control measures to keep traffic flowing smoothly. They use the simplest method of control. Plan for the use of two different control measures. This provides a back-up method, if needed. Control measures can include the following:

- Visual signals to tell traffic when to move, such as hand and arm signals, flashlights, or a handheld flags. Use any technique that shows vehicles when to move. Visual signals work best for small defiles where holding areas are not needed.
- Wireless or wire communications to tell teams in holding areas to hold or start traffic through the defile. Link communications directly between holding areas, or route them through the leader

at the defile site. Use wire communications as the main means of communication. Use wireless communications as a backup or when no other means are available.

- A flag to identify the last vehicle moving through a defile. Give the flag to the last driver or attach it to the last vehicle entering the defile. Another MP removes the flag when the vehicle reaches the end of the defile. This serves as a signal for traffic to start in the opposite direction. This is repeated as often as needed.
- An MP rider to indicate the last vehicle of a column. The rider stays in the last vehicle until the column reaches the opposite side. He dismounts and rides back in the last vehicle returning. This technique ensures that all the vehicles clear the defile.
- MP lead and trail vehicles in the front and rear of a column to guide it through the defile. After the column clears the defile, the vehicles guide a column moving in the opposite direction. MP use this method when movement through a defile is complex and requires an escort. The trail vehicle ensures that all the vehicles clear the defile. A single lead or a single trail vehicle can also be used, depending on the number of vehicles and the complexity of the defile.

5-107. To ensure traffic flow is not interrupted by a disabled vehicle, MP plan for a recovery vehicle to stand by at the defile. If a recovery vehicle is not available, they use field expedient measures.

5-108. Regardless of what special control measure is employed, MP leaders coordinate with the PM for the location, duration and special criteria for the execution of the control measure. Once operational, MP monitor its effectiveness and ensure that force protection and security measures are followed and maintained.

Chapter 6

Area Security

MP conduct AS operations to protect critical functions, facilities, and forces. They synchronize efforts with base and base cluster defense planners within a specific AO to ensure that support and sustainment operations are not interrupted. The HN, when capable, retains responsibility for security of all areas outside US bases. However, US commanders are always responsible for the defense and security of US forces and bases regardless of HN support. AS is conducted by MP across the full spectrum of army operations to protect the force, impose order, and ensure freedom of movement. MP activities that support AS include reconnaissance operations, ADC, base and air base defense (ABD), response force operations, and critical site asset and high-risk personnel security.

RECONNAISSANCE OPERATIONS

6-1. MP plan and conduct area and zone reconnaissance, screening and surveillance missions, and counterreconnaissance.

6-2. MP conduct reconnaissance and screening missions to obtain information about the activities and resources of an enemy or potential enemy or to secure data concerning the characteristics of a particular area. MP reconnaissance, screening, and surveillance efforts include area, zone, and route reconnaissance and counterreconnaissance. These missions may be conducted primarily in the rear area, but may occur anywhere sustaining operations are conducted. Refer to *Chapter 5* for more information about route reconnaissance. MP employ NBC detection equipment to determine the absence or presence and extent of NBC contamination. Refer to *Appendix J* for more information about NBC reconnaissance.

AREA RECONNAISSANCE

6-3. Area reconnaissance is performed to obtain detailed information concerning the terrain or enemy activity within a prescribed area, such as a town, ridgeline, woods, or any terrain critical to the operations. MP conduct area reconnaissance to help guard against unexpected enemy attack in the rear area. Area reconnaissance and surveillance are vital to maintaining AS and contribute to the commander's intelligence collection plan. MP area reconnaissance is a composite of actions. It is initiated from observations and reports gathered over time by MP patrols and information gained through coordination with HN police and other friendly forces. Refer to *FM 7-8*.

6-4. Reconnaissance patrols may differ slightly, depending on the type of reconnaissance to be performed. However, all reconnaissance patrols have a reconnaissance and security team. The size of the patrol is determined by METT-TC. Other considerations to determine the size of the patrol include—

- Size and number of reconnaissance objectives.
- Requirement to secure the objective rally point (ORP) and other points.
- Time allowed for conducting the mission.

6-5. MP leaders plan area reconnaissance based on the IPB and the commander's critical information requirements. Information on enemy activity and likely avenues of approach is coordinated with military intelligence (MI). MP monitor likely enemy avenues of approach and LZ and DZ in critical areas to give early warning of rear-area enemy activity.

6-6. MP area reconnaissance plans include areas near facilities that are designated as critical by the commander, such as—

- NAIs.
- Air bases.
- Bases and base clusters.
- Communications centers.
- Logistic support clusters.
- Key terminals, depots, and bridges.
- Critical terrain features.
- High-value assets.

6-7. When leading an area reconnaissance patrol, in addition to using troop-leading steps and following the general principles for making a reconnaissance, the patrol leader—

- Uses a scheme of maneuver.
- Secures and occupies an ORP.
- Conducts a leader's reconnaissance of the objective area to confirm or change the plan.
- Returns to the ORP, completes the plan, and briefs the soldiers.

6-8. The security elements leave the ORP before the reconnaissance element. The security element leader places security teams at the ORP and on enemy avenues of approach into the objective area. The reconnaissance element conducts the reconnaissance by moving to several vantage points around the objective.

6-9. The reconnaissance element leader may have a small reconnaissance team move to each vantage point instead of having the entire element move as a unit from point to point. This reduces the chances of being spotted.

6-10. After the objective has been reconnoitered for the details outlined in the order, all elements return to the ORP. Teams share their information, consolidate it, and report it, then return to the patrol HQ or continue to the next mission.

ZONE RECONNAISSANCE

6-11. A zone reconnaissance is a directed effort to obtain detailed information concerning all routes, obstacles (to include chemical or radiological contamination), terrain, and enemy forces within a zone defined by boundaries. A zone reconnaissance is normally assigned when the enemy situation is vague or information concerning cross-country trafficability is desired. Zone reconnaissance techniques include the use of moving elements, stationary teams, or a series of area reconnaissance actions. Refer to *FM 7-8* and *FM 17-98*.

6-12. The four methods used to conduct a zone reconnaissance are—

- Box method.
- Fan method.
- Converging routes method.
- Successive sector method.

Box Method

6-13. To use the box method *(Figure 6-1)*, the leader sends his reconnaissance and security teams from the ORP along the routes that form a boxed-in area. He sends other teams along routes through the area within the box. All teams meet at a linkup point at the far side of the box from the ORP.

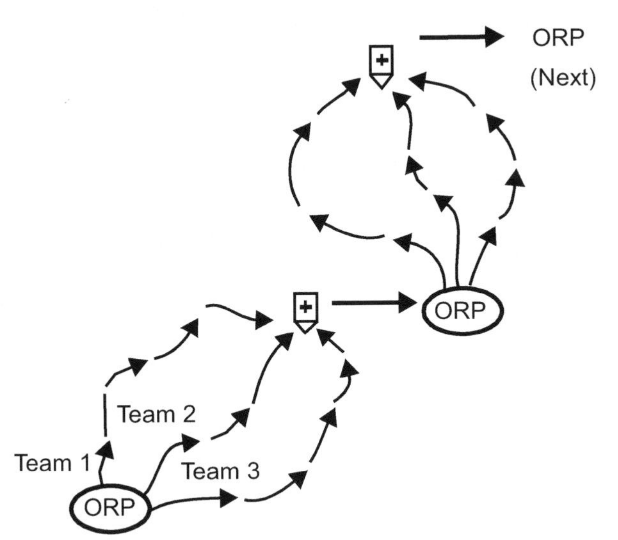

Figure 6-1. Box Method of Zone Reconnaissance

Fan Method

6-14. To use the fan method *(Figure 6-2, page 6-6)*, the platoon leaders selects a series of ORPs throughout the zone. At the first ORP halt and set up security. After confirmation of the patrol's location, the platoon leaders selects reconnaissance routes out from and back to the ORP.

NOTE: These routes form a fan-shaped pattern around the ORP. They must overlap to ensure that the entire area has been reconnoitered.

6-15. Once the routes have been selected, send out reconnaissance elements along the routes. Do not send out all the elements at once. The platoon leader keeps a reserve at the ORP. He sends elements out on adjacent

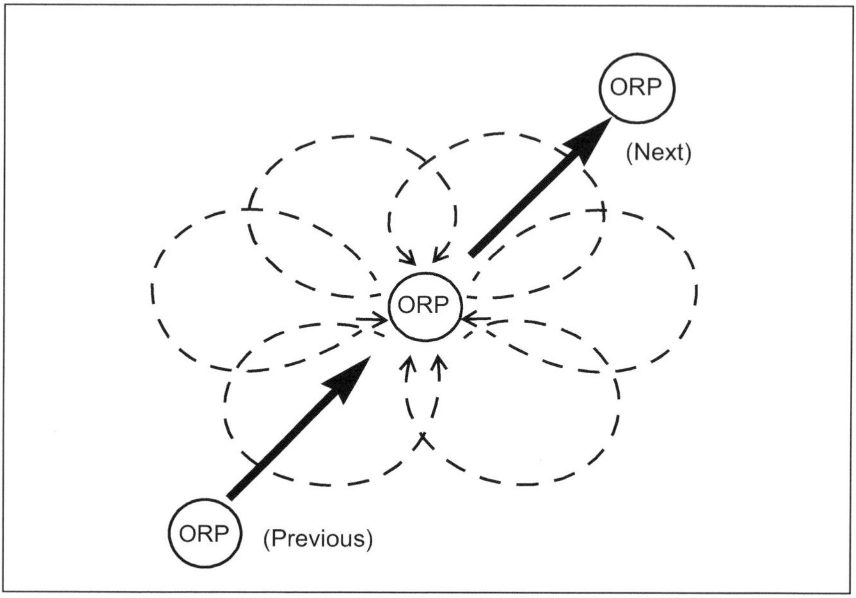

Figure 6-2. Fan Method of Zone Reconnaissance

routes to keep from making contact in two different directions.

6-16. After the entire area (fan) has been reconnoitered, report the information then move the patrol to the next ORP. Repeat this action at each successive ORP.

Converging-Routes Method

6-17. To use the converging-routes method *(Figure 6-3)* (which incorporates the fan method), select an ORP and reconnaissance routes through the zone and the rendezvous point.

NOTE: The rendezvous point is a place where patrol members link up after the reconnaissance.

6-18. Halt the patrol at the ORP and set up security. Confirm the patrol's location. Designate a route for each

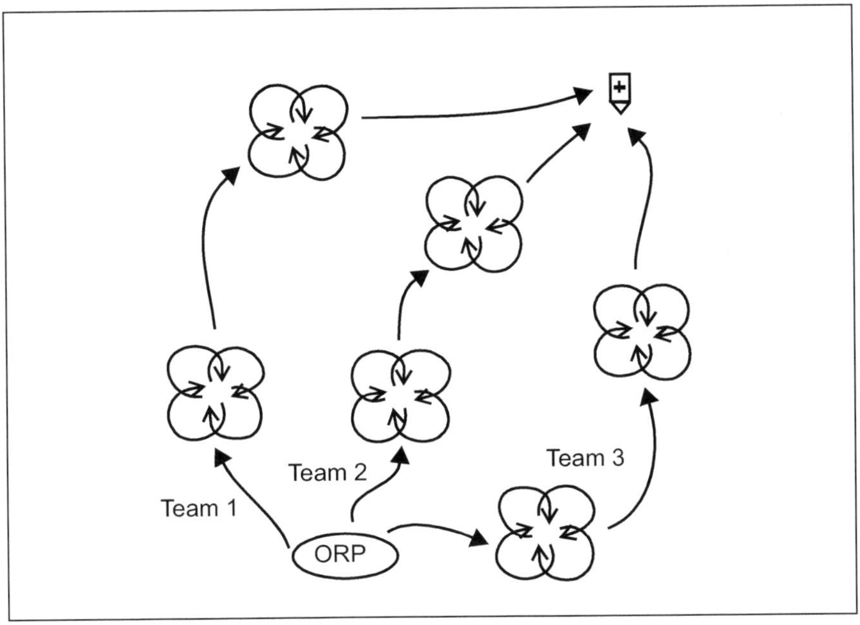

Figure 6-3. Converging Routes Method of Zone Reconnaissance

reconnaissance element, a location for the rendezvous, and a linkup time at the rendezvous point. Send a reconnaissance element to reconnoiter each route (usually using the fan method). The leader moves with the center element.

6-19. At linkup, the patrol secures the rendezvous point as it did the ORP. While at the rendezvous point, information gained by each member is exchanged with all the other members. This provides backup to ensure that all information is passed onto higher HQ. The patrol then returns to friendly lines or continues on to another mission.

Successive-Sector Method

6-20. To use the successive-sector method *(Figure 6-4, page 6-8),* build on the converging-routes method. Select an ORP and a series of reconnaissance routes and rendezvous points. Use the converging-routes method from each ORP to each rendezvous point.

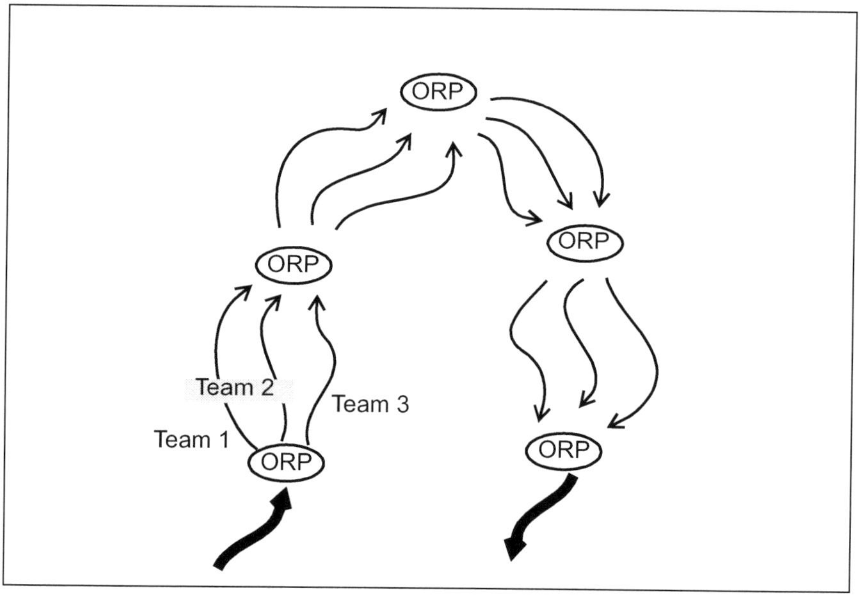

Figure 6-4. Successive Sector Method of Zone Reconnaissance

6-21. Each rendezvous point becomes the ORP for the next phase. Designate reconnaissance routes, a linkup time, and the next rendezvous point when the patrol links up at the rendezvous point. Continue this action until the entire zone has been reconnoitered.

6-22. Regardless of the type of method used, report the information through proper MP channels as soon as possible. Commanders rely on fast, accurate reconnaissance information to plan successful operations.

SCREENING MISSIONS

6-23. Screening missions are defensive in nature and largely accomplished by establishing a series of OPs and conducting patrols to ensure adequate surveillance of the assigned sector. Division cavalry units normally conduct security missions that include a screen for maneuver units during offensive operations.

6-24. MP conduct screening missions for friendly forces in the rear area to provide early warning of enemy approach and to provide real-time information and reaction time for stationary units. In the event of a Level III threat, MP may come under the OPCON of a TCF, which is also referred to as a combined-arms maneuver unit. In this role MP may provide limited security missions, such as a screen to the flank or rear of the main body, with the primary mission of providing early warning and disrupting or destroying enemy reconnaissance vehicles.

6-25. Generally, MP are tasked to observe specific avenues of approach or, more precisely, NAI. The area to observe should be identified in either the reconnaissance and security plan that the platoon leader receives or in the OPORD from higher HQ. If the platoon does not receive an IPB product, the higher OPORD must specifically state where it must focus the screening operation. If the platoon is assigned multiple requirements, the higher HQ must prioritize them.

SURVEILLANCE

6-26. On order, MP maintain continuous surveillance of all the assigned NAIs or enemy reconnaissance avenues of approach into a particular sector. This is accomplished by setting up a series of OPs. MP may conduct active mounted patrols to extend their observation limits or to cover dead space and the area between OPs. Refer to Chapter 4 for more information about setting up OPs/LPs.

6-27. Once the platoon leader understands what his surveillance requirements are, he task organizes the platoon and any assigned assets to achieve the most effective surveillance of the avenue or NAI.

6-28. Unlike a scout platoon, MP focus on providing early warning of enemy reconnaissance elements rather than gaining and maintaining contact with the enemy's main body or destroying it. During screen missions, it is important to understand that an MP platoon by itself does not have enough assets to both acquire and kill an enemy reconnaissance larger than the engaging element. Generally, other assets will be given the specific mission of killing these forces. If MP are ordered to engage enemy forces, they do so by engaging at the maximum effective range of their organic weapons. If available, the platoon leader also plans for and uses CAS and indirect fire.

6-29. During surveillance the platoon's ability to report is critical. Effective early warning requires detailed planning for uninterrupted communications. The platoon leader considers communication distances and significant terrain features to identify potential wireless communication problems. If problems exist, he requests support from the higher HQ.

COUNTERRECONNAISSANCE

6-30. MP contribute to the commander's concept of operations by conducting security and reconnaissance missions designed to detect, disrupt, and impede enemy reconnaissance elements. Counterreconnaissance is not a distinct mission; rather, it is a combination of measures taken by friendly forces to reduce the threat's ability to gather information. It contains both active and passive elements and includes combat action to destroy or repel enemy reconnaissance units.

6-31. Counterreconnaissance prevents enemy reconnaissance forces from observing the main body of friendly forces by defeating or blocking the enemy forces. In the execution of counterreconnaissance, MP operate either offensively or defensively using whatever tactics best accomplish the mission. The principal techniques used are—

- A hasty attack.
- An ambush.
- Indirect fire support.

6-32. MP must task organize to defeat enemy reconnaissance forces. Enemy reconnaissance capabilities in any given situation must be compared to the MP unit's capabilities to determine if additional maneuver or CS assets are required.

6-33. Conventional reconnaissance elements are usually squad-size or smaller. However, special-purpose reconnaissance forces can consist of mechanized forces up to company size. In all counterreconnaissance operations, the goal is to acquire, identify, and kill the enemy reconnaissance force after it has penetrated the initial screen line. Defeating such forces usually requires combined-arms forces, but this is dependent on the type, size, and capabilities of the reconnaissance element.

6-34. MP platoons are not organized or equipped to fight for extended periods or to destroy enemy armor vehicles. MP employ AT weapons, such as AT-4s, for defensive purposes (self-protection and breaking contact). However, MP teams are highly skilled at reconnaissance and surveillance and providing early warning of enemy activity.

6-35. A scout platoon acquires and identifies enemy reconnaissance forces along a screen line, which is a control measure usually named as a phase line, and is an established forward of the main body. MP conduct their counterreconnaissance efforts in a similar manner in the rear area or anywhere sustainment operations are taking place.

6-36. In most cases, the scout platoon cannot be expected to acquire, identify, and defeat enemy reconnaissance elements. As a CS asset, MP can assist a scout platoon by locating the enemy reconnaissance element, freeing the scouts or TCF to perform the killing function of counterreconnaissance on larger mechanized enemy reconnaissance elements. MP activities that contribute to counterreconnaissance include—

- Area reconnaissance.
- Zone reconnaissance.
- Route reconnaissance.
- OP operations.
- Physical security and vulnerability risk assessment.
- Critical asset security.
- OPSEC.
- Deception operations.

6-37. Conventional threat reconnaissance elements push far out in front of their combat unit to gain intelligence on their rear area objective. Unconventional threats such as terrorists, criminals, or gangs may try to observe installations, deep-water ports, or other facilities to obtain information. Conventional threat reconnaissance efforts are concentrated on gaining intelligence on the capability of friendly forces. Unconventional threats try to obtain information to plan sabotage or criminal activity, or to simply disrupt the efforts of friendly forces.

6-38. MP leaders plan measures to counter enemy reconnaissance by coordinating with various staff sections and agencies that include—

- MI for information on enemy capabilities, likely rear area targets and objectives, likely enemy reconnaissance avenues of approach, and the commander's critical information requirements (CCIR).

- The PM and criminal investigation division (CID) for HN police information on local gangs, known criminals, and criminal activity.

6-39. To assist commanders with their counterreconnaissance efforts, friendly forces such as MP, CID, engineers, and MI conduct physical security surveys and vulnerability and risk assessments of bases and base clusters, deep-water ports, and air bases. They advise commanders of these facilities on security measures designed to prevent the threat from gaining access to friendly forces and facilities.

6-40. During AS missions, MP conduct security activities around NAI; critical assets, such as communications nodes; and air bases. MP perform area and zone reconnaissance of all terrain that dominates critical facilities. They concentrate their efforts on locating enemy reconnaissance forces. MP deny the enemy the opportunity to observe friendly forces by reporting their location, maintaining surveillance, and assisting in their destruction if required.

6-41. Enemy reconnaissance forces are not likely to use primary reconnaissance avenues of approach to gather information on friendly forces. MP teams are more likely to come in contact with enemy reconnaissance forces operating on trails, rough terrain, and dead space that allows mounted movement. They use the cover of darkness for their operations. MP must make maximum use of NVDs and illumination to help detect their movement. They put the devices on key terrain and along avenues of approach to critical bases, and cover the area with crew-served weapons. Enemy reconnaissance teams are most vulnerable during the day. MP concentrate daytime mounted or dismounted operations on locating their base camp or hide positions.

Once they are discovered, if ordered to do so, MP can lay ambushes on likely routes to destroy them. Refer to *Chapter 4* for more information about ambush patrols.

6-42. Mounted MP patrols use overlapping search techniques to make it difficult for enemy reconnaissance teams to reach their objectives without being exposed. Overlapping searches provide random coverage not easily predictable by simple observation.

6-43. If contact is made, MP may be directed to maintain contact or surveillance until enough force can be assembled to counter the threat. They disrupt their movement and delay them until a large enough force is available to defeat them. If directed to disrupt or delay, they clearly identify the enemy before engaging them; there are generally numerous friendly forces operating in the rear area. MP must consider fratricide preventive measures. Refer to *Appendix F* for more information about fratricide avoidance.

6-44. MP make good use of terrain and maximum use of weapon systems. They use harassing fire from mounted crew-served weapons (MK19/M2), but do not become decisively engaged. They deceive the enemy with heavy harassing fire from crew-served weapons. If available, MP destroy the threat with indirect fire. MP make the threat believe he has encountered a defense. The enemy must not realize these actions are only delaying tactics.

AREA DAMAGE CONTROL

6-45. ADC is basic to successful rear-area operations. ADC measures are taken before, during, and after hostile actions or natural or man-made disasters to minimize effects and reduce damage. All commanders try to limit the impact of enemy actions and reestablish unit operations as quickly as possible. Commanders at each level plan ADC operations.

6-46. ADC operations integrate the functions of many specialized units. Engineers plan and coordinate ADC operations at the site. Medical teams help sort and treat mass casualties and assist in initial evacuations. Army aviators help evacuate casualties and provide emergency resupply, communications relay operations, area damage assessment, and C^2 actions. MP expedite and control battlefield movement into, around, or through damaged or contaminated areas. Signal elements reestablish the signal system. HN civil efforts, like clearing rubble and providing facilities and services, can greatly aid ADC.

6-47. The amount of MP support needed for ADC depends on the extent of the damage, the importance of the affected area, and the effect of the damage on the movement of troops and logistical supplies. Enemy attacks on key military facilities can leave them unprotected. Heavy damage in urban civilian areas can disrupt local government services.

NOTE: Protecting civilian facilities is a HN responsibility. However, MP preserve law and order in such an area if so doing protects friendly forces and resources.

6-48. Route and area reconnaissance are key to determining the trafficability of the routes into, out of, and around affected areas; obtaining a damage assessment, and having early warning of the continued presence of the enemy. MP determine the level and extent of NBC contamination and identify the location of critical points affected by damage.

6-49. The enemy's damage to the terrain determines the degree and kind of MP support needed and where to place the priority of effort. Downed trees, urban rubble, damaged or destroyed bridges, cratered roads, and contaminated road networks affect circulation control. If the roads can be traveled, MP provide circulation control operations locally in the affected area. However, damage to an area may be so great that roads must be closed and MSR traffic rerouted.

6-50. MP give priority of movement to ambulances transporting wounded and engineers clearing debris. They maintain law and order in the affected area by employing measures to stop looting and unlawful behavior, according to the ROE and the use of force requirements. They disseminate key information such as the location of first aid stations, emergency shelters, and other emergency facilities.

6-51. Security patrols may be needed around key facilities. MP may need to set up an OPs/LPs to observe sectors of the affected area. The OP/LP teams can watch for enemy agents trying to exploit the effects of a conventional attack through arson, sniper fire on firefighters, or other disruptive acts. The OPs/LPs also watch for theft, pilferage, or arson against military property.

6-52. MP ADC operations may be either part of an ongoing operation or a separate requirement. During ADC, MP—

- Perform route and area reconnaissance in affected areas.
- Evaluate the serviceability of the road network.
- Note and report the development of critical points caused by damage to bridges, tunnels, and the like.
- Monitor the flow of DCs from the damaged area.
- Report and block off affected areas.
- Provide AS for involved critical facilities.
- Reroute battlefield movement to alternate road networks.
- Check for and report NBC hazards and contamination.
- Prevent sabotage, looting, and pilferage in the damaged area.

- Protect property, contain panic, and enforce emergency restrictions.
- Direct persons to first aid stations, emergency shelters, and other emergency operations.
- Post temporary signs to prevent entry into unsafe buildings or redirect activity to temporary locations.
- Help establish populace control in affected areas.
- Operate mounted and dismounted mobile patrols, checkpoints, and roadblocks to—
  - Enforce emergency restrictions on movement into, within, and out of the affected area.
  - Direct DCs.
  - Collect stragglers.
  - Enforce curfews, stand-fast orders, and movement authorizations.

BASE DEFENSE

6-53. Base defense is the local military measures, both normal and emergency, required to nullify or reduce the effectiveness of enemy attacks on, or sabotage of, a base to ensure that the maximum capacity of its facility is available to friendly forces.

6-54. The base commander organizes and controls the forces assigned to a base to best capitalize on their capabilities. These forces may be joint or single-service forces operating in the joint rear area (JRA). If a base, base cluster, or air base comes under attack, the base commander responds within his capability. Defense fundamentals, as they pertain to the defense of the bases, include the following:

- Understand the enemy. Defenders must be familiar with the capabilities and limitations of the enemy forces, weapons, equipment, and tactics.

- Use situational awareness. Intelligence operations are key to assembling an accurate picture of the battle space.
- Use the defender's advantages. The defender's advantages may permit a numerically inferior force to defeat a much larger one. These advantages include—
  - The ability to fight from cover.
  - A more detailed knowledge of the local terrain and environment.
  - The ability to prepare positions, routes between them, obstacles, and fields of fire in advance.
  - The ability to plan communications, control measures, indirect fires, and logistic support for contingency operations.
  - The ability to deceive enemy forces about friendly defensive capabilities, dispositions, and execution of operations.
- Concentrate at critical times and places. Conduct the defense of a base along interior lines, permitting the timely and secure movement of forces to engage the most critical threats. The commander must mass combat power at points of decision by economizing in some areas, retaining a reserve, and maneuvering to gain local superiority at critical points.
- Conduct counterreconnaissance and counterattacks. Fixed bases have well-established perimeters that usually have limited depth. Counterreconnaissance and counterattack add depth to the battle, outside the perimeter, allowing the base to continue its primary mission with minimal interference.

- Extend counterreconnaissance patrols 3 to 5 kilometers outside the perimeter, based on terrain. This reduces the risk of fratricide, especially at night.
- Coordinate critical defense assets. Synchronization of indirect fires, air defense resources, tactical aircraft, engineers, dismounted troops, armored vehicles, naval surface fire support, and helicopters can produce a combined-arms effect.
- Balance base security with political and legal constraints. This fundamental is especially critical in a low-intensity conflict environment.
- Know the law of war and ROE. Base commanders and their subordinates must comply with ROE. In joint-service operations, reconcile inconsistencies with ROE.

6-55. A base or base cluster has certain inherent capabilities which allows it a degree of defense. Each base has a BDOC that plans, coordinates, and supervises base defense operations. The BDOC initiates contingency planning that enables the base to—

- Increase the manning posture of the base as the threat level increases.
- Detect and defeat the threat within its capabilities.
- Defend against heavier enemy forces until the response force is committed.
- Maintain control of the fight within the base.
- Support the fire and movement of the response force operating outside the base.

6-56. Each base cluster has a BCOC to monitor base defense plans and establish the base cluster response force. The BCOC—

- Provides C^2 of resources for planning, coordinating, and supervising the defense of the base cluster.

- Coordinates base defense operations.
- Maintains communications with the BDOC within the cluster, MP, and the supporting RAOC.

6-57. MP leaders plan for the interface of their support into the base plan. The plan requires—

- Cover and concealment of personnel and equipment.
- Secure and redundant communications systems at all guard locations (a landline, radio links to the BDOC, and a telephone hookup to the center switch).
- Deception.
- Contingency planning.
- Improvement of base defense positions.
- Coordination with BCOC or RAOC as required.
- OPs/LPs.
- Noise and light discipline.
- Immediate reaction to enemy threat or attack.
- Rehearsals of defense measures.

6-58. All plans and overlays depicting MP support are forwarded through normal MP channels to the BCOC.

6-59. Any threat to the bases must be dealt with swiftly. Few sustainment units can continue their mission while under attack by even a low-level threat.

6-60. Sustaining operations rely on MP to stay apprised of enemy activity near bases. MP forward early warnings of enemy activity to the operations cell at the rear CP. The rear-operations cell immediately notifies the subordinate RAOC and the base and base clusters. It may also alert the response force.

6-61. If a base or base cluster comes under attack, the base commander responds within his capability. However, interruptions of base sustainment operations must be kept to an absolute minimum. If a threat exceeds a base's ability to defeat it, the base commander requests response force support.

DANGER

Base defense forces must be aware of the disposition of other friendly forces in front of them. Fire discipline between the response force, mobile reserve, and TCF is imperative. Positive control must be maintained at all times.

AIR BASE DEFENSE

6-62. The Army is responsible for defending air assets from ground threats outside the boundary of the air base and its area of responsibility (AOR). ABD forces are tasked to defend particularly critical air bases. They may be forces operating in the area specifically on call for ABD or they may be the rear-area response force.

6-63. Air Force security forces (AFSF) are responsible for the internal security of US air bases. HN and US Army forces are primarily responsible for external defense of US air bases. Given the size of an air base and the high priority as a threat target, providing ABD as a dedicated asset requires a force larger than a company. MP do not have the force structure to be dedicated solely to ABD operations. However, an echelon commander could decide to redirect MP mission priorities to ABD and deploy his limited MP assets to serve as part of a larger composite force, such as a TCF. Or an MP response force may be required to respond to threats beyond the base's capabilities.

6-64. MP operating as part of a larger ABD force may provide support that includes —

- Setting up roadblocks and checkpoints to limit access to the base.

- Occupying OPs/LPs and fighting positions on key terrain and likely avenues of approach to the air base.
- Conducting screening missions.
- Conducting combat patrols (security and ambush).
- Conducting DC operations.

6-65. A typical ABD layout consists of a base perimeter that surrounds the runway and the facilities that support air base operations. The tactical perimeter of a base is the outer most ABD tactical AOR. It is where defense forces cover, patrol, or occupy likely avenues of approach to the base and other key terrain. *Figure 6-5* shows a typical ABD layout.

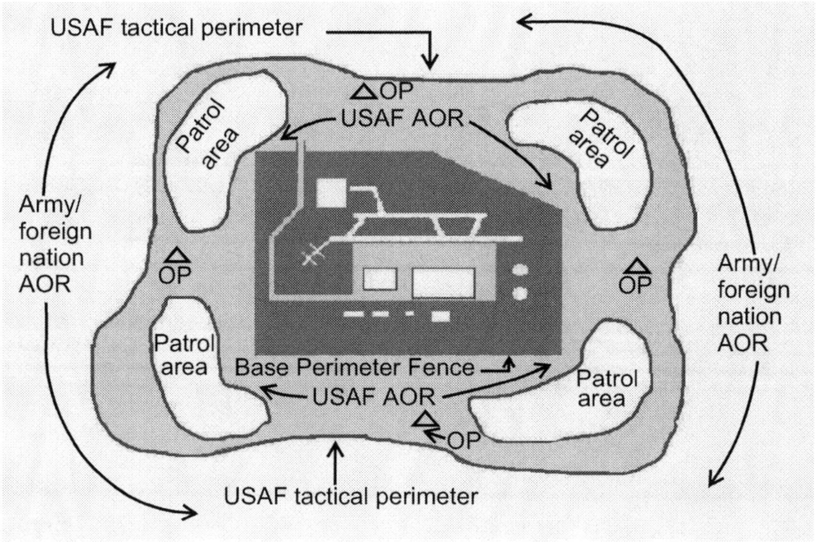

Figure 6-5. Typical ABD Layout

6-66. Commanders develop unit SOPs and contingency plans for ABD. SOPs should address C^2 and fire support issues that are different for ABD. However, AS plans for ABD provide for increased security patrols and static security measures around an air base the same as they would for other critical facilities.

6-67. The depth of the defense area surrounding the air base depends on METT-TC and other defensive planning factors. MP activities that support ABD are concentrated in the tactical perimeter and on key terrain adjacent to the tactical perimeter.

FIRE SUPPORT

6-68. Organic AFSF fire assets typically available are the M29 81-millimeter mortar, the MK19, and the M2 .50 caliber machine gun (MG). These fire systems are incorporated into the base defense plan for both the security force (SF) and MP. This ensures compatibility with all ABD measures in the surrounding area and mutual support to other bases when practical.

6-69. The use of mortars and other indirect-fire weapons located in the defense area is coordinated through a 5-man FDC located at the air base BDOC. The AFSF primarily uses its mortars for illumination, suppression of enemy indirect-fire weapons, and suppression of enemy surface-to-air missile (SAM) capability. AFSF weapons controlled by the FDC vary. The number of weapons available depends on the size of the air base.

6-70. CAS is obtained from aircraft at the base. Aircraft may be sent aloft to provide CAS or targets may be assigned to aircraft already airborne. CAS provides suppressive fire or increased detection capabilities. Among the CAS aircraft that may be available, the AC-130 Spectre gunship and the A-10 Thunderbolt II both have night-flying capability. Both have compatible frequency modulated/modulation communications equipment. Both are especially effective for use against targets considered danger close to ABD forces. However, they are very susceptible to SAM.

DANGER
ABD forces must be aware of the disposition of other friendly forces in front of them. Fire discipline between AFSFs response forces, mobile reserves, and the TCF is imperative. Positive control must be maintained at all times.

COMMUNICATIONS

6-71. Joint communications for all forces supporting ABD must be compatible. Communications must be secure, have redundancy, and possess the ability to integrate all security forces.

6-72. Effective coordination must be an ongoing process. Call signs, frequencies, and SOI are exchanged between all joint, combined, and coalition forces supporting ABD.

MILITARY POLICE RESPONSE TO AIR BASE DEFENSE

6-73. Air bases are normally located in the rear area. Threats that are beyond the capability of the base may require the deployment of an MP response force. In general, the response force responds to an air base the same as any other base or base cluster. However, as a joint service operation there are C^2 and fire support differences. Company and platoon tactical SOPs for ABD should be developed to help speed the response time.

6-74. The response force leader coordinates with the Air Force defense force commander (DFC) for the current tactical situation and threat. When committed, the MP response force may transition to the operational control of the DFC. This ensures that MP actions do not interfere with air sorties being generated by the air base commander.

6-75. The DFC is in charge of all aspects of ABD operations. He provides guidance on where the response force is needed and informs the response force leader of the tactical situation. The response force leader directs the tactical operations of MP elements, keeping the DCF informed. If a TCF must be called to defeat a Level III threat, all units, Air Force and Army, are under the OPCON of the TCF commander until the threat is defeated.

6-76. For the local ground defense of their base, the Air Force provides a BDOC which operates similar to an Army BDOC. It is responsible for combat intelligence, combat operations, logistics, and personnel administration. It collects the ground combat intelligence within the ABD area.

6-77. The AFSF maintains a mobile reserve force. The mobile reserve's purpose is to mass enough firepower to destroy threat forces within the air base boundary or, at least, delay the threat until a larger force can be assembled. The mobile reserve is usually a mounted force. It is under the direct control of the BDOC. The mobile reserve tries to contain any direct landing by threat forces on the airfield. It responds to penetrations of the defense area by forces that have managed to elude external defense forces.

6-78. If the enemy penetrates deeply into the defense area, the response force leader can request that the BDOC commit the mobile reserve to battle. This may prevent the premature commitment of the TCF. Contact points and AAs are established around the base to assist in integrating the mobile reserve during combat operations outside the ground defense area.

6-79. The response force must keep the enemy from destroying resources on the ground and interrupting or stopping air operations.

Night Operations

6-80. Air bases are most vulnerable when visibility is limited. In response to the threat, the bulk of both internal and external ABD operations must be dedicated to detecting threat forces during periods of limited visibility. The large gaps between defensive positions facilitate enemy infiltration through external defenses. To narrow these gaps during limited visibility, external defensive efforts may have to be set up closer to the perimeter. External defensive efforts should integrate the extensive use of—

- Night vision equipment.
- Reconnaissance and surveillance.
- Electronic detection devices.
- Tactical deception.
- OPs/LPs.

Water Obstacles

6-81. Many air bases border water obstacles. Air bases located next to rivers, lakes, or oceans provide unique ABD problems. Threat efforts directed against an air base from the water include infiltration, reconnaissance, and standoff-weapons attacks. As a result, water obstacles that penetrate the defense area may require ABD forces to set up continuous battle positions between the water and any critical air base facility near the water. ABD obstacle plans should include fences, mines, and sensor employment along the portion of the perimeter that borders water. Planned fires are coordinated to defeat watercraft. Floating expedient barricades may have to be assembled. Warning buoys should also be anchored offshore. Underwater approaches provide excellent avenues to air bases. They must be considered during ABD planning. Joint support from the Navy and the Coast Guard may also be needed.

Urban Terrain

6-82. Air bases bordered by urban areas pose a special challenge for ABD operations. Increased security is a necessity. MP must be aware of underground approaches, such as sewers, that could bypass defensive positions. Threat forces, saboteurs, and special-purpose forces will use these avenues of approach to gain access to the air base.

6-83. Infiltrating groups will be very difficult to detect because they avoid direct confrontation and usually move at night. These teams have the greatest potential for disrupting air operations throughout the rear area.

CHECKPOINTS AND ROADBLOCKS

6-84. Checkpoints and roadblocks are set up to control the movement of vehicles, personnel, and material and prevent illegal actions or actions that aid the enemy. They are used to prevent trafficking in contraband and stop the movement of known or suspected belligerents. Checkpoint and roadblock operations contribute to the commander's information and intelligence collection process. Additionally, they assist friendly forces in detecting and establishing the behavioral patterns of the local populace.

6-85. When conducted improperly, checkpoints and roadblocks can foster resentment toward US forces. MP minimize the negative impact that checkpoints and roadblocks may have by treating local nationals with dignity and respect. Whenever possible, checkpoints and roadblocks should be conducted with interpreters, HN police, or other HN authorities.

6-86. In most stability and support operations there exists the potential for threat forces to attempt to exercise some level of force against US forces. Checkpoints and roadblocks may become prime targets for threat forces (both traditional and nontraditional).

METT-TC determines the level of self-protection needed at a checkpoint or roadblock. However, MP leaders must always plan force protection measures that allow their forces to quickly implement adequate self-defense when threatened.

6-87. Checkpoints and roadblocks must be designed so that potential threat forces are discouraged from attempting to disrupt its operation or try to inflict casualties. MP use armored vehicles whenever possible to support the dismounted troops operating the checkpoint or roadblock. The MP security element assigned to protect the site should have an antiarmor capability. The vehicle search area should be located in an area that provides blast protection.

Checkpoints

6-88. Checkpoints are part of a police action. As such, the ROE and the use of deadly force must be clearly understood by all soldiers participating in the operation. MP use the minimum intrusion and imposition necessary to accomplish the mission and protect the force. They allow for a vehicle escape route and plan to destroy a hostile element who uses it. If the checkpoint is completely sealed off, an enemy will have to penetrate it by attempting to run over the barricades. This puts the search team in a position to have to defend itself and fight back.

6-89. MP establish and operate two types of checkpoints—deliberate and hasty.

6-90. **Deliberate Checkpoint.** A deliberate checkpoint is a fixed position set up on a main road in a rural or built-up area. It can be classified as either a *heavy-* or *light-traffic checkpoint*, depending on how much traffic is expected to pass through it. A heavy-traffic deliberate checkpoint normally requires a platoon. An MP squad can only operate a light-traffic checkpoint for a short duration (12 hours or less).

6-91. To operate a heavy-traffic checkpoint, task organize the platoon into—

- A HQ element. The HQ element is responsible for C^2 and maintaining communications with subordinate elements and higher HQ.
- A security element. The security element is an MP squad that provides early warning to the search and assault element, watches for and reports suspicious activity, and monitors traffic flow up to and through the checkpoint. The security element should have an antiarmor capability to protect the site from an enemy armor threat.
- A search element. The search element is an MP squad that halts vehicles at the checkpoint, guides them to the designated search point, conducts personnel and vehicle searches, and directs cleared vehicles on through the checkpoint.
- An assault element. The assault element is an MP squad responsible for destroying (consistent with the ROE) any hostile element that forces its way past the search team. The squad leader places his soldiers beyond the search point and emplaced zigzag obstacles and barriers. The soldiers prepare and occupy fortified fighting positions. When confronted by a threatening vehicle, the search element gets out of the way and allows the vehicle to pass. The vehicle passes through the escape lane, and the assault element makes the decision whether to engage or not. If the assault element has to engage, the battle will occur away from the checkpoint. This reduces the possibility of fratricide to friendly forces or injury to innocent civilians.

6-92. In smaller checkpoint operations, a squad can be organized in a similar fashion. MP elements use handheld portable radios, if available, or wire communications. However, much of the needed signals at a checkpoint or roadblock can be easily accomplished using arm and hand signals.

6-93. A deliberate checkpoint is organized into sections *(Figure 6-6)*. The physical layout and detail of preparation depend on the amount of traffic that will pass through it and the duration of its operation. Normally, a deliberate checkpoint will require engineer support to construct obstacles, barriers, escape lanes, and possibly fighting positions. All checkpoints consists of—

- Obstacles or barriers.
- Search areas (personnel and vehicle).
- Security overwatch and fighting positions.
- Holding areas.

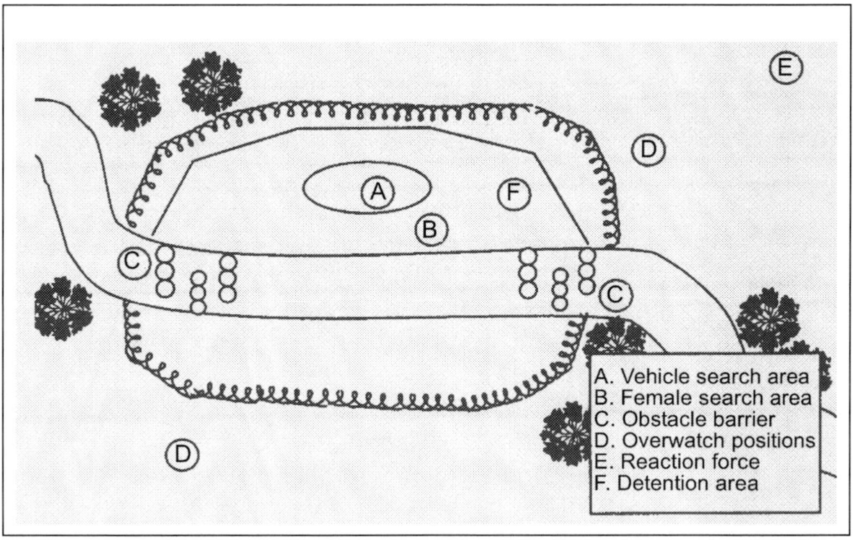

Figure 6-6. Deliberate Checkpoint

6-94. **Hasty Checkpoint.** MP set up hasty checkpoints to achieve surprise. They are temporary and should be moved often. The materials used to construct these checkpoints are carried by the platoon. The platoon or squad uses its vehicles, reinforced with concertina wire, as the obstacle. MP may employ tire deflation devices or road spike stripes. These devices are more effective than concertina wire and may be less intrusive in peace operations. They are commercially available or can be locally fabricated.

6-95. MP position the vehicles to partially block the route *(Figure 6-7)*. MP conduct the search in the area

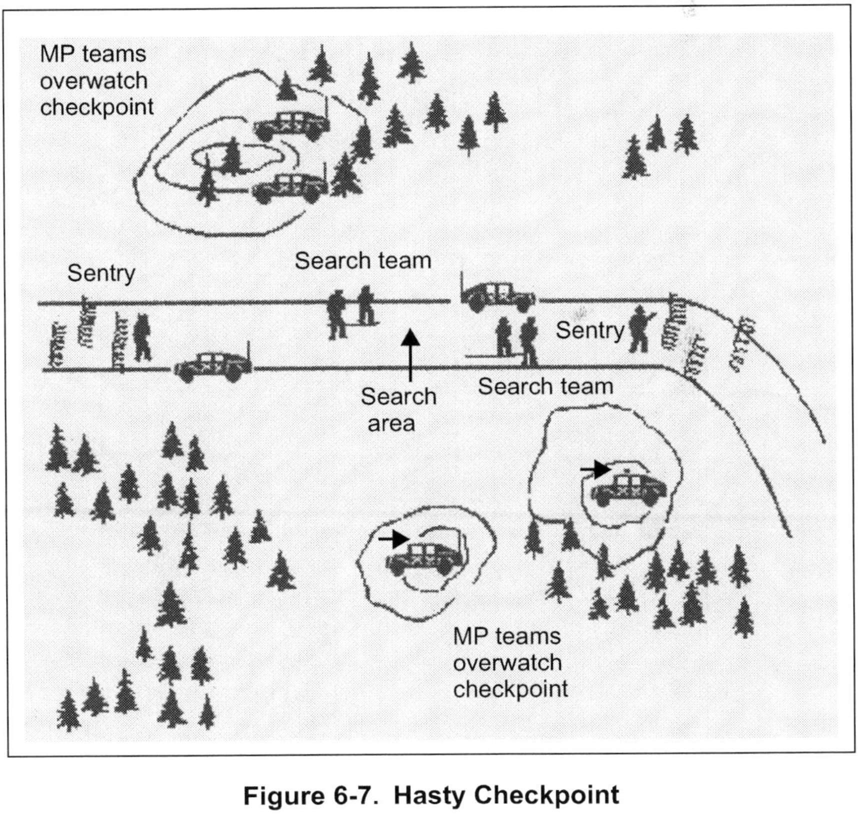

Figure 6-7. Hasty Checkpoint

between the vehicles. MP are positioned at each end of the checkpoint. They cover them by mounted or dismounted automatic weapon positions. MP conceal a reaction force (at least one team) nearby to react in case the site is attacked.

6-96. MP establish hasty checkpoints where they cannot be seen by approaching traffic until it is too late to withdraw. Good locations to set up hasty checkpoints include—

- Bridges.
- Defiles.
- Highway intersections.
- The reverse slope of a hill.
- Just beyond a sharp curve

6-97. **Vehicle Searches**. Two members of the search team position themselves at both rear flanks of the vehicle undergoing a search, putting the occupants at a disadvantage. These soldiers maintain eye contact with the occupants once they exit the vehicle and react to any threat attempts by the occupants during the vehicle search.

6-98. The actual search is conducted by two MP armed with pistols only. One MP conducts interior searches and the other performs exterior searches. They instruct the occupants to exit the vehicle during the interior search and instruct the driver to watch the vehicle search. Once the interior search is complete, they escort the driver to the hood of the vehicle and instruct him to open it. Once the engine compartment has been examined, they instruct the driver to open the other outside compartments (tool boxes, gas caps, trunks, and so forth). The driver removes any loose items that are not attached to the vehicle for inspection. Members of the search team rotate positions to allow for mental breaks.

6-99. MP use MWD teams, mirrors, and metal detectors to thoroughly search each vehicle for weapons, explosives, ammunition, and other contraband. Depending on the threat level, the vehicle search area provides blast protection for the surrounding area.

6-100. **Personnel Searches**. MP may be required to conduct personnel searches at the checkpoints. Every attempt should be made for HN authorities to conduct, or at least observe, searches of local nationals. Additionally, MP leaders must plan for same-gender searches. Personnel searches are conducted only when proper authorization has been obtained, usually from higher HQ, according to the ROE, Status of Forces Agreement (SOFA), or HN agreements. This does not preclude MP from searching individuals that pose a threat to US or other friendly forces.

6-101. MP may have to detain local nationals who become belligerent or uncooperative at the checkpoints. The OPORD and the ROE must address the handling of such personnel. In any case, self-protection measures should be planned and implemented according to the orders from higher HQ.

6-102. Searches of local nationals should be performed in a manner that preserves the respect and dignity of the individual. Special consideration must be given to local customs and national cultural differences. In many cultures it is offensive for men to touch or even talk to women in public. Searchers must be polite, considerate, patient, and tactful. MP leaders must make every effort not to unnecessarily offend the local population. Such situations can have a very negative impact on peace operations and can quickly change popular opinion toward US and other friendly forces.

6-103. Each captive is searched for weapons and ammunition, items of intelligence value, and other inappropriate items.

NOTE: When possible, conduct same gender searches; however, this may not always be possible due to speed and security considerations. Therefore, perform mixed gender searches in a respectful manner using all possible measures to prevent any action that could be interpreted as sexual molestation or assault. The on-site supervisor carefully controls soldiers doing mixed-gender searches to prevent allegations of sexual misconduct.

6-104. MP conduct individual searches in search teams that consist of the following:

- A searcher. A searcher is the MP that actually conducts the search. He is in the highest-risk position.
- Security. Security includes at least one MP to provide security. He maintains eye contact with the individual being searched.
- An observer. The observer is a MP leader that has supervisory control of the search operation. He also provides early warning for the other members of the team.

6-105. The two most common methods that are used to conduct individual searches are the frisk search and the wall search.

- Frisk search. This method is quick and adequate to detect weapons, evidence, or contraband. However, it is more dangerous because the searcher has less control of the individual being searched.
- Wall search. This method affords more safety for the searcher because the individual is searched in a strained, awkward position. Any upright surface, such as a wall, vehicle, tree, or fence may be used.

6-106. If more control is needed to search an uncooperative individual, the search team places the subject in the kneeling or prone position.

6-107. Strip searches should only be considered when the individual is suspected of carrying documents or other contraband on his person. This extreme search method should be conducted in an enclosed area and by qualified medical personnel when available.

Additional Checkpoint Considerations

6-108. The effective use of all task organization elements is vital to the success of checkpoint operations is. Roles and responsibilities must be well defined and rehearsed. Additional considerations when conducting checkpoint operations include—

- Preparing and emplacing signs in the local language instructing drivers what to expect and do at the checkpoint (for example, "You are entering a military checkpoint. Prepare to stop your vehicle, and have your identification papers ready for inspection.").
- Determining if it is necessary to apprehend or detain those who see the checkpoint ahead and turn around to avoid it. If it is, HN police are responsible for this mission, if they are available. If they are not available, it may be necessary to position a respond force close to the approach route to block or detain vehicles that try to avoid the operation.
- Clearing and maintaining control of all buildings and terrain that dominate the checkpoint.
- Staying alert for any change of scenery around the checkpoint. A parked car that was not there before, crowds gathering for no apparent reason, or the media waiting for an event are all indicators that something may happen.

- Using artificial illumination for night operations, arrange the lighting to keep those passing through the checkpoint in the light and our forces in the shadows as much as possible.

6-109. If HN personnel are used to assist, commanders ensure that they do not represent a national, ethnic, or religious group of faction feared and hated by the majority of the local population.

6-110. Leaders avoid setting patterns by moving the checkpoint location and changing the method of operation at random.

6-111. MP can gain valuable police, criminal, and combat information while operating checkpoints. They use a checklist to standardize the information collection effort. The following information is included in the checklist:

- The number and type of vehicles stopped. Report identifying markings, license plate number, and any signs displayed on the vehicle.
- The number of passengers in the vehicle. Report the nationality, the ages, and the sex mixes of passengers.
- The type and quantity of cargo.
- The point of origination and destination of the vehicle.
- The stated reason for travel by passengers.
- A description of arms, ammunition, explosives, and sensitive items found and confiscated from the vehicle.
- Possible or actual sightings of weapons, explosives, or threat forces by the passengers.
- The condition of passengers (general health, dress, attitude).
- Anything unusual reported by the passengers.

6-112. When conducting checkpoint operations, MP will need the support of the following:

- Engineers to build obstacles and barriers to channel traffic.
- Linguists familiar with the local language.
- A civil affairs officer or HN police.
- Trained interrogators.

6-113. Sometimes MP are tasked to operate a checkpoint to assist convoys of friendly forces. These checkpoints are set up at the entrance to the controlled route and will have a vehicle holding area to accommodate large convoys. Refer to *Chapter 5* for more information about holding areas.

6-114. At the convoy checkpoints, MP check convoy vehicle movement credits issued by the local movement control unit to ensure that the convoy is moving on the correct route at the correct time. When convoys are ahead of schedule, MP hold them near the checkpoint in a vehicle holding area until it is their scheduled time to pass. They allow convoys that are behind schedule to proceed if route traffic permits and assist drivers who have taken the wrong route.

Roadblocks

6-115. Roadblocks are set up to stop, slow, or limit movement of vehicles along a route. They are used to limit access to certain areas or roads. A roadblock can be established separate from a checkpoint or used to channel vehicles and personnel to a checkpoint.

6-116. MP use their vehicles with concertina wire, barbed wire, and warning signs as a roadblock *(Figure 6-8, page 6-38)*. If available, armored vehicles make excellent roadblocks. Place a roadblock where unauthorized vehicles or enemy personnel cannot bypass it. If possible, position it so that obstacles like cliffs, swamps, or rivers restrict movement to its flanks or around it. Build man-made obstacles to tie in with and reinforce natural obstacles.

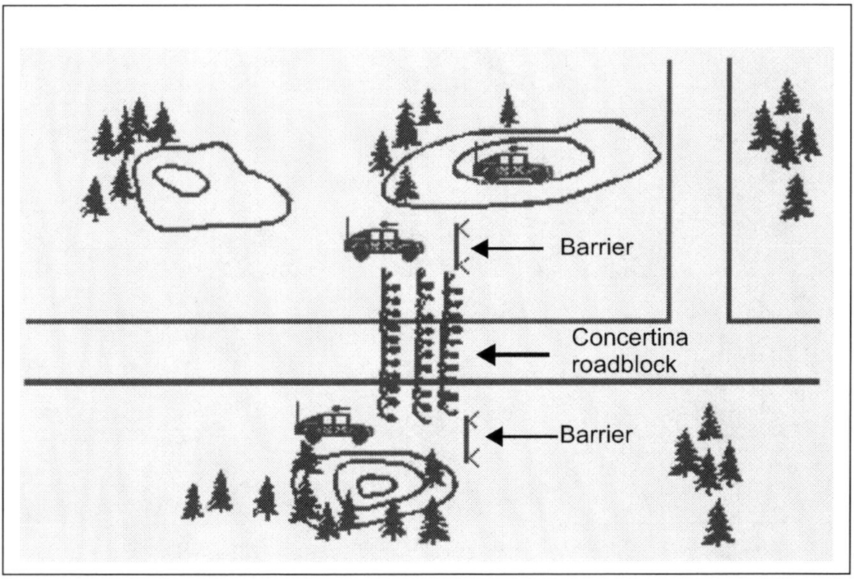

Figure 6-8. Roadblock

6-117. When using a roadblock to channel traffic to a checkpoint, place it where drivers cannot see the roadblock until after they have passed all possible turnoffs. When using a roadblock to close off a road, place it at an intersection to let drivers change to another route with little delay.

6-118. Select a defendable site for the roadblock. Cover the roadblock with automatic weapons. Ensure that defensive positions—

- Include a fighting position for the crew-served weapon to provide overwatch for the roadblock.
- Have fields of fire that cover the approaches to the roadblock to keep it from being breached.
- Are not accessible to an attacker and provide cover and concealment.
- Are placed at an intersection to let drivers change to another route with little delay.

- Are placed near an area where drivers can turn their vehicles around easily.

RESPONSE FORCE OPERATIONS

6-119. MP response forces may respond to threats to a base, a base cluster, or an air base and possibly respond to downed aircraft. MP are normally the commander's designated force to respond to attacks beyond the capabilities of a base or base cluster. The objective of response force operations is to eliminate the threat or make the enemy disengage without requiring the premature commitment of the TCF. MP are organized and equipped to defeat Level II threats. These threats exceed base and base cluster self-defense capabilities but do not require the commitment of a TCF.

6-120. With prior coordination and appropriate supporting fires, MP can defeat the following:

- Small combat units conducting raids, ambushes, and reconnaissance operations.
- Unconventional forces conducting diversionary and sabotage operations.
- Special warfare intelligence collection and operational missions.

6-121. MP response forces are prepared to delay and disrupt Level III threats, protect the flank of a TCF, or allow a base time to establish defenses in greater depth.

Plan

6-122. METT-TC, the number of MP available, and the commander's intent determine the size of a response force. Normally, an MP response force requires at least a full-strength platoon. However, response force operations conducted in built-up areas or in support of ABD may require a much larger force. MP platoons are not organized and equipped to fight for extended periods. The commander task organizes a platoon for response force operations. A response force should have increased antiarmor capability, integrate available fire support, and have CAS available. The result is a force capable of causing an enemy to break off its attack. The response force can then fix and destroy the threat using close combat techniques and the application of artillery, Army aviation, and CAS.

6-123. MP must know the terrain and be able to use it against the enemy. They mass combat power quickly to destroy the enemy or delay them until a TCF can arrive. The rear CP fire support element, along with the main CP fire support cell, establishes procedures by which MP can call for fire support.

6-124. The response force identifies the location of the bases within their AO, and must know which bases are the most critical and which are the most vulnerable. They include this information in the local IPB. In coordination with the rear CP operations cell and the affected RAOC, they position the response force where they can best—

- Detect enemy incursions (near the DZ, the LZ, and the like).
- Interdict enemy forces enroute to their targets.
- Consolidate quickly in response to threatened key assets in the rear area.

6-125. The response force leader must have the following readily available:

- The base defense status of each base.
- Locations of any obstacles or mines near the base.
- Locations and direction of fire of the crew-served weapons.
- The signal for FPF.
- Locations of TRPs and preplanned fires.
- Methods for contacting the BDOC or the BCOC, to include call signs and frequencies.

- Locations of the OPs/LPs and friendly patrols if employed.
- Call signs and frequencies for supporting artillery and Army aviation units tasked to respond.
- Call signs and frequencies for the TCF, RAOC, and the rear CP.
- Fire support targets that are on the approved list.
- Locations of the nearest medical treatment facility (MTF), NBC decontamination site, and ammunition supply point (ASP).

6-126. Continuous communication with the base and base clusters is the key to knowing how and when an MP response force is needed. Bases and base clusters establish 24-hour communications with the RAOC, if they are located in the corps rear area, or the rear CP, if they are located in the division. This liaison and interface allows timely response and information dissemination.

6-127. The response force will be effective only if it can react swiftly and is thoroughly familiar with the base defense plan. They coordinate all response force actions for a base through its BDOC. (Go through the BCOC for base clusters.) They coordinate the following with BDOC:

- Call signs and frequencies.
- Base defense plans and layouts.
- Positions of critical internal assets, external coordination points, and no-fire areas.
- Indirect-fire support.
- Engineer support, if needed, to help prepare defensive positions or for ADC.

6-128. In coordination with the main CP fire support cell, the operations cell sets procedures by which MP can call for fire support.

Execute

6-129. Base response force operations rely heavily on MP tactical skills. The choice of action is based on METT-TC and the base commander's tactical assessment. Once a situation develops that requires a response force, there are several options available to disrupt or eliminate the threat. Response force options include the following:

- Call for fire support to make the enemy break contact.
- Attack into the enemy's flank using the base for fire support.
- Use ambushes along likely escape avenues if it is obvious that the attacking force is not strong enough to overrun the base.
- Augment the base's defense forces and provide support from within a base.

6-130. Once the MP company commander receives the mission to direct response force operations, the company's main efforts shift to supporting the response force. The commander conducts a mission analysis and begins the TLPs, culminating in issuance of an OPORD and a FRAGO. Refer to *Chapter 2* for more information about TLPs.

6-131. The company HQ coordinates for support that includes communications, air evacuation, and logistical support (Class I, III, V, and IX). The company operations section coordinates with follow-on forces and other support that includes the following:

- Fire support to include CAS.
- The location of adjacent units.
- The location of threatened friendly forces.
- The location of noncombatants.
- Coordination with the TCF.

6-132. Once the platoon leader receives the mission to conduct response force operations he initiates TLPs. Normally, the PSG assembles the response force at a prearranged AA near the base under attack.

6-133. The base commander gives the response force an objective and a tactical AO. He may designate a small force to assist the response force. This force would be under the OPCON of the response force leader.

6-134. Normally, the response force acts as a maneuver element against the enemy. Once the platoon leader develops a COA he designates the ORP and selects a scheme of maneuver. The response force moves to the ORP, establishes security, and reports its status to higher HQ. Once the response force has established communications with the friendly forces under attack, the platoon leader reconnoiters the area and assesses the situation. Based on the situation and guidance from higher HQ the platoon leader may decide to—

- Request fire support to destroy the enemy.
- Continue to observe for weaknesses in the enemy's flanks.
- Engage and further develop the situation.
- Attack and destroy the enemy.

6-135. If the decision is made to call for fire to destroy or make the enemy break off the attack, the platoon leader follows the procedures set by the main CP fire support cell. When available, MP can call for and adjust indirect fire provided by artillery. Another option is CAS, if there is a tactical air control party or army aviator trained in joint air attack team operations. Refer to *Chapter 3* for more information about call for fire and fire support. Also refer to *FM 6-20*.

6-136. If the response force is ordered to attack and destroy the enemy, the platoon leader immediately prepares the response force for a hasty attack.

6-137. **Plan a Hasty Attack**. A hasty attack is an immediate action taken without extensive preparations, with the resources at hand and, if by so doing, the attacking element can maintain momentum or take advantage of the enemy situation.

6-138. The platoon leader uses fire support to complement the scheme of maneuver. The goal is to kill as many enemies as possible and suppress the rest to keep them from seeing or shooting the maneuver element as it closes on the objective. The attack plan includes the use of all available direct and indirect fire. The response force leader mainly plans and directs the fire of his organic weapons (MG, M203, AT-4, and MK19). The fire element's makeup depends on how much direct-fire support is needed and what support is to be provided. The initial employment of the fire element includes the following:

- Positioning crew-served and M249 weapons on or near the line of departure (LD), when the objective and route to it can be seen and covered by fire.
- Positioning some crew-served and M249 weapons on or near the LD and some with the moving element, when the objective can be seen and covered by fire from the LD, but only a part of the route is visible, or when the entire route is visible from the LD, but the objective is not visible or is out of range from the LD.

6-139. Plan control measures at the objective and initial employment of the fire element. If time is available before moving to the assault position, make a second reconnaissance to see if the plan needs changing.

6-140. **Conduct the Attack.** The platoon leader moves to where he can lead the force and influence the fight. He moves a fire element into the over watch position and has the attacking element move from the LD to the assault position, using covered and concealed positions. If en route to the objective, the patrol should—

- Receive indirect fire. Move quickly out of the impact area to a preset point.
- Meet obstacles, breach, or bypass them. Report their location if the obstacles might affect follow-on units. Take special care when crossing a danger area like a field, roadway, or creek. Refer to *Chapter 3* for more information about moving in contact.
- Meet direct enemy resistance before reaching the objective and return fire. Call for and adjust fire on the enemy.

6-141. The platoon leader coordinates action so that the response force hits the enemy with its full combat power. He maneuvers the force to strike the enemy's flank or rear. The response force engages the hostile element with the appropriate level of force. He uses overpowering force on armed combat forces. While in contact—

- Immediately assume the best available covered positions, when moving, and at the same time, return fire in the direction of the enemy.
- Locate actual or suspected enemy positions and engage them with well-aimed fire.
- Distribute fires evenly over the objective when no individual positions have been identified.
- Report and monitor the situation and recommend COAs.
- Make visual or verbal contact with soldiers on the left or the right. (Platoon members and team leaders make frequent visual contact with the squad leaders. Squad leaders make frequent visual contact with the platoon leader. Relay all arm and hand signals from the platoon chain of command.)

6-142. When resistance is destroyed, the attacking force continues quickly onward to the assault position. They deploy on line just before reaching the assault position. To maintain the momentum, the attacking force passes through the assault position and attack the objective. They halt at the assault position only to allow all the soldiers to come on line to attack at once. The assault position should be as close to the objective as possible to prevent needless exposure to enemy fire.

6-143. As the moving element attacks, the fire element in its overwatch position will cover the attacking element by shooting at the enemy. Before the attacking element moves beyond support range, the fire element moves up to a position from which it can continue its support and shoot at the enemy.

6-144. With the fire element in overwatch, the platoon leader (commander) gives the command or signal to open fire. The fire leader will control the method and the rate of fire. He and the gunners watch the progress of the attacking element and engage targets that threaten it. When crew-served weapons are close together, the fire leader anticipates the masking of their fire and moves the weapons one at a time. When those weapons are separated, each gunner displaces when its fire is masked or when it can no longer support the attacking element. The platoon leader (commander) times the displacement of weapons in a fire element so that the attacking element has continuous fire support.

6-145. Maneuvering begins when the attacking element makes contact with the enemy. As the attacking element maneuvers near the objective, the fire element in overwatch suppresses the enemy with a high rate of fire.

6-146. When the attacking element reaches the objective and closes with the enemy, the fire element walks its fire across the objective in front of the maneuvering element. Then it shifts its fire to supplementary targets (rear of the objective, escape routes) or lifts its fire to keep from endangering the attack element. Soldiers in the element closing with the enemy move singly, by pairs, by teams, by squads, or by a combination of these, using as much cover as possible. As they close, they fire on selected targets to suppress the enemy. Automatic rifles are fired in short bursts across the forces' front. Rifles, M203s, hand grenades, bayonets, and MG are used to overcome pockets of resistance. The advance continues past the objective far enough to shoot at any withdrawing enemy. Firing continues until the enemy is killed or captured or withdraws.

6-147. Once the threat is eliminated, the platoon leader reconsolidates the response force and assembles it at a secure rally point. Refer to *Chapter 3* for more information about consolidating and reorganizing.

6-148. **Conduct an Attack on Urban Terrain.** In military operations in built-up areas, a response force must be able to find, isolate, and evict the enemy concealed in hardened structures. Normally, responding to a threat in a built-up area will call for a larger response force and should be supported by armored fighting vehicles. When attacking and clearing a building in an urban area, try to—

- Determine and isolate the area around the objective. In urban terrain it is often hard to pinpoint the enemy's position.
- Number the area's buildings and use those numbers as reference points to control movement and for clearing operations.
- Clear buildings from the top down if possible.
- Retain a reserve force to have the flexibility essential for urban combat.

6-149. Refer to *FM 90-10-1* for information about attacking on urban terrain.

Respond to Downed Aircraft

6-150. MP are often required to respond to downed aircraft. This specialized mission requires close coordination with aviation assets. In the event of a downed aircraft, the primary mission is to safeguard the aircraft, render first aid to the crew, evacuate the wounded, and secure the crash site. Normally, this requires MP to be airlifted to the crash site.

6-151. MP leaders plan for downed-aircraft rescue operations as a contingency mission. The primary concern is the safety of personnel onboard the aircraft. The MP leader establishes preplanned lift packages for a mounted or dismounted platoon-size downed-aircraft response force. When organized for downed-aircraft operations, the force must meet the following requirements:

- Possess cross-country capability.
- Be prepared to defend the site against Levels I and II threats.
- Have internal medical support to treat and evacuate injured personnel.

6-152. Engineer support may be necessary to clear lanes to the downed aircraft and medical evacuation (MEDEVAC) LZs.

6-153. Once the response force arrives at the crash site, the platoon leader immediately establishes security and notifies higher HQ of the exact location, the type of aircraft, and the number of wounded or killed personnel. Medical personnel treat the injured and prepare them for evacuation. The platoon leader selects and secures a MEDEVAC LZ. Once all the injured personnel have been evacuated, the response force continues to secure the crash site until relieved.

6-154. Security at the crash site involves MP preventing the handling or disturbance of the wreckage, theft of any parts, or compromising evidence such as gouges or marks on the ground made by the aircraft. Other security measures include—

- Preventing all unauthorized access to the site. This includes media and local authorities.
- Protecting all civil and military property at the site.
- Prohibiting removal of deceased persons until properly identified by military authorities.
- Prohibiting the removal of wreckage.
- Keeping spectators and the media at a reasonable distance from the site for safety.

6-155. All requests for information will be referred to the Public Affairs Office (PAO). MP at the scene will not release information about the crash to the media.

ENEMY DELAY

6-156. A delay is an operation in which a force under pressure trades space for time by slowing down the enemy's momentum and inflicting maximum damage on the enemy without, in principle, becoming decisively engaged. Normally a delay operation is unlikely to be conducted in a sustainment area. However, bypassed enemy forces could create a situation where a MP response force may be required to delay a threat to buy time for a base to establish defenses in greater depth or to delay an enemy until the arrival of the TCF. MP deliberately conduct a delay only when directed to do so by higher HQ. In that case, higher HQ determines the size of the MP force. In the delay, the destruction of the enemy force is secondary to slowing his advance to gain time.

PLAN

6-157. The platoon delays either from successive positions or from alternate positions. When delaying from successive positions, the delaying force will fight rearward from one position to another, holding each position for a given time or until it becomes decisively engaged.

6-158. When delaying from alternate positions, the delaying force will fight rearward with the force deployed so elements can provide overwatch and subsequent maneuver. While the forward element is fighting, the rear element prepares to assume the action. The forward element disengages. It passes through or around the rear element to prepare to assume the action from a position in greater depth. Use the following methods to delay the enemy *(Figure 6-9):*

- Delay forward of a specified line or terrain feature if you have—
  - A narrow sector.
  - Cross-compartment terrain.
  - An area restrictive to armor and the enemy can be canalized into a selected area.
  - Terrain dominating mounted avenues of approach.
- Delay from successive positions when you—
  - Have wide sectors or limited observation, fields of fire, and visibility.
  - Have a mobility advantage simultaneously moving one or two elements at a time.
  - Are in dangerous sectors.
  - Lack good defensible terrain in depth.
- Delay in a sector if you have—
  - An area of responsibility that cannot be adequately covered from one battle position.
  - Multiple enemy avenues of approach.

- Limited fields of fire and observation.
- Delay from alternate positions when —
  - The sector is narrow or there is long-range observation and fields of fire.
  - The enemy is superior in combat power.
  - There is good defensible terrain in depth, and the terrain facilitates movement by bounds.

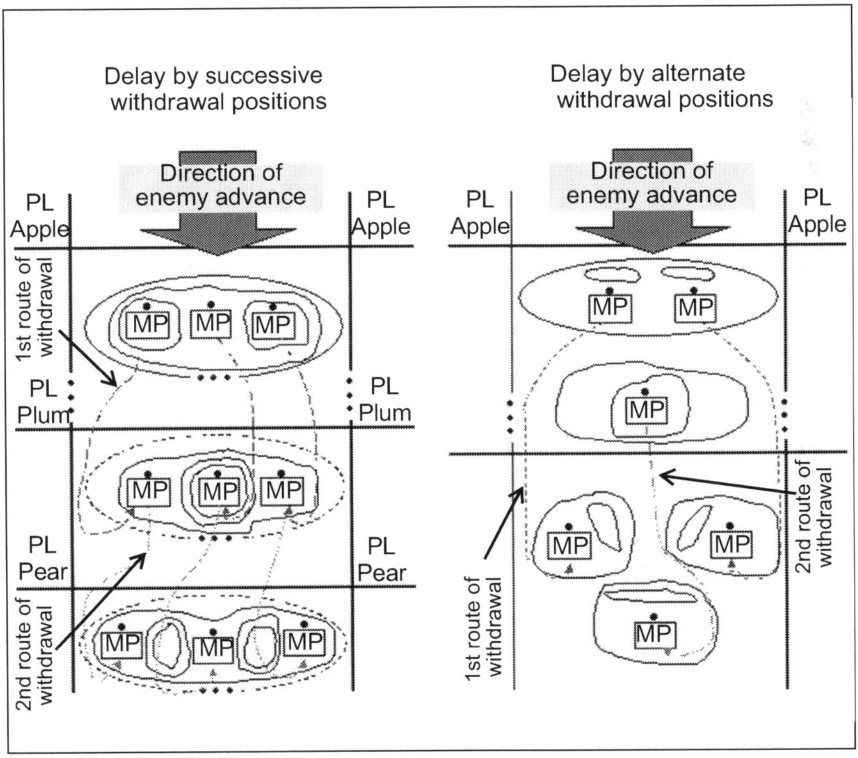

Figure 6-9. Methods of Delaying the Enemy

6-159. The platoon leader develops a plan of action using the factors of METT-TC and the commander's concept and intent. He gives priority to—

- Obtaining a detailed knowledge of the terrain.
- Gearing the terrain analysis to the enemy's intent, avenues of approach, and likely positions and targets.
- Gaining information from aggressive area patrol and reconnaissance.
- Having alternate communications, both sound and sight, ready to be used in case radio communications are disrupted.
- Forecasting time. The response force must consider how fast the enemy will close, how long will it take to move off a position, and how long to get to and occupy the next position. In delaying actions, timing can be critical.
- Requesting fire support along the route.

6-160. In the OPORD to subordinate leaders, includes the—

- Concept of conducting the action.
- Initial delay positions.
- Trigger points identifiable on the ground. ("Squad leader, I plan to initiate fires when the enemy reaches the railroad track.")
- PLs and delay lines. Include the length of time to delay on or forward of the given position or PL.
- Coordination and contact points.
- Indirect-fire plan (smoke, high explosives [HEs] munitions, artillery-delivered scatterable mines).
- Responsibility for adjacent-unit coordination (passage and the like).
- Fire control measures, engagement areas, targets, and rate of fire.
- Disengagement criteria. ("Squad leaders, I plan to start disengagement when the enemy has three armor vehicles across the railroad tracks.")

- Plan for moving after disengagement (signals, breakpoint, use of smoke, and the like).
- Subsequent delay positions and routes and the sequence of disengagement.

6-161. Most of the firepower will be oriented toward the enemy. However, flank and rear security must be provided. The platoon leader plans fire support in front of and on both flanks of the positions. He plans fire to cover the disengagement and movement to subsequent positions. He has fire concentrate on breaking up the enemy's advance at long range.

6-162. Higher HQ will resupply, rearm, and refuel the delaying element. On a platoon action, the platoon sergeant is crucial in ensuring that the platoon has the required support to continue a delaying action. The OPORD provides for a central point where the teams or squads can recover.

EXECUTE

6-163. If directed to delay the enemy, the platoon leader moves to a position where he can best control the action. The platoon leader makes good use of the terrain to mask all movements and provide vantage points for observation and harassing fire.

6-164. The enemy will try to bypass a delaying force. The platoon leader tries to keep the enemy in front of or on an oblique angle to the delaying force's position. He prepares positions as thoroughly as time permits, and reconnoiter routes to subsequent positions.

6-165. When selecting delay positions, look for the following:

- Clear observation and fields of fire onto the avenues of approach.
- Positions offering mutual support and interlocking and flanking fires where possible.
- Restrictions to canalize the enemy.

- Short, covered, concealed, and trafficable routes to rearward positions.
- Positions hidden from enemy observation and fires.
- Defilade positions for HMMWV and ASV.

6-166. If the delaying force has time they—

- Emplace obstacles to slow the enemy while they move to rearward positions.
- Block obvious routes rearward and direct the force to use more hidden routes.
- Continue preparing positions.
- Consider preparing ambush sites along the routes to slow enemy pursuit.
- Select primary and alternate firing positions.
- Have soldiers familiarize themselves with the routes to these positions.
- Place the OPs/LPs, and other security means far enough forward to warn the platoon, particularly if the enemy is using a secondary avenue of approach.

Control Fire

6-167. Fire control is extremely important. The platoon engages the enemy at maximum range with all weapons systems. They fire on a prearranged signal or event (trigger point). They aim for leaders and C^2 vehicles and avoid premature firing. The platoon leader considers not firing until the enemy reaches a preset point on the ground. Then he uses ambush-type fires to quickly destroy lead enemy elements. This also keeps the enemy from learning the friendly force's size and location.

6-168. If the enemy force cannot evade, it will make every effort to envelop and destroy the delaying force. Avoid presenting an exposed flank to the enemy. Use contact points to help ensure flank coordination.

6-169. Each time the delaying force moves back, it will have less time to plan and prepare. Speed, firepower, and maneuver become more important. The platoon leader calls for added fire support or assistance from higher HQ if decisive engagement or envelopment seems likely.

6-170. If additional elements arrive, the platoon leader uses them to augment the element in contact, employing them on the left and right of the element in contact. The platoon leader ensures that as the element in contact moves back, the left and right elements know to also move, but not at the same time. As the element in contact moves back, the flank screens can open harassing fire. Drawing enemy attention to flank elements will allow the element in contact to disengage more easily.

6-171. The platoon leader keeps higher HQ informed of the situation. Forward elements must not become so heavily engaged that they cannot be withdrawn effectively. If contact with higher HQ is lost, the platoon leader uses initiative. He bases each move on METT-TC, the events on the battlefield, and the tactical commander's intent.

Disengage

6-172. Deciding when to disengage is difficult. If the delaying forces remain too long, they risk decisive engagement or envelopment. If they move too soon, they may give up ground unnecessarily and risk pursuit. The decision of which element to move is based on METT-TC. Move the least heavily engaged element first. To disengage—

- Direct supporting elements to engage enemy forces forward of the disengaging element.
- Concentrate all available fire and mutual support.

- Give priority fire, if available, to the disengaging element to rapidly increase its firepower.
- Begin an orderly movement to successive or alternate positions.
- Use all available fire.
- Place the fire far enough forward to avoid impeding personnel leaving the protected positions.
- Place smoke between the delaying force and the enemy.
- Move into subsequent positions from the rear and send a SITREP to higher HQ.
- Report the delaying force's arrival to higher HQ.
- Tell the elements left in contact to disengage and take their next positions.

6-173. If it becomes necessary, replace key leaders, give crew-served weapons a priority of manning, evacuate casualties, and redistribute ammunition. Refer to *FM 17-95* for more information on defensive tactics used for delays.

BATTLE HANDOVER TO A TACTICAL COMBAT FORCE

6-174. When the combined efforts of the base and base cluster defense force and the MP response force are unable to jointly defeat an enemy threat, the operation becomes the responsibility of the TCF. Coordination is essential to determine when, where, and how the response force gives up responsibility for the fight and the TCF takes over. The MP response force and the base-defense force will then hand over the fight to the TCF by conducting a battle handover.

6-175. A battle handover *(Figure 6-10)* is a coordinated operation between two units that transfers responsibility for fighting an enemy force from one unit

ORGANIZATION

Maintain unit integrity for C2.

ORDER OF MOVEMENT

Base the order of movement on the—

- Number of passage points.
- Degree of security required.
- Enemy situation.
- Terrain.
- Prevention of confusion and congestion. Give priorities indicating which units move and when they move.

COMMUNICATIONS

Ensure that the following actions occur:

- Both units have SOI compatibility.
- Both units monitor the rear operations net (critical for recognition signals and fire support).
- Both outgoing and incoming unit elements operate on the outgoing unit's command net.
- The outgoing unit maintains routine radio traffic volume while the incoming monitors it.
- When the responsibility of the battle has changed over, the incoming unit operates on its own net only.

COMMAND AND CONTROL

Ensure that the—

- Number of passage points can determine the method of C2 used.
- Multiple passage points are set up, dictating decentralized control.
- TCF commander decides how he can best influence the action and positions himself accordingly.
- Response force commander collocates with the command group of the TCF during the passage of lines.
- Time or circumstances when responsibility for the battle is transferred are mutually agreed on before hand by the response force and the TCF commanders.
- Responsibility can be passed by radio communications if need be.

Figure 6-10. Planning Considerations for a Battle Handover

COMMAND AND CONTROL POINTS

Ensure that the following occurs:

- All units follow the control measures.
- For a response to force passing through a stationary TCF, control measures include the following:
  - AAs. Used by a unit to prepare for an upcoming mission.
  - BHLs. The single, most important control measure in delineating response force and TCF responsibilities. A PL is placed where the TCF assumes responsibility for the battle from the response force.
  - Attack position. The last position occupied or passed through by the TCF before crossing the LD.
  - Passage lanes. Areas on which the response force moves to avoid the stationary TCF.
  - Passage point. A place where the response force will pass through TCF units. It is located where the commander desires subordinate units to physically execute the passage of lines.
  - Time of passage. The time at which the passage is executed by the leading element of the response force to facilitate control through the passage point.
  - Recognition signals. Established signals to prevent the response force from directing fire on the TCF and vice versa.
  - Contact points. Designated, easily identifiable points on the terrain where two friendly forces are required to meet.
  - RPs. Points on the ground where subordinate units are under the control of their respective commanders while en route to their appropriate destinations.
  - Routes. Routes are the prescribed course to be traveled from a specified point of origin to a specific destination.
- For a TCF passing through a delaying response force, control measures include the following:
  - Battle position. A defensive location established by the response force oriented on the most likely enemy avenues of approach.
  - LD and LC. This control measures places the BHL in that control of the battle is assumed by the TCF once its units cross the PL.

Figure 6-10. Planning Considerations for a Battle Handover (Continued)

to another in close-in battle. It is designed to sustain continuity of the combined-arms fight and protect the combat potential of both forces involved. Battle handover is usually associated with conducting a passage of lines. MP leaders must establish clear tactical SOPs that allow their forces to quickly establish the necessary coordination to preclude a loss of momentum in the attack. In a battle handover, the MP response force will pass rearward through the TCF lines while handing over the fight or the TCF will move forward through the MP response force lines to engage the enemy.

6-176. However the fight is handed off, coordination between the MP response force commander and TCF commander is critical. Use a BHL (refer to *FM 17-95* and *FM 17-98*) and contact points as control measures to set both forces' responsibilities. Both the TCF commander and the response force leader coordinate the BHL location. The BHL is shown on the operations overlay and in the OPORD as a PL or a boundary. It identifies ground as "owned" by the TCF or by the response force. It sets the location where control of the battle passes from one commander to the other. It is where the TCF can use direct fire and observe indirect fire.

6-177. The handover occurs at the time or event coordinated by the TCF commander. The TCF commander decides the method of handover based on METT-TC. Generally, one of the following methods will be used:

- The response force passes rearward through a stationary TCF.
- The TCF passes forward through a delaying response force.
- The response force sets up a final blocking position, with elements of the TCF conducting an attack to destroy the enemy.

6-178. Coordination for battle handover normally flows from the TCF commander to the response force leader. Coordination normally includes—

- Establishing communications.
- Providing updates on both friendly and enemy situations.
- Coordinating passage.
- Collocating C^2.
- Dispatching representatives to contact points.
- Ensuring that there are visual recognition signals for vehicles.
- Ensuring the status of obstacles and routes.
- Coordination of fire support and CSS requirements.

6-179. The most important task for the response force is to maintain contact and continue to engage the enemy. The enemy must continue to see the level of activity that has been established. If the enemy perceives that the handover is occurring, he will attempt to seize the opportunity to destroy the response force.

PASS A RESPONSE FORCE THROUGH A STATIONARY TACTICAL COMBAT FORCE

6-180. The TCF commander coordinates with the response force leader at a designated contact point. The rear CP operations cell dictates the contact point location and notifies both units.

6-181. The contact point should be located forward (50 to 200 meters) of the BHL. It should be near or on easily identifiable terrain features. At a prescribed time, liaison parties from both forces will meet. Normally, a response force representative is sent to the contact point. At the contact point, the TCF commander or his representative briefs the response force representative. Response force teams may remain in the vicinity of the BHL if passage is imminent. TCF security forces screen along the BHL and monitor the response force command net.

6-182. The actions of the elements on the BHL are critical. At the BHL, the TCF scouts, or others assigned the BHL action, pick up the fight from the passing response force elements. If the response force is in contact, its maneuver elements are bound behind the BHL, preferably to a covered and concealed location. At the same time, the TCF elements on the line relieve the pressure on the response force elements.

6-183. At the passage point, where the response force passes through TCF units, response force teams identify passing response force units. They ensure that the passing vehicles are displaying the correct visual recognition signal. TCF scouts on the BHL notify the forward TCF teams by a prearranged signal that contact has been made and friendly forces are at the BHL. Response force elements continue to delay forward of the BHL. Once behind the BHL and covered from enemy direct fire, the response force should—

- Confirm that the proper visual recognition signal is still displayed on all vehicles.
- Orient weapons systems toward the enemy.
- Move quickly, in the directed formation, through the passage points and along designated routes to the RP.

6-184. Stationary TCF elements overwatch this movement. The only time the TCF fires is when positive enemy identification is made. At the completion of the handover, response forces may either be released by the operations cell to resume other priority missions or tasked to remain in OPCON to the TCF and continue the fight.

PASS A TACTICAL COMBAT FORCE THROUGH A DELAYING FORCE

6-185. This technique is similar to the response force passing through the stationary TCF. The response force will have more information on the terrain and enemy situation than the TCF.

6-186. The response force leader sets up contact points, passage points, and routes. At the least, he provides guides. The guides meet the TCF at the contact points and lead them along routes to the passage points and/or RPs near the LD and LC.

6-187. The response force maintains normal radio traffic. The TCF and response force operations elements temporarily collocate. The TCF commander assumes control of the battle handover at this time. TCF company teams may maintain listening silence on their TCF command net. The response force guides notify their commander when the TCF begins moving forward from the contact points. As the TCF company teams deploy across the BHL, the response force leader and the TCF commander transfer responsibility for the battle.

6-188. The TCF fire support officer (FSO) may collocate with the response force forward observer (or the individual appointed to act as forward observer for the artillery unit in the AO). Fire missions are approved by the RAOC supporting the response force until the battle handover has occurred. After that, any fire missions for the response force are cleared through the TCF FSO. Response force units provide direct-fire overwatch according to the TCF commander's scheme of maneuver. It is coordinated directly by the TCF commander and the response force leader at collocated command groups.

6-189. Response force direct-fire support depends on the TCF commander's attack plan and the battle handover.

However, the stationary unit lifts and shifts its direct fire as coordinated by the collocated command groups.

BLOCK FOR A TACTICAL COMBAT FORCE

6-190. When defending from a battle position, the response force positions its elements to achieve the TCF commander's intent. The response force sets up blocking positions to keep the enemy from escaping. It selects battle positions where it can control, maneuver, and concentrate direct fire on the enemy. As blocking positions are set up, the TCF maneuvers to attack the enemy's flank. The TCF commander and the response force commander set up a contact point for coordination and conduct a reconnaissance of this point, time permitting. The TCF commander and response force leader coordinate C^2 and exchange the following information:

- When to lift or shift fire.
- What targets must be destroyed to allow the TCF to close on the enemy.
- The time, location, or activity to pass control of the battle.
- The location of the engagement areas.

6-191. The TCF and blocking force coordinate so that the TCF flanks the enemy. The response force suppresses the enemy from the front allowing time for the TCF to maneuver into position.

6-192. The TCF commander specifies tasks for the response force to achieve the TCF scheme of maneuver. The response force normally will have to orient its weapon systems on the enemy avenue of approach using TRPs or engagement areas (EAs). The response force may be tasked to—

- Destroy a certain enemy force from the battle position.

- Control the terrain or block an avenue of approach by holding the battle position against the enemy assault.
- Reorient weapons systems on a secondary avenue of approach from supplemental positions or avenues of escape for the threat.
- Assist in any other task needed to achieve the TCF mission.

6-193. Refer to *FM 17-95* for more information about battle handover.

CRITICAL SITE, ASSET, AND HIGH-RISK PERSONNEL SECURITY

6-194. MP are often tasked to perform security operations that require integration and prioritization from other MP CS operations. During security operations, MP focus their efforts on measures designed to safeguard resources by conducting special security activities that prevent unauthorized access to installations, equipment, materiel, and documents; and safeguard against espionage, sabotage, damage, and theft. An MP security operation includes protection of CPs, convoys, ASPs, deep-water ports, rail, pipelines, and high-risk personnel.

COMMAND POST SECURITY

6-195. CPs are priority targets for enemy attack, and their security can be crucial to the success of military operations. CPs at division, corps, and EAC are *designated critical sites and assets.* When CPs relocate, MP provide in-transit security. The amount and type of security provided to an established CP depends on whether the CP is dispersed or massed, large or small. In all cases, the focus of MP security efforts is on providing early warning of threat activity. MP possess mobility and communications assets that make them most valuable in operating AS patrols around the CP. Elements do not share a common defensive perimeter in a dispersed CP. Each cell provides its own local security and access control. MP conduct screening operations to provide early warning of enemy activity. In a massed CP where elements share a common perimeter, MP may contribute to perimeter defense with other elements that are collocated at a base.

Division Main CP Security

6-196. All Army of Excellence (AOE) divisions have a main, rear, and tactical CP. Division XXI currently has a division main and tactical CP and a security operations center (SOC). The best use of MP assets to secure any division CP is to employ MP teams to conduct AS around the location of the CP. MP provide early warning of enemy activity by conducting area reconnaissance and surveillance (R&S) of likely enemy avenues of approach to the CP. This security measure also provides the CP with standoff protection from direct enemy fire.

Division Band

6-197. When assigned to augment security operations, the division band should be employed intact, facilitating the commander's ability to quickly return it to its primary mission. The MP commander ensures that the bandmaster and his soldiers know the tactical situation and the defensive plan for the CP when they are assisting in CP security. The band may—

- Assist in perimeter defense of the CP.
- Provide access control on the road to the CP.
- Operate the dismount point for the CP.
- Provide access control to the main CP.

Corps Command Post Security

6-198. Normally, the corps CPs are echeloned into a tactical CP, a main CP, and a rear CP. The corps main

CP is generally located in the general vicinity of the division rear boundary. Traditionally, the corps main CP is massed and is a large organization with attendant mobility and signature problems. The corps tactical CP is small, mobile, survivable, and normally located well forward in the vicinity of a division main CP.

6-199. MP security is provided to both the corps main and tactical CPs. How MP elements deploy depends on the CP dispersion, the particular situation, and the METT-TC. The MP company provides corps CP security. If the CP is massed, one platoon secures the tactical CP and the remainder of the company secures the main CP. Or, depending on the METT-TC,—

- One squad provides access control within the corps tactical operations center (CTOC).
- Two squads serve as a response force.
- Two squads secure the tactical CP.
- Two squads man dismount points or augment other squads.
- The remaining squads operate the OPs/LPs and the security patrols.

6-200. The main security efforts and emphasis of MP are on early warning and screening. Both internal access and screening force measures provide CP security. These measures are enhanced by—

- Internal guard posts.
- Perimeter access control points.
- LPs/OPs and a security post outside the perimeter.

6-201. When a corps main CP is dispersed in cells, focus MP efforts on screening force measures. Allocate MP based on the location of the function and the METT-TC.

CONVOY SECURITY

6-202. MP leaders plan and conduct convoy security (specifically convoy escort) using several different methods to secure them as they move through their AO. Primarily, MP secure vehicle movement by maintaining security in the area that a convoy will pass through. MP also conduct route security on a specific route for a designated period during which multiple convoys move unimpeded.

6-203. The main threats to convoys are mines and ambushes. Convoys are vulnerable to sniper fire, long-range fire from artillery, mortar, threat aircraft, and AT-guided missiles. They are also vulnerable to remotely controlled mines and demolitions. Armored vehicles are better suited to provide convoy escort protection. Leaders carefully evaluate the threat before assigning a convoy escort mission to HMMWV-equipped units.

6-204. When MP conduct convoy escort security missions, platoons normally escort the convoy through their AO and pass responsibility for escort to adjacent platoons as the convoy moves from one platoon's AO to another. However, some situations may require a platoon to escort a convoy from its SP to its final destination.

6-205. Whatever method is used to secure the convoy, MP platoons executing the escort mission operate under the control of the convoy commander. The relationship between the MP platoon and the convoy commander provides for unity of command and effort, if combat operations are required during the course of the mission.

Plan

6-206. The platoon leader plans for security of the convoy in all directions and throughout the length of the convoy. This requires MP teams to be dispersed throughout the convoy formation. When available, locate engineer assets toward the front to respond to obstacles. If a fire-support team (FIST) is attached, locate it near the platoon leader.

6-207. The platoon uses the column formation *(Figure 6-11)* due to its inherent speed and ease of movement.

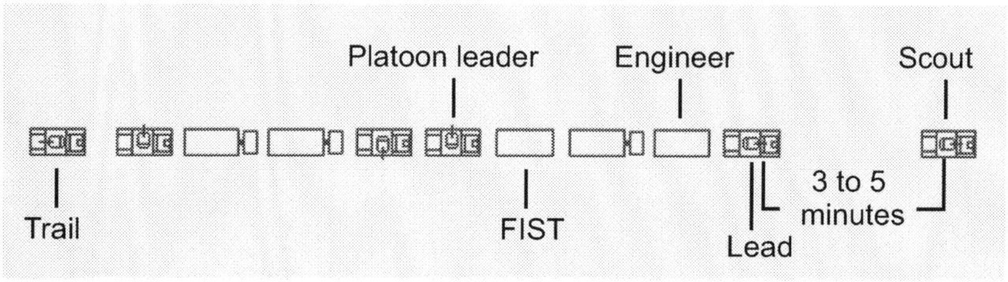

Figure 6-11. MP Platoon Escorting a Convoy

6-208. The platoon leader develops a convoy security plan that includes the following:

- HN support.
- The escorted unit's frequencies and call signs.
- The vehicle number and type and the order of the march.
- Checkpoint identification along the route.
- Security for halts and rest stops.
- Enhanced security at critical points such as, bridges, tunnels, key intersections, and other danger areas.
- An alternate route determination.
- Emergency actions to be taken during ambushes, a sniper attack, an air attack, or an artillery attack.
- Ambush reactions.
- Coordination with the response force responsible for the AO.

6-209. The platoon leader assigns selected teams to remain with the convoy to provide close-in security, if attacked, and other teams to maneuver toward the enemy to deliver a high volume of fire. Use a scout vehicle to travel ahead of the convoy (3 to 5 minutes) to alert the convoy commander of enemy activity, obstacles, or anything that might disrupt the convoys movement.

6-210. If a tracked armored vehicle is not available to lead the escort, the platoon leader assigns one team as the lead security element. Hardening of all vehicles used for convoy escort missions must be SOPs.

6-211. The platoon leader coordinates with the convoy commander to ensure that all convoy personnel are briefed and fully understand the convoy security plan. If time permits, the platoon leader conducts a joint rehearsal with the convoy commander and all security elements. At a minimum, the rehearsal should cover reactions to an ambush and air and artillery attacks.

React to Ambush

6-212. A well-planned ambush can have a devastating effect on a convoy. The reaction to an ambush must be quick, overwhelming, and decisive. The platoon executes its reaction to an ambush as a drill, taking special care to prevent fratricide. Actions to be taken include the following:

- When the ambush is initiated, security vehicles direct their action toward the enemy *(Figure 6-12, page 6-70)*. Selected teams move to covered positions between the convoy and the enemy to suppress the enemy with the highest possible volume of fire and obscure the enemy's view with smoke. At least one team should remain with the convoy and continue close-in security. The platoon leader sends the appropriate reports to higher HQ.

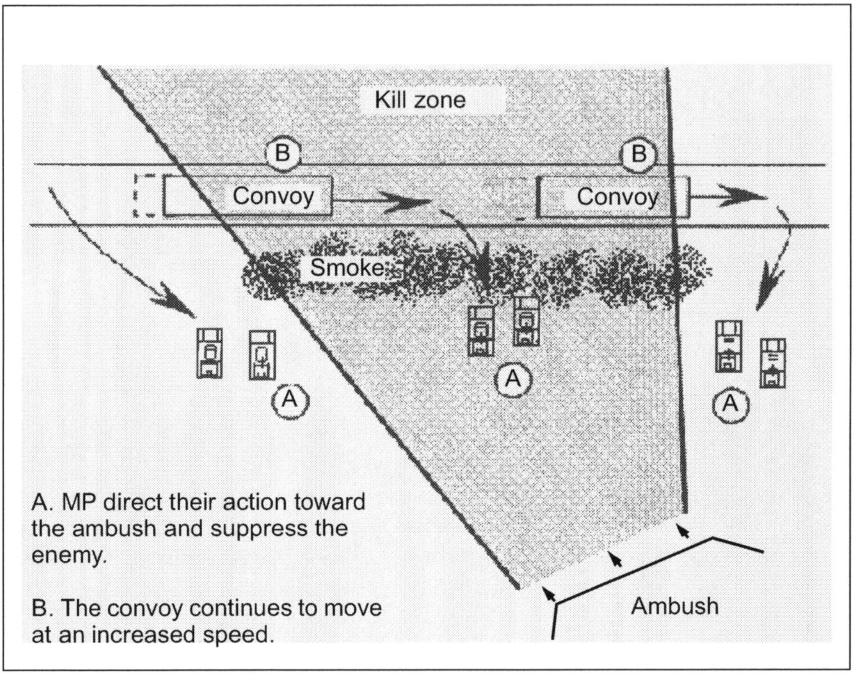

Figure 6-12. Security Vehicles' Actions During an Ambush

- The convoy commander continues to control the convoy and maintains radio contact with the security force (MP platoon leader). Vehicles not yet in the kill zone are stopped before they enter. Vehicles in the kill zone are moved out of the zone at the highest possible speed.
- The scout vehicle immediately selects a safe area for the convoy commander to move vehicles. The vehicles use a Light Vehicle Obscuration Smoke System (LVOSS) or other smoke to mask movement if possible. This safe area will be the linkup location when the convoy resumes.
- The convoy vehicles, if armed, return fire only until the security force positions itself between the convoy and the enemy.

- The damaged or disabled vehicles are abandoned and pushed off the route if they are not mission essential.
- The MP platoon leader submits SPOTREPs. Based on his assessment of the situation he can call for fire or CAS, or request the response force.
- Once the convoy is clear of the kill zone, the MP platoon leader chooses one of the following COAs based on the composition of the escort and the strength of the enemy:
  - Continue to suppress the enemy, while response forces move to support.
  - Assault the enemy.
  - Break contact, move out of the kill zone, and calls for indirect fire.

6-213. Normally, MP equipped with HMMWVs move out of the kill zone as soon as the convoy is clear. ASV-equipped security forces may continue to suppress the enemy or execute an assault. The platoon leader decides when to break contact.

6-214. **Mines**. Command-denoted mines are usually to signal the start of an ambush. When ambush fire comes from one side of the road, the enemy will most likely deploy a second, smaller force on the opposite side. They will deploy mines and obstacles between themselves and the convoy to contain US forces. The ambush force often uses mines to protect its flanks. To avoid damage from mines, drivers must—

- Drive on the track of the vehicle in front.
- Avoid driving on the shoulder of the road.
- Avoid running over foreign objects, brush, or grass in the road.
- Watch local traffic and the reactions of people on foot, which may give away locations of mines and booby traps.

6-215. **Artillery Attack**. If a convoy comes under artillery fire, vehicles must not stop. The convoy continues forward as quickly as possible to clear the area.

6-216. **Air Attack**. If the convoy comes under air attack, vehicles are dispersed, and soldiers attempt to find cover for the vehicles. Personnel begin firing all available weapons in an antiaircraft effort.

6-217. **Obstacles**. Obstacles are a major threat to convoys. Obstacles can be used to harass the convoy by delaying it. If the terrain is favorable, it may stop the convoy altogether. In addition, an obstacle or series of obstacles can be used to channel or stop a convoy to set up an ambush. The purpose of a route reconnaissance ahead of a convoy, is to identify obstacles and either breach or find bypasses. The convoy takes actions to reduce or bypass the obstacle. A convoy becomes more vulnerable when it is stopped and its escort force is occupied with tasks required to overcome or bypass the obstacle. For these reasons, security becomes critical, and actions at the obstacle must be accomplished quickly. Security personnel take the following actions at an obstacle:

- Direct (convoy commander) a halt when the lead security vehicle identifies an obstacle.
- Establish dismounted security and provide security throughout all activities directed by the convoy commander (such as conducting a reconnaissance and breaching the obstacle).
- Establish an obstacle overwatch.
- Ensure that convoy vehicles remain on the road.
- Move escort elements to the flanks to provide security.
- Ensure that all convoy vehicles are aware that the enemy may have buried mines in the area, especially on the road flanks.

6-218. **Halts**. When a convoy has to halt for short breaks it assumes a herringbone formation *(Figure 6-13)*. Security forces take up protective positions 360-degrees around the convoy and orient their weapons outward. They remain alert and establish dismounted local security. The convoy pulls into the protected area in the center of the perimeter. Stagger the convoy vehicles off the road, if there is no threat of enemy mines, but position them so they can easily pull back onto the road.

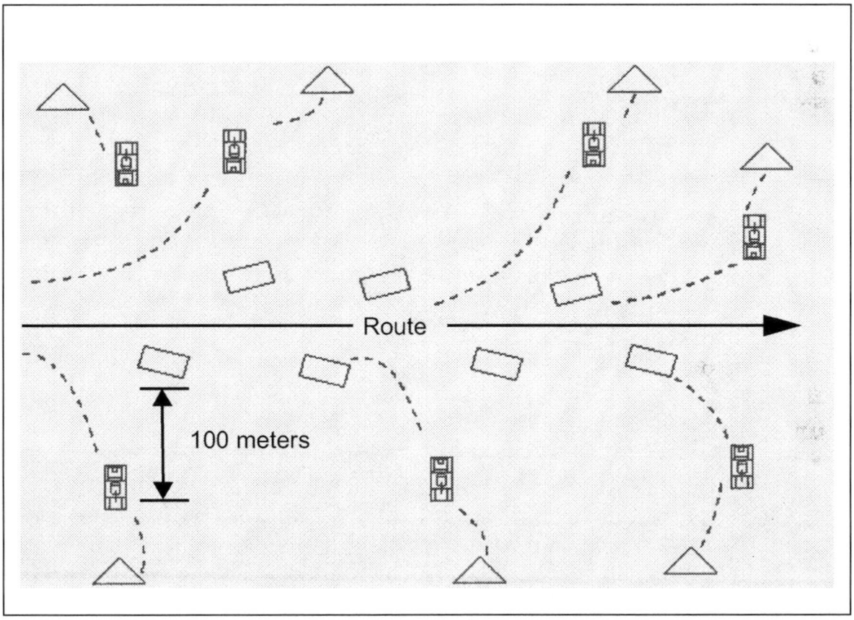

Figure 6-13. Herringbone Formation

6-219. When the convoy is ready to resume, the convoy vehicles reestablish the column formation, leaving space for the security vehicles. Once the convoy is in column, the security vehicles join the column, leaving local security dismounted.

6-220. Once all elements are in column, local security personnel mount and the convoy continues.

CONVENTIONAL AMMUNITION SUPPLY POINT SECURITY

6-221. When an MP platoon has been tasked to provide security for an ASP, the platoon establishes internal and external security. The platoon secures the ASP against enemy or criminal activity so that the ASP is able to maintain normal operations. MP leaders coordinate with ASP personnel to verify communications, distress and duress signals, defense plans, and fratricide prevention.

Internal Security

6-222. When MP conduct internal security, they establish a dismount point and allow only authorized personnel access to the ASP. Access is granted according to the local SOP and access and entry control rosters. MP conduct perimeter patrols ensuring that they do not set a pattern and maintain communications with platoon HQ.

External Security

6-223. When tasked to provide external security for an ASP, MP establish an OP/LP for early warning. They establish a TCP at the intersection of the MSR and the ASP access road. The TCP maintains free traffic flow, screens the traffic entering the access road, and controls stragglers and refugees. To further enhance external security, MP employ movement sensors, establish defensive positions, maintain communications with the platoon HQ, and emplace obstacles and physical security measures.

DEEP-WATER PORT SECURITY

6-224. The very nature of a terminal operation generates some different, and frequently more demanding, aspects of physical security. Ports and harbors are prime targets for enemy and criminal activities. The perimeter areas of these facilities are more vulnerable because of the extensive distance and the exposed beach or pier areas. Terminal areas may include fully developed piers and warehouses or may be an unimproved beach where logistics over the shore (LOTS) or roll-on/roll-off (RORO) operations are conducted.

Security Measures

6-225. When MP provide security for cargo in a port, the main effort is to provide security from the perimeter of the port outward. Security measures focus on aggressive patrolling and reconnaissance designed to detect, report, or destroy enemy threats before they disrupt port operations. Measures may also include—

- Developing police information and intelligence designed to identify criminal threats operating near the port.
- Controlling traffic in the area surrounding the port.
- Conducting mounted or dismounted patrols, with MWD if available, around the port perimeter.
- Establishing an access control and identification section.
- Providing a response force to react to incidents inside the port's perimeter.
- Providing observation and early warning of threat ground and air attacks.

6-226. When providing security for cargo, MP concentrate their efforts on providing AS in the area used to move cargo from the port to its final destination.

6-227. Inside a port's perimeter, MP security forces limit access to cargoes by—

- Operating random mounted or dismounted patrols (with MWD if available).
- Using the combined patrols as a response force for incidents inside the perimeter.
- Controlling access to the most restricted areas.

6-228. The priority is safeguarding the most critical cargoes waiting to be transferred to land transport. To safeguard stored cargo MP—

- Establish access control procedures.
- Search bundles and packages being taken from the area.
- Examine trip tickets and documentation of cargo vehicles.
- Inspect seals on locked containers.

Waterside Security

6-229. If the restricted area is a pier or other maritime environment, access from the water must be controlled as well as from the land. Entry on the landward side of a pier can be limited with fencing, pass control, and aggressive patrolling, but the part of the pier that protrudes over the water is accessible from the sides and below. Methods for securing the pier along its water boundaries include—

- Patrols (both walking on the pier and in small boats).
- Protective lighting.
- Log booms.
- Nets.
- Buoys or floats.
- Anchored or pile-mounted navigational aids and signaling devices.

- Barges.

DANGER
Ports pose special dangers for MP foot patrols. MP leaders must ensure that soldiers are thoroughly briefed on the hazards of patrolling too close to the water's edge.

6-230. While most of the barriers described above will stop or impede access to facilities from boats or swimmers, nets are among the most effective. Well-marked, partially submerged objects are also effective; however, there may be legal prohibitions against the emplacement of barriers that may constitute a hazard to navigation. The barriers should be emplaced only after coordination with and approval by the appropriate legal and HN authorities. Sometimes it is best to just close off the waterside of a pier. A floating boom will keep small boats out. Suspending a cable or chain link net from the bottom of the boom will deny access underwater.

6-231. MP establish at least two security zones on a facility's waterside: the reaction zone and the keep-out or exclusion zone. MP in these zones notify vessels, craft, and swimmers that they are entering restricted waters and should alter their course. MP may stop and search intruders if necessary. Security zones should be extended at least 1,000 meters from the nearest protected asset; however, in some port areas, a security zone this large is not possible. In such cases, other measures, such as boat patrols, must be increased to mitigate the possibility of attack. The two zones are defined as the—

- Reaction zone. The reaction zone extends from the high water mark to a distance beyond the maximum range of anticipated waterborne threats is the reaction zone. Inside the reaction zone, MP will stop and challenge intruders.

- Keep-out zone. This is the zone closest to the protected assets. The keep-out zone extends from the cargo asset to the maximum range of anticipated threat weapons. MP prevent the entry of all unauthorized craft or vessels into this zone. The tactical response force boat, in this case, may be employed if necessary.

Cargo Security During the Transfer

6-232. To keep cargo secure during transfer from one transport method to another, MP control the traffic moving in and out of cargo handling areas by—

- Setting up a single access control point.
- Erecting field-expedient barriers and using truck trailers or other large vehicles to constrict the traffic flow if permanent barriers are not in place.
- Limiting entry to mission-essential personnel, vehicles, and equipment, as designated by the port authority.

6-233. If gates are used by other than cargo vehicles, a *turnout* should be available. Cargo vehicles can pull into it while they are being checked. The turnout must be large enough to handle the volume and size of traffic being inspected. A wooden deck or platform at, or slightly higher than, the level of the truck bed hastens checking; it makes it easier to see the cargo. The platform should be as long as the vehicles being inspected. An empty flatbed trailer could be used.

6-234. Cargo is less likely to be diverted when close observation of cargo documentation and container safety is enforced. Containerized cargo is less likely to be stolen or sabotaged if closely watched as they are filled and sealed. Cargo can be pilfered before the seal is applied. An unsealed container can be moved to a stacking area or someone may apply a false seal, break the seal later, remove cargo, and then apply a legitimate seal. At access control points security personnel—

- Inspect inbound and outbound containers. Look for signs of damage or unserviceability.
- Inspect containers for the presence of seals and/ or locks and hinges. Check their serviceability.
- Verify that the document's transport number, container number, and seal number match those numbers on the transportation control and movement document (TCMD).

NOTE: Check seals by handling them, not simply by visually checking them.

- Allow only containers with valid documents to pass inbound or outbound through the control point.

TRAIN SECURITY FOR CARGO MOVED BY RAIL

6-235. Military resources moving by rail are often vulnerable to theft, pilferage, and sabotage. Depending on the threat, the cargo's sensitivity, vulnerability, and MP availability, the echelon commander may require MP to protect cargo moving by rail.

6-236. Most train operating crews consist of four or five people–the engineer, a conductor, a fireman, a senior brakeman, and a brakeman or flagman. This crew controls the train. The conductor is the train commander unless a transportation railway service officer is assigned to the train. The train commander is responsible for the train's operation and security. He makes all decisions affecting the train. MP integrate their security efforts with the security efforts of the train's crew. MP watch for and report any discrepancies or interruption to normal procedures at any time during the movement. Information about the movement is usually sent along the movement route by the chief dispatcher through a telephone circuit.

Safety Considerations

6-237. Many regions use electric trains. These trains use aboveground, high-voltage lines and travel at greater speeds. When these high-speed trains pass parallel to a stopped cargo train and MP security forces are dismounted, the train can create a vacuum strong enough to pull a security force member under the wheels of the passing train. Some of the slower-moving trains are referred to as floaters. They travel at much slower speeds than the express trains, cause less vibration, and are much quieter. These trains may not be heard until they are too close to react. All trains pose a risk to MP working in and around them.

6-238. Track switches are another danger to MP while they are dismounted. Track switches are used to alter the course of a moving train or car onto another rail line. Track switches are remotely controlled and operated. A member of the rail crew activates some switches manually at some sites. The inherent danger to dismounted MP is getting a foot caught or crushed as the track is being switched.

Security Force

6-239. Normally, an MP squad is an adequate force to secure railway shipments of sensitive freight, but additional MP may be needed for movement of critical cargo. When required, MP conduct railway security to prevent pilferage, detect enemy activity directed against the train and, within their capabilities, defend the train against enemy attack. In addition to MP, the shipper may send specially trained personnel with highly sensitive cargo. The number of MP in a train security force depends on the—

- Sensitivity of the freight.
- Priority of need for the freight.
- Terrain over which the train will pass.
- Length of the train.

- Duration of the trip.
- Degree of enemy threat.

6-240. When planning rail cargo security, MP—

- Review the unit SOP to determine guard force security clearance and weapons qualifications requirements and the use of force and rail security operations training.
- Obtain the time schedule for the rail movement.
- Make a map reconnaissance of the route. If time permits, they conduct an aerial reconnaissance. Bridges and tunnels are especially vulnerable.
- Coordinate with the HN and other agencies, such as the military traffic management command (MTMC) and movement control team (MCT).
- Plan actions at scheduled stops or relief points, deploying MP according to these plans.
- Plot the locations of MP units and other friendly forces along the route and note their radio frequencies and call signs.
- Obtain an intelligence report covering the route to indicate sites where sabotage may occur, or attacks may be expected, and if thefts and pilferage are likely.
- Determine weapons and special-equipment requirements.
- Assess border crossing requirements.
- Develop a security cargo checklist.
- Make a guard force serial-numbered weapons list.

6-241. MP prepare and maintain a record, by car number, of guarded cars in the train. MP may be required to ride in—

- A specific car that requires protection.
- The caboose.

- A special train car provided for the MP to travel in.

NOTE: If only one security car is used, it should be near the center of the train; if more than one is used, space the cars to provide the best protection for the train.

Security Measures

6-242. The shipper is responsible for the security of all carload freight until it is turned over to the transportation railway service (TRS).

6-243. Railway cars are sealed after loading. A seal shows that a car has been inventoried and inspected. The standard method of sealing a railway boxcar door (in addition to padlocks or wires) is by a soft metal strap or cable seal that contains a serial number.

6-244. Maintaining rigid accountability of all the seals is necessary to prevent the undetected replacement of an original seal with another. While sealing does not prevent pilferage, a broken seal is a good indicator that the car and its contents have been tampered with.

6-245. When military vehicles are shipped by railcar, sensitive and high-value items must not be secured in the vehicles. Container express (CONEX) and military-owned demountable containers (MILVANs) are ideal for shipping these and other small items on flatcars since they greatly reduce the chance of pilferage. These containers must be locked and sealed and, if possible, placed door-to-door for additional security.

6-246. Cars containing freight that is likely to be pilfered, high-priority cargo, or special shipments are grouped in the train to permit the most economical use of MP security forces. When flatcars or gondolas are used to transport sensitive or easily pilfered freight, the MP leader positions a security element where it can continuously observe and protect these cars.

6-247. The security force commander conducts a joint inventory and inspection of the seals and locks with the MCT or transportation officer on *DD Form 1911*. The security force commander maintains possession of this document and uses it when transferring custody.

6-248. The MP leader ensures that each seal is checked at every scheduled stop (time permitting) and—

- Reports broken seals immediately to help pinpoint the time and place of possible thefts.
- Dismounts and conducts 360-degree security when the train is stopped.
- Checks for damage to the cars.
- Notes and reports irregularities in procedures, the presence and/or actions of unauthorized personnel, and deficiencies and incidents that occur.
- Conducts a joint inspection of the cars when relieved by another security force while en route. The relief force signs for the train (*DD Form 1911*). The record is kept on the guarded cars.

6-249. Security personnel obtain a receipt for the secured cars when they arrive at their destination. The receipt is attached to the trip report that includes the following:

- The date and time of the SP and the RP.
- Any additional information required by local SOP or command directive.
- Any recommendations for correcting deficiencies or for improving future security on the trains.

6-250. Dock workers unload the cars as soon as possible when arriving at unloading points to reduce the opportunity for loss through pilferage and sabotage. MP assets are normally not available for security of freight in railway yards; however, MP must be prepared to conduct stabling and staging operations.

6-251. Stabling operations are conducted when numerous train shipments are held in a rail yard or other locations, such as a marshalling area, for extended periods pending arrival at the final destination. Stabling operations require continuous security that may include establishing perimeter patrols, access control, and a quick-reaction force to respond to internal and external threats.

6-252. Staging operations are not extended like stabling operations. They occur when parts of a shipment enter a secured or unsecured off-load site. The site may be large enough to accommodate the entire shipment or only several rail cars at a time. Security forces may be split and required to maintain security at two locations. Additional security forces may be required based on the METT-TC. Refer to *FM 55-20* for more information about security of rail cargo.

PIPELINE SECURITY

6-253. Pipeline systems are widely used in a theater of operations to transport bulk petroleum products or other liquids. Such systems are open to a number of security threats from the point of entry to the point of final delivery. Pipeline systems are composed of storage and dispersing facilities, pump stations, terminals, and extended pipelines. They also include discharging facilities for tankers at ports or other water terminals.

6-254. The type and extent of risk to a pipeline varies with the level of conflict in the AO. In a communications zone (COMMZ), the chief hazard is likely to be pilferage. Pipelines can be tapped by loosening the flange bolts that join sections of pipe or cutting holes in the hose line. The risk rises if gasoline is scarce and expensive on the civilian market. Sabotage is a security hazard during all levels of conflict. It is committed by any method such as simply opening pipe flanges, cutting hose line, or setting fires and causing explosions to destroy portions of the line.

6-255. In areas of greater conflict, the likelihood of sabotage and interdiction increase. Pipeline systems are vulnerable to air attacks, especially at—

- Aboveground sections of the pipeline.
- Pump stations.
- Storage facilities.

6-256. Deploy MP in the best manner to provide coverage to the most vulnerable portions of the pipeline which are at the greatest risk to enemy, terrorist, partisan, and ground attack. MP are best able to provide pipeline security by conducting mounted and dismounted patrols; establishing a LP/OP for early warning; and establishing defensive positions around terminals, pump stations, and aboveground pipeline sections. They patrol isolated areas and remote pumping stations and—

- Detect, report, and respond to attacks on or sabotage of the pipeline.
- Monitor critical parts of the pipeline on a routine but random basis.
- Monitor ground sensors and other intrusion detection devices. These are often used at pump stations and elsewhere along the pipeline to detect and identify threats to the system.
- Check line pressure devices in the pipeline and pumping facilities. These devices monitor the flow and detect breaks in the line, which may indicate pilferage of gasoline or other petroleum products.

6-257. Dedicated MP are rarely enough in number for surveillance of an entire pipeline system. Available supporting forces, in the course of their normal duties, assist in observing and reporting items of intelligence for further investigation. Examples of suspicious activities in the pipeline area might include the unusual presence of commercial tank trucks, the appearance of gasoline drums or cans, or the increased use of motor vehicles in fuel-scarce areas.

HIGH-RISK PERSONNEL SECURITY

6-258. Security of high-risk personnel is the responsibility of the US Army Criminal Investigation Division Command (USACIDC). However, due to a limited number of CID special agents, in any theater of operation, MP are often tasked to provide close-in security of protected persons.

Security Considerations

6-259. Normally, the MP company commander designates a platoon leader as the officer in charge (OIC) of the security detail. The platoon leader is given full responsibility for all phases of the security mission, to include the coordination. This allows for a single point of contact for the conduct of the mission. The platoon leader prepares a comprehensive plan to ensure the safety of the designated person. Once the platoon leader receives the designated person's itinerary, he selects an advance team and conducts a reconnaissance of the route and locations the designated person is scheduled to visit.

6-260. Normally, the platoon leader uses organic vehicles and equipment to conduct the mission. If special equipment is required, he makes the necessary coordination through his higher HQ. Personnel selected for the mission are thoroughly briefed on the security plan and their conduct and demeanor during the mission. The security plan should include the following:

- Restrictions on individuals approaching the protected person.

- Evacuation procedures if the protected person is attacked and the threat cannot be immediately reduced.
- Control measures such as designated check-points.
- Crowd control measures.
- Alternate routes.
- Emergency MEDEVAC procedures.
- Emergency actions to be taken in the event of an ambush or indirect fire.

6-261. Depending on the threat level, selected members of the security team are armed with a M4, a M16, or a M203. When dismounted with the protected person, they provide standoff security out to the maximum effective range of their weapons. Shotguns should be considered shotguns if there is a threat of attack in a congested area where there is danger of injuring innocent persons if long-range weapons were used. Members of the team that are selected to physically escort the protected person should be armed with sidearms only. This allows them to have both hands free to react to close-in threats.

6-262. The security plan must be flexible to allow for itinerary and weather changes and mechanical failures. Once the platoon leader has finalized the security plan, he assembles the security team and conducts detailed rehearsals.

Security During the Visit

6-263. Protection may be accomplished by establishing a series of protective cordons around the protected person. The protective measures used must allow the person freedom of movement and must not interfere with the conduct of his duties. Irritating conduct must be avoided at all times.

6-264. MP must remain alert at all times and provide constant security to the protected person. MP provide security by—

- Restricting the circulation of unauthorized personnel around the protected person.
- Remaining a short distance from the protected person.
- Taking positions where they can observe everything and everyone near the protected person.
- Securing the facility of the protected person, but not entering if the protected person does.
- Maintaining constant communications with higher HQ.

6-265. When required to secure a designated person while moving, the mode of transportation must be guarded at all times. Vehicle security normally requires a protected vehicle to transport the designated person, security vehicles to react to threats, and an advance team to travel ahead of the motorcade to provide early warning of threats or hazards before the main body is affected.

6-266. Regardless of the method of transportation, each situation must be evaluated to determine the degree of security that is practical and necessary. Good judgment on the part of the OIC will be necessary in solving the various situations that may exist. Refer to *FM 3-19.30* for more information about the security of designated personnel.

Chapter 7

Internment and Resettlement

I/R consist of those measures necessary to guard, protect, and account for people that are captured, detained, confined or evacuated by US forces. In any military operation involving US forces, accountability and the safe and humane treatment of detainees are essential. US policy demands that all persons who are captured, interned, evacuated, or held by US forces are treated humanely. This policy applies from the moment detainees become the responsibility of US forces and continues until the time they are released or repatriated. (Refer to the *Geneva Conventions* and *AR 190-8, AR 190-14, AR 190-47, FM 3-19.40,* and *FM 27-10*.)

OVERVIEW

7-1. The task areas that support the I/R function are EPW and CI handling, US military prisoner handling, and populace and resource control.

7-2. Captured, detained, and protected persons fall into several different categories that include the following:

- Enemy prisoners of war. EPWs are members of an enemy armed force or militia who must be guarded to prevent escape.
- Civilian internees. CIs are persons who have committed an offense against or poses a threat to friendly forces and must be guarded to prevent escape, but are kept separate from the EPWs.

- Dislocated civilian. DCs are persons that have been removed from their home because of war, disaster, or other reasons. They may be refugees, evacuee, stateless persons, or war victims. DCs are provided sustenance, safety, and humanitarian assistance. They are kept separate from EPWs and CIs. DCs are controlled to prevent interference with military operations and to protect them from combat or to relocate them to safety. DC operations are discussed later in this chapter.
- US military prisoner. US military prisoners are members of the US armed forces being confined, awaiting trial, or waiting transportation to a confinement facility outside the AO. They must be guarded to prevent escape and cannot be confined in immediate association with EPWs and CIs, detainees, or other foreign nationals who are not members of the US armed forces. Refer to *FM 3-19.40* for more information about field confinement of US military prisoners.

7-3. EPWs are more specifically defined in *FM 3-19-40* and the *Geneva Convention Relative to the Treatment of Prisoners of War, August 1949*.

ENEMY PRISONERS OF WAR AND CIVILIAN INTERNEE

7-4. MP receive EPWs and CIs as far forward as possible to prevent maneuvering units from being burdened with large numbers of prisoners. Prisoners are evacuated from the battle area as quickly as possible. The capturing unit is responsible for guarding prisoners until relieved. They field process captives using the *Five Ss-and-T method (Table 7-1)*.

7-5. The capturing unit is usually responsible for delivering the detainees to the collecting point and the

Table 7-1. Five Ss-and-T Methods

Procedure	Description
Search	Search each captive for weapons and ammunition, items of intelligence value, and other inappropriate items. **NOTE: When possible, conduct same gender searches; however, this may not always be possible due to speed and security considerations. Therefore, perform mixed gender searches in a respectful manner using all possible measures to prevent any action that could be interpreted as sexual molestation or assault. The on-site supervisor must carefully control soldiers doing mixed gender searches to prevent allegations of sexual misconduct.**
Silence	Do not allow the captives to speak or let anyone speak to them. Speak only to captives to give orders.
Segregate	Segregate captives by rank, gender, nationally, and status.
Speed	Remove the captives from the battlefield as quickly as possible.
Safeguard	Safeguard the captives according to the Geneva Convention and US policy. Provide medical care as needed.
Tag	Use *DD Form 2745* and include at a minimum the following information: • Date of the capture. • Location of the capture (grid coordinates). • Capturing unit. • Special circumstances of capture (how the EPW was captured, for example, did he resist, did he give up, and so forth).

nonwalking sick or wounded detainees to the nearest medical-aid station for evacuation through medical channels. Medically evacuated EPWs and CIs must be physically segregated from friendly forces. Detainees are normally turned over to MP at the nearest EPW collecting point or holding area. However, MP must be prepared to go forward to accept EPW from capturing units.

7-6. Traditionally, MP operate collecting points in a division AO and holding areas in a corps or EAC AO. However, collecting points and holding areas should be established wherever they are needed. The evacuation chain normally moves from the division forward or the central collecting point to corps holding area, then to internment facilities. When circumstances permit, such as taking advantage of available transportation, EPW evacuation may bypass one or more stations and deliver the detainees directly to a corps holding area or an internment facility.

7-7. At collecting points and holding areas, MP work closely with MI determining if captives, their equipment, or their weapons have intelligence value. MI interrogation teams conduct interrogations during field processing. Other MI interrogations teams conduct interrogations once EPW have been evacuated to more permanent facilities.

OPERATE A DIVISION FORWARD COLLECTING POINT

7-8. The number of MP needed to operate a division forward collecting point is based on the number and rate of captives expected and the METT-TC. A division forward collecting point must be mobile and modular and able to set up, expand, and move quickly with little or no notice. The general location of a forward collecting point is given in the brigade OPLAN or OPORD. It often is located near or in the brigade support area (BSA), but should not allow detainees to observe activities in the BSA. The collecting point should be situated close to an MSR. This makes it easier to get supplies, such as water, food, and barrier material from the BSA. Even a moderate number of detainees will put a strain on the equipment and supplies of an MP company. Minor medical treatment may come from the MP company's combat medical section. However, the company's medical resources are very limited and are primarily used to support medical needs within the company.

Units needed to support the division forward collecting point should be specifically tasked in the brigade OPORD. MP leaders operating the division forward collecting point will—

- Conduct a reconnaissance before selecting an exact location for the collecting point.
- Locate the collecting point far enough from the fighting to avoid minor shifts in the main battle area (MBA) (normally 5 to 10 kilometers from the MBA).
- Notify the BSA TOC and the PM operations section of the selected location. The BSA TOC reports the exact location of the collecting point to the brigade TOC. The brigade TOC notifies subordinate units where the collecting point is located so capturing units with detainees can take them there.
- Coordinate with the MI interrogation team if they are to colocate their interrogation site with the division forward collecting point.
- Request transportation, additional medical supplies, and other support through the forward support battalion.
- Ensure that captives do not remain at the division forward collecting point more than 12 hours before being escorted to the division central collecting point.

7-9. A forward collecting point *(Figure 7-1, page 7-6)* should not be set up near local inhabitants. Existing structures like vacant schools, apartments, or warehouses should be used when possible. This reduces construction requirements and minimizes logistical requirements. If existing structures are not used, detainees, except officers, can be tasked to help construct the collecting point. Prisoners may dig or build cover to protect themselves from artillery, mortar, or air attack. There is no set design for a forward collecting

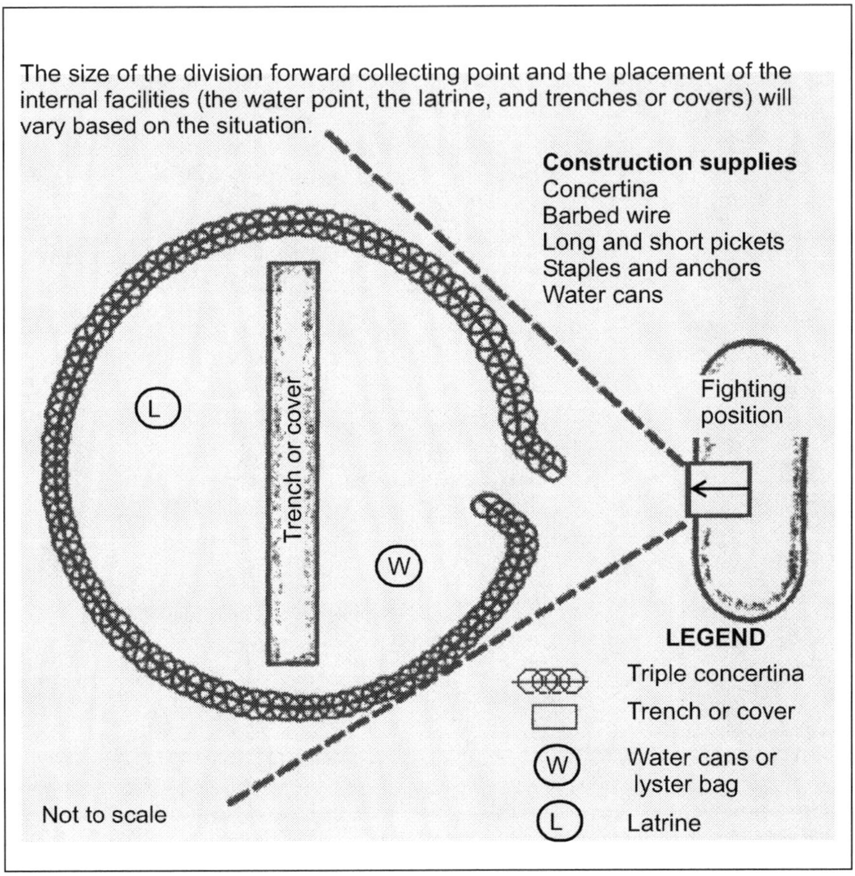

Figure 7-1. Division Forward Collecting Point

point. It can be anything from a guarded, roped off area to a secured, existing structure. The collecting point is built to suit the climate, the weather, and the situation. When selecting a collecting point, consider the following:

- The security of the detainees. The perimeters of the enclosure must be clearly defined and understood by the detainees.

- First aid. Injured or ill detainees require the same treatment that would be given to US casualties.
- Food and water. Detainees may have been without food or water for a long time before capture.
- Latrine facilities.
- Field sanitation. If possible, have detainees wash with soap and water to reduce the likelihood of disease.
- Shelter and cover.
- Language barriers. Provide interpreters and/or instructional graphic training aids (GTAs) in the EPW native language to compensate for the language differences.

7-10. MP at collecting points normally receive detainees directly from the capturing troops. MP then process the detainees using the *stress method*. The six principles of stress are search, tag, report, evacuate, segregate, and safeguard.

7-11. **Search**. Search and inspect every EPW and CI and their possessions. Use males to search male prisoners and females to search female prisoners wherever possible unless, in exceptional situations, an individual of the opposite gender must conduct the search. If this is the case, the search of the opposite sex must be performed in a respectful manner using all possible measures to prevent any action that could be interpreted as sexual molestation or assault. Captives may keep the following items found in a search:

- Protective clothing and equipment (such as helmets, protective masks and clothing) for use during evacuation from the combat zone.
- Retained property, such as identification cards or tags, personal property having no intelligence value, clothing, mess equipment (except knives

and forks), badges of rank and nationality, decorations, religious literature, jewelry, and articles that have sentimental value.

- Private rations of the EPW or the CI (in the early stages of captivity).

7-12. Certain items are confiscated from the EPW or the CI and never returned even if the EPW or the CI is released or repatriated. MP confiscate the following items when searching a captive:

- Weapons and ammunition.
- Items of intelligence value (maps, orders, and so forth).
- Other inappropriate items.

7-13. MP will coordinate with the MI interrogation teams to determine which items that have been confiscated are of intelligence value. Personal items, such as diaries, letters from home, and family pictures may be taken by the MI teams for review, but are later returned to the MP for return to the proper owner.

7-14. Currency will only be confiscated on the order of a commissioned officer (*AR 190-8*) and will be receipted for using *DA Form 4137*.

7-15. Impounded articles are items taken from the EPW or the CI during his internment because the articles make escape easier or compromise US security interests. Items normally impounded are cameras, radios, and all currency and negotiable instruments found on the captives. Refer to *AR 190-8* and *Defense Finance and Accounting Service-Indianapolis (DFAS IN) 37-1* for more information about confiscated and impounded property.

7-16. MP prepare a receipt when taking property from a detainee. The MP leader ensures that both the EPW or the CI and the receiving MP sign the receipt (such as *DA Form 4137*). MP consider bundling a detainee's property or placing it in bags to keep each detainee's property intact and separate. They turn in cleared, confiscated property as far forward as possible. MP maintain a strict chain of custody for all items taken from the EPW or the CI. They ensure that a receipt is obtained for any items you release to any other MP or agency. The escorting MP signs for and transports any remaining property that was taken from the EPW or the CI.

7-17. **Tag**. Each EPW or CI is tagged by the capturing troops using *DA Form 2745* as a way of accounting for them. MP check each tag at collecting points and holding areas for—

- The date and time of the capture.
- The capturing unit.
- The place of the capture (grid coordinates).
- The circumstances of the capture (how the EPW was captured).

7-18. The remaining information on the tag will be included as it becomes available. *DD Form 2745* is a perforated, three-part form which has an individual serial number. It is constructed of durable waterproof, tear-resistant material with reinforced eyeholes at the top of Parts A and C. The capturing unit attaches Part A to the captive with wire, string, or another type of durable material. They maintain Part B in their records and attach Part C to the confiscated property so that the owner may be identified later.

7-19. MP at division collecting points will ensure that *DD Form 2745* has been placed on any captive arriving at the collecting point without it. MP may have to direct the capturing units to complete the capture tag before accepting prisoners into the CP. They ensure that the following is done:

- The tag is filled out with the minimum information listed above (also listed on the back of Part C of the form).

- A statement is on the tag if the captive arrived without a tag.
- The captive is instructed not to remove or alter the tag.
- The capture tag's serial number and the captive's name are annotated on a locally developed manifest.

7-20. MP receive detainees from capturing troops using *DD Form 2708* or a similar document. They ensure that the receipt includes the following:

- The capturing unit.
- The time and date the detainee was received.
- The identification of the detainee. (Use the number on the capture tag when the detainee's name, service number, grade, or date of birth is unknown.)
- The name, service number, grade, unit, and signature of the MP who accepts custody of the detainee.
- A statement in the remarks section about the general physical condition of the detainee. For example, *received without wounds, illness, or injury* or *wounded in upper left arm*.

7-21. **Report**. The number of captives at each collecting point is immediately reported through MP channels. This aids in the transportation and security planning process.

7-22. **Evacuate**. Captives are humanely evacuated from the combat zone through appropriate channels as quickly as possible. MP do not delay evacuation to obtain name, rank, service number, or date of birth. When MP evacuate captives, they give them clear, brief instructions in their own language when possible. Military necessity may require a delay in evacuation beyond a reasonable period. When this occurs, MP leaders ensure that there is an adequate supply of food;

potable water; and appropriate clothing, shelter, and medical attention available.

7-23. MP ensure that EPWs or CIs are not be exposed to unnecessary danger and are protected while awaiting evacuation.

7-24. Medical personnel determine if captives with serious wounds or sickness should be kept in the combat zone. Sometimes prompt evacuation would be more dangerous to their survival than retention in the combat zone.

7-25. **Segregate**. The senior officer or noncommissioned officer in charge (NCOIC) having responsibility for custody of the EPWs or CIs will designate how and at what level to segregate them to ensure their security, health, and welfare. EPWs and CIs are segregated into the following categories:

- Officers, noncommissioned officers (NCOs), enlisted, male, and female.
  - Deserters and those that gave up without a fight may be further segregated for their protection.
  - Nationality, ideologies, and recognized ethnic groups are used for further segregation.
- CIs and/or refugees are physically separated from the EPWs and CIs.
- US military prisoners are physically separated from EPWs, CIs, retained persons (RP), other detainees (OD), and refugees.

7-26. MP do not use coercion of any kind to obtain any information from the captives. This includes basic information, such as name, rank, service number, and date of birth, which they are required to provide under the Geneva Conventions. Coercion or inhumane treatment of any EPW, CI, RP is prohibited and is not justified by the stress of combat or with deep provocation. Inhumane treatment is a serious violation of international law and the Uniform Code of Military Justice (UCMJ).

7-27. MP must not speak to captives except to give orders or directions. Captives must not be allowed to talk to or signal each other. This prevents them from plotting ways to counter security and plan escapes. Uncooperative captives may require a gag in certain tactical situations. However, gags should be used for only as long as needed and should not harm the individual.

7-28. **Safeguard**. In order to safeguard captives according to the Geneva Convention and US policy, MP must—

- Provide first aid and medical treatment for any wounded or sick captive. The wounded and sick will be evacuated separately through medical channels using the same assets as those used to medically evacuate US and allied forces.
- Ensure that the detaining power provides their captives with food and water. These supplies must be the same as to that of US and allied forces.
- Provide firm and humane treatment.

7-29. Protecting detainees from attack, preventing their escape, and quickly removing them from the battle area further safeguards them. Detainees should not remain at the division forward collecting point more than 12 hours, if possible. MP from the division central collecting point move forward to escort detainees back to the central collecting points. When detainees are field processed and ready for evacuation, the MP at the division forward collecting point will—

- Report detainee status to the BSA TOC and through MP channels to the PM.

- Request transport, rations, and water for the detainees from the forward support battalion supply officer (US Army) (S4).
- Ensure that the receipts for the detainees are ready for signing by the escort guards.
- Ensure that items taken from detainees for security or intelligence reasons are signed over to the guards taking the detainees to the rear. Ensure that each item is tagged to identify the owner.

ESTABLISH AND MAINTAIN A DIVISION CENTRAL COLLECTING POINT

7-30. MP in GS are responsible for establishing and maintaining the division central collecting point. They collect detainees from the forward collecting points, then process and secure them until corps MP come forward to evacuate them to the rear. Detainees should be transferred to the corps holding area or directly to an internment facility within 24 hours, if possible. One or more GS MP platoons operate the division central collecting point. The MP platoons are augmented by the division band and/or by the corps MP. Augmentation is based on the number and rate of captives expected.

Band Augmentation

7-31. When necessary, members of the division, corps, or EAC band augment MP for EPW operations. They guard detainees, operate dismount points, and provide perimeter security. When band members are tasked to augment MP for EPW operations they are OPCON to the MP company for the duration of the mission and released at the earliest opportunity to return to their primary mission.

Division Central Collecting Point

7-32. A central collecting point *(Figure 7-2)* is larger than a forward collecting point, but the considerations for setting up and operating the collecting points are

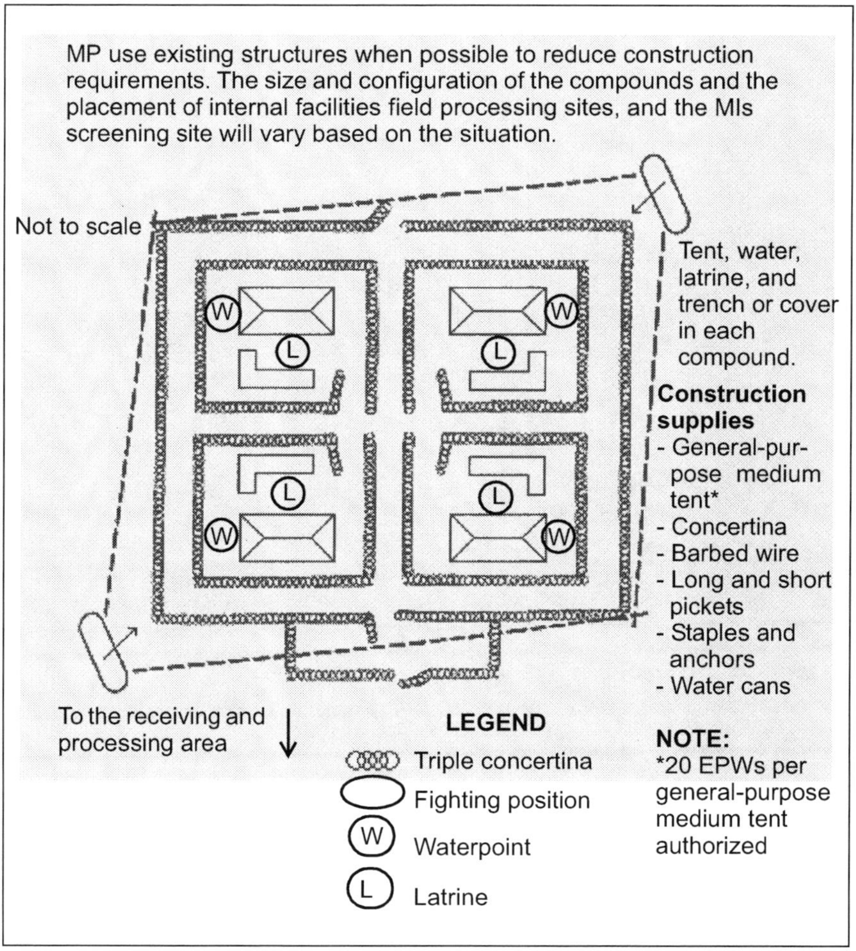

Figure 7-2. Division Central Collecting Point

generally the same. The general location of the central collecting point is given in the division OPORD or OPLAN. It is located near the division support area (DSA), preferably close to an MSR. This makes it easier to obtain supplies, transportation, and additional medical support from the DSA. Non-MP units should be specifically tasked in the coordinating instructions of the division OPORD to provide the support needed for the division central collecting point. MP establishing the collecting point should—

- Coordinate with the unit responsible for the area.
- Conduct a reconnaissance before picking the exact location for the collecting point.
- Notify the PM and the operations cell of the division rear CP (through MP channels) of the collecting point location.
- Coordinate with MI for the location of their screening site.
- Use existing structures when possible.
- Request supplies through the division MP company.

EVACUATE DETAINEES FROM A DIVISION FORWARD COLLECTING POINT TO A DIVISION CENTRAL COLLECTING POINT

7-33. The MP platoon charged with operating the division central collecting point sends MP forward to the division forward collecting point to escort detainees back to the central collecting point. EPWs or CIs must be evacuated from the division forward collecting point as soon as possible, preferably within 12 hours. Before evacuating the detainees, MP checks with MI interrogation teams for any property to be returned to, or evacuated with, the detainees before they are moved.

PROCESS ENEMY PRISONERS OF WAR AND CIVILIAN INTERNEES FOR EVACUATION

7-34. MP consider the physical status of detainees before evacuating them. Categories for consideration are the sick and wounded EPWs and CIs and the able-bodied EPWs and CIs.

Sick and Wounded Enemy Prisoners of War and Civilian Internees

7-35. The MP unit's combat medical section will screen detainees and decide if they will be escorted within MP channels or medically evacuated. Generally, the walking wounded go with the other detainees. Litter patients go through medical channels. US forces provide the same medical care for sick or wounded detainees as that given to US and allied soldiers. Sick and wounded EPWs in the combat zone are either treated and returned to the MP for evacuation or stabilized and moved through medical channels to the rear as far and as quickly as possible. If medically evacuated, MP release the detainees to the medical authorities using *DD Form 2708* or other receipt. The corps medical regulating officer (MRO) and the receiving hospital commander coordinate with the Internment Resettlement Information Center (IRIC) to account for detainees in medical channels.

7-36. MP determine when security is required for sick or wounded detainees. Normally, sick or wounded detainees requiring MEDEVAC are not likely to be a security risk. Detainees well enough to present a security risk can be treated by the combat medical section and evacuated through MP channels as soon as possible.

Able-Bodied Enemy Prisoners of War and Civilian Internees

7-37. Able-bodied detainees are escorted during movement to keep them from escaping. MP planning the movement of detainees consider the following:

- The factors of METT-TC.
- The number of detainees being escorted.
- The condition and morale of the detainees. Fatigued and cooperative detainees may not require as many guards as those who are fresher and more motivated.
- The type of transport to be used. The type of transport may influence the number of guards. A planning consideration is one guard per 5 to 10 detainees. Aircraft is loaded according to the airplane crew's instructions.
- The terrain conditions along the route. Detainees are more likely to attempt escape in close terrain, like dense woods or jungle, and may require more guards than open terrain.
- The level of enemy activity along the route. The more enemy activity in the area, the greater the need for increased security precautions.
- The likelihood or presence of suspected sympathizers and hostile local nationals along the route.
- The scheduled arrival of the transport. Use backhaul transport whenever possible.
- Transportation considerations. Transportation depends on the availability of vehicles delivering cargo in the nearby area.
- The location of MP units or bases and base clusters along the route that could provide assistance during the movement.
- The number and locations of rest stops (based on the type of transportation, distance, and the type of terrain).

7-38. Detainees are evacuated on foot only as a last resort when transport is not available. Transportation for detainees is arranged through the company HQ. At division, the company HQ contacts the local movement control officer.

7-39. Before leaving for the collecting point, the MP in charge of the escort must—

- Conduct a route reconnaissance of the evacuation route.
- Verify the location of the collecting point shortly before departing, since BSAs move often.
- Plan to stop only during daylight and outside towns or installations if possible.
- Designate guards to dismount at halts and supervise the loading of the detainees.
- Segregate detainees by category, if possible.
- Secure the rations and the water. Use captured enemy rations for the detainees, if available. Do not allow the use of utensils or can openers.
- Search detainees and baggage before loading in any transport.
- Use hand irons, leg irons, or special restraining jackets on detainees, if necessary. If hand irons are used, restrain the detainees with their arms in front.
- If prescription drugs are needed, disperse according to the medical officer's instructions.

EVACUATE ENEMY PRISONERS OF WAR AND CIVILIAN INTERNEES FROM A DIVISION CENTRAL COLLECTING POINT

7-40. In order for MP to conduct successful evacuation of EPWs and CIs, MP brief the escorts and the detainees.

Brief the Escorts

7-41. MP or other military personnel may perform as guards in evacuating EPWs and CIs. Escort personnel are briefed on evacuation considerations and escape attempts. Considerations include the following:

- Procedures to ensure that the detainees follow instructions and orders. Escorts must be firm, but will not punish detainees who fail to obey.
- Requirements to inspect passenger areas, latrines, and other places that might be accessible to detainees during transport. Escorts should look for the means of escape or items that could be used as weapons and remove the latches from the latrine doors on transports, if possible.
- The necessity to talk to detainees only to give orders and maintain control.
- Emergency actions to secure and safeguard the detainees in case of enemy contact. Members of the escort element must know in advance which of them will control the detainees and which ones will react to the enemy.

Receive and Brief the Detainees

7-42. The senior MP in the escort element accepts custody of the detainees. Each detainee is tagged and field-processed before being accepted for evacuation. Each detainee is accounted for using *DD Form 2708* or a similar receipt. The senior MP escort ensures that all the detainees (and any equipment) are listed on the custody receipt when custody is accepted. The escort retains a copy of the custody receipt.

7-43. Before moving, the senior MP ensures that the detainees have been briefed on movement discipline in a language understood by them. MP use locally produced GTAs to conduct the briefing. If available, an interpreter should give the instructions to the detainees. They are told—

- The meaning of the word *halt*.
- That the "silence rule" applies at all times (no talking to the guards, no talking to each other).
- The actions to take during an emergency.

COLLECT ENEMY PRISONERS OF WAR IN OTHER OPERATIONS

7-44. Some offensive operations are executed so rapidly that combat forces completely overwhelm the enemy forces. These operations create special considerations when planning EPW operations.

River Crossing Operations

7-45. During river crossing operations, an EPW collecting point is established on the nearside of the river far enough to the rear to prevent interference with tactical operations and to afford reasonable protection against hostile fire. The EPWs are evacuated from the bridgehead area as soon as possible to prevent congestion. The movement of EPWs from the farside of the river must be coordinated with traffic control personnel at the crossing sites. Secondary crossing sites must be used if available.

Amphibious Operations

7-46. During amphibious operations, initially, the shore party or helicopter support teams operate EPW collecting points in the beach support areas or LZs. EPWs are evacuated from the collecting points to designated ships by landing craft, a helicopter, or amphibious vehicles. MP must coordinate with the support force for the handling of EPWs once they have been evacuated from the beachhead. EPWs are retained in the objective area when facilities, supplies, and personnel permit, consistent with the reasonable safety of EPWs from enemy action.

Airborne Operations

7-47. During an airborne operation, the METT-TC considerations for collecting EPWs include the geographical location of the airhead, the tactical plan, the availability of transportation, and plans for linkup with ground forces. EPWs are evacuated primarily by air, especially during the early stages of the operation. The EPW collecting point should be located near a LZ. Plans should provide for the attachment of MP escort guard units from the area EAC to the airborne force to guard EPW during evacuation.

Armored Operations

7-48. Armored units are able to quickly penetrate deep into hostile territory, possibly bypassing enemy strong points. They may leave isolated enemy groups, which would hinder the normal evacuation of EPWs. In this circumstance, it may be necessary to hold EPWs in the area of capture until they can be safely evacuated.

Air Assault Operations

7-49. During an air assault operation, organic military police elements accompany assault elements to the objective areas so that they can collect and evacuate captured EPWs. Collecting points are established as required near heliports or airfields. Arrangements must be made for nondivisional MP to accompany designated incoming or resupply aircraft to guard EPWs during their evacuation from the division.

UNITED STATES MILITARY PRISONER HANDLING (FIELD DETENTION FACILITIES)

7-50. Use field detention facilities *(Figure 7-3, page 7-22)* to hold US soldiers in custody until they can be tried. Use pretrial confinement only to ensure that an accused appears at trial or when the seriousness of the offense or the threat of violence makes confinement essential. Whenever possible, soldiers awaiting trial remain in their units. Only when they are a hazard to themselves or others are they detained in pretrial confinement under MP control.

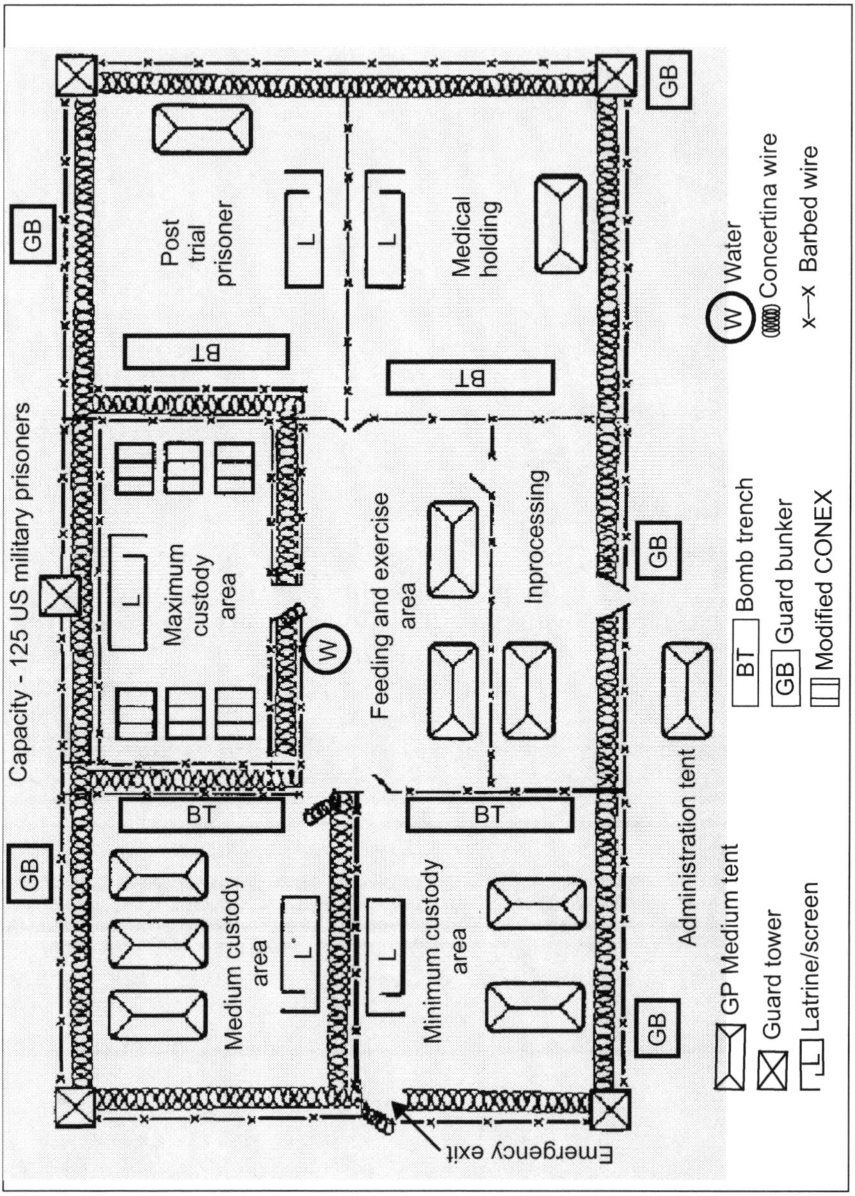

Figure 7-3. Field Detention Facility

7-51. Use field detention facilities to hold sentenced prisoners waiting for transfer to a theater's field confinement facility (FCF) or the continental US (CONUS). After trial, move convicted military prisoners, whenever possible, to confinement facilities outside the combat zone.

7-52. Each echelon commander sets procedures and policies for detaining and confining soldiers. Often US military prisoners in a combat zone are placed under the control of an MP unit operating an EPW collecting point. When small numbers of US prisoners are on hand, a squad operating an EPW collecting point can best take responsibility for the security of US prisoners. US military prisoners must be kept physically apart from EPWs. The policy and procedures for the care and treatment of prisoners and the safeguarding of their personal effects remain the same as that set for other Army confinement facilities.

7-53. When prisoners are retained in-theater, separate temporary detention facilities maybe set up in the corps or division areas. US military prisoners should be held in the division rear area for the shortest possible time. At a division facility MP must—

- Safeguard US prisoners.
- Coordinate for their food and medical care.
- Sustain them until they can be evacuated to a corps facility.
- Transfer them to the corps facility as quickly as possible.

7-54. When the situation permits, MP from a detention facility at the corps come forward to pick up the prisoners at the request of the division's detention facility commander. From the corps, the prisoners are evacuated to the theater confinement facility.

7-55. If a temporary detention facility is set up in the corps, it usually is operated by confinement teams from the confinement battalion in a personnel command (PERSCOM). These teams are organized and trained to perform confinement operations. But when corps detention operations are limited to prisoners being evacuated to a confinement battalion in the PERSCOM, elements from a combat support company can operate a temporary facility.

7-56. A field detention facility usually is located near the MP company CP for food, transport, and supply support. MP request construction materials from the engineers to set up and run a facility. Equipment and supplies must include the following:

- Barbed wire (roll and concertina).
- Fence posts.
- Gates and doors.
- Floodlights and spotlights, complete with wiring.
- An emergency generator.
- Mess equipment and equipment for cleaning mess gear.
- Water cans or lyster bags.
- Computers.
- First aid equipment and supplies. Spare clothing and bedding.
- Hand restrains or leg irons.
- Heating equipment (cold climate).
- Field sanitation supplies.

7-57. MP leaders ensure that the facility is large enough to separate prisoners by prisoner status, custody grade, sex, and rank. They locate the facility away from a base's perimeter or any other area of increased risk.

7-58. The size of the facility is based on the number of prisoners being detained. It may be a room or a tent, as long as it provides shelter equal to that offered to other soldiers in the combat zone. The physical criteria for permanent and temporary structures are the same. MP use existing structures if you can. Otherwise, they use tents. Field-expedient facilities must be approved and periodically inspected by a medical corps officer or a designated representative.

7-59. A team or a squad can operate a field detention facility. A team operating a facility may organize so that the team leader controls the operation and relieves the guards. The other team members perform guard duty in alternating 12-hour shifts. When operating a field detention facility, the team—

- Accepts sentenced soldiers on the written order (currently a *DD Form 2707*) of a court martial convening authority (or the authorized representative).
- Ensures that the order states the name, grade, social security number, organization of the prisoner, offense for which convicted, and the sentence.
- Accepts the accused soldiers on the written order (currently a *DD Form 2707*) of the accused's commanding officer.
- Signs a receipt for each prisoner and his property on the correct and current form.

POPULACE AND RESOURCE CONTROL (PRC)

7-60. PRC operations are the responsibility of the Assistant Chief of Staff, (Civil Affairs) (G5), the Civil Affairs Officer (US Army) (S5), and/or the HN authorities. (Refer to *FM 41-10* for more information about PRC.) PRC is often conducted in stability and support operations where national authority has broken down and the government cannot control the population. MP support PRC by conducting L&O operations designed to restore order and protect the people and property.

7-61. Insurgent organizations often emerge in unstable regions. The aim of such groups is normally to overthrow the established government. The less control the government has, the greater the chance for insurgents to succeed. These organizations try to exploit the population, often through threat and intimidation. When insurgent organizations pose a threat to the population, US forces employ PRC operations that are designed to deny support and assistance to insurgents by controlling the movement of people and goods and restricting access to key facilities. Police activities, such as roadblocks, cordons, curfews, access control, and checkpoints are an important measure in counterinsurgency, but have a high potential for harm if used excessively or incorrectly. MP are specially trained to conduct these operations, as a force focused on security, protection, and assistance. MP are continually trained on the prudent use of force, crisis management, and operations requiring restrictive ROE.

7-62. PRC measures deprive insurgent organizations support and aid in identifying their supporters. Appropriate psychological operations (PSYOP) help make these measures more acceptable to the population by explaining their necessity. The government informs the population that, although its actions may cause inconvenience, the threat posed by the insurgents makes them necessary.

7-63. PRC is often conducted in urban areas. The best use of PRC comes before an organized insurgent movement has the capability for armed conflict. MP intelligence operations support PRC programs. Criminal acts, such as robberies, kidnappings, terrorism, and extortion, may accompany insurgent propaganda or money-raising activities.

7-64. MP employ special control measures to aid populace control that include the following:

- Enforcing curfews.

- Enforcing movement restrictions.
- Verifying travel permits and registration cards.
- Assisting with rescue and evacuation operations.
- Assisting with crowd control.

7-65. MP also employ control measures to aid in resource control. They include—

- Operating roadblocks and checkpoints.
- Inspecting facilities.
- Enforcing local regulations and guidelines.
- Controlling rations.
- Assisting with amnesty programs.

7-66. PRC operations play a vital role in winning support away from the insurgent threat and encouraging support for the goals of the legitimate government.

7-67. MP leaders responsible for supporting PRC must plan for the protection of their forces as these types of operations have a history of turning violent very quickly.

DISLOCATED CIVILIAN RESETTLEMENT

7-68. International law recognizes the humanitarian practice of providing temporary refuge to anyone, regardless of nationality, who may be in imminent physical danger for the duration of the threat. It is US policy to grant temporary refuge in a foreign country to nationals of that country, or of a third nation, solely for humanitarian reasons when extreme or exceptional circumstances exist.

7-69. Civil affairs (CAs) units are trained to plan, coordinate resources for, and monitor the handling of DCs. (Refer to *FM 41-10* for more information about DC operations.) Whenever possible, resources and control should be arranged with the HN, other governmental agencies, and nongovernmental and private organizations.

7-70. To assist in properly identifying DCs, the definition of the five subdivided categories is provided. These subcategories are defined by legal and political considerations as the following:

- Displaced person. A civilian who is involuntarily outside the national boundary of his country in time of conflict.
- Refugee. A civilian who, because of real or imagined danger, has left home to seek safety.
- Evacuee. A civilian removed from his place or residence by military order.
- Stateless person. A civilian who has been denationalized, whose country of origin cannot be determined, or who cannot establish his right to the nationality claimed.
- War victim. A classification that describes civilians suffering injuries, loss of a family member, or damage to or destruction of his home resulting from war.

7-71. DC operations are a special category of PRC. The goal of DC operations is to minimize civilian interference with military operations, relieve suffering, and protect civilians from combat operations or other threats. When the HN cannot or is unwilling to control DCs, MP may be required to collect, evacuate, and resettle them.

7-72. When directed to conduct DC operations, MP leaders coordinate with CA and the PM to establish a traffic control plan. The plan includes the primary and alternate routes used to move DCs and the location for—

- The TCP.
- Holding areas.
- Roadblocks and checkpoints.

7-73. Temporary route signing is not normally an effective control measure during DC operations. Even when the signs are posted in the local language, DCs usually ignore them.

7-74. MP collect and process DCs in the same manner as EPWs with regard to the difference in their status–they are detained personnel, not prisoners. As such, security of the I/R facility should not give the impression that it is a prison.

7-75. MP collect DCs at assembly points located away from the MSR and areas where combat operations are heaviest. They move DCs along selected routes that have the least military traffic. DCs will normally require frequent rest stops. When possible, holding areas for rest stops should be near a source of water and provide protection from the elements or hostile fire.

7-76. MP provide close-in and standoff security for DCs while en route to a civilian camp or collection point. When required, MP provide external and internal security at a DC camp. Within limits, DCs are allowed freedom of movement once they have been resettled. MP and CA must continue to closely monitor DCs at collection points and camps. Recent operations that involve large numbers of DCs have been known to turn violent very quickly. Such a situation can be avoided when DCs are treated with respect and dignity.

EVACUATION OPERATIONS

7-77. In addition to EPWs, CIs, and DCs, MP are often required to evacuate other selected individuals or groups from collecting points, holding areas, or areas of hostility. The safe evacuation of noncombatants, US diplomats, or US military prisoners requires close coordination and extensive planning. This type of operation is referred to as noncombatant evacuation operations (NEO). Such missions require contingency plans and unit SOPs that are specifically designed for these special operations. Unit SOPs should cover the evacuation of designated personnel by ground, air, rail, and water. When directed to conduct evacuation operations, MP leaders task organize based on METT-TC and the availability of MP. They perform extensive rehearsals to successfully conduct evacuation operations.

7-78. When a HN can no longer ensure the safety of US civilians in a foreign nation, US military forces may evacuate them. MP are often required to provide security and escort for evacuees from their point of origin to their destination. A NEO is usually a joint operation conducted with other US and HN government agencies and CA.

7-79. An MP platoon is most likely employed for these operations. The platoon leader conducts direct coordination with CA and the higher HQ. The platoon leader begins TLP, issues a WO to the platoon, and requests information that includes the—

- Screening and identification system being used.
- Number of evacuees, their point of origin, and their anticipated direction of movement and destination.
- Location of marshalling areas, collection points, and AAs.

7-80. If time permits, the platoon and squad leaders conduct a reconnaissance of the evacuation route. The purpose of the reconnaissance is to prepare an overlay which is then used to plan control and special security measures at critical areas along the route. If helicopter support is available, platoon and squad leaders conduct an aerial reconnaissance of the evacuation route. An aerial reconnaissance will normally take less time and the observation of the terrain around the route is much better by air than from the ground. At a minimum, a map reconnaissance must be conducted.

7-81. During the reconnaissance, the platoon leader selects possible rest areas and identifies friendly units along the route. He identifies critical areas, such as defiles, bridges, and areas where enemy activity is likely. He marks these areas on the overlay and plans for special security measures at these locations.

7-82. The platoon leader coordinates with the movement control center to ensure that the evacuation route does not interfere with the movement of friendly forces. If available, he coordinates for fire support along the route, normally at critical sites that were identified during the reconnaissance.

7-83. The platoon leader coordinates for MEDEVAC and vehicle recovery. His plan includes emergency reaction to the following:

- An air attack.
- An artillery attack.
- An ambush.
- A riot.

7-84. NEO are normally carried out according to the guidelines established by CA and US policy. MP may be required to screen for authorized personnel to determine who may actually be evacuated. There are generally three groups of personnel. They are—

- Group I. Group I includes US citizens, officials, dependents, tourist, business persons, and non-US family members when the father, mother or wife is a US citizen.
- Group II. Group II includes foreign nationals holding diplomatic papers, visas, or passports who receive Department of state approval.
- Group III. Group III includes all others, to include HN citizens, who do not fit into the first two categories.

7-85. Persons who fall into any of these groups and require immediate medical attention are always evacuated first. MP should give special consideration to the elderly and children.

7-86. MP brief the evacuees on discipline and the actions to take during emergencies. The briefing must be in a language that all the evacuees understand. The CA unit or HN agency should provide an interpreter. The briefing should cover all aspects of the evacuation. This will help calm the evacuees and instill confidence and cooperation.

7-87. The platoon provides security of the evacuees at the marshalling, evacuation, and holding areas and the reception station. The level of protection depends on the level of the threat. Methods of security depend on the type and location of the facilities used. At a minimum, MP must be prepared to provide interior guards for group areas, establish perimeter security, and operate a dismount point to restrict access to the evacuees.

7-88. During movement, the platoon escorts the evacuees by providing close-in security at the lead, middle, and end of the convoy or in front of and behind a single transportation vehicle. MP use a scout vehicle that travels 3 to 5 minutes ahead of the convoy to alert the main body to danger or delays. If a threat tries to disrupt the evacuation operation or destroy the evacuation vehicle, selected MP teams protect the evacuees as other teams counter the threat within the ROE. Throughout the evacuation operation, MP maintain all-around security, protection, and evacuee accountability.

Chapter 8

Law and Order

This chapter addresses the L&O function across the full spectrum of military operations. Refer to *Appendix K* for more information about civil disturbance control measures.

OVERVIEW

8-1. L&O operations consist of those measures necessary to enforce laws, directives, and punitive regulations; conduct criminal investigations; and control populations and resources that ensure commanders the existence of a lawful and orderly environment. MP enforce laws and appropriate commander directives. They maintain liaison and coordinate joint L&O operations with other DOD police organizations; HN military and civilian authorities; multinational police organizations; and US federal, state, and local police agencies. A coordinated law enforcement effort removes the conditions and opportunities that promote crime, thereby preventing diversion of military resources and maintaining military discipline.

8-2. The evolving criminal threat affects military operations and requires commanders to minimize that threat to preclude negative impacts on forces, resources, and operations. The importance of the criminal threat to military operations is a current, as well as future reality. Nationalist ideologies, the instability of a government to effectively govern and control its population, and the breakdown of government infrastructures will foster the linkage between criminal organizations and the government and its armed forces.

8-3. MP provide the capability to train foreign MP or assist in the reorganize of indigenous constabulary forces as part of stability and support post conflict operations. Under the provisions of and exceptions to *Section 660 of the Foreign Assistance Act*, MP provide initial assistance and training to foreign military and civilian police forces or assist in the creation of these forces where national authority has broken down. Additionally, MP forces provides short-term emergency L&O capabilities until the foreign military and civilian forces are functional.

8-4. MP and the USACIDC are the primary collectors of police information and criminal intelligence. They gather information through contacts made with the local populace and from conducting combined and joint patrols with HN military and civilian police agencies. MP and CID conduct field interviews and gather police information from surveillance operations. They investigate serious offenses and maintain a close liaison with the HN or allied civilian and military police agencies. This police information and criminal intelligence is collected, analyzed, and shared with the intelligence community and contributes to the police information assessment process (PIAP).

8-5. The police activities that support L&O operations include the following:

- Criminal investigations.
- Police information collection and dissemination of the information.
- Traffic accident investigations.
- Antiterrorism force protection support.
- Crowd control.
- US customs operations.
- Use of MWDs.

8-6. The enforcement of military laws, orders, and regulations is a command responsibility, as well as an MP responsibility. Each commander is responsible for maintaining order and discipline in the unit. To support commanders, the PM plans the use of MP assets to help enforce military laws, orders, and regulations.

8-7. MP performing L&O operations enhance and extend the tactical commander's C^2 by—

- Aiding commanders in maintaining combat strength.
- Helping prevent diversion of military resources.
- Suppressing opportunities for criminal behavior by US and non-US personnel or elements.
- Assisting and protecting military forces.
- Helping ensure the discipline of US forces.
- Assisting intelligence organizations in obtaining a true tactical intelligence picture by providing criminal and operational data and intelligence.

LAW AND ORDER AUGMENTATION DETACHMENT

8-8. The technical and supervisory expertise to support L&O operations in a mature theater is provided by L&O teams organized under an L&O augmentation detachment. The L&O teams are designed to give the commander the additional flexibility and capability in any environment to conduct a wide range of force protection mission requirements, to include split-based operations. This enables the supported commander to perform needed L&O missions while other MP assets are conducting MMS, AS, or I/R.

8-9. When these teams are not available or have not arrived in theater, L&O missions are prioritized with other MP combat support missions and are usually not full-scale dedicated operations. The theater commander determines when he can afford to dedicate MP assets to L&O operations.

8-10. A complete 45-person L&O augmentation detachment includes a detachment C^2 team, an operations team, a desk team, five MP investigation (MPI) teams, five traffic accident investigation (TAI) teams, and two force protection teams. Refer to *Figure 8-1*.

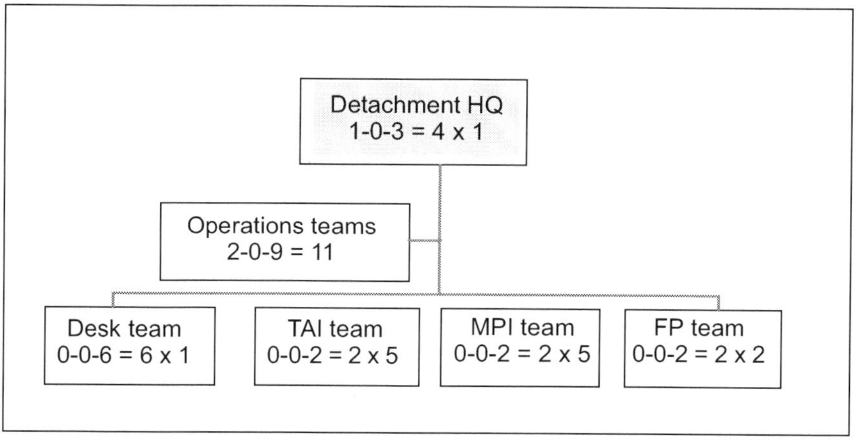

Figure 8-1. L&O Augmentation Detachment

8-11. When deployed, detachment and teams are under the overall operational C^2 of the PM deployed for the operation. Those teams may be attached to MP battalions and companies, as the situation requires. In war, the L&O augmentation detachment is assigned to an MP brigade and further attached to MP battalions and division MP companies as required. The L&O augmentation detachment consists of the following teams:

- Operations. The operations team provides C^2; supervises traffic, force protection, and investigative operations; and provides an evidence custodian to ensure that the chain of custody for evidence is maintained.

- Desk. A desk team establishes and operates the MP desk and coordinates law enforcement patrol activity.
- MPI. MPI teams provide the technical personnel to investigate crimes that do not fall within the investigative purview of the CID and conduct surveillance operations. Each MPI team consists of two MP with additional skill identifier (ASI) V5. These teams have no organic vehicles or communications capabilities.
- TAI. TAI teams provide the technical personnel to conduct TAIs. Based on operational requirements, TAI teams augment PM and MP organizations to ensure unimpeded movement and regulation enforcement on the MSRs. Each TAI team consists of two MP with ASI Q9.
- Force protection. Force protection teams provide technical personnel to assist in the safeguarding of personnel and equipment according to the command policies and field operating procedures. The role of the force protection team is to assist, inspect, and educate. Force protection teams provide expertise for safeguarding personnel and PRC for refugees, displaced persons, and civil disturbance operations. They enhance the unit's ability to ensure that soldiers and units are aware of appropriate personal security safeguards. Additionally, force protection teams assist customs operations teams in the establishment of sterile areas. Each force protection team consists of MP NCOs.

8-12. MP brigade commanders employ L&O augmentation teams based on the METT-TC to support the US forces operating within the brigade AO. While the number of teams is based on the troop population supported, whenever possible the team AO should coincide with the MP battalion and the division AO. The detachment HQ, operations, desk, and force protection teams should be collocated with the supported battalion HQ or division PM. If transportation requirements for these teams exceed the capabilities of the operations team, the supported unit must provide transport.

COMPANY AND PLATOON LEVEL LAW AND ORDER OPERATIONS

8-13. As previously stated, L&O augmentation teams may not arrive until a theater matures and L&O operations become the priority. Corps and division MP units must be prepared to conduct full-scale L&O missions. When ordered to establish L&O operations, the company commander determines mission requirements that include the following:

- Special equipment (vehicles, communications, and barricades or traffic cones).
- Specialized personnel support (a linguist, HN police, and PSYOPS or civil affairs personnel).
- CCIR.
- Threat assessments from MI and the CID.

8-14. The commander and platoon leaders conduct a map reconnaissance of the AO and determine the platoon areas of responsibility. When the tactical situation allows and adequate structures are available, the commander selects a building that is centrally located within the AO to establish the MP operations center. The facility should, at a minimum, offer areas for the following:

- An MP station desk.
- Offenders processing.
- Good radio transmissions.
- Arms, evidence, and property storage.
- Suspect detainment.

- Latrine facilities.

8-15. The company operations section operates the MP operations center and organizes it into MP sections that include—

- MP operations.
- Administration.
- Force protection.
- An MP desk.
- MP investigations.
- TAI.
- MWDs (if attached or assigned).
- Customs (if attached or assigned).

8-16. MP that are school-trained and have been awarded an ASI should be in charge of the appropriate section.

MILITARY POLICE OPERATIONS AND ADMINISTRATION OPERATIONS

8-17. The company operations sergeant is in charge of the day-to-day activities of the MP operations center. He instructs platoons regarding patrol patterns and distribution and coordinates for specialized equipment and personnel. He reviews all MP reports and the MP blotter. The center maintains close liaison with HN and allied military and civilian police agencies. The operations sergeant directs and supervises the administration section and provides guidance to other sections.

FORCE PROTECTION

8-18. The company physical security NCO (ASI H3) coordinates and directs the efforts of force protection for the commander. The force protection section is responsible for reviewing the security measures of all critical facilities and assets within the AO. The force protection NCO reviews points designated as critical by the senior commander and identifies criminal and operational threats. He develops a draft mission-essential or vulnerable area (MEVA) list and submits the list to higher HQ for approval. Once the list is approved, he conducts vulnerability assessments of points on the MEVA list. He applies risk analysis procedures according to *AR 190-51, AR 190-11*, and *DA Pamphlet 190-51* and ensures that appropriate army antiterrorism force protection standards are met according to *AR 525-13*. He reassesses assets for addition to or deletion from the MEVA list or changes in priority. The role of the physical security NCO is to assist, inspect, and advise commanders on force protection measures.

MILITARY POLICE DESK

8-19. The MP desk coordinates MP patrol activity, documents those activities, and reviews incident reports to ensure accuracy and completeness. The MP desk is also responsible for referring incidents to the appropriate investigative agency for further action.

8-20. The MP desk operates continuously during the conduct of L&O operations. It should be located as near as possible to the center of the L&O operational area, and have communications throughout the AO. When possible, existing facilities should be used for the MP station, but requirements for communications, transportation, and proximity to troops and facilities may rule out existing structures. The MP desk should be located within a reasonable distance of the HN police desk during stability, support, or other operations involving interaction with HN authorities. It may be necessary to establish substations if the area covered or the volume of activity is large. The MP station should have, at a minimum, the following:

- Areas and lighting for preparing reports and other paperwork.

- Communications with the MP elements performing L&O and the command and populace that the L&O operations are supporting.
- Detention areas where persons can be secured by a guard and/or physical constraints, such as a detention cell.
- Latrine facilities.

8-21. MP establish joint service, multiagency, or multinational operations when appropriate for the populace being provided L&O support or as directed. The PM establishes agreements before initiating joint operations to prevent conflicts in jurisdiction, documentation, and communication. Those agreements may be informal at first but should be formalized as soon as the situation permits. Refer to *FM 19-10* for more information about operating a MP desk.

MILITARY POLICE INVESTIGATIONS

8-22. Crimes occurring within an AO may require an in-depth investigation, depending on the seriousness of the offense and the availability of investigative personnel. Commanders report minor offenses within the unit area to the MP for statistical purposes, but may still investigate the offense themselves.

8-23. MPI investigators handle the most criminal offenses not investigated by CID or the unit commander. MPI normally investigate those offenses cited in *AR 190-30*. The commander ensures that only school trained (ASI V5) personnel are assigned to the MPI section to investigate crimes. Outside continental US (OCONUS) areas, MPI investigate off-post incidents according to SOFA and/or the US and HN agreements.

8-24. CID investigates serious offenses. It has elements in support of all echelons down to division level. CID operates across the full spectrum of army operations, with an emphasis on logistics security (LOGSEC) during wartime operations.

8-25. When MPI has purview to investigate a crime, they use the following investigation measures:

- Take control of the crime scene, ensuring that the crime scene is protected and secure.
- Identify the personnel involved, both suspects and witnesses.
- Identify the type of offense or offenses that have been committed.
- Process the crime scene using notes, sketches, and photographs.
- Collect and secure the physical evidence.

8-26. Once the investigator has processed the crime scene, he pursues leads and conducts investigation activities that include—

- Interviewing victims, suspects, subjects, and witnesses, and obtaining written statements from them.
- Preparing case documents and the required reports.
- Collecting related documents necessary to support the investigation.
- Coordinating with the Staff Judge Advocate (SJA).
- Submitting the final draft report for review by the MPI supervisor.

8-27. The commander appoints an evidence custodian to maintain evidence obtained during criminal investigations. The evidence custodian coordinates the needed laboratory examinations and ensures the proper disposition of evidence. It may be necessary to coordinate with the CID for this support.

8-28. Joint investigative activities, such as drug or black market suppression teams, may involve investigators from CID, MPI, or other US services or agencies. Such operations may require close liaison and cooperation with HN or allied civil or military agencies.

8-29. Refer to *FM 19-20* for more information about criminal investigation techniques and procedures.

TRAFFIC ACCIDENT AND INVESTIGATION

8-30. MP help reduce nonbattle casualties and the loss of equipment by ensuring that vehicles are operated according to regulations and determining the cause of accidents. Traffic enforcement supports the commander's intent in HNs by reducing the likelihood that military traffic will be seen as a threat to the native population. Thorough investigations of traffic accidents can—

- Identify the personal, environmental, and equipment factors that caused or contributed to the accident.
- Document the facts of the incident for future criminal or civil actions.

8-31. Traffic enforcement measures vary according to the operational and political environment. Those measures may include the following:

- Operator license and dispatch checks.
- Safety inspection checkpoints.
- Vehicle load and route restrictions enforcement.
- Speed control measures.

8-32. Traffic accidents are investigated on the scene whenever the tactical situation permits. MP investigate traffic accidents involving military personnel or vehicles within their AO. When the HN or other authorities have primary responsibility for an accident investigation involving US forces or equipment, MP conduct a concurrent investigation.

8-33. The first MP that responds to a traffic accident is responsible for securing and protecting the accident scene. When necessary, they administer medical attention to the injured persons and implement traffic control measures. MP safeguard classified materials and take the appropriate action to identify and contain hazardous materials at the scene.

8-34. The TAI team takes control of the accident scene when it arrives. Whenever possible a school-trained (ASI Q-9) traffic accident investigator processes the accident scene.

8-35. Once the TAI team receives a briefing from the senior person present, it takes the following steps to investigate the accident:

- Collect physical evidence to include evidence of drug or alcohol involvement.
- Photograph the scene.
- Interview victims and witnesses.

8-36. Once the TAI team has the necessary information, it clears the scene. To do this it ensures that the vehicles and debris are removed and reestablishes traffic flow. The TAI team then prepares a final report and submits it to the proper authority.

8-37. Accidents involving fatalities or life-threatening injuries require a serious incident report (SIR) according to *AR 190-40*.

8-38. Refer to *FM 19-25* for more information about traffic enforcement techniques and conducting TAIs.

MILITARY WORKING DOGS

8-39. MWD teams further enhance L&O and customs operations with capabilities to detect explosive devices and residue and controlled substances, and detect, track, control, and apprehend personnel.

8-40. Mission support requirements determine the number of functional MWD teams that make up a wartime, mission-oriented MWD team. These mission-oriented MWD teams are assigned to the PM. MWD teams conducting explosive or narcotic detection are unable to provide security for themselves and require security by the supporting unit.

8-41. Explosive, narcotics, and patrol teams each consist of three handlers and three working dogs. This allows each team to provide 24-hour support for a mission that requires one MWD or up to three short-duration missions. There is also a kennel master team, which provides technical supervision and is responsible for establishing kennel operations.

8-42. The explosive or patrol team provides the capability to detect explosive devices or residue in support of personal protection, MOUT, health and welfare, crime scene, and customs operations. When not required for explosive detection, this team functions as a patrol team.

8-43. The narcotic and patrol team provides the capability to detect controlled substances in support of the crime scene, health and welfare, and customs operations. When not required for narcotics detection, the team functions as a patrol team. The patrol team provides the capability to search for, track, detect, and control personnel and augment security in support of personal protection, MOUT, criminal apprehension, force protection, EPW, refugee or displaced person, and crowd control operations.

8-44. Refer to *AR 190-12* for more information about the MWD program, to include standards for selecting and retenting handlers and dogs, training teams, constructing kennels and training areas, maintaining and disposing of dogs, using teams, and using force. Refer to *DA Pamphlet 190-12* for more information about the MWD program.

UNITED STATES CUSTOMS SUPPORT

8-45. Units (and individuals) returning to the customs territory of the US (CTUS) have the responsibility to ensure that their vehicles and equipment meet US customs, other governmental agencies, and US Department of Agriculture (USDA) standards. Units request and receive customs support to ensure that they comply with all the established requirements. Trained military personnel, normally from the unit deploying to CTUS, conduct military customs inspections in the absence of US customs personnel. Senior military customs inspectors (SMCIs) train and supervise unit personnel on custom's responsibilities and requirements. SMCIs supervise the customs operations. The SMCIs supporting the theater deploy as part of the customs operations team.

CUSTOMS OPERATIONS TEAM

8-46. The customs operations team supports the commander's requirement to ensure that personnel, equipment, and materiel meet US customs, immigration, USDA, and other federal agency requirements for redeployment back to the CONUS. A complete customs operations team includes an SMCI supervisor team and two 5-member SMCI inspection teams.

8-47. The customs operations team, when deployed, is assigned to the Army service component command (ASCC) PM. It may be under the OPCON of or attached to the theater support command (TSC) PM.

8-48. The SMCI supervisor provides technical supervision, staff planning, and coordination for mission requirements with the Army staffs and agencies, redeploying units, and US federal agencies.

8-49. Two SMCI teams provide the capability to support two ports of embarkation (POEs). Before arrival at the

POE, these teams advise deploying units on requirements and procedures to preclude delays or rescheduling of unit and strategic lift transportation. Team members train and supervise selected redeploying unit personnel as inspectors to augment the SMCI team.

REDEPLOYMENT OPERATIONS

8-50. *AR 700-93* provides guidance on processing and shipping DOD-sponsor retrograde cargo destined for the CTUS. In order for a unit to redeploy successfully, it must—

- Coordinate for site requirements and standards.
- Establish wash and holding areas.
- Determine the existing USDA preclearance requirements.
- Use the expertise of assigned SMCI.

Site Requirements

8-51. A site includes those areas necessary to getting the unit's vehicles and equipment ready for the customs inspection and shipment. A suitable site includes the following areas:

- Staging.
- Download.
- Vehicle or equipment wash area.
- Inspection.
- Holding (sterile).

8-52. **Staging Area.** A staging area is an open area large enough to handle the number of vehicles and equipment being processed. Units often underestimate the amount of space needed for the staging area. An accurate count of vehicles and personnel redeploying to the CTUS must be received to accurately determine the staging area requirement.

8-53. **Download Area.** The download area is an area that must have a base of cement, asphalt, or clean or coarse gravel at least 6 inches deep. Vehicles arriving from the staging area download equipment here. The download area breaks out equipment and vehicles for wash area activities. Vehicles are taken to the vehicle wash area and equipment is taken to the equipment wash area.

8-54. **Vehicle and Equipment Wash Area.** A vehicle and equipment wash area or wash site consists of a hardstand area with adequate drainage, water outlets, high-pressure steam cleaners, and ramps. Most wash sites resemble a birdbath using long tubes with holes in them. A vehicle passes by the tubing while it pressure sprays the sides and undercarriage of the vehicle. Unit personnel clean vehicles to free them of dirt, vegetation, insects, and so forth, and check for and clean up fluid spillage. Equipment is cleaned meticulously, ensuring that there is no debris or soil present. When vehicles and equipment are assumed to be clean, they are reloaded and moved to the inspection area.

8-55. **Inspection Area.** The inspection area is where the vehicles and equipment are inspected for cleanliness. If vehicles and equipment are clean, they are moved to sterile holding areas. If they are dirty, they are moved to a spot-wash area and spot-cleaned. Vehicles and equipment that were spot-washed are reinspected, and if clean, go to the sterile holding area.

8-56. **Holding Area.** Cleaned vehicles and equipment are placed in the holding area. Unit personnel guard the holding area to ensure that unit members or other personnel do not contaminate or place restricted or contraband items in vehicles or equipment. The holding area must be free of USDA concerns. Some of these concerns include the following:

- Weeds. The download area must be vegetation free.

- Soil. The download area must be completely free of soil and soil particles.
- Pests. Insects and their debris or excrement are a major concern in the download area. Ensure that insects are kept out of the download area.

NOTE: SMCIs assist units in identifying and eliminating USDA concerns.

8-57. Vehicles and equipment in the holding area will be loaded onto rail, ship, or other means of transport for redeployment.

MULTINATIONAL LAW AND ORDER OPERATIONS

8-58. Multinational police operations may involve joint patrols with allied forces, HN military or civilian police, or a combination of these options. Commanders may also choose to operate certain patrols or stations with MP forces of one nationality, with all police operations answering to a combined, multinational operations cell. Because each nationality has its own statutes, regulations, policies, and traditions, police forces should enforce laws only on persons normally under their own jurisdiction. This reduces confusion about authority and the potential ill feelings that can result when citizens are controlled by other foreign forces. If a situation occurs involving HN or third-nation citizens, the MP should contain or monitor the situation until the appropriate police agency can respond to control the situation. MP may be required to control the situation (consistent with ROE or ROI) if the appropriate agency is not reasonably available or the situation is so severe that a delay would endanger lives.

8-59. *The Foreign Assistance Act of 1961* prohibits the military from training HN civilian police. However, MP may be directed to provide initial assistance and training to foreign military or civilian police forces or assist in the creation of these forces, where the local national authority has broken down. This is an exception to *Section 660 of the Foreign Assistance Act.* MP, when directed, conduct training for foreign military police and assist in reorganizing local constabulary forces during stability and support operations. MP provides short-term emergency L&O capabilities until the foreign military or civilian police forces are functional.

8-60. US military training assistance of foreign military or civilian police normally includes the following:

- Special forces. These detachments teach basic skills needed to perform security duties such as small arms training, unit security tactics, radio procedures, and human rights observation and safeguard.
- Military police. MP teach investigative techniques, law enforcement skills, and confinement operations.
- Civil affairs. CAs coordinate with the HN to effectively integrate training with the established political system. They also provide language interpreter support.
- Psychological operations. PSYOPS personnel work to gain and maintain acceptance of the training effort by the civilian population and encourage their acceptance of the new or reconstituted civilian police force.

8-61. The US ambassador and country team provide guidance for implementing all US Department of Justice (USDOJ) training, including police training. US military forces conduct training under the direction of the civilian authorities, and only until the International Criminal Investigative Training Assistance Program (ICITAP) staff or another agency can assume training. Contractors working for ICITAP, USDOJ, or another international agency may also provide instruction and assistance, and frequently assume those duties from the military as the situation matures. US military involvement in training foreign military or civilian police should not normally exceed 180 days following the end of a conflict.

8-62. MP train HN military or police forces to maintain L&O, using current operational and investigative techniques. They may also act as the agents to equip those forces with the material needed in police operations and train HN personnel on the use and maintenance of that equipment. The training provided may be formal or informal instruction, or consist of on-the-job training with MP working with, and monitoring the performance of, HN police authorities. Classroom instruction is based on training packages developed locally or by higher HQ, or exportable training packages from a service school.

8-63. Once the basics of law enforcement have been learned or reinforced, new or reconstituted police may benefit most from working together with MP in joint operations. ROE and ROI may vary by situation, but some basic guidelines for working with newly constituted or reconstituted police agencies are—

- Allowing the HN authorities to handle situations involving local nationals. US forces assists HN police and monitor the situation. US forces should not appear to be giving orders to the HN agents.
- Ensuring that US forces take care that they do not embarrass or demean HN agents, especially in view of the populace. Soldiers must understand that negative behavior toward their HN counterparts could seriously damage the relationship between US forces and the HN authorities.
- Ensuring that US forces are aware of and sensitive to the cultural traditions and

standards of the HN. What may be acceptable in one culture may be, at best, ill-mannered and possibly, a serious insult or even criminal behavior in another culture.

- Ensuring that US forces appear publicly to be respectful of its agents and leadership and be confident in their abilities. This helps the HN force be effective and respected by the local populace.

8-64. As HN authorities develop professionally and gain the confidence of the HN population, the role of US forces changes. MP shift from being the sole providers of L&O, to being role models and advisors for local authorities, to being monitors of HN police activity.

8-65. Combined police operations reinforce training and provide HN police with the mobility, security, and communications to operate more effectively while conserving critical personnel resources. MP provide support through training and the following, usually combined, operations:

- Law enforcement, security, and criminal information support to the indigenous police force to include L&O operations and administration in a democratic society.
- Patrol and desk.
- Traffic control.
- PRC and civil disturbance.
- MWD.
- Physical security.
- Personnel security.
- Area and route security.
- Counterdrug.
- Antiterrorism.

8-66. In most cases, US forces will continue to exercise jurisdiction over their own forces, especially for offenses of a purely military nature. Local government, once reestablished, may choose to assume jurisdiction in serious or high interest offenses involving US personnel or according to the applicable SOFA or other agreement. MP, in the absence of competent civil authority, establish control over the movement of personnel and supplies and guard critical food supplies and material during production and storage. As the local police and paramilitary forces become effective and assume more security responsibilities, MP elements reduce their activity. They withdraw their support gradually, ensuring that the HN has adequate resources to carry out ongoing programs.

Chapter 9

Police Intelligence Operations

The PIO support, enhance, and contribute to the commander's protection program, situational awareness, and battlefield visualization by portraying relevant threat information that may affect the operational and tactical environment. MP and the CID gather police, criminal, or combat threat information during the performance of their functions.

OVERVIEW

9-1. MP collect and share information during the conduct of their day-to-day operations. This information, which may be police, criminal, or combat, is provided as input to the intelligence collection effort and turned into action or reports. The PIO function ensures that information collected during the conduct of MMS, AS, I/R, and L&O is reported up through the proper channels so that it can be analyzed. MP perform PIO while conducting combat support tasks, such as—

- Checkpoints and roadblocks.
- TCPs.
- Field interviews.
- Criminal investigations.
- Reconnaissance (zone, area, and route).

9-2. Collecting police information during MP activities under the functions of MMS, AS, and I/R may result primarily in combat information, but may lend to police or criminal information. Likewise, the police collection efforts during full-scale L&O operations may result in combat information. MP apply lessons learned from peacetime, conflict, and war environments to the PIO function to gain the most advantage for performing the function. The MP platoon is capable of operating in all functions, day or night, and in various terrain, weather, and visibility conditions. An MP platoon operates independently over large, dispersed areas. MP, along with MI, engineers, and NBC reconnaissance, are key to the IPB (refer to *FM 34-130* for more information about IPB). They are collectors of information in the rear area during sustaining operations or anywhere throughout the battlefield. MP gather police information from contacts that are often very valuable in substantiating or verifying other sources of information. These sources include the following:

- Daily contact with the local populace.
- Combined police patrols with HN military and civilian police agencies.
- Close liaison with local, HN, and multinational police agencies.
- Field interviews.
- Nongovernmental organizations (NGOs).
- Private volunteer organizations (PVOs).

9-3. Refer to *FM 3-19.1* for more information about the processes discussed in this chapter.

POLICE INFORMATION ASSESSMENT PROCESS

9-4. The PIAP is a tool used to contribute to the PIO function. Information gained through the PIAP may contribute independently or simultaneously to the all-source analysis product (ASAP) and the IPB process. The PIAP, the ASAP, and the IPB enhance and support the commander's force protection program, situational awareness, and battlefield visualization. However, the PIAP independently or collectively—

- Provides the commander with information necessary to improve measures to protect the forces.
- Provides information that clarifies the threat and operational situation.
- Reduces opportunities for threat forces to disrupt military operations and inflict US or friendly casualties.

9-5. The commander and his staff continually monitor the environment at the tactical level consistent with the METT-TC. They apply the military aspects of terrain (OCOKA) as a means of protecting the force. PIO clarify the evolving criminal threat picture for commanders through the PIAP. This helps planners predict threat courses of action against our forces or protected populations.

9-6. MP use PIAP to continuously collect, organize, interpret, and report police and criminal information in support of the IPB. The PIAP consists of 6 steps. Refer to *Table 9-1, pages 9-4 and 9-5*.

POLICE INFORMATION

9-7. MP gather police information actively or passively. Active collecting efforts result from a direct tasking, and passive collecting efforts result from normal, daily MP operations.

ACTIVE MODE

9-8. MP perform the PIO function in the active mode when directed by the higher HQ. In this mode, the MP platoon conducts specific missions with the *intent* to actively collect information. Specific MP activities (such as setting up a checkpoint or roadblock) are performed to specifically fulfill a requirement (such as looking for individuals who are in possession of or have knowledge of others in possession of weapons, US military property,

Table 9-1. Police Information Assessment Process

Step	Action
1.	Determine the scope of the PIAP by— • Conducting a detailed mission analysis. • Reviewing the mission of the higher HQ and the commander's intent. • Reviewing the CCIR. • Reviewing mission priorities. • Determining the AO and area of interest (AI). • Reviewing the IPB estimates of the higher HQ. • Determining the required information products.
2.	Assemble the working aids and— • Post the applicable maps. • Acquire crime statistics and other related data. • Obtain language aids, such as cultural references and interpreters. • Acquire the necessary automation equipment.
3.	Determine information requirements and coordinate access to data by— • Coordinating with the CID for access to available criminal information developed by CID programs. • Determining police and criminal gaps in the designated geographical area. • Determining the requirement for information collectors to fill police and criminal gaps. • Coordinating with the Intelligence Officer, US Army (S2), the Assistant Chief of Staff, G2 (Intelligence), PSYOP, and other agencies to determine if the information is already available.

Table 9-1. Police Information Assessment Process (Continued)

4.	Recommend and supervise police and criminal collection efforts by— • Coordinating with the S2, the G2, the SJA, the CID, and other applicable police agencies before initiating a collection effort. • Determining which police and criminal information tasks will be assigned to the MP and the CID. • Determining the criteria to satisfy information requirements. • Providing collectors with reporting instructions (such as the reporting frequency and the report format). • Monitoring information collection efforts to prevent duplication of efforts.
5.	Process police and criminal raw data by— • Assembling and assessing the reliability of the data (according to the assessment criteria established above). • Integrating information from the collectors (the MP and CID agents in the field). • Evaluating the data to determine if it meets the requirements. • Developing criminal trends and indicators.
6.	Report and disseminate police and criminal information assessment by— • Recommending MP and CID actions to improve and focus future collection efforts. • Reporting police and criminal information assessments to the S2, the G2, and other MP units. • Reviewing and ensuring that the release of police and criminal information assessments do not violate established guidelines and constraints. **NOTE: Refer to *Appendix E of FM 3-19-1* for a further discussion of PIAP.**

and so forth). MP gain valuable police, criminal, and combat information while conducting these operations. A standardized checklist enhances the information collection effort and aids in the analysis of the information collected. The checklist indicates a pattern in the behavior of the local nationals. It shows what the local nationals are transporting, to where they are transporting the items, and so forth. The checklist may include—

- The number and types of vehicles stopped. Identifying marks, license plate numbers, and any signs displayed on the vehicles are recorded and reported.
- The number of passengers in the vehicle. The nationality, age, and sex mix of the passengers are recorded and reported.
- The type and quantity of cargo.
- The vehicle's point of origin and destination.
- The stated reason for travel by the passengers.
- The description of arms, ammunition, explosives, and sensitive items found and confiscated from the vehicle.
- The possible or actual sightings of weapons, explosives, or threat forces by the passengers.
- The condition of the passengers.
- The reporting of anything unusual by the passengers.

PASSIVE MODE

9-9. Every MP conducts the PIO function in the passive mode during their normal day-to-day operations. In the passive mode, PIO are not a stand-alone function and, as such, it cannot be separated from the other MP functions. If while performing MMS, AS, I/R, and L&O, MP receive, observe, or encounter police, criminal, or combat information, they immediately submit a SALUTE, SPOTREP, or other appropriate report to relay information up the chain of command. The information is then integrated into the on-going PIAP and forwarded to the higher echelon S2 and G2 for IPB applications.

RESPONSIBILITIES

9-10. The responsibilities for the PIO function start with the MP company. The company has the overall responsibility of supervising the collecting and reporting of information to the higher HQ. (Refer to *FM 34-2* for more information about the collection management process). Training scenarios can be developed to enhance the PIO information collecting process of both the active and passive mode. Additionally, the scenarios stress collective training (such as processing police information) and individual training (such as improving interpersonal communication and interview skills).

PLATOON LEADER

9-11. When the platoon leader is tasked to conduct PIO collecting and reporting, he initiates TLPs for the mission. The platoon leader—

- Coordinates with the higher HQ for the CCIR, police and criminal information requirements (PCIR), and threat estimates. The PCIR allows additional focus on local threats.
- Establishes liaison with civil and MP forces and law enforcement agencies in the AO.
- Coordinates with the local and HN police to determine the existence of organized crime in the AO and the identification of current and emerging criminal leaders and associates.
- Coordinates with the local and HN police and the populace to identify the types of criminal activity (such as smuggling, counterfeiting, narcotics, extortion, and so forth) in the AO.
- Reports information of potential intelligence value by—

- Gathering collected information from squads and teams.
- Consolidating original reports and sketches of potential intelligence value and forwarding them through intelligence channels for analysis and a determination of their intelligence value.

PLATOON MEMBERS

9-12. The platoon sergeant, squad, and team leaders, down to the junior member of the platoon share in seeking out sources of information by—

- Collecting R&S information that supports the commander's overall PIOR and PCIR collection efforts while on a routine patrol. They accomplish this by—
  - Noting terrain information pertaining to streets, roads, canals, subterranean systems, built-up areas, cities and villages, and the impacts of weather on the terrain.
  - Collecting information on progovernment and antigovernment individuals and groups who might disrupt L&O during protests, strikes, riots, and other spontaneous or organized efforts.
  - Identifying private establishments that may be a target or whose presence or operations contribute to the disruption of L&O (such as gun shops, pawnshops, and liquor stores).
  - Identifying critical infrastructures such as power stations, water works, radio and television stations, telephone and communication facilities, public transportation, and other establishments that may be critical to the sustenance of the community.

- Identifying EPWs, stragglers, and DCs who may have information of potential intelligence value and reporting it to the chain of command. Use a SPOTREP, SITREP, or a SALUTE report or a format directed by the chain of command to report information. SOPs may also be used. Refer to *Appendix D* for report formats

Appendix A

Metric Conversion Chart

A-1. Use *Table A-1* to convert from US to metric measurement and metric to US measurement.

Table A-1. Metric Conversion Chart

US Units	Multiplied By	Equals Metric Units
	Length	
Feet	0.30480	Meters
Inches	2.54000	Centimeters
Inches	0.02540	Meters
Inches	25.40010	Millimeters
Miles (statute)	1.60930	Kilometers
Miles (nautical)	1.85320	Kilometers
Yards	0.91400	Meters
	Area	
Square inches	6.45160	Square centimeters
Square feet	0.09290	Square meters
Square yards	0.83610	Square meters
	Volume	
Cubic inches	16.38720	Cubic centimeters
Cubic feet	0.02830	Cubic meters
Cubic yards	0.76460	Cubic meters
Gallons	3.78540	Liters
Fluid ounces	29.57300	Milliliters
Quarts	0.94600	Liters

Table A-1. Metric Conversion Chart (Continued)

US Units	Multiplied By	Equals Metric Units
Weight		
Ounces	28.34900	Grams
Pounds	453.59000	Grams
Pounds	0.45359	Kilograms
Short tons	0.90700	Metric tons
Long tons	1.01600	Metric tons
Metric Units	**Multiplied By**	**Equals US Units**
Centimeters	0.39370	Inches
Meters per second	2.23700	Miles per hour
Millimeters	0.03937	Inches
Kilometers	0.62137	Miles (statute)
Kilometers	0.53960	Miles (nautical)
Meters	3.28080	Feet
Meters	39.37000	Inches
Meters	1.09360	Yards
Area		
Square centimeters	0.15500	Square inches
Square meters	10.76400	Square feet
Square meters	1.19600	Square yards
Volume		
Cubic centimeters	0.06100	Cubic inches
Cubic meters	35.31440	Cubic feet
Cubic meters	1.30790	Cubic yards
Milliliters	0.03380	Fluid ounces
Liters	1.05700	Quarts
Liters	0.26420	Gallons
Weight		
Grams	0.03527	Ounces
Kilograms	2.20460	Pounds
Metric tons	1.10200	Short tons

Appendix B

Media Relations

Military operations are carried out, especially in the initial phases, under the full glare of public scrutiny. The press can distribute reports and pictures faster than the news can be released by the command by using satellites and modern communications technology. Incidents, sometimes embroidered or slanted towards a partisan viewpoint, are screened on the television the same day and in the countries that are parties to the dispute and their allies.

OVERVIEW

B-1. Journalists fall back on speculation when information is withheld. Such speculation, although usually inaccurate, is often near enough to the truth to be accepted as such by large sections of the public and even by the governments. Belligerents find it advantageous to leak part of a story to the press to build public support for their position. On occasion, such activities can grow into a fully orchestrated press campaign.

MEDIA INTERACTION

B-2. Establish procedures for media interaction and incorporate them into SOPs. Know what (such as a simple theme which they can tie their responses back to) to discuss with the media and what to refer to the PAO. Ensure that deploying units receive a predeployment briefing from the installation PAO to clarify what may or may not be discussed with the media. Ensure that this action occurs before any exposure to the media. Do not make off-the-record statements in briefings or discussions with the media. Public or media knowledge of any classified activity associated with an operation does not imply or mean that the information is unclassified or may be released or confirmed.

B-3. Ensure that MP are courteous, respond candidly, speak effectively, and continue the mission when interacting with broadcast and print reporters and photographers. This relationship with the media prevents any misgivings. The media will report more accurately and not feel that information is being withheld. The media provides its audiences with real-time information of varying accuracy and completeness.

B-4. The media can be a valuable tool or a lost asset during military operations. It can potentially have a quick and pervasive impact on the plans and operations of commanders, and can change the public's opinions and perceptions about military operations. It can target audiences whose support is crucial to the desired end state of an operation. MP may be the centerpiece for humanitarian assistance or during operations to relieve suffering. A reporter or a photographer may show up at a checkpoint or a TCP where local nationals (belligerents) are refusing to cooperate. Ensure that the media is not only documenting the activities and behavior of the belligerents, but is also recording the military's response to the incident.

Appendix C

Training Execution Model

The TEM allows numerous rehearsals and identification of the TTP required for mission accomplishment. It causes each soldier, regardless of position, to understand the unit leader's plan with contingencies and the TTP required. It allows junior leaders and soldiers to see the big picture and understand the tactical environment. Place the company as the OC for the subordinate platoons, and place the platoon leader as the OC for the subordinate squads during internal evaluations.

OVERVIEW

C-1. The TEM ensures that soldiers are trained to standard, not to time. It is derived from train-up or certification exercises (CERTEXs) before deploying to the National Training Center (NTC).

TASK IDENTIFICATION

C-2. The TEM takes place after the METL is approved. The approved METL, coupled with an assessment process (such as external or internal evaluations), determines the collective and individual tasks to be covered during training exercises. Once these tasks are identified, implement the following eight-step TEM:

Step 1. Back brief number 1.

Step 2. Back brief number 2.

Step 3. Rock drill.

Step 4. Subunit leaders walk-through.

Step 5. Subunit leaders walk-through with troops.

Step 6. Dry run on the actual lane.

Step 7. Execution.

Step 8. After-action review (AAR).

C-3. The TEM steps complement but do not replace the TLPs. The following outline explains the eight-step model applicable to an MP platoon. The TEM assumes that the platoon's collective task lists are approved at the appropriate level, the OCs are certified and trained, and all resources are available.

STEP 1: BACK BRIEF NUMBER 1

C-4. During step 1 *(Figure C-1),* the OC issues the mission order to the platoon leader. He briefs the platoon's mission using a sand table, which the platoon built. The platoon leader provides a mission back brief to the OC. The back brief eliminates any confusion or misunderstanding at the very beginning of the mission or plan cycle. The platoon leader's back brief is only with the OC. This allows the interaction between the OC and the platoon leader to take place while leaving the platoon leader's credibility intact. With no subordinate present, the environment to learn and ask questions is established.

C-5. In a nearby AA, leaders train soldiers on critical individual soldier tasks that support the platoon's collective tasks. Here, the PSG checks on the focus and progress of those individual tasks that have been assessed as weak. Additionally, AA actions in step 1 focus on PCIs.

STEP 2: BACK BRIEF NUMBER 2

C-6. During step 2 *(Figure C-2, page C-4),* the platoon leader returns to the sand table area where the mission order is received from the OC. He briefs the tentative

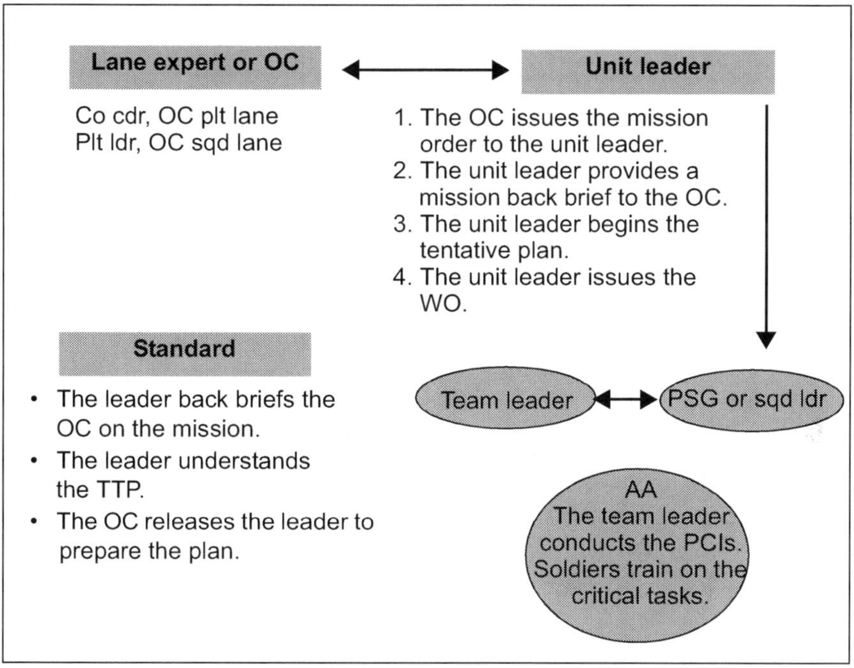

Figure C-1. Back Brief Number 1

plan to the OC. Once the OC verifies the MTP standards and pertinent TTPs, the platoon leader briefs squad leaders on the same sand table. The OC remains present to hear the interaction among the platoon's leadership. After the squad leaders are briefed, the platoon leader conducts the leader's reconnaissance. If any modification to the tentative plan is required based on the reconnaissance, the platoon leader briefs the OC. He briefs the squad leaders again if changes are approved. The interaction and planning get more focused after the leader's reconnaissance.

STEP 3: ROCK DRILL

C-7. During step 3 *(Figure C-3, page C-5)*, the platoon leader and squad leaders conduct a rock drill on a

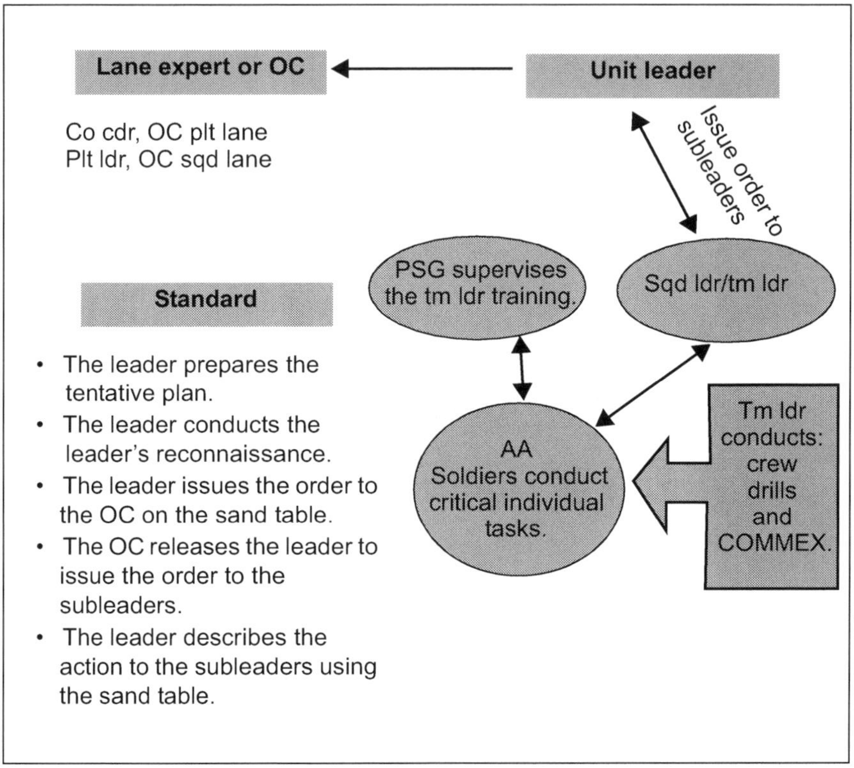

Figure C-2. Back Brief Number 2

terrain model that the platoon constructed of the actual training lane. The OC is in an overwatch position. He always assesses the interaction of the leadership and serves as the OPFOR for the rock drill. The platoon leader talks or walks through the plan on the terrain model with all the squad leaders.

C-8. In reaction to the OPFOR, leaders solidify contingencies based on the interaction on the terrain model. The OC's "what if" drills will usually trigger detailed planning. The credibility of the platoon leadership is still intact because the interaction and pertinent questions are not being asked in front of the soldiers. The soldiers are in the AA continuing their individual training and preparing equipment based on the WO received. Locate the PSG in the AA to coach team leaders in their AA activities. Once the OC is sure that the platoon leaders understand the plan, he releases them to conduct a walk-through.

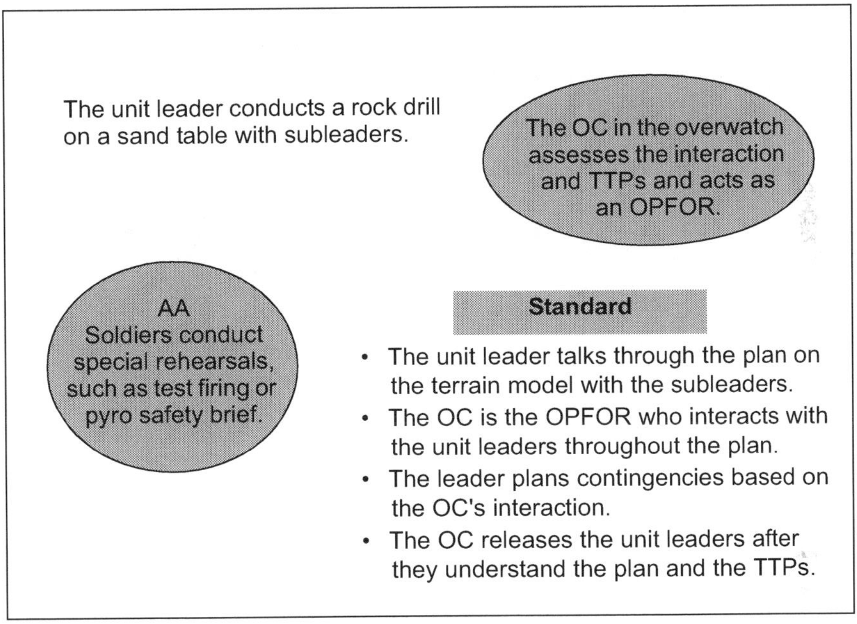

Figure C-3. Rock Drill

STEP 4: SUBUNIT LEADER'S WALK-THROUGH

C-9. During step 4 *(Figure C-4, page C-6)*, the platoon leader displays all terrain features and control measures. During the walk-through, the platoon leadership imitates the movement selected, such as traveling, traveling overwatch, and bounding overwatch. The platoon leadership uses actual radio procedures and call signs according to the platoon's plan. The platoon's leadership demonstrates its grasp of the plan and the mission without troops, showing that credibility is still sound.

C-10. Once the OC is satisfied that the platoon's leadership knows the tasks, he releases them to brief the soldiers on the sand tables previously mentioned. Squad leaders ensure that their team leaders and soldiers understand the mission by showing them on the sand table what their part of the mission is and how it fits into the overall scheme of maneuver.

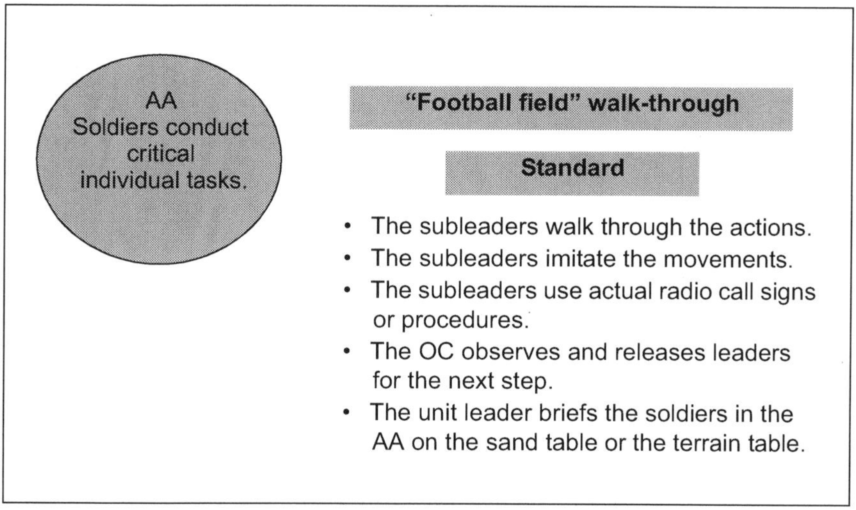

Figure C-4. Subunit Leader's Walk-Through

STEP 5: SUBUNIT LEADER'S WALK-THROUGH WITH THE TROOPS

C-11. During step 5 *(Figure C-5),* the soldiers are assembled and briefed on the mission using sand tables. The MP teams walk and talk through the mission. The platoon walks through its action of the plan; and the soldiers react and dress on their team, squad, and platoon leaders. Shout commands out so that all can hear. Imitate movement as well as radio procedures, call signs, fire commands, calls for suppression, and so forth. The OC who plays the OPFOR and interacts with the platoon through its leaders observes the platoon. Often, while conducting the walk-through with the troops it will be necessary to stop and inform soldiers of any risk hazards and individual responsibilities. The platoon's interaction on the field is a critical component of mission accomplishment. The OC only releases the unit for the next step if he is satisfied that the unit understands the leader's plan and the required TTP.

Issue warning OPORD

Subleaders walk-through

All now in final rehearsal

- The subleader walks through the actions with troops on the sand table.
- The troops act on the leader's commands.
- The subleader imitates movements.
- The subleader uses actual radio call signs or procedures.
- The OC observes and releases the leaders for the next step.

Figure C-5. Subunit Leader's Walk-Through with the Troops

STEP 6: DRY RUN ON THE ACTUAL LANE

C-12. During step 6 *(Figure C-6, page C-8),* the unit follows the leader's plan and conducts a dry run of the mission on the actual training lane. The OC observes to ensure that the platoon follows its plan and conducts the mission to MTP standards. The dry run identifies weak subtasks and allows the platoon to retrain or repeat the dry run. This is a full dress rehearsal and ensures that the platoon is meeting the standard. If the platoon performs satisfactorily on the dry run, the OC releases it to execute at full combat speed.

The unit conducts a dry run of the mission following the leader's plan on the actual lane.

Actual lane dry run.	Standard
	• The unit conducts the mission to the MTP standard. • The unit identifies the weak subtasks. • The unit identifies the terrain-induced weakness to the plan. • The unit retrains the weaknesses. • The unit does the lane again. • The OC releases the unit to execute at the full combat speed if satisfied with its performance.

Figure C-6. Dry Run on the Actual Lane

STEP 7: EXECUTION

C-13. During step 7 *(Figure C-7),* the unit executes the mission at full combat speed against an unrestricted OPFOR that is controlled by the OC. This phase involves force on force, controlled by the OC, with a complete simulation system (such as multiple-integrated laser engagement system [MILES]). The OC ensures that soldiers adhere to MTP standards and TTPs. The OC calls for an end of exercise (ENDEX) when the training lane has run to culmination (such as the mission is accomplished and all the training effects took place). The OC has enough information to assess the platoon's performance and initiate the next step—the AAR.

The unit executes the mission.

Standard

Actual lane

- The unit conducts the mission to the MTP standards.
- The unit has full use of OPFOR and MILES.
- The unit uses force on force or is unrestrained.
- The OC controls the OPFOR.
- The OC calls for an ENDEX.

Figure C-7. Execution

STEP 8: AFTER-ACTION REVIEW

C-14. During step 8 *(Figure C-8, page C-10),* the unit conducts an AAR of the lane. If possible, it conducts the AAR on the objective site that overlooks the lane. The OC facilitates the AAR. He or the platoon leader states the tasks, condition, and standards for the training objective. The OC ensures that the AAR becomes a review that is driven from the soldier level up toward the top. He elicits the soldiers to identify their actions, both right and wrong, and gets the leaders to do the same. The OC states what was right and wrong with the mission according to the MTP and the appropriate TTPs, always striving to leave the AAR on a positive note to improve soldier and unit morale and performance.

The unit conducts an AAR of the training event.

Standard

- The OC facilitates the AAR.
- The unit members state the training objectives (task, condition, standards).
- The unit members have a bottom-up discussion.
- The unit members have the soldiers identify their mistakes.
- The unit members have the leaders identify their mistakes.
- The OC assesses the task according to the MTP.

NOTE: The assessment dictates whether to go to the next training or redo the training.

Figure C-8. AAR

Appendix D

Orders and Reports

Orders are written or oral communications directing action. They are based on plans or the receipt of a new mission. Use reports extensively to provide information to higher, lower, and adjacent commands. Although there are many, this appendix is limited to combat orders and reports. Refer to *FM 101-5* for more information about orders.

ORDERS

D-1. Combat orders pertain to strategic, operational, or tactical operations and their service support. The combat orders used by MP are WOs, OPORDs, and FRAGOs.

WARNING ORDERS

D-2. Refer to *FM 101-5* for an example of a WO. WOs provide advance notice of an action so that MP can use available time for preparations, and they—

- Are issued at each level down to the squad.
- Are issued to subordinates in as much detail as possible.
- Are issued as brief, oral, or written messages.
- Are a part of planning the use of available time. WOs should be kept simple.
- Describe the operation and the preparations to be made before the OPORD is issued.

- Include the situation, the mission, special instructions, and the time and place for issuing the OPORD.

OPERATION ORDERS

D-3. OPORDs coordinate actions to carry out the commander's plan for an operation, and they—

- Explain how leaders at different levels want the operation conducted. To ensure that maneuver and other non-MP units carry out actions or provide a form of support that MP plans and operations depend on, state the required actions or support in the coordinating instructions paragraph of the division and brigade OPORDs.
- Have a great impact on how subordinate leaders employ units and perform missions. For example, the MP platoon leader's latitude to employ the unit could be restricted, based on how the company commander wants the operation conducted.
- May be written, oral, graphic (such as traces and overlays), or a combination of these forms.
- Are usually written when prepared at company level and above.
- Have a standardized system of designating days and hours in relation to an operation or an exercise.
- Follow a prescribed written format, that contains a classification, heading, body, and ending.
- Have a prescribed five-paragraph format for the body of the order, which includes the following:
  - Situation.
  - Mission.
  - Execution.
  - Service support.
  - Command and signal.

D-4. The risk management process is greatly enhanced by incorporating the risk assessment into the OPORD. The commander integrates risk management into the order, rather than treating it as an afterthought. Leaders at all echelons assess the effectiveness of their units by reviewing how well hazards are identified and risk controls are specified in oral and written orders. Refer to *Figure D-1, pages D-5 through D-10,* for an example of an OPORD format. (Refer to *FM 101-5, Appendix H*, for further discussion of an OPORD.)

D-5. FRAGOs issue supplemental instructions to a current OPORD or OPLAN while the operation is in progress, and they–

- Contain missions of immediate concern to subordinate units.
- May be either written or oral.
- Provide brief, specific, and timely information without loss of clarity.
- Have no prescribed format. Prevent confusion OPORD.
- May be issued to change an OPORD that has already been issued. As such, only those items from the original OPORD that have changed are included in the FRAGO, as long as clarity is not sacrificed.

STANDING OPERATING PROCEDURES

D-6. SOPs prescribe routine methods to be followed in operations, and they–

- Supplement other combat orders.
- Reduce the number, length, and frequency of other orders. Because the SOP is a standing order, do not repeat the information contained therein in other orders unless emphasis is desired.

- Have no prescribed format. However, ensure that the subordinate unit SOPs follows the format of the next higher HQ SOP if possible.
- Prescribe actions of a recurring nature that lend themselves to definite or standardized procedures. Examples include the following items:
  - Troop safety matters
  - Methods of reporting unit locations.
  - Measures for handling captured personnel or equipment
  - Distribution of supplies.
  - Standard communication procedures for exercising C^2.
  - Other items that lend themselves to standardization.

NOTE: These items are generally the constants in what is otherwise a frequently changing set of circumstances.

REPORTS

MP contribute to the commander's situational awareness by providing timely information to higher HQ. They collect and report information in the form of administrative, operational, or intelligence reports. These reports ensure that the commander receives continuous current information. *Tables D-1* through *D-3, pages D-11 through D-18,* are examples of reports used by MP when conducting CS operations. Refer to *FM 101-5-2* for a complete listing of the standardized Army report and message formats.

____
(Classification)

Place the classification at the top and bottom of every page of the OPORD.

(Change from oral orders, if any.)

This statement is applicable only if an oral order is issued by the commander. The phrases "No change from oral orders" or "No change from oral orders except paragraph ___ " are necessary here.

Copy ___ of ___ copies
Issuing HQ
Place of issue (coordinates)
DTG of signature

Show the place of issue (location of the issuing HQ) on each copy. Show the name of the town or the place in capital letters, the coordinates in parentheses, and the country in capital letters. Both may encode.

The effective time for implementing the plan or order is the same as the DTG, unless coordinating instructions state otherwise. When orders apply to units in different time zones, use time zone Zulu (Z). In OPORDs and service support orders, list the time zone applicable to the operation in the heading of the order following the references. When an order does not specify the actual date and hour for beginning an operation, apply the proper reference designations.

Message Reference Number

Message reference numbers are internal control numbers that the unit signal officer issues and assigns to all orders. The unit's SOP normally describes the number's allocation and use. Using the number allows an addressee to acknowledge receiving the message in the clear.

OPORD ________ (code name)
(Number)

Orders normally contain a code name and are numbered consecutively within a calendar year.

Reference(s): The heading of the order includes a list of maps, charts, datum, or other related documents the unit will need to understand the plan or order. The user does not need to reference the SOP but may refer to it in the body of the order. The user references a map using the map series

Figure D-1. OPORD Format

number (and country or geographic area, if required), sheet number, name, edition, and scale, if required. Datum is the mathematical model of the earth used to calculate the coordinates on any map. Different nations use different datums for printing coordinates on their maps. Reference the datum in the marginal information of each map.

Time Zone Used Throughout the Order: The time zone used throughout the order (including annexes and appendixes) is the time zone applicable to the operation. Operations across several time zones use Z time.

Task Organization: Describe the allocation of forces to support the commander's concept. Show task organization in one of two places: preceeding paragraph one or in an annex, if the task organization is long and complicated.

OPORD ________ - ________________________

(Number) (Issuing HQ)

(Place this information at the top of the second and any subsequent pages of the OPORD.)

1. **SITUATION.**

a. **Enemy Forces.** Express this information down to two enemy echelons below yours (for example, battalions address platoons or companies address squads). Describe the enemy's most likely and most dangerous COA. When possible, provide a sketch of the enemy COA in lieu of verbiage (Appendix ___ [sketch] to Annex ___ [title]). Include an assessment of terrorist activities directed against US government interests in the AOs. Reference more sources using the final subparagraph to refer the reader to the documentation.

b. **Friendly Forces.** Include the mission, the commander's intent, and the concept of operations for HQ one and two levels up. Subparagraphs state the missions of the flank units (left, right, front, and rear) and other units whose actions would have a significant bearing on the issuing HQ.

c. **Attachments and Detachments.** State when, and for how long, units are attached or detached to the operation. Do not repeat information already listed under task organization or in a task organization annex. State "See Annex ___ (title)" here if a task organization annex is necessary.

Figure D-1. OPORD Format (Continued)

2. **MISSION**. State the mission derived during the planning process. This statement describes the task and purpose of the operation and clearly indicates the action to be taken and its reason. There are no subparagraphs in a mission statement. The mission statement covers on-order missions.

3. **EXECUTION**.

Intent: State the commander's intent derived during the planning process. It is a clear, concise statement of what the force must do to succeed with respect to the enemy and the terrain. It provides the link between the mission and the concept of the operation, which provides the basis for subordinate initiative and decentralized mission execution. Always include the desired end state.

a. **Concept of Operations**. The concept of operations may be a single paragraph or divided into two or more subparagraphs. It is based on the COA statement from the decision-making process and, at a minimum, should address close, deep, rear, and security operations as well as designating the main effort. The commander uses this subparagraph when he feels he must supply enough detail to ensure appropriate action by subordinates in the absence of additional communications or further instructions. When an operation involves two or more clearly distinct and separate phases, the concept of operations may be prepared in subparagraphs describing each phase. Ensure that the concept statement is concise and understandable. The concept describes the following:

- The employment of maneuver elements in a scheme of maneuver.
- A plan of fire support or *scheme of fires* supporting the maneuver with fires.
- The integration of other major elements or systems within the operation, including reconnaissance and security elements, intelligence assets, engineer assets, and air defense.

NOTE: Depending on what the commander considers appropriate, the level of command, and the complexity of any given operation, the following subparagraphs are examples of what may be included within the concept of operations:

(1) **Maneuver**. State the scheme of maneuver derived during the planning process. Ensure that this paragraph is consistent with the operation overlay. Ensure that this paragraph and the operation overlay is

Figure D-1. OPORD Format (Continued)

complementary and adds to the clarity of, rather than duplicating, each other. Do not duplicate information to be incorporated into the unit subparagraphs and coordinating instructions.

(2) **Fires**. State the scheme of fires to support the concept, and include priorities of and restrictions for fire support.

(3) **Reconnaissance and Surveillance**. Specify the reconnaissance and surveillance (R&S) plan and explain how it ties in with the basic concept of operations.

(4) **Engineer**. Clarify the scheme of engineer support to the plan, paying particular attention to the integration of engineer assets and obstacles, including the priorities of effort.

(5) **Air Defense**. State the overall concept of air defense in support of the scheme of maneuver if necessary.

(6) **Information Operations**. State the overall concept of information operations in support of the scheme of maneuver. Refer to the appropriate annexes if necessary.

NOTE: Units required to accomplish specific tasks for information operations and R&S are specified in the appropriate subparagraphs of 3b (tasks to maneuver [subordinate] units).

b. **Tasks to Maneuver (Subordinate) Units.** Clearly state the missions or tasks for each maneuver (or subordinate element) unit that reports directly to the HQ issuing the order. List the units in the same sequence as in the task organization, including attachments. Use a separate subparagraph for each subordinate element. State the tasks that are necessary for comprehension, clarity, and emphasis. Place tactical tasks that commonly affect two or more elements in subparagraph 3c (coordinating instructions).

c. **Coordinating Instructions.** List only those instructions applicable to two or more units (subordinate elements) and not routinely covered in unit SOPs. This is always the last subparagraph in paragraph 3.

NOTE: The following are examples of subparagraphs that are generally included as coordinating instructions. Subparagraphs 1 through 5 below are mandatory.

Figure D-1. OPORD Format (Continued)

(1) Time or condition when an order becomes effective.

(2) CCIR.

(a) Priority intelligence requirements (PIR).

(b) Essential elements of friendly information (EEFI).

(c) Friendly force information requirements (FFIR).

(3) Risk-reduction control measures may include such items as MOPP, operational exposure guidance, vehicle recognition signals, and fratricide prevention.

(4) ROE. When lengthy and complicated, include ROE as a separate annex.

(5) Environmental considerations.

(6) Force protection.

(7) Any additional coordinating instructions.

4. **SERVICE SUPPORT**. Address service support in the areas shown below as needed to clarify the service support concept. Refer to any annexes, where necessary. Subparagraphs can include the following:

a. **Support Concept**. State the concept of logistics support necessary to support the operation. Include the concept for—

- Support HQ or support area locations, including locations of logistical bases if not clearly conveyed in the overlay.
- Commander's support priorities.
- HN support.
- Any significant sustainment risks.
- Unique support requirements in the functional areas of manning, arming, fueling, fixing, moving, and sustaining the soldier and the systems.

b. **Materiel and Services**.

c. **Medical Evacuation and Hospitalization**. At a minimum, include frequencies, call signs, and locations of medical support facilities and units, as well as the plan for casualty evacuation (CASEVAC).

d. **Personnel Support**.

e. **EPW**.

f. **Transportation**.

Figure D-1. OPORD Format (Continued)

5. **COMMAND AND SIGNAL**.

a. **Command**. State the map coordinates for the CP locations and potential future locations for each echelon CP applicable to the operation. Identify the chain of command if not clearly addressed in unit SOPs.

b. **Signal**. List signal instructions not specified in unit SOPs; and identify the specific SOI addition in effect, required reports and formats, and the times the reports are submitted. List primary and alternate means of communications, as well as pyrotechnic signals and their meanings.

Acknowledge: Include instructions for the acknowledgement of the order by addressees. The word *acknowledge* may suffice or a message reference number. Acknowledgement of an order means that it has been received and understood.

Commander's last name
Rank

The commander or his authorized representative signs the original copy. If the representative signs the original, add the phrase "For the Commander." The signed copy is the historical copy and remains in the HQ files. It is always designated "Copy 1 of ______ copies."

OFFICIAL:
(Authentication) Use only when applicable. If the commander signs the original, no further authentication is required. If the commander does not sign, authorization is required by the signature of the preparing officer or individual and only the last name and rank of the commander appear in the signature block.

Annexes: List annexes by letter and title and in the sequence by which they were referenced in the order.

Distribution: Furnish distribution copies either for action or for information. List in detail those who are to receive the order. If extremely lengthy, refer to an annex containing the distribution list or to a standard distribution list or SOP.

____
(Classification)

Place the classification at the top and bottom of every page of the OPORD.

Figure D-1. OPORD Format (Continued)

Table D-1. Administrative Reports

Types	Who	What	When	Where	How	Content
Casualty	The unit or the section with casualties	The number of dead, wounded, or sick	Upon experiencing casualties or as required by the SOP or HQ	Higher HQ	By the most secure means and encoded if by unsecure radio	*DA Form 1156* or per SOP
Personnel	Submitted at company level and higher (input is needed from the squad or section)	Personnel strength accounting and status	Daily	Higher HQ	*DA Form 5367-R*	*DA Form 5367-R*
Periodic logistic	Submitted at company level and higher (input is needed from the squad or the section)	Supported strength and status of critical supplies	As supplies become depleted or as required by the SOP	Higher HQ	By the most secure means and encoded if by unsecure radio	• Logistical situation • Supply • Maintenance • Transportation • Service

Table D-1. Administrative Reports (Continued)

Type	Who	What	When	Where	How	Content
Journal	Units or sections operating independent of the parent organization	Events about a unit or section during a given or specified period	As events occur over the given or specified period	Maintained locally (provide to higher HQ on request)	*DA Form 1594*	*DA Form 1594* • Item • Time • Incident, message, or order • Action taken • Initials
Closing	The unit leader	Notice of change of the location	Upon arrival at the new site	Higher HQ	Secure the radio	• Unit • Date and time • CP location • Vehicles and radios • Personnel and weapons

Table D-2. Operational Reports

Type	Who	What	When	Where	How	Content
Situation and status	Commander or leader closest to the situation	Tactical situation or status	Immediately after a significant event or as specified	Higher HQ.	By the most secure means and encoded if by unsecured radio	• Enemy • Own situation • CS • General • Commander's evaluation
Force tracking (diversion, holding, or passing)	Designated TCP and mounted patrols	Unit movement along the MSR	As requested or designated in the SOP	Local movement control unit	Secure voice or encoded	Division transportation office (DTO) or the Provost Marshal Office (PMO) adopted SOP format
Offense	TCP and mounted patrols	Violation of MSR regulations	When tasked to conduct MMS and MSR regulations are in force	Through military channels to the driver's commander	Written report. (refer to *FMs 55-10* and *101-5*)	• Date • Driver's name • Name of vehicle commander • Particulars of the offense observed

Table D-2. Operational Reports (Continued)

Type	Who	What	When	Where	How	Content
EPW	Platoon, squad, and team operating the collecting points	The number of EPWs collected and evacuated	As required	Higher HQ	By the most secure means and encoded if by unsecured radio	Per SOP
Intention of laying a minefield	Platoon, squad, or team leader preparing to lay the minefield	Tactical objectives and characteristics of the minefield	Before emplacing the mines	Higher HQ	Written or by a secure radio (refer to *FM 20-32)*	• Purpose of the minefield • Estimated number and type • Location • Proposed start and completion times
Initiation of laying a minefield	Platoon, squad, or team leader of the force laying the minefield	Emplacement of the mine	When emplacement begin	Higher HQ (mandatory)	Written or by a secure radio (refer to *FM 20-32)*	• Time begun • Location and target number

Table D-2. Operational Reports (Continued)

Type	Who	What	When	Where	How	Content
Completion of laying a minefield	Platoon, squad, or team leader of the force laying the minefield	Completion of the minefield	Upon completion of the minefield	Higher HQ	Written or by a secure radio (refer to *FM 20-32*)	• Field is complete and functional (expedite the report and follow up with a hasty protective minefield report)

Table D-3. Intelligence Reports

Type	Who	What	When	Where	How	Content
Spot	Unit, section, or individual observing the enemy (all the echelons)	Enemy activity and area information of immediate value	Upon contact or as requested	Higher HQ	By the quickest means and encoded if by unsecured radio	• Reporting unit • Date and time of the event • Location and grid coordinates (encoded) • Event (SALUTE) • Original source • Remarks

Table D-3. Intelligence Reports (Continued)

Type	Who	What	When	Where	How	Content
Meaconing, intrusion, jamming, interference (MIJI)	Soldier experiencing electronic warfare	Correct report according to the type of interference	As soon as possible after the incident	Through signal channels to the higher HQ.	SOI defines the report and how to prepare it	• Type of report • Frequency or channel affected • Victim designation and call sign • Type of emission or audio characteristics • Coordinates of the affected station
NBC 1	Observer	Observation concerning suspected NBC attacks and resulting hazardous areas	Upon observing a suspected enemy NBC attack	Higher HQ.	Secure radio	• Observer's location • Direction from the observer • Date-time group (DTG) of the attack • Type of burst

Table D-3. Intelligence Reports (Continued)

Type	Who	What	When	Where	How	Content
NBC 4	Element in contact with radiation	Radiation dose rates	Upon contact with radiological contamination or as directed	To higher HQ	Secure radio	• Location • Dose rate • DTG of the measurement
Patrol	Patrol leader	Pertinent information pertaining to the patrol's activity	Upon completion of the patrol	To higher HQ	Written	• Designation of the patrol • Maps • Terrain • Enemy • Results of encounters with the enemy • Condition of the patrol
Enemy minefield location	Soldiers encountering an enemy minefield	Characteristics of the minefield	Upon locating a minefield	To higher HQ	Secure radio (refer to *FM 20-32*)	• Type of minefield • Location and depth • Enemy weapons or surveillance • Routes for bypassing • Coordinates of lane entry and exit • Width of lanes, in meters

Table D-3. Intelligence Reports (Continued)

Type	Who	What	When	Where	How	Content
						• Map sheet designation • Date and time information collected • Coordinates of minefield boundaries • Estimated time to clear the minefield • Estimated material and equipment required to clear the minefield • Other, such as the type of the mine, new mines, or booby traps
SALUTE	Observer	Enemy activity; for example, convoy, patrol vehicles, and aircraft	Upon encountering the enemy's activity	Higher HQ	By the most secure means and encoded if by unsecure radio (refer to *FM 101-5*)	S - Size A - Activity L - Location U - Unit T - Time E - Equipment

Appendix E

Precombat Inspections

Combat missions start with PCIs. These PCIs are an essential part of every mission. The objective of PCIs is to confirm the combat readiness of the unit. Subordinate leaders conduct detailed checks, allowing the unit leader to conduct PCIs in an atmosphere of total preparedness. After the PCI, the unit is prepared to move on the receipt of orders.

MODIFIED PRECOMBAT INSPECTION CHECKLIST

E-1. A modified PCI checklist *(Figure E-1, page E-2)* ensures that the team has the required equipment to conduct its mission when the unit is short on time. The modified PCI allows the team leader to focus on preparing the team for the mission, while the platoon leader issues the mission to the squad leader. Use the checklist as a recovery checklist when the team returns from the mission. Immediate, proper recovery ensures that the team is ready for the next mission. This checklist is only a guide and can be modified based on METT-TC and the unit SOP.

Steps	Actions
1	The squad leader and the team leader receive the WO.
2	The team leader prepares the squad for movement, while the squad leader is receiving the mission from the platoon leader.
3	The team leader follows the same checklist, while the squad leader debriefs the platoon leader if recovering from a mission.
4	The team leader inspects the communications equipment and ammunition in HMMWVs. He— • Conducts a communications check with the platoon CP. • Ensures that proper frequencies are set and that call signs are available. • Ensures that there is a Single-Channel, Ground-to-Air Radio System (SINCGARS), a manpack, accessories, and a battery case. • Ensures that there are batteries for the radios (automated net control device [ANCD], precision lightweight global positioning system receiver [PLGR], OE-254). • Ensures that the Class V basic load is accounted for.
5	The driver conducts preoperation checks and services on the vehicle. He ensures that— • The correct paperwork is completed, such as the dispatch, folder, and signature. Ensures that the correct TM is on hand. • He has a valid driver's license. • Preventive-maintenance checks and services (PMCS) are performed to a -10 standard. • A current *DA Form 2404* is on hand. • The fire extinguisher is sealed, tagged, and dated. • The engine's coolant, oil, transmission fluid, and washer fluid are at the correct level. • The battery level is correct and that cables are secured. • Lights are operational. • Fuel and water cans are filled and that there are meals, ready-to-eat (MREs) available. • There are no exhaust leaks. • Basic-issue item (BII), on-vehicle material (OVM), pioneer tools, locks, and chains are present on the vehicle. • The load plan is followed and that the vehicle's interior is clean.

Figure E-1. Modified PCI Checklist

Steps	Actions
6	The gunner conducts preoperation checks and services on the crew-served weapon and NBC equipment. He ensures that— • A TM for the equipment (-10) is available. • The weapon is oiled, cleaned, and functional. • There is a spare barrel and BII. • There is a T&E mechanism. • There is a wrench. • There is a tripod mount with an adapter. • There are weapon cleaning kits present. • There are night vision devices (NVDs) present. • All the NBC equipment is present and serviceable.

Figure E-1. Modified PCI Checklist (Continued)

PRECOMBAT INSPECTION CHECKLIST

E-2. This PCI checklist is extensive and time-consuming. Use it when the team has a lot of downtime and has a secure environment in which to work. Once combat operations begin, the secure environment and the downtime will be lost. *Figure E-2, page E-4,* is a draft checklist, which may be modified to meet the needs of the unit.

Steps	Actions
1	Soldiers. Soldiers should ensure that they— • Have a seasonal uniform. The uniform should include the following: ▪ Load-carrying equipment (LCE) that is complete. ▪ First aid packet. ▪ Canteens that are full and have a M17 drinking cap. ▪ M40 with a hood, M258A1 kit, M13 kit, and optical inserts (if required). ▪ Kevlar with a camouflage cover and band. ▪ Individual weapon with 6 magazines. ▪ ID tags around the neck and ID cards. ▪ Ear plugs. • Have been briefed on the current situation and the mission. • Possess a current driver's license. • Have been briefed on hot- and cold- weather injuries. Leaders. Leaders should conduct a safety briefing, to include— • Checking for hazards, such as ticks, chiggers, and spiders, and taking preventive first aid measures. • Checking for snake hazards and ensuring that there is first aid. • Checking for poisonous plants, such as oak, ivy, and sumac. • Checking for hot- and cold- weather injuries and preventing them. • Ensuring that correct driving practices are followed, to include— ▪ Speed in relation to the weather. ▪ Ground guides. ▪ Seat belts. ▪ Blackout driving. ▪ Gunner positioning (goggles). • Ensuring that equipment is set up, such as camouflage nets or tents. • Ensuring that there are fire-prevention vehicles and tents. • Taking appropriate actions during severe weather. • Using pyrotechnics. • Using the buddy system for safety and during the mission.

Figure E-2. Extensive PCI Checklist

Steps	Actions
	Vehicle commander, squad leader, or team leader. The vehicle commander, squad leader, or team leader ensures that— • There is a map with a current overlay. • There is a current SOI. • Includes the following in the leader packet: ▪ A unit tactical standing operating procedure (TACSOP). ▪ Map markers. ▪ Overlay sheets. ▪ A notebook and a pen or a pencil. ▪ A route reconnaissance GTA card. ▪ A bridge classification GTA card. ▪ A call for-fire GTA card. ▪ A rigging GTA card. ▪ An NBC GTA card. ▪ An *EPW GTA card (GTA 19-7-1).* ▪ Flex cuffs for EPWs. ▪ *A DD Form 1408.* ▪ *A DA Form 4137.* ▪ *A DA Form 3975.* ▪ *A DD Form 2708.* ▪ *A DA Form 3881.* ▪ *A DA Form 2823.* ▪ A compass and a wrist watch. ▪ Batteries for all the equipment. ▪ Binoculars. ▪ NVGs. ▪ A sensitive item serial number list (radio, weapons, NVGs, and so forth).
2	Vehicles. • Tie down all the items according to the load plan, and ensure that— ▪ There is space left for ammunition. ▪ The TA-50 is loaded. ▪ The MOPP chemical protective overgarment (CPOG) is stored for ready access. ▪ The gas tank is topped off. ▪ There are POL package products and weapon oil. ▪ Water cans are full.

Figure E-2. Extensive PCI Checklist (Continued)

Steps	Actions
	▪ Spare fuel cans are full and tied down. ▪ MRE rations are inventoried and stored. ▪ There are weapon-cleaning kits. ▪ The OVM is stored and serviceable. ▪ There are chock blocks. ▪ That a complete first aid kit is available. ▪ A TM for the equipment (-10) on hand. ▪ A vehicle dispatch packet and a *DA Form 2404* are available. ▪ There are tools and tool kits. ▪ There are tire chains. ▪ The spare tire is properly stowed. • Perform a premovement inspection and ensure that— ▪ There is a current TM (-10) on hand. ▪ There are no fuel leaks. ▪ The fire extinguisher is sealed, tagged, and dated. ▪ The engine coolant level is correct. ▪ The oil level is correct. ▪ The transmission fluid level is correct. ▪ The battery level is correct and the cables are tight. ▪ The air filter is clean and secure. ▪ The lights are operational. ▪ There are no exhaust leaks. ▪ The fuel filter is drained. ▪ Warning triangles are on hand. ▪ The canvas is attached and tied down. ▪ The gauges are functional. ▪ The tire pressure is correct. ▪ The vehicle service is current. ▪ The interior of the vehicle is clean and orderly.
3	Crew-served weapons. Ensure that— • All mounting equipment is present. • The TM for the weapon (-10) is available. • There is a spare barrel and BII (such as gloves, a T&E mechanism, a wrench, a blank adapter, a tripod mount with an adapter, and a cleaning kit). • The NVDs are operational. • The tripod is clean and serviceable. • The weapons are mounted with pins.

Figure E-2. Extensive PCI Checklist (Continued)

Steps	Actions
4	Communications equipment. Ensure that— • Radios (vehicles or manpacks)— ▪ Are operational. ▪ Are secure. ▪ Have the proper frequency setting. ▪ Have a matching unit operational/set. ▪ Have an antenna with the tip protector tied down. ▪ Have connectors that are clean and serviceable. ▪ Have batteries available. ▪ Have the TM for the equipment (-10) is on hand. ▪ Have cleaning materials available (such as cotton swabs, alcohol, steel wool, and so forth). • The field telephone (TA 312) is operational. • The digital nonsecure voice terminal (DNVT) with wire is operational (communication). • The appropriate phone listing is available. • Radio antennas (OE 254s) are complete. • Communication test sets, including tools, are on hand. • Frequencies, call signs, challenges, and passwords are disseminated. • Each team has one SINCGARS manpack for quick-dismount operations.
5	NBC equipment. • Ensure that protective clothing is complete and serviceable. • Ensure that the M40 is complete, such as— ▪ M13 deconnaissance kit. ▪ M258A1 deconnaissance kit. ▪ M256 detector kit. ▪ M8 detector paper. ▪ M9 detector paper. ▪ Antifogging kit. ▪ Hood. ▪ Optical inserts, if needed. ▪ Waterproof bag.

Figure E-2. Extensive PCI Checklist (Continued)

Steps	Actions
	• Ensure that the mask and hood are serviceable. • Implement the proper MOPP level. • Brief the threat condition. • Ensure that nerve agent antidotes are available. • Issue and mount M11 deconnaissance apparatus, to include the following: ▪ Two DS2 cans per vehicle. ▪ An M8 alarm. ▪ An M229 refill kit for each M8. • Ensure that NBC defense equipment is available, such as— ▪ An M256 agent detector kit. ▪ An M 174 RADIAC meter. ▪ Two sets of batteries for the IM 174. ▪ An IM 93/147 RADIAC meter (dosimeter). ▪ A PP 157A RADIAC charger 1/IM 93/147. ▪ NBC contamination signs (marking kits).

Figure E-2. Extensive PCI Checklist (Continued)

Appendix F

Fratricide Avoidance

Fratricide is defined as the unforeseen or unintentional death, injury, or damage to friendly personnel or equipment. It is committed during the employment of friendly weapons and munitions with the intent to kill the enemy or destroy his equipment or facilities. Fratricide prevention is the commander's responsibility, and all leaders assist him across all operating systems to accomplish this mission.

FRATRICIDE

F-1. Friendly forces operating in the sustainment area are particularly vulnerable to fratricide due to the high concentration of soldiers. Most MP missions are conducted in the sustainment area in support of these friendly forces. During MP operations, it is critical that MP know where other friendly elements are operating. With this knowledge, they anticipate dangerous conditions and take steps to either avoid or mitigate them. The platoon leader is always aware of changes and developments in the situation that may place friendly forces in danger from MP long-range weapon systems.

F-2. It is important that other friendly forces operating in the AO know the location of MP to prevent them from becoming victims of fratricide. MP leaders constantly report the location of their forces to higher HQ. Higher HQ disseminates this information to other friendly forces. When the platoon leader perceives potential for fratricide, he quickly responds to prevent it.

FRATRICIDE EFFECTS

F-3. Fratricide results in unacceptable losses and increases the risk of mission failure; it usually affects the unit's ability to survive and function. Units experiencing fratricide suffer the following consequences:

- Loss of confidence in the unit's leadership.
- Increased self-doubt among the leaders.
- Hesitancy to employ deadly force.
- Oversupervision of units.
- Hesitancy to conduct night operations.
- Loss of aggressiveness.
- Loss of initiative.
- Disrupted operations.
- General degradation of unit cohesiveness, morale, and combat power.

FRATRICIDE CAUSES

F-4. The following paragraphs discuss the primary causes of fratricide. MP leaders must identify any of the factors that may affect their soldiers and strive to eliminate or correct them.

TARGET IDENTIFICATION

F-5. The primary cause of fratricide is the lack of positive target identification. MP must positively identify targets before engaging.

LAND NAVIGATION FAILURES

F-6. MP may stray into unknown locations, report wrong locations, and become disoriented. In this situation, they may unexpectedly encounter an errant unit, that may fire their weapons at the friendly force.

REPORTING AND COMMUNICATIONS FAILURES

F-7. Units at all levels may fail to generate timely, accurate, and complete reports as locations and tactical situations change. This distorts the tactical picture available at each level and can lead to erroneous clearance of supporting fire.

WEAPON ERRORS

F-8. Lapses in individual discipline can result in fratricide. These incidents include charge errors, accidental discharges, mistakes with explosives and hand grenades, and incorrect gun data use.

OPERATION HAZARDS

F-9. A variety of explosive devices and materiel may create danger on the battlefield, such as UXO, unmarked or unrecorded minefields (including scatterable mines), and booby traps. Failure to mark, record, remove, or otherwise anticipate these threats will lead to casualties.

FRATRICIDE RISK ASSESSMENT

F-10. *Table F-1, page F-4,* shows an example of a work sheet for evaluating fratricide risk in the context of mission requirements. The work sheet lists six mission accomplishment factors that affect the fratricide risk, along with related considerations for each factor. Platoon leaders assess the potential risk in each area (low, medium, or high), and assign a point value to each (one point for low risk, two for medium risk, three for high risk), and add the point values to calculate the overall fratricide assessment score.

F-11. They use the resulting score only as a guide. The platoon leader's final assessment is based on observable risk factors, such as those listed on the work sheet, and on his feel for the intangible factors affecting the operation. Note that descriptive terms are listed only in the low- and high-risk columns of the work sheet. The assessment of each factor determines whether the risk matches one of these extremes or lies somewhere between them as a medium risk.

Table F-1. Fratricide Risk Assessment Work Sheet

Factors Affecting Fratricide	Potential Risk Categories (With Variable Conditions and Point Values)		
	Low Risk (1 point)	**Medium Risk (2 points)**	**High Risk (3 points)**
Understanding of the Plan			
Commander's intent	Clear	⟷	Vague
Complexity	Simple		Complex
Enemy situation	Known		Unknown
Friendly situation	Clear		Unclear
ROE/ROI	Clear		Unclear
Environmental Factors			
Intervisibility	Favorable	⟷	Unfavorable
Obscuration	Clear		Obscured
Battle tempo	Slow		Fast
Positive target ID	100 percent		None
Control Measures			
Command relationships	Organic	⟷	Joint/ combined
Audio communications	Loud/clear		Jammed
Visual communications	Easily seen		Obscured
Graphics	Standard		Not understood
SOP	Standard		Not used
Liaison personnel	Proficient		Not trained
Location/navigation	Sure		Unsure

Table F-1. Fratricide Risk Assessment Work Sheet (Continued)

Factors Affecting Fratricide	Potential Risk Categories (With Variable Conditions and Point Values)		
	Low Risk (1 point)	**Medium Risk (2 points)**	**High Risk (3 points)**
Equipment (Compared to US Equipment)			
Friendly Enemy	Similar Different	⟷	Different Similar
Training			
Individual proficiency Unit proficiency Rehearsals Habitual relationship Endurance	MOS qualified Trained Realistic Yes Alert	⟷	Untrained Untrained None No Fatigued
Planning Time (Based on the 1/3–2/3 Rule)			
Higher HQ Own unit Subordinate elements	Adequate Adequate Adequate	⟷	Inadequate Inadequate Inadequate
Overall risk assessment (by total point value)	26-46 points	42-62 points	58-78 points
NOTE: Point values alone may not accurately reflect fratricide risk. The commander tailors the assessment to the unit's requirements. Refer to *FM 100-14*.			

PREVENTIVE MEASURES

F-12. These measures provide the platoon leader with a guide to actions he can take to reduce and prevent fratricide risk. These guidelines are not directive in nature, nor are they intended to restrict initiative. All MP leaders must learn to apply them as appropriate based on the specific situation and METT-TC factors.

Platoon leaders must educate their soldiers on the following reduction and prevention of fratricide principles:

- Identifying and assessing the potential fratricide risks in the estimate of the situation. Express the risks in the OPORD and applicable FRAGOs.
- Maintaining situational awareness. Focusing on such areas as current intelligence, unit locations and/or dispositions, denial areas (minefields and scatterable mines; contaminated areas, such as improved conventional munitions [ICM] and NBC), SITREPs, and METT-TC factors.
- Ensuring positive target identification. Review vehicle and weapons ID cards. Become familiar with the characteristics of potential friendly and enemy vehicles, including their silhouettes and thermal signatures.
- Establishing a command climate that emphasizes fratricide prevention. Enforce fratricide prevention measures, placing special emphasis on the use of doctrinally sound TTP. Ensure supervision in the execution of orders and in the performance of all tasks and missions to standard.

F-13. Additional considerations for fratricide reduction and prevention include the following:

- Reporting MP patrol locations accurately. The TOC and CP track the patrols' locations in relationship to friendly forces.
- Recognizing the signs of battlefield stress. Maintain unit cohesion by taking quick, effective actions to alleviate the battlefield stress.
- Conducting individual, leader, and collective (unit) training covering fratricide awareness.
- Giving complete and concise mission orders.

- Simplifying mission orders by using SOPs that are consistent with the doctrine. Periodically review and update SOPs as needed.
- Striving to provide maximum planning time for leaders and subordinates.
- Planning and establishing effective communications.
- Ensuring that ROE are clear and understood.
- Conducting rehearsals when the situation allows.
- Being in the right place at the right time. If an MP patrol becomes lost or misoriented, leaders must know how to contact higher HQ immediately for instructions and assistance.
- Including a discussion of fratricide incidents in all AARs.

FRIENDLY FIRE INCIDENTS

F-14. MP patrol's could become involved in a friendly fire incident in one of several ways: as the victim of the fire, as the firing element, or as an observer observing an attack of one friendly element on another. If a patrol encounters such a situation, it should follow the procedures for the various situations listed below:

FALL VICTIM TO FRIENDLY FIRES

F-15. When falling victim to friendly fire, react to contact until recognizing friendly fire, then—

- Cease fire.
- Take immediate actions to protect soldiers and vehicles.
- Use a visual recognition signal directing the firing unit to cease fire.
- Report the following on the next higher unit net:

  - Announce that friendly fire has been received.
  - Request medical assistance as needed.
  - Give the location and direction of the firing vehicles.
  - Warn the higher unit not to return fire if the firing unit has been identified as friendly.

ENGAGE FRIENDLY FORCES

F-16. When engaging friendly forces, cease fire and report the following on the next higher unit net:

- The ID of the engaged friendly force (if the unit is unidentified, report the number and type of vehicles).
- The location of the incident.
- The direction and distance to the engaged force.
- The type of fire.
- The target effects.

OBSERVE A FRIENDLY FIRE INCIDENT

F-17. When observing a friendly fire incident, seek cover for protection, then—

- Use a visual recognition signal directing the firing unit to cease fire.
- Report the following on the next higher unit net:
  - The ID of the engaged friendly force (if the unit is unidentified, then report the number and the type of vehicles).
  - The location of the incident.
  - The direction and distance to the victim and the firing unit.
  - The type of fire.
  - The target effects.

- Provide assistance as needed (when safe to do so).

LEADER RESPONSIBILITIES

F-18. In all situations involving the risk of fratricide and friendly fire, leaders must be prepared to take immediate actions to prevent casualties and equipment damage or destruction. Recommended actions in fratricide situations include—

- Identifying the incident and ordering the parties involved to cease fire.
- Conducting an in-stride risk assessment.
- Identifying and implementing the controls to prevent the incident from recurring.

Appendix G

MK19 Qualification Tables

The tables in this appendix are designed to enable the MK19 gunner and assistant gunner to qualify with the weapon.

NOTE: These tables are not for ASV qualification. The ASV tables are under development.

PRIMARY GUNNER, MK19 QUALIFICATION AND ZERO/PRACTICE TABLES

G-1. *Table G-1, page G-2,* is a zero/practice table for the MK19 primary gunner.

Table G-1. Primary Gunner Zero/Practice Table

Task	Condition	Standard	Ammunition	Time
Mount the MK19 on a vehicle or a M3 tripod	Given a mounted MK19, an M3 tripod, an M64 gun mount, a T&E mechanism, an organic carrier, and a selected firing position	The gun crew must be able to mount the MK19 on a vehicle or a tripod.	0	2 minutes
Zero the MK19 using a 400-meter vehicle target	Given a mounted MK19, a 400-meter vehicle target, a selected firing position, and 4 40-millimeter training practice (TP) rounds	The gunner must impact on the target with at least one of the 4 40-millimeter TP rounds.	4	No time limit
Engage a stationary vehicle target at 1,100 meters	Given a mounted MK19, a vehicle target at 1,100 meters, a selected firing position, and 4 40-millimeter TP rounds	The gunner must impact on the vehicle target with at least 1 of the 4 rounds.	4	No time limit
Engage a stationary vehicle target at 1,500 meters	Given a mounted MK19, a vehicle target at 1,500 meters, a selected firing position, and 4 40-millimeter TP rounds	The gunner must impact on the vehicle target with at least 1 of the 4 rounds.	4	No time limit

Table G-1. Primary Gunner Zero/Practice Table (Continued)

Task	Condition	Standard	Ammunition	Time
Engage a stationary vehicle target at 600 meters	Given a mounted MK19, a vehicle target at 600 meters, a selected firing position, and 4 40-millimeter TP rounds	The gunner must impact on the vehicle target with at least 1 of the 4 rounds.	4	No time limit
Dismount the MK19 from the vehicle or the tripod	Given a mounted MK19 and a selected firing position	The gunner must take the MK19 out of operation from the carrier or tripod.	0	1 minute

PRIMARY GUNNER, INDIVIDUAL MK19 QUALIFICATION (DAY), DISMOUNTED

G-2. *Table G-2* is the MK19 qualification table for the primary gunner, dismounted. The commander may decide to fire using this table (mounted) based on the METT-TC.

Table G-2. Primary Gunner (Day), Dismounted

Task	Condition	Standard	Ammunition	Time
Mount the MK19 on an M3 tripod or a vehicle	Given a mounted MK19, an M3 tripod, an M64 gun mount, a T&E mechanism, an organic carrier, and a selected firing position	The gun crew must be able to mount the MK19 on a vehicle or a tripod.	0	120 seconds
Engage a stationary vehicle target at 800 meters	Given a mounted MK19, a selected firing position, 6 40-millimeter TP rounds, and a vehicle target at 800 meters	The gunner must impact on the vehicle target with at least 2 of the 6 rounds.	6	90 seconds
Engage a stationary vehicle target at 600 meters	Given a mounted MK19, a selected firing position, 4 40-millimeter TP rounds, and a vehicle target at 600 meters	The gunner must impact on the vehicle target with at least 1 of the 4 rounds.	4	90 seconds
Engage a stationary vehicle target at 1,100 meters	Given a mounted MK19, a vehicle target at 1,100 meters, a selected firing position, and 6 40-millimeter TP rounds	The gunner must impact on the vehicle target with at least 2 of the 6 rounds.	6	120 seconds

Table G-2. Primary Gunner (Day), Dismounted (Continued)

Task	Condition	Standard	Ammunition	Time
Engage a stationary vehicle target at 1,500 meters	Given a mounted MK19 vehicle target at 1,500 meters, a selected firing position, and 8 40-millimeter TP rounds	The gunner must impact on the vehicle target with at least 2 of the 8 rounds.	8	180 seconds
Dismount the MK19 from the vehicle or the tripod	Given a mounted MK19 and a selected firing position	The gunner must take the MK19 out of operation from the carrier or tripod.	0	60 seconds

PRIMARY GUNNER, MK19 QUALIFICATION, MOUNTED

G-3. *Table G-3, page G-6,* is the primary gunner's individual MK19 qualification table, mounted.

Table G-3. Primary Gunner, Mounted

Task	Condition	Standard	Ammunition	Time
Mount the MK19 on a vehicle or a M3 tripod	Given a mounted MK19, an organic carrier, an M3 tripod, an M64 gun mount, a T&E mechanism, and a selected firing position	The gun crew must be able to mount the MK19 on a vehicle or a tripod.	0	120 seconds
Engage a stationary vehicle target at 400 meters	Given a mounted MK19, a selected firing position, 4 40-millimeter TP rounds, and a vehicle target at 400 meters	The gunner must impact on the vehicle target with at least 2 of the 4 rounds.	4	90 seconds
Engage a stationary vehicle target at 1,100 meters	Given a mounted MK19, a selected firing position, 8 40-millimeter TP rounds and a vehicle target at 1,100 meters	The gunner must impact on the vehicle target with at least 2 of the 8 rounds.	8	120 seconds

Table G-3. Primary Gunner, Mounted (Continued)

Task	Condition	Standard	Ammunition	Time
Engage troops at 600 meters	Given a mounted MK19, a selected firing position, 4 40-millimeter TP rounds, and a troop target at 600 meters	The gunner must impact on the target with at least 1 of the 4 rounds	4	60 seconds
Engage a stationary vehicle target at 1,500 meters	Given a mounted MK19, a selected firing position, 6 40-millimeter TP rounds, and a vehicle target at 1,500 meters	The gunner must impact on the vehicle target with at least 1 of the 6 rounds	6	120 seconds
Engage a stationary vehicle target at 1,100 meters and troops in the open at 600 meters	Given a mounted MK19, a selected firing position, 10 40-millimeter TP rounds, a vehicle target at 1,100 meters, and a troop target at 600 meters	The gunner must impact on the vehicle and troop targets with at least 2 rounds per target	10	180 seconds
Dismount the MK19 from a vehicle or a tripod.	Given a mounted MK19 and a selected firing position	The gunner must take the MK19 out of operation from the carrier or tripod	0	60 seconds

PRIMARY GUNNER, MK19 QUALIFICATION (NIGHT)

G-4. *Table G-4* is the primary gunner's MK19 qualification table for night firing.

Table G-4. Primary Gunner (Night)

Task	Condition	Standard	Ammunition	Time
Mount the AN/ TVS-5 to the MK19	Given a complete AN/TVS-5, an MK19, an M3, and a T&E mechanism or organic carrier	The gunner must mount the AN/ TVS-5 on the MK19.	0	120 seconds
Use a zeroed MK19 with an AN/ TVS-5 mounted for observation and engage a 400-meter vehicle target	Given a complete AN/TVS-5, a mounted MK19, a selected firing position, 6 40-millimeter TP rounds, and a vehicle target at 400 meters.	The gunner must impact the target with at least 2 of the 6 rounds.	6	60 seconds

Table G-4. Primary Gunner (Night) (Continued)

Task	Condition	Standard	Ammunition	Time
Engage a stationary vehicle target at 600 meters	Given a mounted MK19, a selected firing position, 8 40-millimeter TP rounds, and a vehicle target at 600 meters	The gunner must impact the vehicle target with at least 4 of the 8 rounds.	8	120 seconds
Engage troops in the open at 800 meters	Given a mounted MK19, a selected firing position, 8 40-millimeter TP rounds, and troops in the open at 800 meters	The gunner must impact the troop target with at least 4 of the 8 rounds.	8	120 seconds
Dismount the AN/ TVS-5 from the MK19	Given an MK19 and an AN/ TVS-5	The gun crew must take the AN/ TVS-5 out of operation.	0	120 seconds

ASSISTANT GUNNER, MK19 FIRING TABLE, MOUNTED

G-5. *Table G-5* is for familiarization only.

Table G-5. Assistant Gunner, Mounted

Task	Condition	Standard	Ammunition	Time
Mount the MK19 on a vehicle	Given a zeroed, mounted, MK19; a T&E mechanism; an MK64 gun mount; an organic carrier; and a selected firing position	The gun crew must be able to mount the MK19 on a vehicle or a tripod	0	120 seconds
Engage a stationary vehicle target at 400 meters	Given a mounted MK19, a vehicle target at 400 meters, a selected firing position, and 4 40-millimeter TP rounds	The gunner must impact the vehicle target with at least 2 of the 4 rounds	4	90 seconds
Engage troops in the open at 600 meters	Given a mounted MK19, a troop target at 600 meters, a selected firing position, and 4 40-millimeter TP rounds	The gunner must impact the troop target with at least 1 of the 4 rounds.	4	60 seconds

Table G-5. Assistant Gunner, Mounted (Continued)

Task	Condition	Standard	Ammunition	Time
Engage a stationary vehicle target at 1,100 meters	Given a mounted MK19, a vehicle target at 1,100 meters, a selected firing position, and 8 40-millimeter TP rounds	The gunner must impact the vehicle target with at least 2 of the 8 rounds.	8	120 seconds
Engage a stationary vehicle target at 1,100 meters and troops at 600 meters	Given a mounted MK19, a vehicle target at 1,100 meters, troops in the open at 600 meters, a selected firing position, and 10 40-millimeter TP rounds	The gunner must impact the vehicle and troop targets with at least 2 rounds per target.	10	180 seconds
Dismount the MK19 from a vehicle or a tripod	Given a mounted MK19 and a selected firing position	The gunner must take the MK19 out of operation from the carrier or tripod.	0	NA

ASSISTANT GUNNER, MK19 FIRING TABLE, DISMOUNTED

G-6. *Table G-6* is a dismounted firing table. The commander may decide to fire using *Table G-5* based on the METT-TC.

Table G-6. Assistant Gunner, Dismounted

Task	Condition	Standard	Ammunition	Time
Mount the MK19 on a vehicle or an M3 tripod	Given a mounted MK19, an M3 tripod, an M64 gun mount, a T&E mechanism, an organic carrier, and a selected firing position	The gun crew must be able to mount the MK19 on a vehicle or a tripod.	0	120 seconds
Engage a stationary vehicle target at 800 meters	Given a mounted MK19, a vehicle target at 800 meters, a selected firing position, and 6 40-millimeter TP rounds	The gunner must impact the vehicle target with at least 2 of the 6 rounds.	6	60 seconds
Engage a stationary vehicle target at 600 meters	Given a mounted MK19, a vehicle target at 600 meters, a selected firing position, and 4 40-millimeter TP rounds	The gunner must impact the vehicle target with at least 1 of the 4 rounds.	4	90 seconds

Table G-6. Assistant Gunner, Dismounted (Continued)

Task	Condition	Standard	Ammunition	Time
Engage a stationary vehicle target at 1,100 meters	Given a mounted MK19, a vehicle target at 1,100 meters, a selected firing position, and 6 40-millimeter TP rounds	The gunner must impact the vehicle target with at least 2 of the 6 rounds.	6	90 seconds
Engage a stationary vehicle target at 1,500 meters	Given a mounted MK19, a vehicle target at 1,100 meters, a selected firing position, and 6 40-millimeter TP rounds	The gunner must impact the vehicle target with at least 2 of the 6 rounds.	6	120 seconds
Engage a stationary vehicle target at 1,500 meters	Given a mounted MK19 vehicle target at 1,500 meters, a selected firing position, and 8 40-millimeter TP rounds	The gunner must impact the vehicle target with at least 2 of the 8 rounds.	8	120 seconds
Dismount the MK19 from a vehicle or a tripod	Given a mounted MK19 and a selected firing position	The gunner must take the MK19 out of operation from the carrier or tripod.	0	60 seconds

ASSISTANT GUNNER, MK19 FIRING TABLE (NIGHT)

G-7. *Table G-7* is for night firing familiarization only.

Table G-7. Assistant Gunner (Night)

Task	Condition	Standard	Ammunition	Time
Mount the AN/TVS-5 to the MK19	Given a mounted AN/TVS 5 and a T&E mechanism	The gunner must mount the AN/TVS-5 on an MK19.	0	120 seconds
Engage a stationary vehicle target at 600 meter	Given a mounted MK19, a vehicle target at 600 meters, a selected firing position, and 6 40-millimeter TP rounds	The gunner must impact on the vehicle target with at least 1 of the 6 rounds.	6	120 seconds
Engage troops in the open at 800 meters	Given a mounted MK19, troops in the open at 800 meters, a selected firing position, and 6 40-millimeter TP rounds	The gunner must impact on the vehicle target with at least 1 of the 6 rounds.	6	120 seconds
Dismount the AN/TVS-5 from the MK19	Given an MK19 and an AN/TVS-5	The gun crew must take the AN/TVS-5 out of operation within 2 minutes.	0	120 seconds

MK19 SCORECARD

G-8. *Figure G-1* is the scorecard for primary and assistant gunner qualification.

MK19 Qualification Scorecard

1. Name (Last, First, Middle Initial)	2. SSN	3.Grade	4. Unit	5. Date

Mounted Table

Rounds	Range (Meters)	Target	Time (Seconds)	Required Hits	Hits
4	400	Vehicle	90	2	
8	1,100	Vehicle	120	2	
4	600	Troops	60	1	
6	1,500	Vehicle	120	1	
10	1,100/600	Vehicle/ troops	180	2/2	
				Total hits	

Rating:
Distinguished = 5 of 5 qualified engagements.
Qualified = 4 of 5 qualified engagements.
Unqualified = 3 and below.

Dismounted Table

Rounds	Range (Meters)	Target	Time (Seconds)	Required Hits	Hits
6	800	Vehicle	90	2	
4	600	Vehicle	90	1	
6	4,100	Vehicle	120	2	
8	1,500	Vehicle	180	2	
				Total hits	

Rating:
Distinguished = 4 of 4 qualified engagements.
Qualified = 3 of 4 qualified engagements.
Unqualified = 2 and below.

Night-Firing Table

Rounds	Range (Meters)	Target	Time (Seconds)	Required Hits	Hits
6	400	Vehicle	60	2	
8	600	Vehicle	120	4	
8	800	Troops	120	4	
				Total hits	

Rating:
Distinguished = 3 of 3 qualified engagements.
Qualified = 2 of 3 qualified engagements.
Unqualified = 1 and below.

Standards

All standards achieved within table = distinguished.
Missed 1 standard within table = qualified.
two or more missed standards = unqualified.

Overall Rating:
Three distinguished tables = distinguished.
Two distinguished tables = qualified.
One distinguished table = unqualified.

Figure G-1. MK19 Qualification Tables

Appendix H

Countermine Operations

Military operations are often conducted in areas where warring factions have left unrecorded mines and minefields scattered across the landscape. In these environments, emphasize mine awareness at all levels of command and plan countermine operations into every military operation.

OVERVIEW

H-1. Undertake countermine operations to breach or clear a minefield. All the tasks fall under breaching or clearing operations and include detecting, reporting, reducing, proofing, and marking.

H-2. MP efforts in countermine operations are in detecting, reporting, and marking. They conduct countermine operations in conjunction with their other missions. Countermine operations include–

- Denying infiltrators, insurgents, and terrorist groups the opportunity to use mines.
- Using mine detection equipment.
- Detecting mines visually or by probing.
- Reporting the location of suspected mines and UXO.
- Marking the location of these devices.

H-3. MP patrols look for suspicious persons along MSRs; watch approaches to critical points like bridges or defiles; and maintain surveillance of MSRs, key terrain, and critical facilities. Check with the HN police and local nationals for information on unusual activity in a particular AO. Mines are usually emplaced at night. Use night vision and early warning devices to maintain surveillance and detect enemy activity.

H-4. Mine and UXO awareness involve soldier and leader skills. Soldier skills are a mix of individual and collective tasks that are required for an element to maintain its combat effectiveness in and around a mined environment. Soldier skills involve individual and collective tasks that are required for basic survival in a mined environment. They include minefield indicators, probing techniques, mine detector operations, extraction drills, survival rules, casualty treatment, and evacuation drills. The soldier's basic mine awareness skills are critical to his and the unit's survival. Leader skills involve planning missions, assessing situations, and tracking and disseminating mine information. Soldiers must be proficient in all mine awareness skills to effectively operate in a mined environment (refer to *FM 20-32*).

DETECT

H-5. Detection is the actual confirmation and location of the mines and may be accomplished through reconnaissance or unintentionally (such as a vehicle running into a mine). Use mine detection in conjunction with intelligence-gathering operations, minefield bypass reconnaissance, and breaching and clearing operations. Use the following three methods to detect mines and UXO:

- Visual.
- Physical (probing).
- Electronic (mine detector).

VISUAL INDICATORS

H-6. Mine and UXO indicators are part of all combat operations. Understanding and recognizing mine indicators could determine whether or not a soldier becomes a casualty. The following may indicate the presence of mines and UXO:

- Trip wires.
- Signs of road repair (such as new fill or paving, road patches, ditching, or culvert work).
- Signs, usually understood only by local populace, placed on trees, posts, or stakes. Threat forces mark their minefields to protect their own forces.
- Dead animals.
- Vehicles that are damaged.
- Disturbances in previous tire tracks or tracks that stop unexplainably.
- Wires leading away from the side of the road. They may be firing wires that are partially buried.

H-7. Check for odd features in the ground or patterns that are not present in nature. Plant growth may wilt or change color, rain may wash away some of the cover, the cover may sink or crack around the edges, or the material covering the mines may look like mounds of dirt. Civilians may know where mines or booby traps are located in the residential area. Civilians staying away from certain places or out of certain buildings are good indications of the presence of mines or booby traps. Question civilians to determine the exact location of these devices.

H-8. Pieces of wood or other debris on a road may be indicators of pressure or pressure-release firing devices. These devices may be on the surface or partially buried. The enemy uses mines that are fired by command, so search road shoulders and areas close to the objects.

PHYSICAL INDICATORS

H-9. Physical detection (probing) is very time-consuming. Use it primarily for clearing operations, self-extraction, and covert breaching operations. Detection of mines by visual or electronic methods should be confirmed by probing. Use the following procedures and techniques when probing for mines:

- Roll up your sleeves, and remove any jewelry to increase sensitivity. Wear a Kevlar helmet, with the chin strap buckled, and a protective fragmentation vest.
- Stay close to the ground, and move in a prone position to reduce the effects of an accidental blast. When moving into a prone position–
  - Squat down without touching your knees to the ground.
  - Scan forward up to 2 meters and to the sides up to 3 meters for mine indicators.
  - Probe the area around your feet and as far forward as possible.
  - Kneel on the ground after the area is found to be clear. Continue probing forward until you are in a prone position.
- Use sight and touch to detect trip wires, fuses, and pressure prongs.
- Use a slender, nonmetallic object as a probe, and—
  - Probe every 5 centimeters across a 1-meter front.
  - Push the probe gently into the ground at an angle that is less than 45 degrees.
  - Apply just enough pressure on the probe to sink it slowly into the ground.
  - Check the probe for resistance. If the probe encounters resistance and does not go into the ground freely, carefully pick the soil

away with the tip of the probe and remove the loose dirt by hand. Take care to prevent functioning the mine.

- Stop probing when you touch a solid object, and use two fingers from each hand to carefully remove the surrounding soil and identify the object. If the object is a mine, remove enough soil to show the mine type and mark its location.

DANGER
Do not attempt to remove or disarm the mine.

H-10. Probing is extremely stressful and tedious. The senior leader sets a limit to the time a prober can actually probe in the minefield. To determine a reasonable time, the leader considers the METT-TC factors, weather conditions, the threat level, the unit's stress level, and the prober's fatigue level and state of mind. As a rule, 20 to 30 minutes is the maximum amount of time that an individual can probe effectively.

DANGER
Use extreme caution when probing. If the probe is pushed straight down, its tip may detonate a pressure fuse.

ELECTRONIC INDICATORS

H-11. Electronic detection is effective for locating mines, but this method is time-consuming and exposes personnel to enemy fire. Confirm suspected mines by probing.

H-12. The AN/PSS-12 mine detector can only detect metal, but most mines have metal components in their design. The detector locates and identifies plastic or wooden mines by a slight metallic signature. Refer to *TM 5-6665-298-10* for more information about the employment and operation procedures for the AN/PSS-12. The detector is handheld and identifies suspected mines by an audio signal in the headphones.

H-13. As in probing, take consideration for the maximum amount of time an individual can operate the detector. The leader considers the METT-TC factors, weather conditions, the threat level, the unit's stress level, and the individual's fatigue level and state of mind. As a rule, 20 to 30 minutes is the maximum amount of time an individual can use the detector effectively.

REPORT

H-14. Intelligence concerning enemy mines and UXO is reported by the fastest means available. Report sightings of these devices using a SPOTREP format. SPOTREPs originate from patrols that have been sent on specific reconnaissance missions or from MP patrols that discover mine information in the course of their normal route operations.

MARK

H-15. When mines and UXO are detected, mark the location to prevent friendly follow-on forces and local nationals from accidentally encountering them. Mark the mines and UXO with standard North Atlantic Treaty Organization (NATO) markers (a red triangle with white letters). Use concertina wire to construct a perimeter around the mined area and place markers on the fencing about waist-high.

Appendix I

Route Classification and Signing System

The military route classification system helps in planning and executing battlefield movements. Military engineers develop route classifications using a route classification formula. This formula consists of a series of numbers and letters that express, in standard sequence, the route width, the route type, the lowest MLC, overhead clearance, obstructions to the traffic flow, and special conditions on a given route. Findings are based on information extracted from route reconnaissance reports.

OVERVIEW

I-1. The first step in completing a route reconnaissance overlay is to understand what information must be included on it. As a minimum, include the following information on the overlay:

- The route classification formula.
- The name, rank, and social security number of the person preparing the overlay.
- The unit conducting the reconnaissance.
- The DTG that the overlay was conducted.
- The map name, edition, and scale.
- Any remarks necessary to ensure the complete understanding of the information contained on the overlay.

I-2. This appendix focuses on the route classification system and the methods necessary to determine the classification of a road. Refer to *FM 5-170* for more information about route classification and reconnaissance.

ROUTE CLASSIFICATION FORMULA

I-3. The route classification is derived from the information gathered during the route reconnaissance. The formula is recorded on the route reconnaissance overlay *(Figure I-1)* and consists of the following:

- The route width (in meters).
- The route type (based on the ability to withstand weather).
- The lowest MLC.
- The lowest overhead clearance (in meters).
- Any obstructions (OB) to traffic flow, if applicable.
- Any special conditions, such as snow blockage (T) or flooding (W).

I-4. The following are examples of route classification formulas:

- 6.1/Z/40/∞. A fair-weather route (Z) with a minimum traveled way of 6.1 meters and an MLC of 40. Overhead clearance is unlimited (∞), and there are no obstructions to traffic flow. This route accommodates both wheeled and tracked, single-flow traffic without obstruction.
- 7/Y/50/4.6 (OB) (W). A limited, all-weather route (Y) with a minimum traveled way of 7 meters, an MLC of 50, an overhead clearance of 4.6 meters, and an obstruction. This route width is not suitable for double-flow traffic (wheeled or tracked). This route is subject to regular, recurrent flooding.

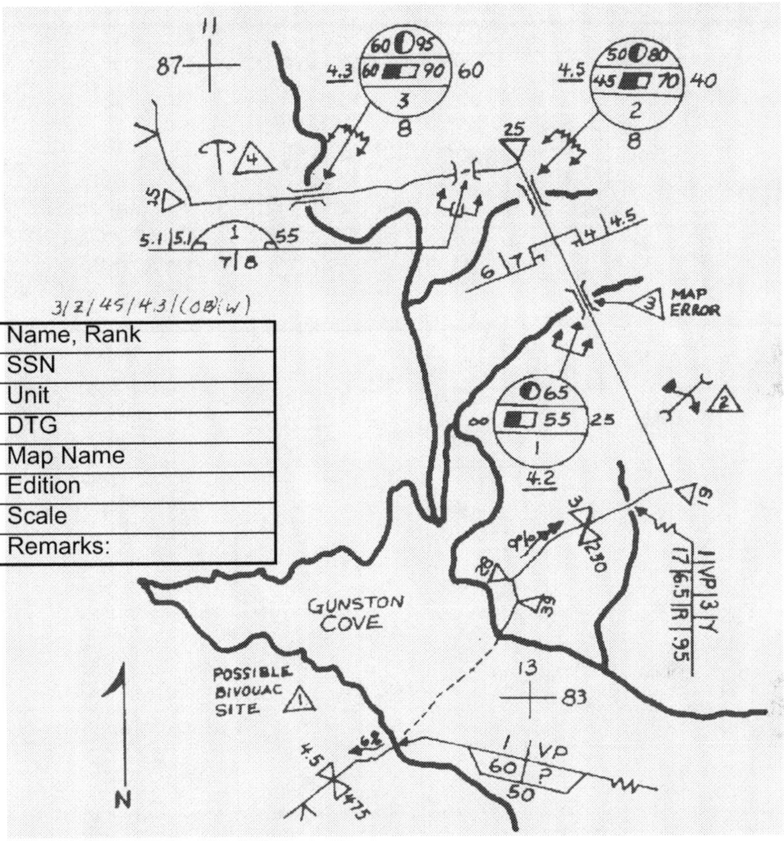

Figure I-1. Route Reconnaissance Overlay

I-5. *Table I-1, page I-4,* gives route reconnaissance symbols used on a route reconnaissance overlay.

Table I-1. Route Reconnaissance Symbols

Explanation	Symbol	Remarks
1. Abbreviated bridge symbol	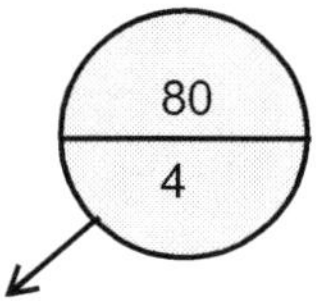	Use this symbol only when the map scale does not permit the use of the full NATO bridge symbol. If this symbol is used, submit DA Form 1249. Draw an arrow to the map location of the bridge. Show the bridge's serial number in the lower portion of the symbol and the MLC for single-flow traffic in the upper portion. If there are separate load classifications for tracked or wheeled vehicles, show the lesser classification. Underline the classification number if the width or overhead clearance is below the minimum standard.
2. Axial route	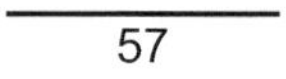	Use a solid line and identify the route by an odd number.
3. Bypass difficult	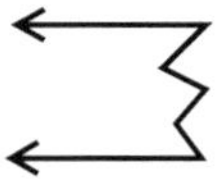	Use this symbol when the obstacle can be crossed in the immediate vicinity, but some work is necessary to improve the bypass.
4. Bypass easy	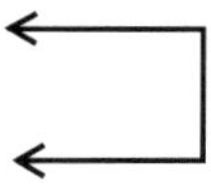	Use this symbol when the obstacle can be crossed in the immediate vicinity by a US 2 1/2-ton truck (or NATO equivalent) without work to improve the bypass.
5. Bypass impossible	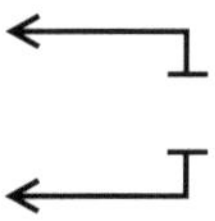	Use this symbol when the obstacle can be crossed only by repairing or constructing a feature or by detouring around the obstacle.
6. Civil or military route designation		Write the designation in parentheses along the route.

Table I-1. Route Reconnaissance Symbols (Continued)

Explanation	Symbol	Remarks
7.Concealment	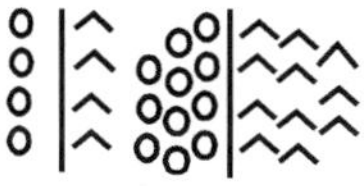	Show roads lined with trees by a single line of circles for deciduous trees and a single line of inverted Vs for evergreen trees. Show woods bordering a road by several rows of circles for deciduous trees and several rows of inverted Vs for evergreen trees.
8. Critical points	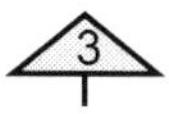	Number, in order, and describe critical points on *DA Form 1711-R*. Use critical points to show the features not adequately covered by other symbols on an overlay.
9. Damage or destruction	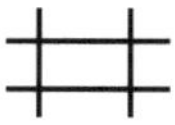	Damage or destruction that prevents movement along the route.
10. Ferry *Ferry type* P–pedestrian V–vehicular	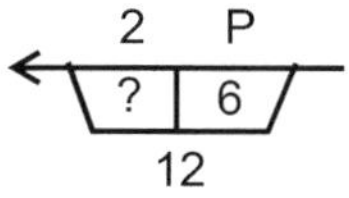 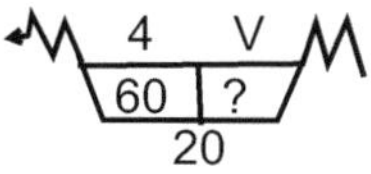	Draw an arrow to the ferry location. The data above the symbols shows, in order, the left approach, the ferry's serial number, the ferry type, and right approach. The data inside the symbol shows, from left to right, the MLC and the dead weight capacity in tons. The number below the symbol shows the turnaround time, in minutes. A question mark indicates unknown information. Show the different approaches by a zigzag line, and show easy approaches by a straight line.

Table I-1. Route Reconnaissance Symbols (Continued)

Explanation	Symbol	Remarks
11. Ford *Ford type* P–pedestrian V–vehicular	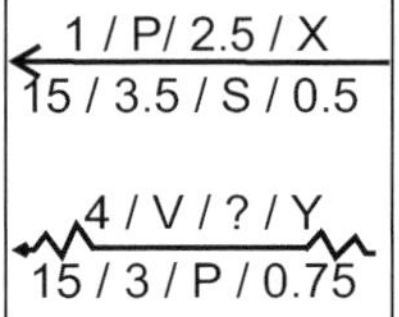	Draw an arrow to the ford location. The data above the line shows, in order, the left-bank approach, the ford serial number, the ford type, the stream velocity (in meters per second), seasonal limitations, and the right-bank approach. A zigzag line corresponding in position to the shore where the approach is located represents difficult approaches. Straight lines identify an easy approach. The left and right banks are determined by looking downstream. The data below the line shows, in order, the length, the width, the bottom type, and the depth. All measurements are in meters. Seasonal limiting factors: X – none; Y – significant; ? – unknown information. Bottom type: M – mud; C – clay, S – sand; G – gravel; R – rock; P – artificial paving.
12. Full NATO bridge symbol	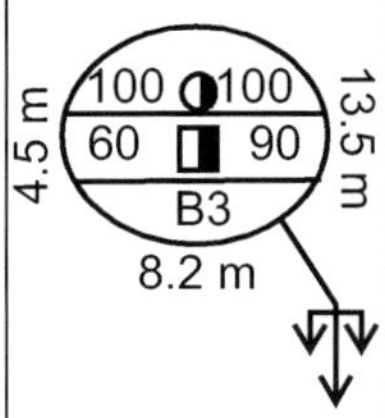	Indicate wheeled vehicles in the upper third of the symbol with the two-way wheeled classification on the left and the one-way wheeled classification on the right. Show tracked vehicles in the center third of the symbol with the two-way tracked classification on the left and the one-way tracked classification on the right. Place the bridge serial number in the lower third of the symbol. Draw an arrow to the bridge location and show the bypass conditions on the arrow shaft. Place the traveled-way width below the symbol, the overhead clearance to the left of the symbol, and the overall length to the right of the symbol.

Table I-1. Route Reconnaissance Symbols (Continued)

Explanation	Symbol	Remarks
13. Grades	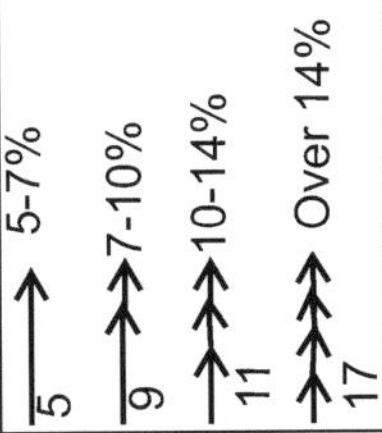	Show the actual percent of grade to the right of the symbol. Any grade of 7 percent or more is an obstruction. Include the obstruction in the route classification formula. Arrows should point uphill. The length of the arrow represents the length of the grade if the map scale permits.
14. Lateral route	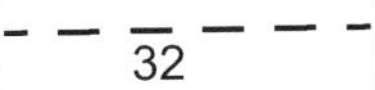	Use a broken line, and identify the route by an even number.
15. Limits of sector		Show the beginning and ending of a reconnoitered section of a route or road with this symbol.
16. MSR		Label the route as MSR and assign a code name.
17. Obstacle a. Proposed block b. Prepared but passable c. Completed block	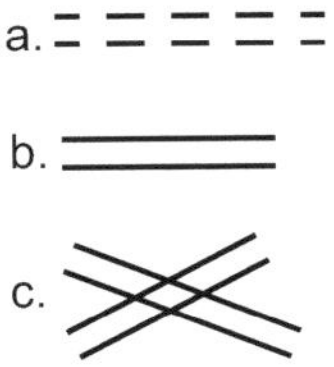	Place the center of the symbol over the location of the blocked part of the route. Use parallel broken lines for a proposed block, parallel lines for a prepared but passable block, and crossed lines for a completed block.
18. Clearance unlimited		Use this symbol when overhead clearance is unlimited.
19. Parking area		Designate parking areas.
20. Railroad crossing	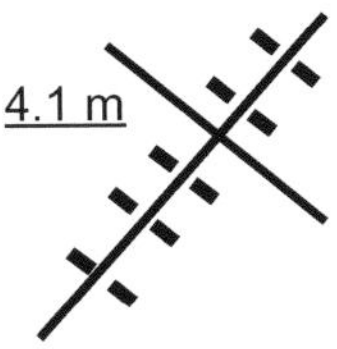	Use this symbol to show a level crossing where passing trains would interrupt the traffic flow. If there is a power line present, show its height, in meters, from the ground. Underline the overhead clearance if it is less than 4.3 meters.

Table I-1. Route Reconnaissance Symbols (Continued)

Explanation	Symbol	Remarks
21. Railway bridge symbol	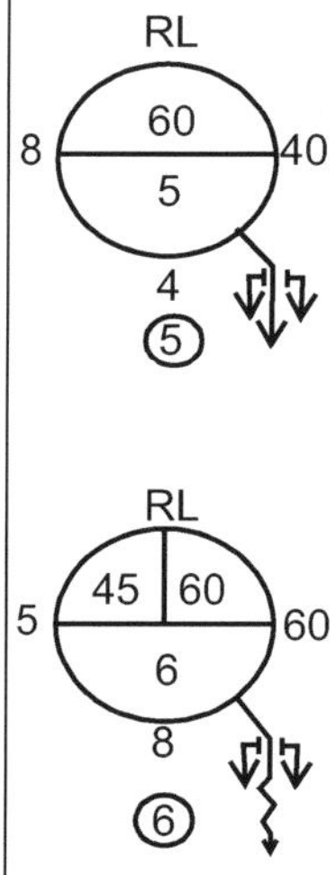 RL = railroad bridge	Place RL above the symbol to indicate a railway bridge. At the left of the symbol, show the overhead clearance. Show the overall length of the bridge on the right of the symbol. Indicate the traveled-way width below the symbol, and underline it if it is below standard for the classification. Inside the symbol, show the bridge classification in the upper half. If the class is different for single- and double-flow traffic, show single flow on the left and double flow on the right. Place the railway bridge serial number in the lower half of the symbol. Draw an arrow to the bridge location. On the arrow shaft, indicate the ease of adapting the bridge for road vehicle use. A zigzag line means it would be difficult to adapt, and a straight line means it would be easy to adapt. Place the bypass symbol on the arrow shaft to indicate the bypass conditions.
22. Route classification formula	6 / Z / 30 / 4 / (OB)	Express the formula in the order of the route width, the route type, MLC, minimum clearance, obstructions (if present), and special conditions. All measurements are in meters. Route types: X – all weather; Y – limited all-weather route; Z – fair-weather route. Special conditions: (T) – regular snow blockage; (W) - regular flooding.
23.Series of sharp curves	 5 / 15	Write the number of curves and the radius of the sharpest curve of the series to the outside of the triangle.

Table I-1. Route Reconnaissance Symbols (Continued)

Explanation	Symbol	Remarks
24. Sharp curve		Point the vertex of the triangle to the curve location and indicate the radius of the curve, in meters, outside the triangle. A curve of 45 meters or less must be reported on the overlay, and a curve of 25 meters or less is an obstacle.
25. Traffic control HQ	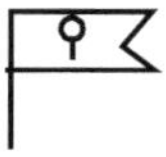	None
26. Traffic control post		Manned traffic control post.
27. Tunnel	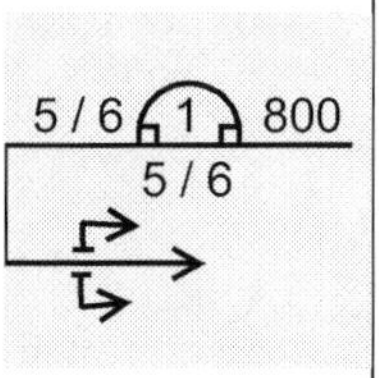	Draw an arrow to the tunnel location. Place the bypass condition symbol on the arrow. Show the minimum and maximum overhead clearances to the left of the symbol, the tunnel serial number inside the symbol, and the total tunnel length to the right of the symbol. Below the symbol, show the traveled-way width. If sidewalks are present, follow with a slash and the total traveled way, including sidewalks. Underline the traveled way if the road entering the tunnel is wider than the traveled way of the tunnel. Use a question mark to show unknown information.

Table I-1. Route Reconnaissance Symbols (Continued)

Explanation	Symbol	Remarks
28. Turnout; the symbol may be amplified as follows: a. Wheeled vehicle b. Tracked vehicle c. A length of road exceeding 1 kilometer	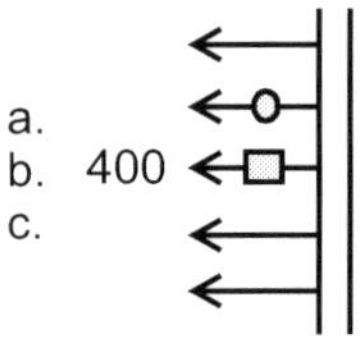	Use this symbol to show the possibility of driving off the road. Draw the arrow in the direction of the turnout (right or left of the road). For wheeled vehicles, draw a small circle on the shaft of the arrow. For tracked vehicles, draw a small square on the shaft of the arrow and place the length of the turnout, in meters, at the tip of the arrow. When the turnout is longer than 1 kilometer, use double arrows.
29. Underpass constrictions (arched or rectangular)	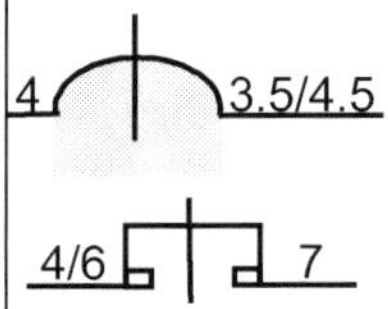	Draw the symbol over the road. Place the width of the traveled way, in meters, to the left of the symbol. If sidewalks are present, follow the traveled-way width with a slash and the total width, including sidewalks. Underline the traveled-way width if the road entering the underpass is wider than the underpass traveled way. Show the overhead clearance, in meters, to the right of the symbol. Show both minimum and maximum overhead clearances, if different.
30. Unknown information	?	Use a question mark with almost any symbol when information required for the symbol is unknown or doubtful.
31. Width constriction	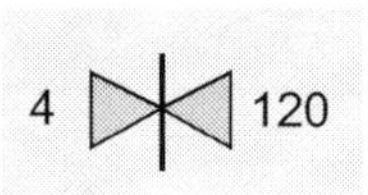	The minimum width is to the left of the symbol. The length of the minimum width is to the right of the symbol.

ROUTE WIDTHS

I-6. The route width is the narrowest width of a traveled way on a route *(Figure I-2)* and is expressed in meters. It may be the width of a bridge, a tunnel, an underpass, or any other constriction that limits the traveled-way width. The establishment of the traveled-way width will help determine the number of lanes of traffic that the route may accommodate.

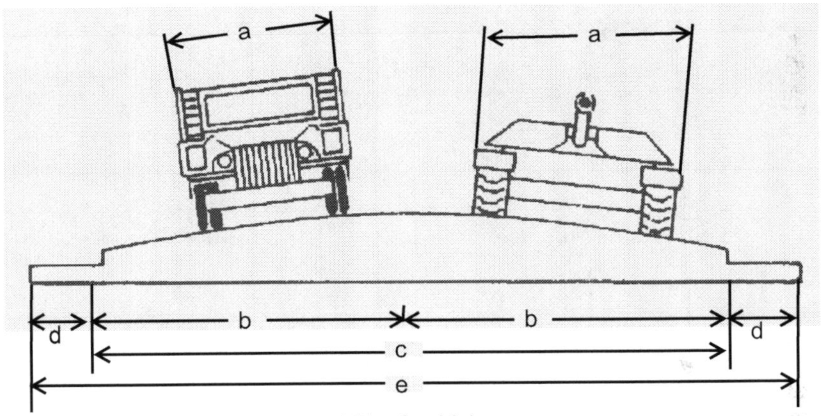

a = width of vehicle
b = width of lane
c = width of traveled way
d = width of hard shoulder
e = width of grading

Figure I-2. Road Width

I-7. Based on the number of lanes of traffic, categorize the route as either—

- Limited access. Limited access permits passage of isolated vehicles of the appropriate width in one direction only.

- Single lane. A single lane permits use in only one direction at any one time. Passing or movement in the opposite direction is impossible.
- Single flow. Single flow permits the passage of a column of vehicles and allows isolated vehicles to pass or travel in the opposite direction at predetermined points only.
- Double flow. Double flow permits two columns of vehicles to proceed simultaneously.

ROUTE TYPES

I-8. Route types are designated by their ability to withstand the effects of weather, and are determined by the worst section of the route. The three types of routes are—

- Type X. Type X is an all-weather route that, with reasonable maintenance, is passable throughout the year to its maximum volume of traffic. This type of route is normally formed of roads having waterproof surfaces, is only minimally affected by precipitation or temperature changes, and is never closed from the effects of weather.
- Type Y. Type Y is a limited, all-weather route that, with reasonable maintenance, is passable throughout the year, but at times having a volume of traffic considerably less than maximum capacity. This type of route is normally formed of roads that do not have waterproof surfaces and are affected by weather or temperature changes. This type of route is closed for short periods (up to one day at a time) by adverse weather conditions during which heavy use of the road would probably lead to complete collapse.
- Type Z. Type Z is a fair-weather route that may become impassable in adverse weather. It may

be closed for long periods, and may require major maintenance or construction efforts to reopen it.

MILITARY LOAD CLASSIFICATION

I-9. A route MLC is a class number representing the safe load-carrying capacity and indicating the maximum vehicle class that can be accepted under normal conditions. Usually, the lowest bridge MLC (regardless of the vehicle type or conditions of traffic flow) determines the route MLC. If there is not a bridge on the route, the worst section of road determines the route overall classification.

I-10. Whenever possible, the basic military road network is composed of average routes and includes a number of heavy traffic routes and a few very heavy traffic routes. Individual routes are grouped and identified in broad categories, such as—

- Average traffic routes are Class 50.
- Heavy traffic routes are Class 80.
- Very heavy traffic routes are Class 120.

OVERHEAD CLEARANCE

I-11. Overhead clearance is the vertical distance between the road's surface and any overhead obstacle (power lines, overpasses, tunnels, and so forth) that denies the use of the route or road to all vehicles or loads that exceed this height. This distance is represented in the route classification formula, in meters. Use the infinity symbol (∞) in the formula for unlimited overhead clearance. Any point along the route where the minimum overhead clearance is less than 4.3 meters is considered an obstruction.

ROUTE OBSTRUCTIONS

I-12. Route obstructions are conditions that restrict the type, amount, or speed of the traffic flow. They are indicated in the route classification formula as OB. If an obstruction is encountered and reported, depict its exact nature on the route classification overlay. Reportable obstructions include the following:

- Overhead obstructions, such as tunnels, underpasses, power lines, and so forth, with a clearance of less than 4.3 meters.
- Reductions in traveled-way widths that are below the standard minimums prescribed for the type of traffic flow *(Table I-2)*. This includes reductions caused by bridges, tunnels, craters, and lanes through mined areas, protruding buildings, or rubble.
- Slopes (gradients) of 7 percent or greater.
- Curves with a radius of 25 meters or less. Curves with a radius of 25.1 to 45 meters are not considered obstructions; however, report them on the reconnaissance overlay.
- Ferries.
- Fords.

Table I-2. Traffic Flow

Type of Vehicles	Limited Access (In Meters)	Single Lane (In Meters)	Single Flow (In Meters)	Double Flow (In Meters)
Wheeled	At least 3.5	3.5 to 5.5	5.5 to 7.3	Over 7.3
Tracked or combination vehicles	At least 4.0	4.0 to 6.0	6.0 to 8.0	Over 8.0

SPECIAL CONDITIONS (SNOW BLOCKAGE AND FLOODING)

I-13. Conditions, such as snow and water, are not normally classified as obstructions, except where these conditions are regular, recurrent, and serious. In cases where snow accumulation is excessive and is either blocking traffic or has the potential to block traffic, the symbol following the route classification formula is T. Similarly, where flooding poses the same problem, the symbol following the formula is W.

CALCULATIONS

I-14. Make calculations for the various curves, slopes, and fords that may affect vehicle movement along the route.

CURVES

I-15. The speeds that vehicles move along the route and security planning measures are affected by sharp curves in the roadway. As previously stated, curves with a radius of 25 meters and less are obstructions to traffic and are indicated by OB in the route classification formula and identified on *DA Form 1248*. There are several ways to measure the radius of a curve, but the two easiest are the tape measure and formula methods.

Tape Measure Method

I-16. The quickest way to estimate the radius of a sharp curve is by using a tape measure to find the radius *(Figure I-3, page I-16)*. Imagine the outer edge of the curve as the outer edge of a circle. Find (estimate) the center of this imaginary circle; then measure the radius using a tape measure. Start from the center of the circle and measure to the outside edge of the curve. The length of the tape measure from the center of the imaginary circle to its outer edge is the curve radius. This method is practical for curves having a radius up to 15 meters and located on relatively flat ground.

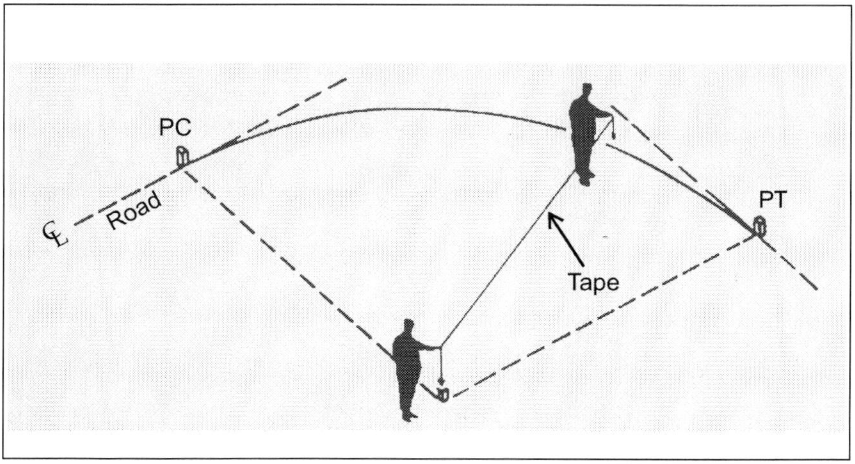

Figure I-3. Find the Radius with a Tape Measure

Formula Method

I-17. The other, and often more practical, method of determining a curve radius *(Figure I-4)* is based on the following formula:

$$R = (C^2/8m) + (M/2)$$

where—

R = radius of the curve

C = distance from the centerline of the road to the centerline of the road at the outer extremities (beginning and end) of the curve

M = perpendicular distance from the center of the tape to the centerline of the road

I-18. When conditions warrant, set M at 2 meters from the centerline, then measure C at 2 meters from the centerline. Use this method when there is a time limitation or because natural or man-made restrictions prevent proper measurements.

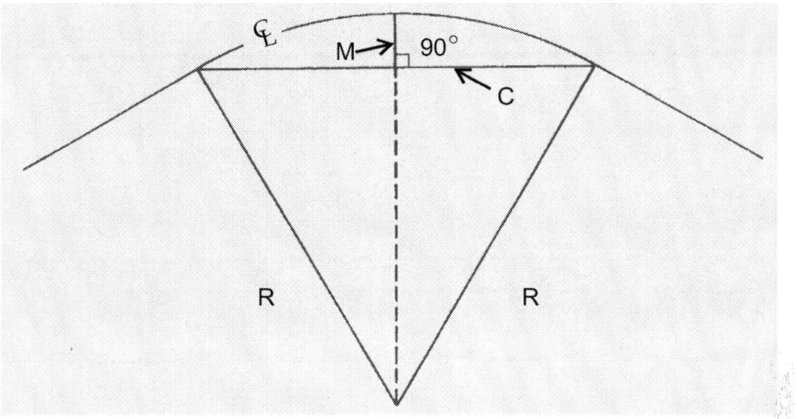

Figure I-4. Determine a Curve Radius

I-19. If C is 15 meters and M is fixed at 2 meters, the formula becomes:

$R = (15^2/16) + 2/2$

I-20. The result of this calculation would be an obstruction to traffic flow, and OB would be placed in the route classification formula.

SLOPES

I-21. The rise and fall of the ground is known as the slope or gradient (grade). Roadways with a slope of 7 percent or greater affect the movement speed along the route and are considered an obstruction. The percent of slope is the ratio of the change in elevation (the vertical distance to the horizontal ground distance) multiplied by 100 *(Figure I-5, page I-18)*. It is important to express the vertical and horizontal distance in the same unit of measure. Although not an obstruction, report all hills with a slope greater than 5 percent on the reconnaissance overlay.

I-22. There are numerous methods and instruments available for determining the percent of slope, but the

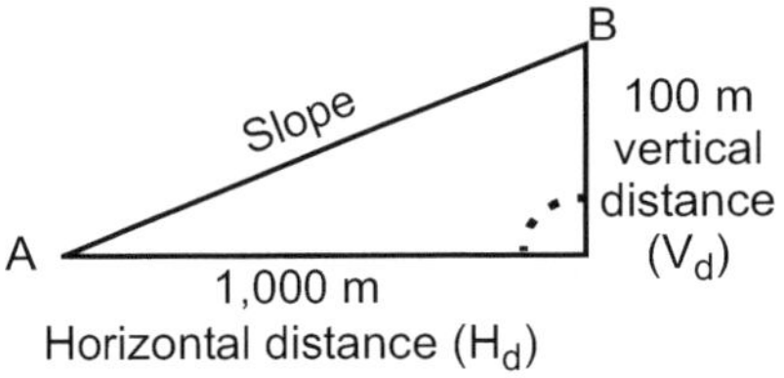

$$Percent\ of\ slope = \frac{V_d}{H_d} \times 100$$

$$= \frac{100\ m}{1,000\ m} \times 100$$

$$= +10\ percent$$

Figure I-5. Calculate the Percent of Slope

two hasty and most practical methods are the map and pace methods.

Map Method

I-23. Use a large-scale map (such as 1:50,000) to estimate the percent of slope quickly. After identifying the slope on the map, find the difference in elevations between the top and bottom of the slope by reading the elevation contours or spot elevation. Then, measure and convert the horizontal distance (usually road distance) to the same unit of measurement as the elevation difference. Substitute the vertical and horizontal distances in the percent-of-slope formula and compute the percent of slope *(Figure I-6)*. This is a hasty method and is the less accurate of the two methods.

Pace Method

I-24. The pace method is another quick way to estimate the percent of slope. Before using this method, accurately determine the height and pace of each member of the reconnaissance team. As a general rule of thumb, the eye level of the average soldier is 1.75 meters above the ground *(Figure I-7, page I-20)*. The pace of the average soldier is 0.75 meters. Perform the following steps to determine the percentage of slopes:

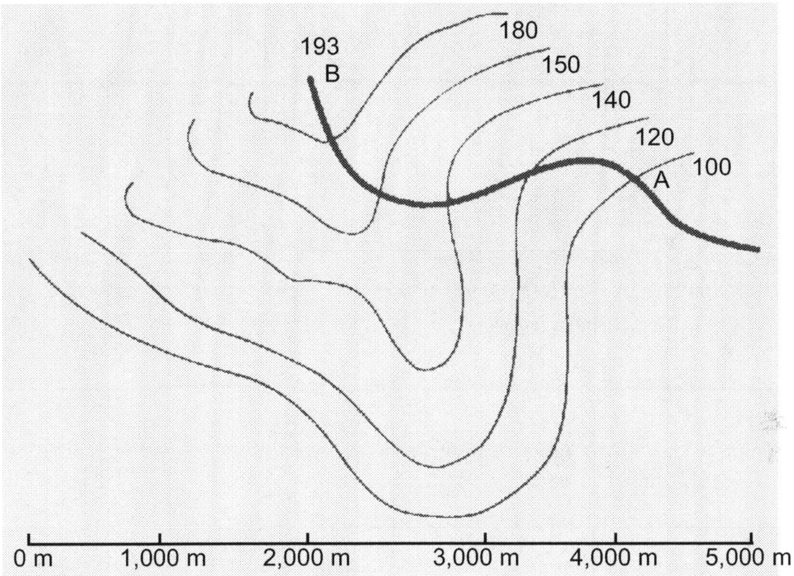

Find the difference in elevation (V_d)

Spot elevation B	=	193 m
Spot elevation A	=	<u>100 m</u>
Difference in elevation (V_d)	=	93 m

Find the horizontal distance (H_d)

Road distance A to B (use a piece of paper and a graphic scale as shown in FM 21-26) = 3,720 m

Use the percent-of-slope formula.

$$\frac{V_d}{H_d} \times 100$$

$$= \frac{93}{3,720} \times 100$$

$$+ 2.5 \textit{ percent}$$

Figure I-6. Determine the Slope with a Map

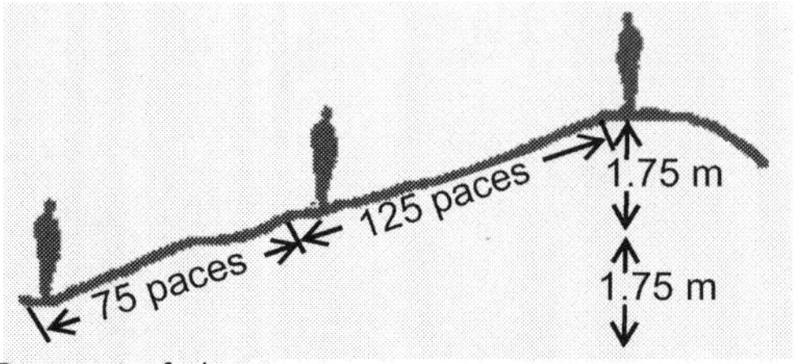

Find:	Percent of slope	=	
Given:	Eye-level height	=	1.75 m
	Pace	=	0.75 m
Answer:	Vertical distance	=	2 x 1.75 m
		=	3.5 m
	Horizontal distance	=	75 paces +125 paces x 0.75
		=	150 m

$$Percent\ of\ slope = \frac{V_d}{H_d} \times 100 = \frac{3.5}{150} \times 100$$

Figure I-7. Determine the Slope Using the Pace Method

Step 1. Stand at the bottom of the slope with your head and eyes level.

Step 2. Sight an easily identifiable spot on the slope, horizontal to eye-level.

Step 3. Walk forward and stand on the spot, recording the number of paces. Repeat this procedure until you reach the top of the slope. Estimate fractions of eye level.

Step 4. Compute the vertical distance by multiplying the number of sightings by the eye-level height of 1.75 meters. Compute the horizontal distance by totaling the number of paces and converting them to meters by multiplying by the known pace-to-meter conversion factor of 0.75 meters.

Step 5. Calculate the percent of slope by substituting the values into the percent-of-slope formula *(Figure I-7)*. Because this method considers horizontal ground distance and incline distance as equal, the reasonable accuracy is only for slopes less than 30 percent. This method requires repeated practice to achieve reasonable accuracy.

FORDS

I-25. A ford is a location in a water barrier where the current, bottom, and approaches allow personnel and vehicles to cross and remain in contact with the bottom during the crossing. While engineers are fully trained and equipped in this area, MP are also capable of performing this task. In the course of a route reconnaissance, MP identify potential fording sites as a contingency for bridges that may become unusable. When identifying a ford on the reconnaissance overlay, include the following information in the ford symbol:

- The ford location, as indicated by an arrow from the ford symbol to the site on the overlay.
- A serial number assigned to each ford for reference. Follow the unit SOP in assigning serial numbers, ensuring that the same number is not duplicated on any one overlay.
- The type of ford, as determined by bottom composition, width, and water depth. Use V for vehicle, P for pedestrian, or VP for both. Approaches are not considered in determining the ford type.
- The stream's normal current velocity is expressed in meters per second. Seasonal limiting factors follow the stream velocity notation and are shown by the following letters:
  - X - No seasonal limitations except for potential sudden flooding of limited duration (flash flooding).
  - Y - Serious, regular, or recurrent flooding or snow blockage.

NOTE: If the Y symbol is used, the route type in the route classification formula automatically becomes type Z.

- The length of the ford (distance from nearshore to far shore), in meters.
- The nature of the bottom surface, using the most appropriate letter symbol—
  - M – mud.
  - C – clay.
  - S – sand.
  - G – gravel.
  - R – rock.
  - P – artificial paving.
- The normal stream depth at its deepest point, expressed in meters.

I-26. Separate all elements of the ford symbol by slashes. Substitute a question mark for any item of information that is unknown. Record all the information on the route reconnaissance overlay.

Stream Width

I-27. There are many methods of determining (estimating) the stream's width. The most common and practical is the compass method. Begin by using a compass to take an azimuth from a point on the nearshore and close to the water's edge to a point on the opposite shore and close to the water's edge *(Figure I-8).* On the nearshore, establish another point that is on a line and at a right angle to the azimuth selected. The azimuth to the same point on the far shore is plus or minus 45 degrees (800 mils) from the previous azimuth. Measure the distance between the two points on the nearshore. This distance is equal to the distance across the stream.

Current Velocity

I-28. Current velocities vary in different parts of a stream. Velocity is usually slower near the shore and

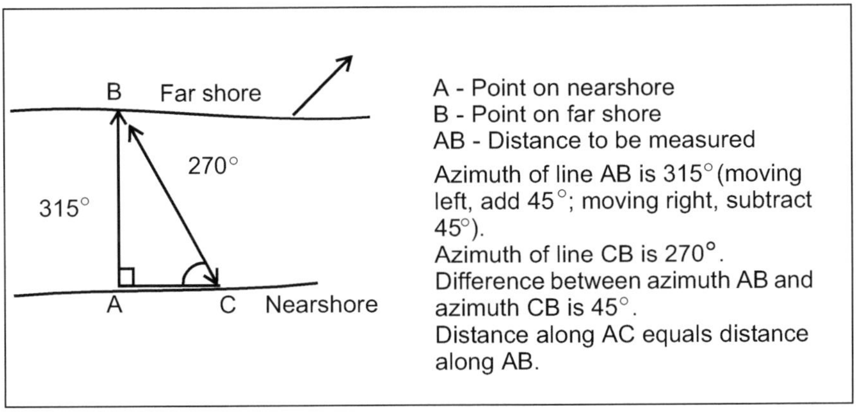

Figure I-8. Determine Stream Width

faster in the main channel of the stream. To estimate stream velocity—

- Measure a distance along the nearshore.
- Throw a light, floating object that will not be affected by the wind into the stream.
- Record the time that is takes for the object to travel measured distance.
- Repeat this procedure at least three times to establish an accurate average.
- Use the average time of the test in the following formula *(Figure I-9, page I-24)* to determine the stream's velocity.

Velocity (in meters per second) = measured distance (in meters) / average time (in seconds)

TEMPORARY ROUTE SIGNING

I-29. A military route sign system, like the US highway sign system, can enable road users to reach their destinations by following route signs and road markings displayed along the roadside. MP patrols monitor signs on a routine basis, checking specific signs before critical

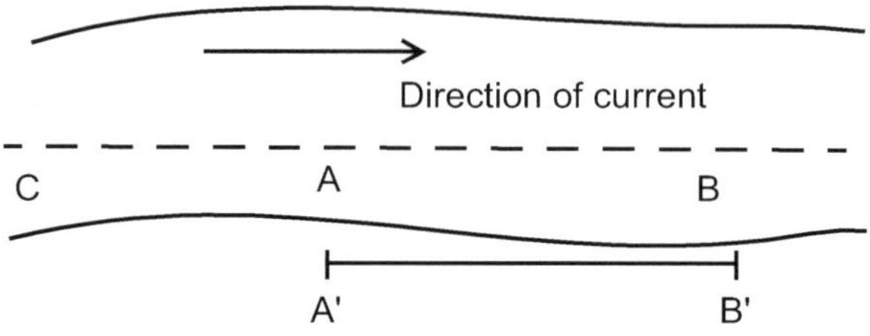

$$Velocity = \frac{A'B' \text{ (meters or feet)}}{\text{Time to float from } A \text{ to } B \text{ (seconds)}}$$

Measure distance A'B'.
Throw a floating object into the stream at C.
Determine the time required for the object to float to distance from A to B.

NOTE: This should be done three times, with the final velocity being the average of the three.

Figure I-9. Determine Stream Velocity

moves. Engineers erect permanent signs, but these signs can be damaged, destroyed, or moved by weather, saboteurs, and battle.

I-30. When MP on patrol encounter immediate and temporary MSR obstructions, like blown bridges or NBC contamination, they construct and erect signs quickly to guide vehicles around the obstruction. Prepare and post temporary signs to—

- Identify routes.
- Reroute traffic around problem areas.

- Help convoys and units move quickly and easily to their destinations even on an unfamiliar route.
- Show the drivers the locations of staging areas, tactical AAs, detours, key units, and facilities.
- Give directions, distances, and general information.
- Help lost military personnel find their way to the closest MP element.

TEMPORARY SIGNING

I-31. Obtain signing materials, signs, paint, and wire through the Army supply system. In an emergency, use boards, shingles, or cardboard. Use a portable sign-making kit to prepare signs when the signs will not conflict with standardization agreements (STANAGs).

I-32. Preplanned route signing for convoys and units traveling long stretches of MSRs and link routes must be planned well in advance. It is a time-consuming and manpower- and material-intensive operation. Signs used for this type of route are built by engineers or by MP battalion or company supply personnel. If tasked to take part in such an activity, ensure that after the signs have been constructed and coded for a particular route or operation, they are stored in a secure place to avoid compromise.

TEMPORARY ROUTE SIGNING

I-33. Place signs where they will support the traffic control plan and the traffic circulation plan. Specific sign locations are shown on the traffic control plan overlay prepared at the PM operations section. Changes in sign locations are reflected on the traffic control plan overlay as signs are added and deleted.

I-34. **Plan**. When MP encounter immediate and temporary MSR obstructions, like blown bridges or NBC contamination, they use the squad route sign kit or construct and erect hasty signs quickly to guide vehicles around the obstructions. Planning considerations include the following:

- Identifying routes.
- Rerouting traffic around problem areas.
- Helping convoys and units move quickly and easily to their destinations, even on unfamiliar routes.
- Showing drivers the locations of staging areas, tactical AAs, detours, key units, and facilities.
- Giving directions, distances, and general information.

I-35. **Execute**. Place signs where they will support the traffic control plan and the traffic circulation plan. Specific sign locations are shown on the traffic control plan overlay. Changes in sign locations are reflected on the traffic control plan overlay as signs are added or deleted.

I-36. Often, one three-man team can place signs along a route. To post a sign, a team member dismounts and walks 50 to 100 meters up the road. The other team members provide security and check and confirm the sign's placement to ensure that the drivers will be able to see the sign.

I-37. MP squads use the route signing kit to place signs along a route within the squad's AO. One team posts the signs while the other teams provide security. The squad leader checks and confirms the sign placement by traveling the entire route. He develops an overlay with the location of each sign in his squad's AO. Squad overlays are consolidated and contribute to a platoon traffic control plan.

I-38. More teams, up to a squad, may be needed for signing tasks in urban areas, in areas where a thorough reconnaissance has not been conducted, and in areas where the threat is unknown or is thought to be great. One team erects signs while following teams provide overwatch security. Use the following guidelines when placing signs:

- Place hazard signs about 150 meters before the hazard.
- Place regulatory signs where a regulation takes effect.
- Place all signs on the side of the road facing the traffic flow, about 1 meter off the traveled roadway. Conceal them from air view. If no cover is available, slant the sign stake forward.
- Place signs 1 to 2 meters above road level. Place all the signs at the same height if possible. Sign height is governed by roadside foliage, by whether the route is in an urban or a rural area, and by day or night use. In urban areas, place signs so that they are not hidden by vehicles or pedestrians, do not hinder pedestrians, and can be seen at night with street lighting or vehicle headlights. In open country, a good sign height is between thigh and knee height. This usually makes signs visible by day and night. Be sure that signs are not obscured by foliage.
- Use the least number of signs needed to be effective. Ensure that every sign is necessary and specific.
- Use more signs in urban areas than in rural areas.
- Use more signs on night routes than on routes primarily used during the day.
- Use signs to inform drivers to follow the common route when one road is used for two signed routes. Use signs to inform drivers when the routes diverge.

I-39. Conceal all the signs so that they are seen only from the direction from which they are approached.

There is no exact rule stating the distance from which a sign should be visible. However, the distance should be no greater than security allows and not less than is reasonable for those receiving directions.

I-40. Carefully conceal illuminated signs. Ensure that the light source is just strong enough to light the sign, but not strong enough to be seen from the air. This entails masking and covering the light sources. Consider placing chemical light sticks on top of the signs.

I-41. Place temporary route signs where they will provide warning and reaction time for the drivers. Do not block existing civilian signs. Place warning panels at convenient distances from where a route regulation takes effect. This distance can be shown on the panel (for example, BLACKOUT 500 METERS). In areas where blackout drive is the rule and drivers are using night-vision equipment, employ IR emitters on signs along routes to help guide the units to their destination. Just like any other light source, ensure that the IR emitters cannot be seen from the air.

I-42. Place guide signs at road junctions to prevent confusion. Put signs on both sides of the road if needed. Place confirmation signs *(Table I-3)* 150 meters beyond the critical road junctions to let drivers know that they are on the correct route.

Table I-3. Signs Identifying Military Routes and Locations

Type	Examples	Description	Purpose	Placement
Direction indicator		White, 30-centimeter diameter disk with black directional arrow (ID number or name is mounted below the disk) or White disk on black rectangular board with the identification number above	Indicates the direction and identifies the route or the name	At intersections where roads merge and where routes separate

Table I-3. Signs Identifying Military Routes and Locations (Continued)

Type	Examples	Description	Purpose	Placement
Guide signs (use together at important road junctions)		Rectangular with the symbols in white on black background; includes directional arrow and route number, name, and/or symbol	Indicates the locations, distances, directions, and routes	Where needed
Warning signs (use together at important road junctions)		Same as above	Shows the correct direction to take at route junctions	A sufficient distance (50 to 100 meters) before a junction to allow drivers to make the turn safely (On roads where speed is restricted, place the signs 25 meters before the junction.)

Table I-3. Signs Identifying Military Routes and Locations (Continued)

Type	Examples	Description	Purpose	Placement
Confidence signs		Same as above	Reassures the drivers that they are still on the correct route (Use them in urban areas to assure drivers that they are following the correct route. Use them on long stretches of road where it is unnecessary to use warning and confirmation signs for a considerable distance.)	Where needed
Confirmation signs		Same as above	Lets drivers know that they are on the correct route after changing direction	Just after turns, but visible while making the turn, if possible

Table I-3. Signs Identifying Military Routes and Locations (Continued)

Type	Examples	Description	Purpose	Placement
Countdown signs		Same as above	Warns of significant locations, such as SPs, RPs, starts and ends of routes, link routes, MSR junctions, and blackout areas—anything requiring a major change to the movement	Series of three signs at 100-meter intervals before the designated location
Regulatory signs		Military equivalent of civilian signs like STOP and YIELD	Regulates and controls the traffic on a route	Where needed (posted by engineers and considered permanent)

Table I-3. Signs Identifying Military Routes and Locations (Continued)

Type	Examples	Description	Purpose	Placement
Hazard signs		Yellow, diamond-shaped background with information printed in black	Indicates traffic hazards, such as dangerous corners, steep hills, or crossroads (rarely used in COMMZ, as civilian signs usually suffice. Use military signs in combat zones.)	Where needed (posted by engineers and considered permanent)
Military casualty evacuation route signs	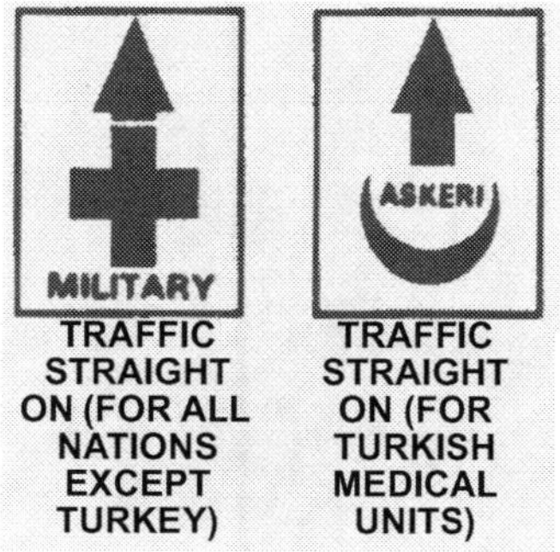	Rectangular white background with a red directional arrow, cross, or crescent; the word MILITARY; the unit or subunit designation; and other information like national markings or A directional disk with four segments cut out to form a cross or a directional disk with a crescent cut out (include the same information as above.)	Indicates evacuation route for military casualties	Where needed

Table I-3. Signs Identifying Military Routes and Locations (Continued)

Type	Examples	Description	Purpose	Placement
Civilian casualty evacuation route signs	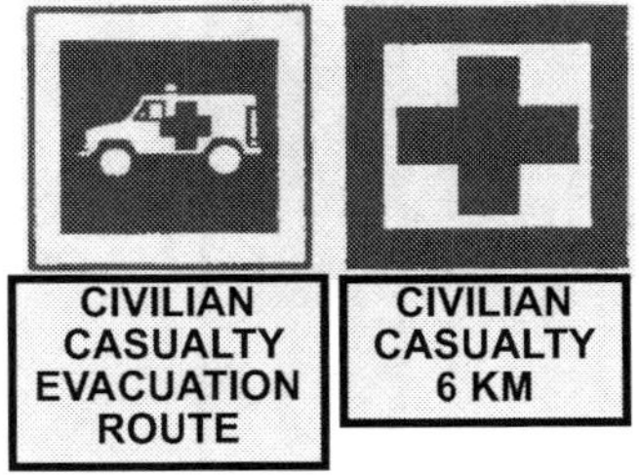	Blue Geneva Conventions information sign (includes ambulance in white with a red cross or crescent and the words CIVILIAN CASUALTY EVACUATION ROUTE beneath the sign in the HN language)	Indicates civilian casualty evacuation routes	Along routes for civilian traffic (designated by the HN)
Blackout warning signs		Based on the Geneva Conventions hazard warning sign (Place the legend and the distance on a rectangular plaque beneath the warning sign.)	Indicates the beginning of a blackout area	Same as warning signs

Table I-3. Signs Identifying Military Routes and Locations (Continued)

Type	Examples	Description	Purpose	Placement
Blackout enforcement signs		Geneva Conventions prohibitory sign; plaque beneath says VEHICLE LIGHTS FORBIDDEN	Indicates that a blackout is in effect	Every 100 meters along the blackout route
Blackout relaxation signs		Same as a blackout warning sign	Indicates the end of a blackout area	At the end of the blackout route

I-43. Place detour signs next to general traffic signs to identify the detour. Place the detour sign to the side (left or right) of the general sign that corresponds to the new direction to be taken.

I-44. Use signs to mark the entrance to a HQ or an installation along with a halt sign or other regulatory signs. Signs marking turnoffs and roads or tracks should include a directional disk or a directional arrow. Place signs to—

- Indicate where vehicles leave a signed route to get to the HQ or the installation.
- Mark the road or track leading to the HQ or the installation.
- Mark the entrance to the HQ or the installation.

SIGNING FOR KEY POINTS ON ROUTES

I-45. Place countdown signs *(Table I-3, page I-29)* at the beginning and end of a route. Clearly mark SPs and RPs with the appropriate countdown signs. Feeder routes to the SP may require signs to help the convoys find it.

I-46. Mark route detours with countdown signs showing the distance to where the detour begins. The signs clearly indicate the route to be detoured.

I-47. Place countdown signs so that they give clear warning of the end of the detour. Mark the end of the detour with a sign reading DETOUR END. Erect a warning sign at the end of the detour to show how to return to the original route.

I-48. Position detour signs ahead of time where terrain will require a defile or the like. Position the signs off the road, facedown on the ground. To implement the detour, erect the signs.

NOTE: The point where a link route meets the circuit is an ideal place for MP control. It is the point where the circuit and the link route begin and end.

I-49. Place signs at junctions of axial and lateral MSRs from all four directions. Place countdown signs 300, 200, and 100 meters before the junction.

I-50. To keep the number of signs to a minimum, battalion-size and smaller groups use a directional sign. The space for the military symbol may contain the UIN or the unit map-marking symbol. Print the UIN or the symbol so that it can be read when the arrow is vertical. This allows the sign to be used as a disk direction.

I-51. Therefore, only one type of sign is needed for all purposes. Put unit signs in place immediately before a unit movement. Remove them as soon as possible after the move.

MAIN SUPPLY ROUTE SIGNS

I-52. MSR signs identify MSRs by number, a pictorial symbol, or a name. Names and pictures—

- Are easier to identify and to remember than numbers.
- Prevent confusion with link-route signs marking the routes of units having three-figure identification numbers.
- Prevent security compromises by removing the chance use of an identification number already used on a map overlay.

I-53. There are two types of MSRs, axial and lateral, in a theater of operations. On the theater of operation's traffic circulation plan—

- Axial MSRs run to and from the forward edge of the battle area (FEBA) and are identified by odd numbers, like 87 or 215. Axial MSRs are shown as solid lines on the traffic circulation plan overlay. Axial routes are represented by a pictorial symbol on the route sign and are marked on the overlay as pictorial; for example,

MSR CLUB (PIC). On axial MSRs, *up* is toward the FEBA. It is shown on overlays and signs as a plain arrow. *Down* is away from the FEBA. It is shown as an arrow with a bar on the tail end.

- Lateral MSRs run parallel to the FEBA and are identified by even numbers, like 86 or 214. Lateral MSRs are shown as broken lines on the traffic circulation plan overlay. A named route sign represents lateral routes. The names are short, three- or four-letter words like *fox*, *ant*, or *hen*. *Up* and *down* on lateral MSRs show only general directions of travel. The general direction shown by *up* or *down* varies with the theater of operation and with the location of the FEBA. *Up* is usually to the north or east; d*own* is usually to the south or west. A plain arrow on an overlay or sign indicates *up*; a barred arrow shows *down*. To avoid confusion on lateral MSRs, use the letters N, E, S, W, NE, SE, NW, and SW on route signs to show the general direction of movement.

This paragraph implements STANAG 2010.

I-54. Ensure that all route signs are large enough to be read easily in poor light. Size is not specified, but—

- Signs for international use cannot be less than 40 centimeters by 33 centimeters.
- Bridge classification signs must conform to STANAG 2010.

LINK ROUTE SIGNS

I-55. There is no set requirements for the design of link route signs *(Figure I-10)*. (Sometimes showing direction with a white arrow on a black background is enough.) A link route connects a unit or an activity to an MSR. A departing convoy follows the link route signs to the MSR. The convoy follows the MSR until guided off the MSR by signs warning of the need to exit and again follow link route signs until arriving at its destination. That part of a traveled route coinciding with an MSR will not have link route signs.

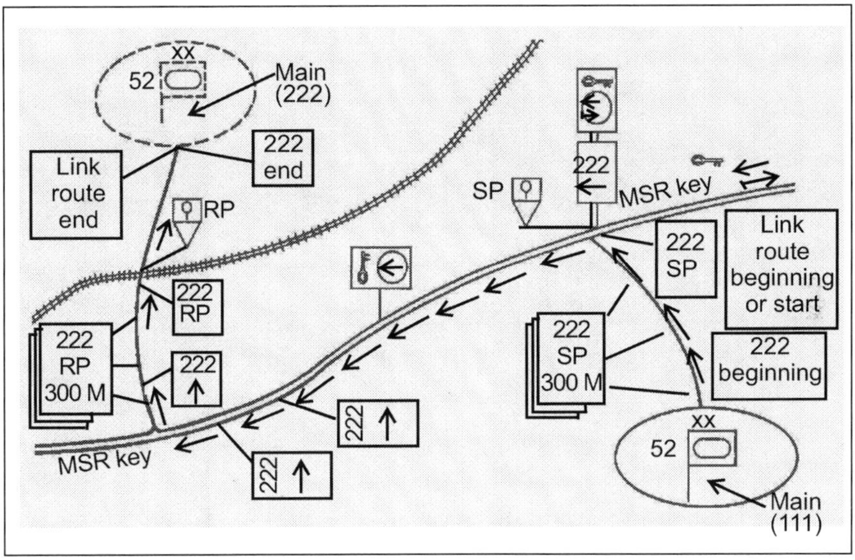

Figure I-10. Link Route Signing

HEADQUARTERS AND INSTALLATION SIGNS

I-56. All signs must be used in HQ and parts of HQ, down to battalion or equivalent level. Examples are—

- Medical installations.
- Water points.
- Ammunition depots.
- TCPs.
- Decontamination stations.
- Fuel installations.

This paragraph implements STANAG 1059.

I-57. HQ and installation signs show the military symbol for the HQ or the installation. For HQ at brigade level and above, include the national distinguishing letters given in STANAG 1059. The sizes of the sign and the symbol are not standardized. The symbol and the background must be clearly contrasting colors. The actual colors used are left to the discretion of each nation.

DIRECTIONAL SIGNS

I-58. A directional sign *(Table I-3, page I-29)* must be—

- A white disk displaying a black directional arrow accompanied by a route identification number or name that can be mounted beneath the disk.
- A white disk mounted or superimposed on a black rectangular board with the number or name on the board above the disk.

I-59. Directional disks cannot be less than 30 centimeters in diameter. Drill eight holes in the disks at equal intervals around the circumference so that they can be erected with the arrow pointing in the correct direction.

I-60. Directional signs show directions for axial and lateral routes on hastily prepared routes or very temporary routes like detours. The initial sign shows the use of disk directional signs and orders the drivers to follow the disk direction.

I-61. An MSR may require hundreds of signs. Make the signs in quantity without the black arrow. Stencil or stick the arrows on later. The superimposed signs are easiest to store and use. It is hard to secure the two elements of the circular two-piece signs, and it is hard to position that sign for display.

I-62. Use yellow instead of white for signs used during prolonged snowfall conditions or for signs permanently erected where there can be prolonged snowfalls.

PORTABLE SIGN-MAKING KIT

I-63. The new portable sign-making kit (national stock number [NSN] 9905-01-478-3723) is lightweight and simple to use and contains tools and materials that, when used in conjunction with vehicle BII and locally available materials, produce temporary signs in support of MP operations. Since the kit is not fully self-contained, the user must plan ahead. The new, lighter kit gives MP units the ability to pack according to the METT-TC. Even though a significant difference exists between signs made with the kit and specifications in STANAGs (STANAG route signs generally have a black background with white letters, and the kit's sign background material is white), it does not differ from the kit it replaces. The soft-sided sign kit comes with white plastic bags that can be used to make signs up to 21 inches wide by 23 inches high. *Figure I-11,* shows a sample sign; and *Table I-4, page I-42,* shows a component listing for sign making.

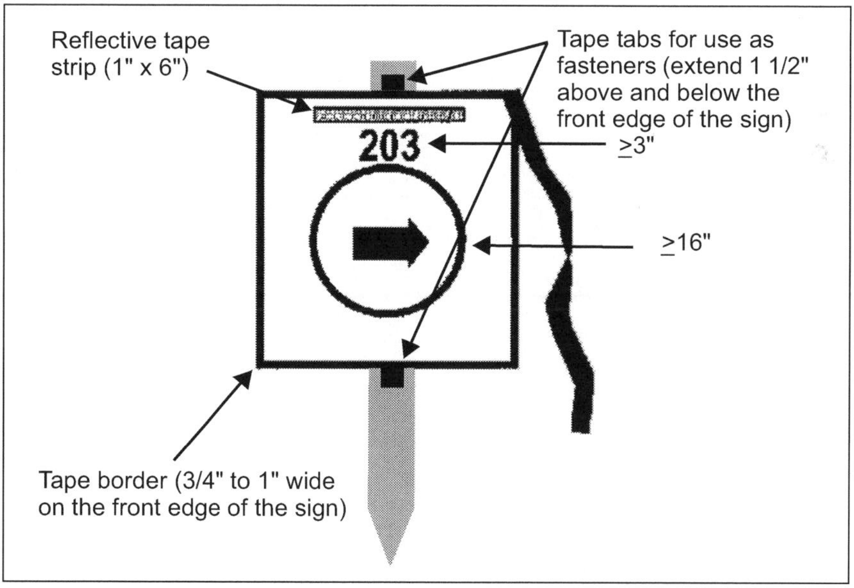

Figure I-11. Sample Sign

Table I-4. Component Listing

Tools	**NSN**	
Tool bag	5140-01-481-4460	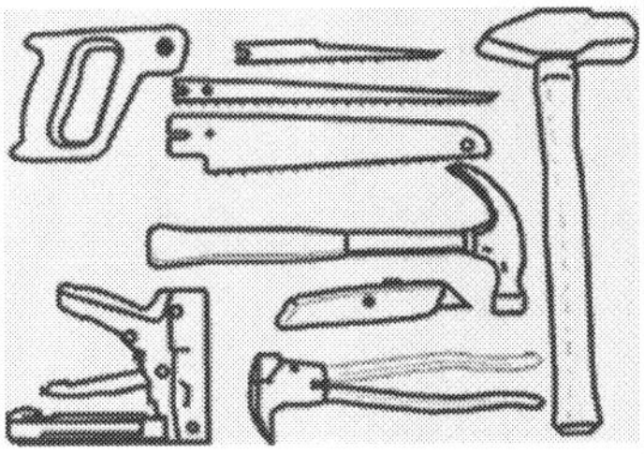
Carpenter's hammer	5120-01-112-8350	
Utility knife Replacement blades	5110-00-892-5071 5110-00-293-2865	
Linemans tool	5120-00-293-3485	
Blacksmith's/engi-neer's hammer (3 lb)	5120-00-900-6103	
Nested saw set	5110-00-293-0090	
Staple gun	5120-00-889-1796	
Fasteners	**NSN**	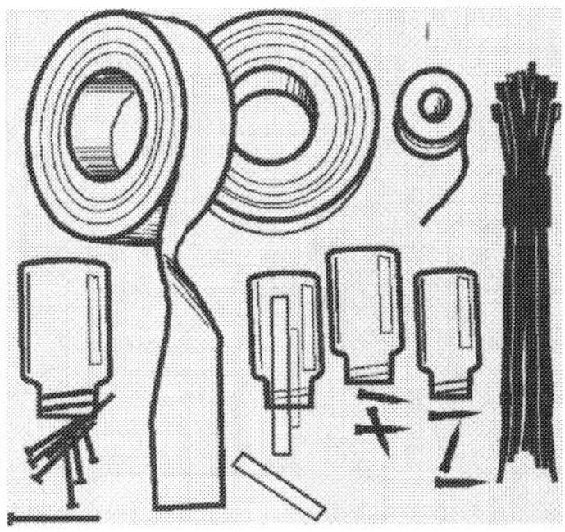
One pound of 1.75" steel screw-shank nails (concrete)	5315-00-543-3023	
Five pounds of 3.5" (10-12d) common nails	5315-00-753-3885	
One-half pound of 1.5" steel barbed-shank nails (roofing)	5315-01-381-9609	
Tape, pressure sensitive (2" x 160', 100 miles an hour)	7510-00-074-5124	
Staples, 1/2", box of 5,000	5315-00-889-2605	
Cable ties, pkg of 100	5975-00-156-3253	
Wire, nonelectrical, 1-lb roll	9505-00-293-4208	

Table I-4. Component Listing (Continued)

Marking Materials	NSN
Black tube-type poster markers (2 each)	7520-01-336-2856
Red tube-type poster marker	7520-01-336-2857
Blue tube-type poster marker	7520-01-336-2858
Bags, white plastic, trash, 8 gal, roll/box of 20 to 30 (2 each)	8105-01-481-4225
Reflective tape (white 1" x 150')	9390-00-949-8044
Ribbon, flagging, surveyors (arctic-glo pink)	9905-00-194-9703

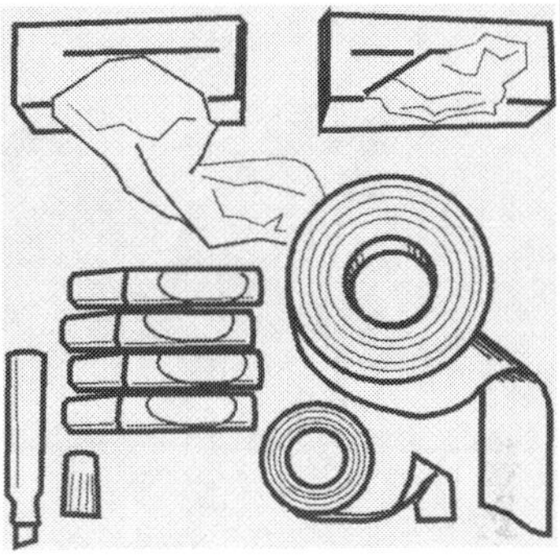

Appendix J

Nuclear, Biological, Chemical Defense

The technology to produce and deliver chemical and biological agents and weapons of mass destruction is widely available. This technology provides military and nonmilitary leaders in unstable regions with a means to deter more technically advanced nations. Weapons of mass destruction can radically change the flow of battle, shift the balance of combat power, and place a superior force at risk. Areas targeted with these weapons may remain hazardous for days, even weeks after an attack. Additionally, nuclear reactor and chemical production facility accidents or sabotage can release radiological or chemical contaminants that can threaten friendly forces and military operations.

OVERVIEW

J-1. Enemy forces may threaten sustainment areas to seize and maintain the initiative, while degrading or eliminating a unit's flexibility and capability to support operations. Sustainment area activities are lucrative targets for enemy NBC attacks. Attacking sustainment nodes weakens the main battle area's force effectiveness without adversely affecting an enemy's immediate maneuver. To achieve these aims, threat activities in sustainment areas target key critical support and logistic facilities. These facilities include—

- Special-weapons storage sites and delivery systems.

- C^2 facilities.
- Air defense artillery sites.
- Air bases.
- Seaports.
- MSRs.

HAZARD DETECTION AND REPORTING

J-2. The best units to perform NBC reconnaissance are those units with the M93 NBC Reconnaissance System (FOX). Planners consider the disadvantages of conducting NBC reconnaissance with HMMWV-equipped units. The disadvantages of using a HMMWV include the following:

- The crew conducts operations in MOPP 4 while in contamination.
- The vehicle stops to conduct survey and detection.
- The vehicle has difficulty keeping up with the maneuver force over rugged terrain.
- The readings obtained using an M256A1 kit, detector paper, and a chemical agent monitor (CAM) are not as reliable as readings from the M93 (FOX).

J-3. Planners use the IPB and the METT-TC to determine the type of NBC reconnaissance technique and the type of equipment to be used for the NBC reconnaissance.

J-4. Early detection of NBC hazards is required for timely warning of units and personnel in the affected areas. NBC reconnaissance is an active contamination avoidance measure that provides commanders with information on NBC hazards in an AO.

J-5. NBC reconnaissance contributes to the overall intelligence collection effort and consists of the following five critical tasks: detecting, identifying, marking, reporting, and sampling. The four types of NBC reconnaissance include search, survey, sample, and surveillance.

J-6. MP may be tasked to conduct NBC reconnaissance while conducting other reconnaissance operations. They are equipped to detect and identify radiological and chemical hazards, mark the contaminated area, and report the hazard. They are not equipped to collect or handle samples.

J-7. NBC surveillance activities are planned to support the intelligence collection effort. MP conduct NBC surveillance by monitoring MSRs, critical facilities, and key terrain. They employ NBC detector equipment and measures that include the following:

- M8A1 automatic chemical agent alarm system.
- AN/VDR-2 RADIAC set.
- CAM.
- Visual observation for unexplained artillery or spray.
- M256A1, M8, and M9 paper.

J-8. If MP observe the indications of an NBC attack, they may be ordered to conduct an NBC survey to define the boundaries of the contamination, locate and mark clean bypass routes, and occupy OPs to observe the designated area. MP conduct an NBC survey, on order, to obtain detailed information concerning a contaminated area, including the location of the general boundaries of the contaminated area and the intensity of the contamination (nuclear) or the type of agent (chemical).

J-9. MP locate the boundaries and place warning markers around the contaminated area at all the entry points. They report information using an NBC 4 report and inform the road users of the contaminated area, and direct the traffic to alternate routes to avoid the contaminated area.

J-10. At all echelons, MP provide continuous information to higher HQ using NBC 1 and NBC 4 reports *(Figure J-1)*. These report formats provide a rapid means of disseminating information. Use the NBC 1 report to record the initial use and subsequent data concerning enemy NBC attacks. The initial NBC 1 report precedence is flash and all others are immediate. *Table J-1, pages J-6 through J-10,* gives an explanation of the various lines in an NBC report.

J-11. Use the NBC 4 report for radiation dose rate measurements. Usually, the unit submits two reports—one on initial contact and another for peak dose rate. Measure radiation dose rates in the open, 1 meter above the ground. If radiation dose rates are taken inside a vehicle or a shelter, at least one outside reading is necessary to determine the correlation factor. Other items of the report are optional. Refer to *FM 3-3* for more information about NBC reports.

J-12. As MP move throughout an AO, they monitor for the presence of an NBC hazard, mark contaminated areas, send NBC reports through operational channels, and direct traffic around or through hazards to ensure that the troops and the supplies get where they are needed on the battlefield.

CONTAMINATION MARKING

J-13. Mark contamination so that unsuspecting personnel will not be exposed to it. When detecting or suspecting NBC hazards, mark all likely entry points into the area and report the contamination to higher HQ. The only exception to this policy occurs when marking the area would help the enemy. If the commander makes this exception, report the hazard to protect friendly units.

NBC 1 Report (Observer's Report)		
NBC 1 (Nuclear) Report	**NBC 1 (Biological) Report**	**NBC 1 (Chemical) Report**
B. NB062634	B. LB206300	B. LB200300
C. 90-degree grid	D. 200410Z	D. 201405Z
D. 201405Z	E. 200414Z	E. 201412Z
G. Aircraft	F. LB206300, actual	F. LB206300, estimated
H. Surface	G. Aerial spray	G. Bomblets
J. 60	H. Unknown	H. Nerve
L.15 degrees		V. Airburst
NOTE: Line items B, D, H, and either C or F should always be reported. Use other line items if the information is known.		

NBC 4 Report (Reconnaissance, Monitoring, and Survey Results)	
NBC 4 (Nuclear) Report	**NBC 4 (Chemical) Report**
Q. LB123987	H. Nerve V
R. 35	Q. LP200300, liquid
S. 201535Z	S. 170610Z
NOTES: **1. Line items H, Q, R, and S may be repeated as often as necessary.** **2. In line R, descriptive words, such as initial, peak, increasing, decreasing, special, series, verification, or sumary may be added.**	

Figure J-1. Example of NBC 1 and NBC 4 Reports

Table J-1. Explanation of Line Items in NBC Report

Line	Nuclear	Chemical and Biological	Remarks
A	Strike serial number	Strike serial number	Assigned by the NBC center
B	Position of the observer	Position of the observer	Grid coordinates (universal transverse mercator [UTM] or place).
C	Direction of the attack from the observer, to include the unit of measure	Direction of the attack from the observer	Nuclear: Use degrees magnetic north (DGM), degrees true north (DGT), degrees grid north (DGG), mils magnetic north (MLM), mils true north (MLT), or mils grid north (MLG) Chemical: Measure clockwise from the grid or magnetic north (in degrees or mils)
D	DTG of detonation	DTG for the start of the attack	Nuclear: Use Zulu time. Chemical: Designate time zone used.
E	NA	DTG of the end of the attack	Designate time zone used.
F	Location of the area attacked	Location of the area attacked	Use grid coordinates (or place). State whether the location is actual or estimated.
G	Suspected or observed event and the means of delivery or kind of attack	Kind of attack	State whether the attack was by artillery, mortars, rockets, missiles, bombs, or spray.

Table J-1. Explanation of Line Items in NBC Report (Continued)

Line	Nuclear	Chemical and Biological	Remarks
H	Type of burst	Type of agent and type of burst P = persistent NP = nonpersistent	Nuclear: Specify air, surface, or subsurface. Chemical: State whether by air, ground, or spray attack.
I	NA	Number of munitions or aircraft	If known.
J	Flash-to-bang time	NA	Use seconds.
K	Crater present or absent and the diameter	Description of terrain and vegetation	Nuclear: Sent in meters. Chemical: Sent in NBC 6.
L	Cloud width at H+5 minutes	NA	State whether measured in degrees or mils.
M	Stabilized cloud top or cloud bottom angle or cloud top or bottom height at H+10 minutes	NA	Nuclear: State whether the angle is cloud top or cloud bottom and whether it is measured in degrees or mils. Chemical: State whether the height is cloud top or cloud bottom and whether it is measured in meters or feet.
N	Estimated yield	NA	Sent as kilotons.
O	Reference DTG for estimated contour line not H+1	NA	Use when the contours are not plotted at H+1.
P	For radar purposes only	NA	None

Table J-1. Explanation of Line Items in NBC Report (Continued)

Line	Nuclear	Chemical and Biological	Remarks
PA	NA	Predicted hazard area (coordinates)	If the wind speed is 10 kilometers per hour or less, this item is 010 (the radius of the hazard area in kilometers).
PAR	Coordinates of external contours of the radioactive cloud	NA	Six-digit coordinates. Letter R identifies RADAR set.
PB	NA	Duration of the hazard in the attack and hazard area	State in days, hours, minutes, and so forth
PBR	Downwind direction of the radioactive cloud and the unit of measure	NA	DGM, MLM, DGT, MLT, DGG, or MLG. The letter R identifies a RADAR set.
Q	Location of the reading	Location of the sampling and the type of sample	Nuclear: UTM or the place. Chemical: UTM or the place. State whether the test was air or liquid.
R	Dose rate or actual value of decay exponent	NA	State the dose rate in centigray per hour (cGyph).
S	DTG of the reading	DTG contamination detected	State the time of the initial identification test sample or when the reading was taken.

Table J-1. Explanation of Line Items in NBC Report (Continued)

Line	Nuclear	Chemical and Biological	Remarks
T	H+1 DTG	DTG of the latest contamination survey of the area	NBC 5 and NBC 6 reports only.
U	1,000-cGyph contour line	NA	Plot in red.
V	300-cGyph contour line	NA	Plot in green.
W	100-cGyph contour line	NA	Plot in blue.
X	20-cGyph contour line (30 cGyph contour line is used by other NATO forces)	Area of actual contamination	Nuclear: Plot in black. Chemical: Plot in yellow.
Y	Direction of the left and right radical lines	Downwind direction of the hazard and the wind speed	Nuclear: Direction measured clockwise from grid north (GN) to the left and then right radial lines (degrees or mils, state which), four digits each. Chemical: Direction is four digits (degrees or mils) and the wind speed is three digits (kilometers per hour).

Table J-1. Explanation of Line Items in NBC Report (Continued)

Line	Nuclear	Chemical and Biological	Remarks
Z	Effective wind speed Downwind distance of Zone 1 Cloud radius Unit of measure for each category	NA	Three digits—effective wind speed (kilometers per hour) Three digits—downwind distance of Zone 1 (kilometers or nautical miles) Two digits—cloud radius (kilometers or nautical miles) (the wind speed is less than 8 kilometers per hour, this line contains only a three-digit radius of Zone 1 [kilometers])
ZA	NA	Significant weather phenomena	Air stability: Two digits Temperature in centigrade: Two digits Humidity: One digit Significant weather phenomena: One digit Cloud cover: One digit
ZB	Used to transmit correlation factors or transmission factors	Remarks for additional chemical and biological attacks only.	Include any additional information.

PROCEDURES FOR MARKING

J-14. Face markers away from the contamination. For example, if markers are placed 50 meters from the outside edge of a contaminated area to mark a radiological hot spot, they face away from the point of the highest contamination reading. Place markers at roads, trails, and other likely points of entry. When time and mission permit, use additional markers. The distance between the signs varies. In open terrain, place them farther apart than in hilly or wooded areas. Soldiers should be able to stand in front of a marker and see the markers to the left and right of it.

J-15. Units discovering a marked, contaminated area do not have to conduct elaborate, time-consuming surveys. The new unit checks the extent of the contamination and alters its plans if necessary. If the size of the hazard has expanded or decreased, relocate the signs. If the hazard is gone, remove the signs and report changes to higher HQ.

TYPES OF MARKERS

J-16. US forces use standard NATO markers to make it easier for allies to recognize the hazards. These markers are in the standard NBC marking set. Colors and inscriptions on a marker indicate the type of hazard. Other contamination information is written on the front of the sign.

RADIOLOGICAL CONTAMINATION DETECTION AND MONITORING

J-17. Conduct radiological monitoring using an AN/VDR2, and start on the order of the commander or as set by the SOP. Mark the area with radiological contamination markers. Report radiation dose rates and the times and locations of the readings to higher HQ using the NBC 4 nuclear report. All units start continuous monitoring when–

- Moving from one area to another on the battlefield.

- Getting a fallout warning.
- Knowing that a nuclear burst was reported, seen, or heard.
- Detecting radiation of 1 centigray per hour by periodic monitoring.
- Being ordered by the unit commander.

J-18. Continuous monitoring stops on order from the higher HQ or when the dose rate falls below 1 centigray per hour, except for units on the move.

TECHNIQUES FOR MONITORING

J-19. Use both direct and indirect techniques when conducting radiological monitoring. Refer to *FM 3-3* for more information on radiological monitoring.

Direct Monitoring

J-20. The direct technique is the simplest and the most precise. The soldier must—

- Stand at the desired location.
- Hold the RADIAC meter waist high and turn it slowly 360 degrees. Refer to *TM 11-6665-251-10*.
- Record the highest reading on *DA Form 1971-R*.
- Take the reading in the open, at least 10 meters from buildings or large structures, if possible.

Indirect Monitoring

J-21. Use the indirect technique inside shelters or vehicles. The soldier must—

- Stand at the center of the shelter.
- Hold the RADIAC meter 1 meter above the floor and slowly rotate it 360 degrees.
- Record the highest reading.
- Take all readings from one selected position when monitoring from inside a vehicle or shelter.

SURVEY TEAM

J-22. MP may be tasked to perform as part of a survey team. When essential contamination information cannot be obtained from monitoring reports, a radiological survey may be required. Radiological surveys are directed efforts to learn the extent and intensity of radiological contamination. A survey requires a control team and one or more survey teams. The HQ directing the survey usually provides the control team. In a team assigned to a survey—

- The team leader monitors the RADIAC meter and provides C^2.
- The driver operates the vehicle.
- The gunner provides security.

J-23. The control team instructs the survey teams on what to do. The MP team leader ensures that each team member receives the following information from the control team:

- Purpose. The survey determines the presence and the level of fallout or induced radiation near ground zero, in a specific area, along a route, at a specific location, or any combination of these.
- Start and finish times. The times during which a survey will be conducted.
- Survey execution. The route to be followed and the locations where readings are to be made or the distances needed between the readings are given.
- Radiation safety precautions. The control team tells the survey team the turn-back dose. The dose rate is monitored on the AN/VDR2. If the turn-back dose rate is reached as the survey team travels forward, the team immediately leaves the fallout area by the same route used to enter it, unless instructed otherwise. If the dose rate decreases as the survey team moves ahead, the team continues to perform its mission.

- Recording limitations. The survey team will not convert inside readings to outside readings. Record readings in the dose rate column only.
- Communications. The method and times to be used to communicate survey data to the chain of command.
- Special instructions. Any special instructions for marking the contaminated areas or special security precautions.

CHEMICAL AGENT DETECTION

J-24. During routine operations, MP may detect chemical agents while making periodic checks for contamination. Or as previously discussed, teams may be tasked to conduct a chemical reconnaissance along an MSR or in a given area after an enemy attack. Some chemical agents are odorless, colorless, tasteless, and invisible. Teams must use chemical agent detection equipment and materials to be certain that all chemical agents are detected.

PREPARATION

J-25. When assigned a mission or a task to locate and mark a suspected contaminated area, MP leaders must ensure that they prepare properly for the mission. Preparation for any NBC reconnaissance mission begins with the inspection of personnel and equipment. As a minimum, each squad must have the following equipment on hand:

- M8 paper.
- M9 paper.
- M256 and M256A1 detector kits.
- M8A1 alarm.
- CAM.
- Marking kit.
- M13 decontamination apparatus (DAP).

- M291 decontamination kit.
- MOPP gear.
- Mark 1 nerve agent auto injector kit (NAAK).
- VS-17 marking panels.

J-26. In addition to ensuring that the proper equipment is on hand, leaders must ensure that alarms and paper are properly mounted and functional and that all external equipment is stowed. The platoon leader includes a rehearsal of NBC reconnaissance techniques when preparing the mission. The platoon leader coordinates with the unit chemical officer for any special instructions, ensuring that thorough decontamination support is available at the conclusion of the mission.

RECONNAISSANCE

J-27. When detecting for chemical contamination during NBC reconnaissance, the MP leader—

- Ensures that all members of the party wear MOPP 4 with M9 paper attached to their protective clothing and vehicle.
- Drives along the route until the chemical agent alarm sounds. By the time the alarm sounds, the team may be some distance from where the agent was first encountered. Keep this in mind when selecting the area to be searched.
- Gives special attention to shell craters, low-lying patches of woods, defiles, ravines, streambeds, and areas covered with high grass or underbrush.
- Has one MP remain in each vehicle to provide security and to monitor the radio.
- Has other patrol members dismount and check for contaminants.
- Ensures that one person uses M8 detector paper on the tip of a stick.

- Ensures that one person operates the M256 chemical agent detector kit. (Refer to *TM 3-6665-307-10* for detailed operating instructions for the M256, and refer to *FM 3-6* for more information.)

J-28. When MP find chemical contamination, they mark the area with NBC markers. They service the automatic chemical agent alarm to ready it for further operation. MP must be outside the contaminated area when the alarm is serviced. *TM 3-6665-312-12&P* provides instructions for operators and organizational personnel. MP remount the vehicles and continue the reconnaissance. While moving, they send an NBC 4 report by radio and follow this procedure each time a chemical agent is detected. The goal of the reconnaissance is to define the contaminated area only to the degree necessary to provide friendly forces with the information needed to avoid the area.

BIOLOGICAL AGENT DETECTION

J-29. For warning of biological attacks or contamination outside a unit's immediate area, the unit relies on warnings relayed through its chain of command from division level or higher units. MP help provide warning of biological hazards by their reports of area activity passed on from their route, zone, and area reconnaissance.

J-30. Be alert to any sign that biological agents are being used. Promptly report any unusual occurrences of sickness in troops or civilians. Report any unusual actions of animals, birds, or large numbers of dead animals or birds not likely to have been caused by combat. Observe the types of weapons and munitions used during attacks. Compare them to the known characteristics of enemy biological attacks. Refer to *FM 3-3* for detailed information on biological agent detection, identification, and defense.

J-31. Biological attacks are most likely to occur at night or during extended twilight to avoid toxin exposure to direct sunlight. (Direct sunlight degrades biological agents.) Cloudy and foggy days are also favorable for the use of biological agents.

J-32. Watch for the following indications that a biological attack has occurred:

- A low-flying aircraft that appears to be producing a mist or a spray.
- The use of any type of spray device.
- The use of ammunition that does not seem to have any immediate effect, such as a bomblet.
- Unusual types of bomblets.
- Insect swarms that suddenly appear after an aircraft drops containers that do not seem to have any immediate effects.

SELF-DEFENSE MEASURES

J-33. MP not only provide early warning for units in the sustainment area, but they also employ NBC defense measures to protect their own forces. NBC defense is a balance of the following three principles:

- Avoidance.
- Protection.
- Decontamination.

AVOIDANCE

J-34. Avoiding NBC attacks and hazards is the key to NBC defense. Avoiding contamination involves bypassing contaminated areas and avoiding detection by the enemy. If forces avoid detection, they are less likely to be targeted for attack and will not need to apply protection and decontamination. *FM 3-3* provides detailed guidance on conducting contamination avoidance procedures.

J-35. NBC contamination avoidance is especially critical for light forces. Leaders conduct risk assessments to determine what protective measures they will direct. Refer to *FM 3-4* for detailed information on conducting a risk assessment.

PROTECTION

J-36. A unit may become contaminated because of direct NBC attack or because the mission causes them to enter a contaminated area. In either case, NBC protection is an integral part of operations. Avoidance and protection are closely related. Techniques that work for avoidance also provide protection. There are four broad groups of activity that comprise protective measures and include the following:

- Hardening positions and protecting personnel.
- Assuming MOPP.
- Reacting to attack.
- Using collective protection.

J-37. Refer to *FM 3-4* for detailed guidance on conducting NBC protective procedures.

DECONTAMINATION

J-38. The extent and timing of decontamination depend on the tactical situation, the mission, the degree, the type of contamination, and the resources available. Refer to *FM 3-5* for detailed guidance on conducting decontamination operations.

J-39. The primary purpose of decontamination is to stop erosion of combat power and to reduce the number of casualties. Combat power drops as soon as personnel don MOPP gear. The mask restricts the field of vision and reduces observation and target acquisition ability. Communication is more difficult. Mobility is reduced because personnel reduce their physical work rate to avoid heat stress. The longer a unit remains contaminated, the greater the chance of NBC casualties. The three levels of decontamination are immediate, operational, and thorough.

Immediate Decontamination

J-40. Immediate decontamination is a basic soldier survival skill carried out by soldiers as soon as possible after they discover they are contaminated. Its basic purposes are to minimize casualties, save lives, and limit the further spread of contamination. Personnel treat any contact between chemical or biological agents and bare skin as an emergency. Some agents can kill if they remain on the skin for longer than a minute. The best technique for removing or neutralizing these agents is to use the M291 skin decontamination kit. Leaders must ensure that their soldiers are trained to execute this technique automatically, without waiting for orders.

J-41. Begin personal wipe down within 15 minutes of contamination. The wipe down removes or neutralizes contamination on the hood, mask, gloves, and personal weapon. For chemical and biological contamination, use mitts from the M295 individual equipment decontamination kit (IEDK). For radiological contamination, wipe off the contamination with a cloth or simply brush or shake it away.

J-42. Begin the operator's spray down of the equipment immediately after completing the personal wipe down. The spray down removes or neutralizes contamination on the surfaces that operators must touch frequently to perform their mission. For chemical and biological contamination, use on-board decontamination apparatuses, such as the M11, the M13, or the M295 IEDK to decontaminate surfaces where DS2 cannot be applied.

NOTE: Wash decontamination solution number 2 (DS2) off surfaces no more than 30 minutes after application. If necessary, use 5-gallon water cans or other water sources to assist in removing DS2.

J-43. For radiological contamination, brush or scrape away the contamination with whatever is at hand or flush it with water and wipe it away.

Operational Decontamination

J-44. Operational decontamination allows a force to continue fighting and sustain its mission after being contaminated. It limits the hazard of transferring contamination by removing most of the gross contamination on equipment and nearly all the contamination on soldiers. This speeds the weathering process and allows clean areas (people, equipment, and terrain) to stay clean. Following operational decontamination, soldiers who have removed sources of vapor contamination from their clothing and equipment unmask temporarily in hazard-free areas to eat, drink, and rest.

J-45. Accomplish operational decontamination using the assets of the parent unit. Make use of the following decontamination techniques:

- Vehicle wash down.
- MOPP gear exchange.

J-46. Conduct these procedures at the same time. They are best performed at squad level. Vehicles and personnel that are not contaminated should not go through either technique. Conduct vehicle wash down as far forward as possible. It is normally performed using the power-driven decontamination equipment (PDDE) crew with assistance from the squad decontamination crew. It is most effective if started within 1 hour after contamination.

J-47. The two steps in vehicle wash down are as follows:

Step 1. Button up the vehicle and secure the equipment.

Step 2. Wash down the vehicle and the equipment with hot, soapy water for 2 to 3 minutes.

J-48. Because speed is important, do not check vehicles for contamination after vehicle wash down. Remove only gross contamination.

Thorough Decontamination

J-49. Thorough decontamination reduces contamination to negligible risk levels. It restores combat power by removing nearly all contamination from the unit and individual equipment. This allows troops to operate equipment safely for extended periods at reduced MOPP levels. A contaminated unit conducts detailed troop decontamination (DTD) under the supervision of the chemical unit. Ordinarily, the chemical unit selects a site, sets it up, and performs the detailed equipment decontamination (DED) with assistance from the contaminated unit. A small risk from residual contamination remains, so make periodic contamination checks after this operation.

J-50. Conduct thorough decontamination as part of an extensive reconstitution effort in brigade, division, and corps support areas. Support sites at lower levels cannot provide the quantities of decontamination resources (water, decontaminants, and time) required for such an extensive process. In some cases, a contaminated unit could conduct a thorough decontamination operation with organic decontamination assets, but support from a chemical unit is usually required.

J-51. After thorough decontamination, the unit moves into an adjacent AA for reconstitution. Support elements from the brigade, division, or corps support areas replenish combat stocks, refit equipment, and replace personnel and equipment. The newly reconstituted unit leaves the AA fully operational and fit to return to battle.

J-52. Thorough decontamination does the best job of getting rid of contamination and its hazards, but it is often not possible. It requires large quantities of valuable resources that may not be immediately available. The next best solution is to decontaminate only what is necessary to sustain the force and continue to fight.

MILITARY POLICE LEADERS' RESPONSIBILITIES

J-53. MP leaders ensure that all soldiers understand the concepts of NBC defense and the skills necessary to survive an attack. Individual NBC skills include–

- Recognizing hazards and taking cover.
- Using protective measures and MOPP gear.
- Knowing avoidance measures, first aid, and decontamination.

J-54. MP commanders also designate and train special-purpose teams. These special-purpose NBC teams include the following:

- Chemical agent detection teams.
- Radiological survey and monitoring teams.
- Decontamination teams.

J-55. To assist with NBC defense, MP commanders rely on the unit NBC officer and the NBC NCO. They—

- Provide technical assistance to the commander on NBC defense training and operations.
- Coordinate unit NBC defense activities.
- Give NBC defense instructions to achieve basic operating standards of proficiency for the unit and individuals.
- Plan and supervise NBC defense training aspects of operational training exercises and maneuver.
- Supervise the preparation of unit NBC defense SOPs.
- Supervise the operation and maintenance of NBC material.
- Determine by dosimetry or calculation (as appropriate) the total dose of radiation and the time spent in radiologically contaminated areas to avoid exceeding command exposure guidance.
- Prepare fallout prediction patterns.

- Plan NBC reconnaissance and advise commanders on the best routes to cross or bypass an NBC contaminated area.
- Plan and coordinate decontamination within the unit.
- Maintain records of the unit's radiation exposure.
- Estimate downwind hazards for chemical attacks.
- Report NBC data to the next higher HQs.
- Analyze the unit's vulnerability to an NBC attack.
- Evaluate individual and unit competence in NBC defense, and advise the commander on the unit's ability to survive and continue operations in an NBC environment.
- Advise the commander on all matters pertaining to unit NBC defense.

MISSION-ORIENTED PROTECTION POSTURE LEVELS, ALARMS, AND SIGNALS

J-56. Soldiers on the integrated battlefield face a combination of NBC and conventional attacks. Individual and unit protection against chemical attack or contamination hinges on effective use of MOPP gear and on individual proficiency in basic NBC skills. All MP must be familiar with the standard MOPP levels shown in *Figure J-2, page J-24,* and *Table J-2, page J-25.*

J-57. When an NBC attack is recognized, every soldier receives the warning and assumes the appropriate MOPP level. Those in immediate danger need warnings they can see or hear. The alarm or signal must be simple and unmistakable for quick and correct reaction. Units not immediately affected need the information to prepare for the hazard or to change plans.

Figure J-2. Illustrated MOPP Levels

J-58. Mark the contaminated area if an NBC hazard has been located. The NBC Warning and Reporting System (NBCWRS) and the contamination markers contribute to the warning procedures for follow-on forces. In the immediate area of contamination, several methods (or a combination of methods) will allow quick reaction by all platoon members.

J-59. These methods include vocal alarms (shout of *GAS*), the M8A1 alarm, nonvocal alarms (horn blast or banging of metal-to-metal objects), and visual alarms (most commonly, hand-and-arm signals). The tactical situation may not allow for audible alarms; therefore, the platoon SOP should clearly detail the visual signals for contamination.

BIOLOGICAL DEFENSE

J-60. The key protective measure against a biological attack is maintaining a high order of health, personal

Table J-2. Standardized MOPP Levels

MOPP Equipment	MOPP Levels: MOPP Ready	MOPP Levels: MOPP Zero	MOPP Levels: MOPP 1	MOPP Levels: MOPP 2	MOPP Levels: MOPP 3	MOPP Levels: MOPP 4	Command: Mask Only
Mask	Carried	Carried	Carried	Carried	Worn[1]	Worn	Worn
Overgarment	Ready[3]	Available[4]	Worn[1]	Worn[1]	Worn[1]	Worn	Available
Vinyl overboots	Ready[3]	Available[4]	Available[4]	Worn	Worn	Worn	Available
Gloves	Ready[3]	Available[4]	Available[4]	Available[4]	Available[4]	Worn	Available
Helmet protective cover	Ready[3]	Available[4]	Available[4]	Worn	Worn	Worn	Available
Chemical protective undergarment[2]	Ready[3]	Available[4]	Worn[2]	Worn[2]	Worn[2]	Worn[2]	Available

[1]In hot weather, the coat or hood can be left open for ventilation.

[2]The chemical protective undergarment (CPU) is worn under the battle dress uniform (BDU) (applies to the special operation forces and armor vehicle crewman).

[3]Must be available to the soldier within 2 hours, and the second set must be available within 6 hours.

[4]Within arms reach of the soldier.

NOTE: Refer to *Figure J-2* for an illustration of the MOPP levels.

hygiene, and sanitation discipline. Biological attacks are hard to detect. If an attack occurs, chances of survival are better if soldiers are healthy, physically fit, keep all immunizations up to date, and maintain good personal hygiene. Keep your body clean to prevent ingestion of biological agents. Keep small cuts or scratches covered and germ-free by using soap, water, and first aid measures. Since insects carry biological agents, keep clothes buttoned and your skin covered to prevent insect bites.

NUCLEAR ATTACK DEFENSE

J-61. MP can defend themselves against nuclear attack before, during, and after the event.

DEFENSE BEFORE A NUCLEAR ATTACK

J-62. The best defense against a nuclear attack is to dig in. Prepare unit defensive positions, which vary from individual foxholes to improved positions, whenever the tactical situation permits. Keep individual weapons, equipment, clothing, and other issue items in vehicles. Secure equipment because the blast wave will convert unsecured items into lethal missiles. Disperse and protect supplies, explosives, and flammable materials. Turn off all electronic equipment and remove antennas from vehicles.

J-63. Reverse slopes of hills and mountains give some nuclear protection. The initial radiation, heat, and light from the fireball of a nuclear blast tend to be absorbed by hills and mountains. Using of gullies, ravines, ditches, natural depressions, fallen trees, and caves can reduce nuclear casualties.

DEFENSE DURING A NUCLEAR ATTACK

J-64. Defense during a nuclear attack is either dismounted or mounted. When taking dismounted defensive actions, never run for cover. Immediately drop flat on the ground (face down) or to the bottom of a foxhole, with your head toward the blast. Cover exposed skin as much as possible and close your eyes. Remain down until the blast wave has passed and debris has stopped falling. Stay calm, check for injury, check weapons and equipment for damage, and prepare to continue the mission.

J-65. MP in vehicles should exit and move away from the vehicle and take dismounted defensive actions.

DEFENSE AFTER A NUCLEAR ATTACK

J-66. Once the attack ends, forward an NBC 1 nuclear report, organize survivors, secure and organize equipment, assist casualties, improve protection against possible fallout, and begin continuous monitoring. If the radiation dose rate reaches a hazardous level after fallout is complete, be prepared to move, on order, to a less hazardous area.

Fallout Warning

J-67. The first person to detect the arrival of fallout is usually the radiological monitor operating a RADIAC meter. As soon as he notes a dose rate of 1 centigray per hour or higher, he warns unit personnel. All personnel hearing the warning relay it to others. If the mission allows, move into a shelter with overhead cover and stay there until given an *all clear* signal or until directed to move. If the mission does not allow the unit to take cover, decontamination becomes more important and perhaps more difficult.

Supervision of Radiological Monitoring

J-68. Perform radiological monitoring routinely with a AN/VDR-2 RADIAC meter to determine the presence and intensity of a radiation hazard. MP leaders must ensure that their teams are properly trained on this equipment. The two types of monitoring is periodic and continuous.

J-69. **Periodic Monitoring**. Periodic monitoring (once every hour) assures the platoon that the area is not contaminated or, if applicable, provides a warning when contamination is detected after the platoon arrives. Periodic monitoring is initiated under the following conditions:

- After the first use of nuclear weapons in the theater of war.
- When the platoon is out of contact with higher HQ.
- When ordered by higher HQ.
- When the platoon stops continuous monitoring.

J-70. **Continuous Monitoring**. Continuous monitoring is the surveillance for radiation in the platoon's area or position. Initiate continuous monitoring when any of the following situations occur:

- When a nuclear detonation is observed or reported in the area of operations.
- When an NBC 3 report is received and the platoon is in the predicted area of contamination.
- When ordered by higher HQ.
- When a dose rate of 1 centigray per hour is recorded in periodic monitoring.

Supervision of Tactical Dosimetry Operations

J-71. Each MP team is issued a dosimeter. Selected soldiers wear them. Check all dosimeters to be used for the operation. Turn in dosimeters for recharging if they do not read zero. If a charger is not available, note the original reading. Ensure that the dosimeter readings are reported accurately. Collect readings at least once daily. Average these readings, round to the nearest 10, and report the average and the radiation exposure status (RES) to higher HQ.

J-72. Use individual actions for nuclear defense when operating in or crossing radiologically contaminated areas. Close vehicles tightly and cover cargoes with tarps or tenting. Mission permitting, keep the speed down to prevent dust. Vehicles should maintain adequate following distances to stay out of the dust raised by preceding vehicles.

J-73. After the unit exits a contaminated area, check personnel, equipment, and cargo for contamination and decontaminate them, if necessary. Monitor dose rates closely to ensure compliance with operational exposure guidance (OEG). Update the RES if appropriate.

CHEMICAL ATTACK DEFENSE

J-74. MP must be prepared to defend themselves before, during, and after a chemical attack.

DEFENSE BEFORE A CHEMICAL ATTACK

J-75. The best defense before a chemical attack is preparation. Use protective procedures. Ensure that all personnel have protective masks available, and ensure that each mask fits and functions properly. All personnel should wear the proper protective clothing according to the MOPP level designated by the commander. Protect all equipment and supplies from liquid chemical contamination by keeping them organized and covered.

J-76. Emplace an M8A1 automatic chemical agent alarm. The M8A1 is the primary means of detecting an upwind chemical attack. The system provides two essential elements of survival: detection of a toxic agent cloud and early warning to the troops in the monitored position. The platoon leader decides where to place the chemical alarm. Place the detector units no more than 150 meters upwind from the platoon's perimeter or position. Space the available detector units about 300 meters apart, and ensure that each detector unit is connected to the alarm unit by telephone cable (WD-1).

Position the alarm units near radiotelephone assets; this makes it easy to alert the unit of an attack. Blowing sand or dust, rain, sleet, snow, temperatures below 40 degrees Fahrenheit, and tropical conditions can affect the operation of the alarm.

J-77. If an NBC hazard cannot be avoided, soldiers must be prepared to protect themselves and equipment from the effects of exposure. Base the type and degree of protection required on the unit's mission and the hazard. Note that the line between contamination avoidance and protection is not distinct; many actions contribute equally to both.

DEFENSE DURING A CHEMICAL ATTACK

J-78. Give the alarm. All unmasked soldiers put on their protective masks and other MOPP gear. All personnel move inside their vehicles and close all doors and hatches, if applicable. This aids in the protection from gross liquid contamination. The platoon leader directs the use of M256A1 detector kits and the CAM to determine the type of agent, and he submits an NBC 1 report. The platoon then continues the mission.

DEFENSE AFTER A CHEMICAL ATTACK

J-79. Forward an NBC 1 chemical report, treat casualties, perform immediate decontamination as required, and mark the contaminated area.

SYMPTOMS AND TREATMENT OF CASUALTIES

J-80. Potential adversaries may have access to a wide variety of biological and chemical agents on the modern battlefield. These agents can be dispensed alone or with other carriers or agents. Casualties resulting from exposure to biological or chemical agents require medical treatment as quickly as possible.

J-81. The first step in the treatment process is usually appropriate self-aid and buddy-aid measures. These vary depending on the agent. Soldiers mask to prevent inhaling or ingesting additional agents and remove agents from exposed skin by washing with soap and water or using the M291 kit. They use buddy-aid procedures to observe each other for early symptoms of toxic exposure and provide medical assistance.

J-82. The platoon leader selects separate casualty collection points for contaminated and noncontaminated casualties to prevent cross contamination. Decontaminate all contaminated casualties as thoroughly as the situation allows before being evacuated. The platoon includes the number of contaminated patients in its casualty evacuation request. This allows the evacuation team to send the proper number of vehicles for pickup.

J-83. Chemical agents fall into four major categories: nerve, blister, blood, and choking. The primary routes of attack on the body are through the respiratory system and the skin. These agents are especially dangerous because they can kill or incapacitate quickly. The first, and most important, step in dealing with them effectively is to recognize symptoms so that proper treatment can be administered. *Table J-3, page J-32,* shows protection and detection measures, symptoms, and treatment and decontamination procedures for chemical agents.

UNMASKING PROCEDURES

J-84. Soldiers unmask as soon as possible except when a biological or chemical attack is expected. Use the procedures outlined in the following paragraphs to determine if unmasking is safe.

J-85. If an M256 or M256A1 detector kit is available, use it to supplement unmasking procedures. The kit

Table J-3. NBC Protection and Detection Chart

Agent Type	Protection	Detection	Symptoms	Effects	First Aid	Deconta-mination
Nerve	Protective mask and suit	M8A1, M256A1, CAM, M8 paper, or M9 paper	Difficult breathing, drooling, nausea, vomiting, convulsions, or blurred vision	Incapacitates	Use Mark 1 NAAK and convulsant antidote for nerve agents (CANA).	Use an M291 kit and flush eyes with water.
Blister	Protective mask and suit	M256A1, M8 paper, M9 paper, or CAM	Burning eyes, stinging skin, and irritated nose (no symptoms with mustard or nitrogen mustard)	Blisters the skin and damages the respiratory tract	Use the same procedures as for second- and third-degree burns. Refer to FM 21-11.	Use an M291 kit and flush eyes with water.
Blood	Protective mask	M256A1	Convulsions and coma	Incapacitates.	None	None
Choking	Protective mask	Odor only (resembles newly-mown hay or green corn).	Coughing, choking, nausea, headache, and tightness in the chest	Floods and damages the lungs	Avoid move-ment and keep warm.	None

does not detect all agents; therefore, use the proper unmasking procedures, which take about 15 minutes. If all the tests with the kit (including a check for liquid contamination using M8 detector paper) have been performed and the results are negative, use the following procedures:

- The senior person selects one or two soldiers to start unmasking procedures. If possible, they move to a shady place. Bright, direct sunlight can cause pupils to constrict, giving a false symptom.
- The selected soldiers unmask for 5 minutes and then clear and reseal their masks.
- The leaders observe the unmasked soldiers for 10 minutes. If no symptoms appear, request permission from higher HQ to signal *all clear*.

J-86. The leaders watch the soldiers for possible delayed symptoms. Always have first aid treatment immediately available in case it is needed.

J-87. If an M256 or M256A1 kit is not available, unmasking procedures take about 35 minutes. When a reasonable amount of time has passed after the attack, find a shady area and use M8 paper to check the area for possible liquid contamination. Conduct unmasking using the following procedures:

- The senior person selects one or two soldiers. They take a deep breath and break their mask seals, keeping their eyes wide open.
- The soldiers clear and reseal their masks after 15 seconds. The leaders observe them for 10 minutes.
  - If no symptoms appear, the same soldiers break the seals, take two or three breaths, and clear and reseal their masks. Observe them for 10 minutes.

- If no symptoms appear after 10 minutes, the same soldiers unmask for 5 minutes and then remask.
- If no symptoms appear after 10 minutes, the leaders request permission from higher HQ to signal *all clear*. He continues to observe soldiers in case delayed symptoms develop.

J-88. Give the all-clear signal by word of mouth through the chain of command. Higher HQ initiates the all-clear signal after testing for contamination proves negative. If required, use standard sound signals, such as a continuous, sustained blast on a siren, a vehicle horn, or a similar device. When *all clear* is announced on the radio, it must be authenticated before compliance. The commander designates the specific all-clear signal and includes it in the SOP.

Appendix K

Civil-Disturbance Measures

This appendix discusses the crowd control measures used when there is a civil disturbance.

OVERVIEW

K-1. The mission of the military forces in a civil disturbance is to apply the minimum force necessary to help local or HN authorities restore law and order. Most often, military personnel disperse unauthorized assemblages and patrol disturbed areas to prevent unlawful acts. Other missions assigned to MP or other military forces include—

- Maintaining the mechanics of essential distribution, transportation, and communications systems.
- Making a show of force.
- Setting up roadblocks.
- Cordoning off areas.
- Dispersing crowds.
- Releasing riot control agents (RCAs).
- Serving as security forces or reserves.
- Initiating needed relief measures, such as distributing food or clothing or establishing emergency shelter.
- Employing nonlethal munitions and equipment.

K-2. Military control force commanders must know what options are available to them. They select the option that is best for the specific physical and psychological environment. They must be able to reduce the intensity of the confrontation and restore order.

K-3. The commitment of military forces to civil-disturbance control operations does not automatically give these forces police power. There are legal and commonsense reasons to restrict the police power of military forces. All military leaders and planners must be familiar with laws, regulations, and policies that govern military involvement in civil disturbances.

K-4. In all contacts with the civilian population and the participants of the disturbance, military forces must display fair and impartial treatment and must adhere to the principle of minimum force. Whenever possible, civil police apprehend, process, and detain civil law violators. Military forces perform these functions only when necessity dictates and to the minimum extent required. The return these functions to civil authorities as soon as possible.

K-5. As the disturbance subsides, the commander takes steps to restore control to civil authorities. The control force gradually reduces the number and scope of its operations. It takes care not to give the impression that all controls have been removed. Withdrawal is not immediate. The disturbance may flare up again if the participants get the impression that authorities have abandoned the operations. The control force gradually withdraws in a phased return of control to civil authorities.

CIVIL DISTURBANCES ON DEPARTMENT OF DEFENSE INSTALLATIONS OUTSIDE THE CONTINENTAL UNITED STATES

K-6. On DOD installations overseas and at US embassies and consulates, HN forces generally control disturbances targeted at US facilities because of the possible international political ramifications of foreign civil disturbances. Status of Forces Agreements define the legal considerations that guide and constrain actions by US military commanders. Commanders must have an effective liaison with HN authorities. Through close coordination with HN authorities, US commanders can determine the level of visibility and the involvement, if any, of US forces.

CROWD BEHAVIOR

K-7. The presence or absence of social factors like leadership, moral attitudes, and social uniformity may influence crowd behavior. Leadership has a profound effect on the intensity and direction of crowd behavior. When blocked from expressing its emotions in one direction, a crowd's frustration and hostility can be redirected elsewhere. The first person to give clear orders in an authoritative manner is likely to be followed. Agitators can exploit a crowd's mood and convert a group of frustrated, resentful people into a vengeful mob. Skillful agitators using television, radio, and other communications media can reach large portions of the population and incite them to unlawful acts without having direct personal contact. In a civil-disturbance environment, any crowd can be a threat to law and order because it is open to manipulation.

K-8. Additionally, crowd behavior may be affected by emotional contagion or panic.

- Emotional contagion provides the crowd psychological unity. The unity is usually temporary, but it may be long enough to push a crowd to mob action. When emotional contagion prevails, normal law and authority are suppressed, increasing the potential for violence.
- Panic can occur during a civil disturbance when—
  - Crowd members perceive their safety is at risk and attempt to flee the area.

- Crowd members cannot disperse quickly after exposure to RCAs.
- Escape routes are limited.
- Escape routes are blocked or congested.

K-9. Control force members are also susceptible to crowd behavior. They may become emotionally stimulated during a tense confrontation. Commanders must counteract this. Control force members must exercise restraint individually and collectively. Rigorous training, firm and effective supervision, and complete awareness and understanding of ROE and ROI are necessary to offset the effect of crowd contagion on the control force.

CROWD TACTICS

K-10. In civil disturbances, crowds employ any number of tactics to resist control or to achieve their goals. Tactics may be unplanned or planned and nonviolent or violent. The more purposeful the disturbance, the more likely the possibility of well-planned tactics.

K-11. Nonviolent tactics may range from name-calling to building barricades. Demonstrators may converse with control force members to distract them or to gain their sympathy. They may try to convince control force members to leave their posts and join the demonstrators. Demonstrators may use verbal abuse. Expect obscene remarks, taunts, ridicule, and jeers. Crowd members want to anger and demoralize the opposition. They want authorities to take actions that later may be exploited as acts of brutality.

K-12. Sometimes women, children, and elderly people are placed in the front ranks. This plays on a control force's sympathy to try to discourage countermeasures. When countermeasures are taken, agitators take photographs to stir public displeasure and embarrass the control force. Demonstrators may form human blockades to impede traffic by sitting down in roads or at the entrances to buildings. This can disrupt normal activity, forcing control personnel to physically remove the demonstrators. Demonstrators may lock arms, making it hard for the control force to separate and remove them. It also makes the control force seem to be using excessive force.

K-13. Groups of demonstrators may trespass on private or government property. They want to force mass arrests, overwhelm detainment facilities, and clog the legal system. Demonstrators may resist by going limp and forcing control force members to carry them. They may chain or handcuff themselves to objects or to each other. This prolongs the demonstration. Agitators may spread rumors to incite the crowd and try to force the control force to use stronger measures to control or disperse the crowd. The agitators want to make the control force appear to be using excessive force. Terrorist groups may try to agitate crowds as a diversion for terrorist acts. They also try to provoke an overreaction by the control force.

K-14. Violent crowd tactics, which may be extremely destructive, can include physical attacks on people and property, fires, and bombings. Crowd use of violent tactics is limited only by the attitudes and ingenuity of crowd members, the training of their leaders, and the materials available to them. Crowd or mob members may commit violence with crude, homemade weapons, or they may employ sophisticated small arms and explosives. If unplanned violence occurs, a crowd will use rocks, bricks, bottles, or whatever else is at hand. If violence is planned, a crowd can easily conceal makeshift weapons or tools for vandalism. They may carry—

- Balloons filled with paint to use as bombs.
- Bolt cutters to cut through fences.
- Picket signs to be used as clubs.

- Pipes wrapped in newspapers to throw as deadly missiles.
- Firecrackers dipped in glue and covered with BBs or small nails to use as deadly grenades.
- Plywood shields and motorcycle helmets to protect against riot batons.
- Safety goggles to protect against tear gas.

K-15. A crowd may erect barricades to impede troop movement or to prevent a control force from entering certain areas or buildings. They may use vehicles, trees, furniture, fences, or any other material. In an effort to breach barriers, rioters may throw grapples into wire barricades and drag them. They may use grapples, chains, wire, or rope to pull down gates or fences. Long poles or spears may be used to keep control forces back while removing barricades or to prevent the use of bayonets. They also may crash vehicles into gates or fences to breach them.

K-16. Rioters can be expected to vent their emotions on individuals, troop formations, and control-force equipment. Rioters may throw rotten fruits and vegetables, rocks, bricks, bottles, improvised bombs, or other objects from overpasses, windows, and roofs. In the past, troops, firefighters, and utility workers on duty during a civil disorder have been beaten, injured, or killed. Vehicles have been overturned, set on fire, or otherwise damaged.

K-17. Rioters may direct dangerous objects like vehicles, carts, barrels, and liquids at troops located on or at the bottom of a slope. On level ground, they may drive wheeled vehicles at the troops, jumping out before the vehicles reach the target. This tactic is also used to breach roadblocks and barricades.

K-18. Rioters may set fire to buildings and vehicles to block the advance of troops. Fires are also set to create confusion or diversion, destroy property, and mask looting and sniping. Rioters may flood an area with gasoline or oil and ignite it. On the other hand, they may pour gasoline or oil down a slope or drop it from buildings and ignite it.

K-19. Weapons fire against troops may take the form of selective sniping or massed fire. The fire may come from within the ranks of the rioters or from buildings or other adjacent cover. The weapons used can vary from homemade one-shot weapons to high-powered rifles. Snipers may try to panic control force members into firing a volley into the crowd. Innocent casualties make a control force appear both undisciplined and oppressive.

K-20. Explosives may be used to breach a dike, levee, or dam. Bombs can be exploded ahead of troops or vehicles so rubble blocks a street. They can be used to block an underpass by demolishing the overhead bridge. In extremely violent confrontations, bombs placed in buildings may be timed to explode when troops or vehicles are near. Demolition charges can be buried in streets and exploded as troops or vehicles pass over them. Explosive-laden vehicles can be rolled or driven at troops. Rioters may attach explosives to animals and force them toward troops. They then detonate the explosives by remote control. Even harmless looking objects like cigarette lighters and toys can be loaded with explosives and used as weapons.

COMPANY LEVEL OPERATIONS

K-21. Normally, civil-disturbance operations are conducted at company level. The company operations section coordinates for special equipment that includes the following:

- Kevlar and face shield.
- Body armor.
- Shields, work gloves, and batons.
- Protective masks.

- Elbow pads.
- Shin guards.
- Weapons and bayonets.
- CS/OC spray.
- Smoke grenades.
- Breaching ramps.
- Scaling ladders.
- Portable radios.
- Portable lighting.
- Bullhorns.
- Video home system (VHS) cameras to videotape individuals in the crowd for identification.
- Transportation assets to move people to detention cells or detention areas.
- Hand and leg irons and flex cuffs.
- NVDs.

K-22. The MP operations center also coordinates for support, to include medical and emergency operations center (EOC) personnel, SRTs, hostage negotiators, MWD teams, PSYOP, civil affairs, local or HN law enforcement personnel, and apprehension or detention facilities for civilian personnel.

K-23. Rehearsals are imperative to the overall success of suppressing civil disturbances even before they occur. Rehearsals should cover the use and employment of nonlethal munitions, practice situations in which ROE or the use of force is applied, the use of riot dispensers, hand and arm signals or voice commands, and formations.

K-24. MP use appropriate crowd control measures that include formations such as—

- Using the line formation to push or drive a crowd straight back.
- Using the echelon formation to turn or move crowds away from buildings, fences, or walls.

- Using the wedge formation penetrate and split a crowd.
- Using the diamond formation to enter a crowd.

K-25. When a small unit, normally a squad, moves a dignitary through a crowd or an apprehension team into a crowd, they generally use the diamond formation. The other formations are trained at squad level and above but are normally performed by a platoon or company. The line, echelon left or right, and wedge formations are the basis for platoon and company formations. A squad must be skilled in the basic formations before practicing in platoon-size or larger formations.

K-26. Refer to *FM 19-15* for more information about civil-disturbance techniques and procedures.

NONLETHAL MUNITIONS

K-27. Nonlethal munitions are explicitly designed and primarily employed to incapacitate personnel or material, while minimizing fatalities, permanent injury to personnel, and undesired damage to property and the environment. Unlike conventional lethal munitions that destroy their targets through blast, penetration, and fragmentation, nonlethal munitions employ means other than gross physical destruction to prevent the target from functioning.

K-28. The doctrine and concepts of operation for nonlethal weapons are designed to reinforce deterrence and expand the range of options available to commanders. They enhance the capability of US forces to accomplish the following objectives:

- Discourage, delay, or prevent hostile actions.
- Limit escalation.
- Take military action in situations where the use of lethal force is not the preferred option.
- Protect our forces better.

- Disable equipment, facilities, and personnel temporarily.

K-29. Nonlethal munitions are not required to have a zero probability of producing fatalities or permanent injuries, and complete avoidance of these effects is not guaranteed or expected. When properly employed, nonlethal weapons should significantly reduce these effects when compared with using lethal weapons to physically destroy the same target.

K-30. Nonlethal capabilities provide a wider range of options that augment but do not replace traditional means of deadly force. The option to resort to deadly force must always remain available when the commander believes it is appropriate to the mission.

K-31. MP normally use the lowest level of force necessary to control most incidents that involve noncombatants. RCAs, such as CS and MWD, and crowd control devices, such as the riot baton, are examples of nonlethal means currently used by MP forces. These options are currently listed in *AR 190-29*.

K-32. Training for the use of nonlethal munitions requires leaders and soldiers to understand the limited use of these systems in environments with restrictive ROE. Training will be continuous at all levels to ensure that nonlethal munitions are properly employed.

K-33. The countermeasures for thwarting virtually all nonlethal options are usually apparent, quickly learned, and readily available. Because they do not kill, nonlethal options teach an adversary what to avoid in the future. Commanders must be prepared to stay one step ahead of motivated belligerents. Many nonlethal options have both maximum effective and minimum safety ranges. Individuals struck short of the minimum safety range often suffer severe injuries or death, while the effects of most nonlethal devices are greatly mitigated at longer ranges. In order to be effective, an adversary must be engaged within the effective zone (beyond the minimum safety range and short of the maximum effective range).

K-34. Nonlethal munitions and equipment currently available are shown in *Appendix L*. Refer to *Figure 3-2, page 3-11,* for the range of the munitions.

NONLETHAL MUNITIONS EMPLOYMENT CONSIDERATIONS

K-35. Nonlethal munitions employment must be well documented in ROE. Leaders must constantly ensure that soldiers understand when and how to effectively employ them. Incorrect application of these munitions can have significant operational and political ramifications. Employment considerations include the following:

- **Individual.** When possible, do not change individual weapons. Designate individuals as nonlethal shooters. Nonlethal shooters carry lethal munitions only for personal protection. Carry lethal rounds in a place to avoid confusing nonlethal rounds with lethal rounds.
- **Squad.** Squad leaders carry stun grenades and maintain their weapon loaded with lethal ammunition. Ideally, the squad does not change their task organization to accommodate the addition of nonlethal equipment. They should designate nonlethal shooters instead.
- **Patrols.** Commanders do not plan a nonlethal patrol, but they plan a combat and security patrol with a nonlethal attachment when the mission dictates. Carrying a shotgun limits the flexibility an individual has because of the time it takes to transition from nonlethal to lethal. Use shotguns only from a fixed position where adequate coverage is available.
- **Static position.** Individuals on static positions have their weapons loaded with lethal

ammunition, and additionally they carry nonlethal munitions. Both shotguns and M203s work well from static positions and are used together when possible.

- **Contact teams.** Teams of personnel whose primary responsibilities are to make personal contact with the adversary are established before deployment. Train team members in unarmed self-defense, open-hand control, handcuffing, and flex cuffing. Equip contact teams with personal protection gear, 9-millimeter pistols, and flex cuffs. Due to the physical nature of contact teams, do not arm contact team members with long rifles. If security for the contact team is an issue, attach a security element. MWD teams may also augment contact teams to help locate the adversary. Refer to *FM 90-40* for more information on nonlethal techniques and procedures.

Appendix L

Weapons and Equipment

MP must be rapidly deployable, versatile, and as lethal as the force that they support. MP are organized and equipped to provide functional battlefield capabilities that range from the deliberate attack (area security) to civil-disturbance control operations (L&O). Although not all inclusive, the listed data is intended to aid in planning MP operations. In any tactical situation, it is important that MP be able to properly identify threat weapons, vehicles, and aircraft. When threat is observed, MP report their location, activity, and direction of travel to higher HQ by the fastest means available.

FRIENDLY VEHICLES

L-1. *Table L-1, page L-2,* gives MP leaders a quick look at the capabilities of the various vehicles in the Army inventory.

Table L-1. Vehicle Capabilities

Vehicle Capabilities	M998	M1025/1026	M1114	M1117	Light Medium Tactical Vehicle (LMTV)
Weight (gross vehicle weight [GVW]) (in pounds)	4,950	5,250	12,100	29,500	16,500
Height (in inches)	69	73/73	73	102	112
Length (in inches)	180	180/185	190.5	237	251
Width	85	85/85	85	101	96
Fuel capacity (in gallons)	25	25	25	50	58
Maximum speed (in miles per hour)	55	55	55	63	58
Range (in miles)	350	300	275	440	400+

Table L-1. Vehicle Capabilities (Continued)

Vehicle Capabilities	M998	M1025/1026	M1114	M1117	LMTV
Fording Capabilities					
With kit (in inches)	60	60	60	NA	60
Without kit (in inches)	30	30	30	60	60
Pay load (maximum pounds)	2,500	2,500	NA	3,360	5,000

FRIENDLY WEAPONS

L-2. *Table L-2* gives MP leaders a quick look at the capabilities of the weapons used by MP units.

Table L-2. Weapon Capabilities

Weapon Capabilities	M2	M4	M9	M16A2	MK-19	M136 (AT4)	M203	M249
Weight (in pounds)	84	7.5	2.6	8.7	76	14.8	11	15.5
Length (in inches)	66	29.75 closed, 33 open	8.5	39	43	40	39	41.1
Range								
Maximum (in meters)	6,765	3,600	1,800	3,600	2,212	2,100	400	3,600
Arming (in meters)	NA	NA	NA	NA	18 to 30	10	14	NA
Minimum safe (in meters)	NA	NA	NA	NA	28	NA	31	NA

Table L-2. Weapon Capabilities (Continued)

Weapon Capabilities	M2	M4	M9	M16A2	MK-19	M136 (AT4)	M203	M249
Effective Range								
Area target (in meters)	1,830	600	NA	800	2,212	NA	350	800
Point target (in meters)	1,200	500	50	580	1,500	NA	160	600
Moving target (in meters)	NA	NA	NA	200	NA	NA	NA	NA
Rate of Fire (in Rounds Per Minute)								
Cyclic	500	700 to 970	NA	700 to 800	375	NA	NA	800
Rapid	40*	NA	NA	NA	60	NA	35	200*
Sustained	40*	12/15	60	16	40	NA	35	85

*With barrel change.

FRIENDLY NONLETHAL EQUIPMENT AND MUNITIONS

L-3. The following descriptions and illustrations describe nonlethal equipment and munitions that are currently available.

NONLETHAL EQUIPMENT

L-4. The following nonlethal equipment provides bodily protection for soldiers involved in a nonlethal operation and allows soldiers to capture the threat with a minimum of force.

Nonballistic and Ballistic Riot Face Shields

L-5. The riot face shield *(Figure L-1)* provides soldiers with improved facial protection from thrown objects. The face shield is lightweight, adaptable to the current helmet, transparent, and scratch-resistant. It is adjustable to up and down positions. The mechanism for attaching and removing the shield from the helmet is robust and simple and requires no tools in the field.

Figure L-1. Riot Face Shield

L-6. The ballistic riot face shield is similar in design, but it is heavier and provides facial protection up to a 9-millimeter full metal jacket (FMJ) or 124 grains at 1,400 feet per second. This face shield is primarily used by SRTs in force-entry scenarios and can also be used for MOUT operations.

Nonballistic and Ballastic Riot Body Shields

L-7. The nonballistic riot body shield *(Figure L-2)* is 24 inches by 48 inches by .157 inches. It provides soldiers with improved protection from frontal, side, and overhead assaults. This shield is lightweight, transparent, and scratch-resistant. The ballistic riot body shield *(Figure L-3, page L-8)* is 24 inches by 36 inches with a 4- by 16-inch window for viewing the threat. It is similar in design to the nonballistic shield but is heavier. It provides ballistic protection up to a 9-millimeter FMJ or 124 grains at 1,400 feet per second. This shield is primarily used by SRTs in forced-entry scenarios and can also be used for selected MOUT operations.

Figure L-2. Nonballistic Riot Body Shield

Figure L-3. Ballistic Riot Body Shield

Nonballistic and Ballistic Shin Guards

L-8. Nonballistic shin guards *(Figure L-4)* provide soldiers with improved protection from thrown objects. They are lightweight and black in color.

Figure L-4. Nonballistic Shin Guards

L-9. Ballistic shin guards are similar in design to nonballastic guards, but they are heavier and provide protection up to a 9-millimeter FMJ or 124 grains at 1,400 feet per second. They are primarily used by SRTs in forced-entry scenarios and can also be used in selected MOUT operations.

Baton

L-10. The standard 36-inch wooden riot baton *(Figure L-5)* is currently in use for riot control. It is used for self-defense and to keep rioters out of arm's reach of the soldiers conducting crowd control tactics.

Figure L-5. Standard 36-inch Wooden Riot Baton

Portable Bullhorn

L-11. The portable bullhorn *(Figure L-6, page L-10)* is a critical communication device when conducting crowd control tactics. The bullhorn can facilitate communication with the crowd in conjunction with linguist or PSYOP support. It also assists in communicating commands to troops engaged in crowd control by projecting over the crowd's noise.

Individual Voice-Amplification System

L-12. The individual voice-amplification system *(Figure L-7, page L-10)* is a critical communication device for conducting crowd control tactics when using RCAs and wearing a protective mask. This device facilitates oral communications and increases the user's ability to be heard on radios and other devices.

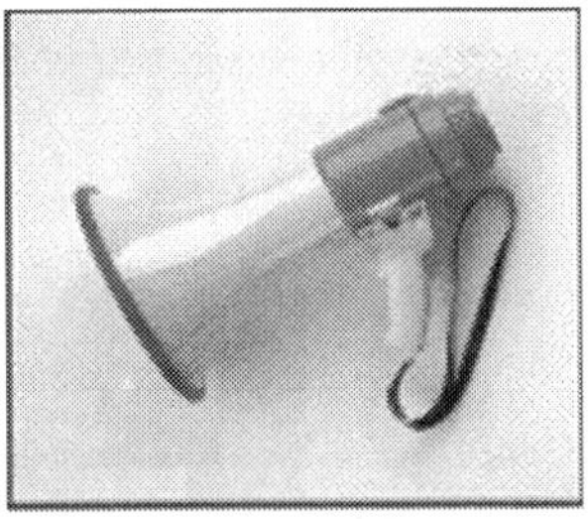

Figure L-6. Portable Bullhorn

Figure L-7. Individual Voice-Amplification System

Individual Oleoresin Capsicum and M36 Individual Chlorobenzyl Malononitrile Dispersers

L-13. These individual RCA dispersers *(Figure L-8)* are used primarily for self-defense and to keep rioters out of arm's reach of soldiers conducting crowd control tactics or engaged in missions were noncombatant threat exists.

Midsize Riot Control Dispersers of Oleoresin Capsicum and Chlorobenzylidene Malononitrile

L-14. This RCA disperser *(Figure L-9)* is primarily used by formations conducting crowd control tactics, law

Figure L-8. Individual OC and CS Dispersers

enforcement, and I/R operations. It is lightweight, can be operated by one person, and is easily refilled and pressurized with available maintenance equipment or 2½- or 5-ton truck compressors. It is intended to provide a small unit with self-defense capabilities from large crowds out to 10 meters (more range is possible based on the wind) or offensively to clear crowds from critical areas (toward preplanned escape routes).

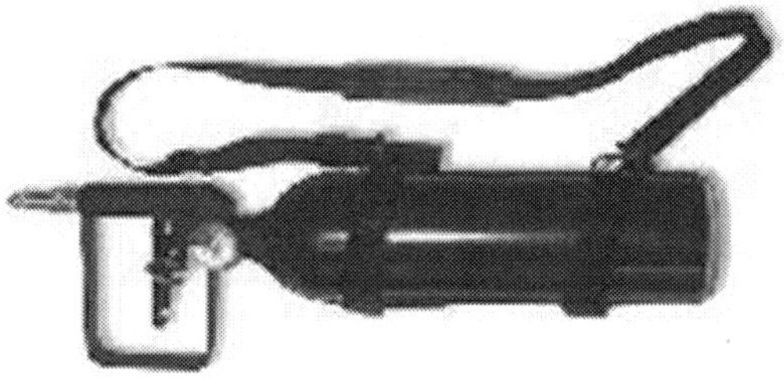

Figure L-9. Midsize Riot Control Dispersers of OC and CS

M33A1 Riot Control Disperser Filled With Chlorobenzylidene Malononitrile or Dibenz (B, f)-1, 4-Oxazepine (CR)

L-15. This RCA disperser *(Figure L-10)* is primarily used by formations conducting crowd control tactics. It is twice as heavy as a midsized disperser, carries a larger payload, and dispenses powdered CS or liquid CR. One individual can operate it, and it can be refilled and pressurized with available M4 compressors. It is intended to provide a small unit with self-defense capabilities from large crowds out to 15 meters (100 meters is possible based on wind speed and direction). Use the M33A1 offensively to clear crowds from critical areas.

Figure L-10. M33A1 Riot Control Disperser Filled With CS or CR

L-16. Some training is required to operate the dispenser. The unit NBC NCO may conduct this training. The device requires up to 2,000 pounds per square inch of pressure to function properly, and it is maintained with an M254 maintenance kit.

High-Intensity Xenon Searchlight

L-17. Use this individual, handheld searchlight *(Figure L-11)* for illumination in crowd control operations during darkness. Use it for general illumination of the operational area; to pinpoint agitators or threat, to reduce the ability of rioters to see troop formations and actions, and to enhance tactical deception techniques for units conducting crowd control operations.

Figure L-11. High-Intensity Xenon Searchlight

Lightweight Disposable Restraints

L-18. Lightweight disposable restraints *(Figure L-12, page L-14)* are also known as flex cuffs. Individual soldiers can carry large quantities of these restraints to immobilize individuals being detained. When freeing a detainee, cut the restrains off him with the safe cutting device supplied with the restraints or with utility shears. A reusable, red restraint training device is available.

Shotgun Munitions Carrier

L-19. The shotgun munitions carrier *(Figure L-13, page L-14)* is a 12-gauge ammunition carrier that straps to the stock of the M12 shotgun. It allows the firer to carry nonlethal ammunition that is readily available.

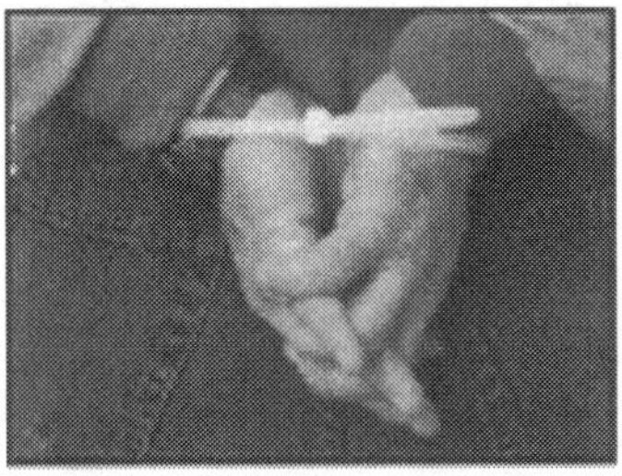

Figure L-12. Lightweight Disposable Restraints

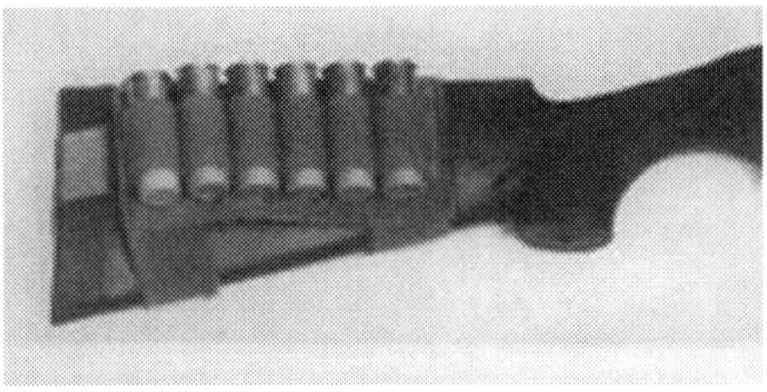

Figure L-13. Shotgun Munitions Carrier

Portable Vehicle-Arresting Barrier (PVAB)

L-20. The PVAB *(Figure L-14)* is designed to assist with short-term physical security of critical facilities and for use at checkpoints. Use the PVAB to augment a vehicle checkpoint. It adds the ability to stop any light vehicle (up to 7,500 pounds) that attempts to flee without killing the occupants. This device may be set up by 2 or 3 soldiers in less than an hour.

L-21. When emplaced, the PVAB resembles a standard speed bump. When armed, the device operates within 1.5 seconds and deploys a high-tensile net that catches the vehicle and slows it to a stop with internal braking

Figure L-14. Vehicle Entrapped in a PVAB

mechanisms. The net and other features prevent occupants from fleeing the vehicle. The PVAB is best suited for mobile, short-term vehicle checkpoints in areas under US control where there is a threat of terrorist-type activity and where the threat uses the cover of noncombatants to infiltrate US AOs.

NONLETHAL MUNITIONS

L-22. The following munitions provide soldiers with a nonlethal way to break contact, enforce a buffer zone, or stun an individual.

12-Gauge Nonlethal Point Target Cartridge Round (M1012)

L-23. The point target cartridge round *(Figure L-15, page L-16)* stuns individuals by delivering a strong blow to the body without penetrating it. This munition allows soldiers to enforce a buffer zone (standoff distance) with a violent crowd, break contact, or stun an individual target for possible detention by snatch teams. Fire the round at the center mass of an adult subject at ranges of 10 to 30 meters. Beyond 30 meters, the projectile loses accuracy and may no longer have the velocity required to stun an individual. This round has applications in law enforcement, I/R facilities, and US military detention facilities.

DANGER
Shots fired at subjects closer than 10 meters or shots to the head or groin may cause serious injury or even death.

Figure L-15. 12-Gauge Nonlethal Point Target Cartridge Round

12-Gauge Aerial Diversionary-Device Round

L-24. The aerial diversionary-device round *(Figure L-16)* provides multishot nonlethal capability to distract individuals or crowds. In crowd control, it delivers a flash bang projectile over the heads of a violent or potentially violent crowd, and is used to distract the crowd (in combination with other distraction devices and troop maneuvers). It allows other troop formations to maneuver to positions that are more advantageous

L-25. The round is designed to be fired at ranges of 75 to 100 meters and is placed about 5 meters above the crowd.

DANGER
Shots fired directly at subjects or in enclosed areas may cause serious injury.

Figure L-16. 12-Gauge Aerial Diversionary-Device Round

12-Gauge Nonlethal Area Target Cartridge Round (M1013)

L-26. The area target cartridge round *(Figure L-17, page L-18)* provides the capability to stun or deter two or three threats by delivering a strong blow to the body without penetrating it. This round has a wide range of capabilities for tactical, law enforcement, I/R, and US military detention operations.

L-27. The round is designed to be fired at the center mass of an adult threat at ranges of 10 to 30 meters.

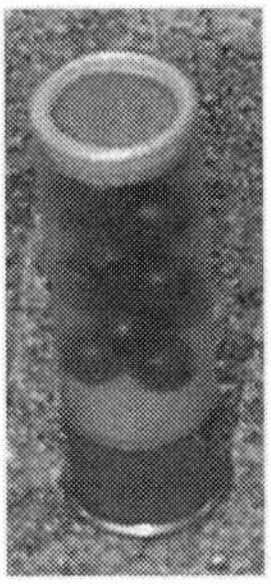

Figure L-17. 12-Gauge Nonlethal Area Target Cartridge Round

Shots fired closer than 10 meters may cause serious injuries. The projectile loses accuracy when shot beyond 30 meters and may no longer be effective.

40-Millimeter Sponge Round (Point) (M1006)

L-28. The 40-millimeter sponge round *(Figure L-18)* delivers a strong, stunning blow to a threat's body without penetrating it. This round has a wide range of capabilities for tactical, law enforcement, I/R, and US military detention operations.

Figure L-18. M1006 40-Millimeter Sponge Round

L-29. The round is designed to be fired at the center mass of an adult threat at ranges of 10 to 50 meters.

WARNING
Shots fired closer than 10 meters may cause injuries.

40-Millimeter Crowd Dispersal Round (Area) (XM1029)

L-30. The 40-millimeter crowd dispersal round *(Figure L-19)* delivers a strong, stunning blow to a threat's body without penetrating it. This round has a wide range of capabilities for tactical, law enforcement, I/R, and US military detention operations. In crowd control, it provides a nonlethal capability that can be used to break contact and enforce a buffer zone (standoff distance) with a violent crowd.

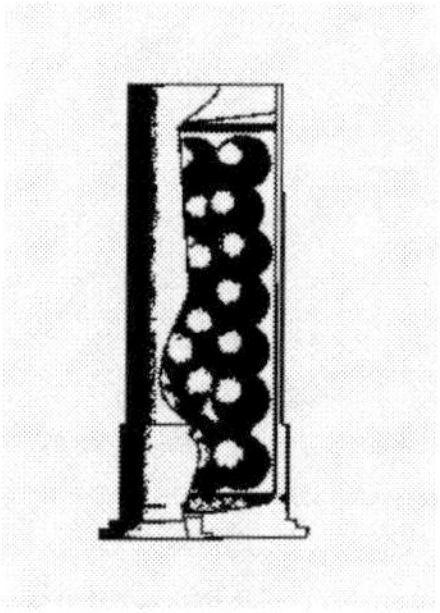

Figure L-19. 40-Millimeter Crowd Dispersal Round (Area)

L-31. The round is designed to be fired at the center mass of an adult threat at ranges of 10 to 30 meters.The projectile loses accuracy when shot beyond 30 meters and may not stun or deter the threat.

WARNING
Shots fired closer than 10 meters may cause injuries.

40-Millimeter Carrying Pouch

L-32. The 40-millimeter carrying pouch is slung over the shoulder of a soldier. It provides the ability to carry nonlethal ammunition separate from lethal rounds.

M84 Stun Grenade (Diversionary Device, Hand-Thrown)

L-33. The M84 *(Figure L-20)* is a hand-thrown, flash bang, stun device used primarily by SRTs in forced-entry scenarios. It is used for selected MOUT or crowd control operations.

Figure L-20. M84 Stun Grenade

L-34. The M84 is designed to be thrown into a room (through an open door, a standard glass window, or other opening). It delivers a loud bang and a brilliant flash that temporarily disorients and detracts the occupants. Because of its reusable metal body, do not throw it into a crowd, as it may be returned to friendly troops in the form of a projectile. Instead, throw it into a controlled area in conjunction with other deception and distraction techniques.

NONLETHAL, TURRET-MOUNTED, 66-MILLIMETER GRENADE LAUNCHER (M315) AND LIGHT VEHICLE OBSCURATION SMOKE SYSTEM, 66-MILLIMETER, NONLETHAL GRENADE (L96/97 XM98/99)

L-35. The LVOSS/M315 *(Figure L-21)* is a 66-millimeter smoke grenade-launching platform designed to give HMMWVs ASV M1117 the ability to obscure their position in the same manner as armored vehicles. These launchers can launch any of the smoke or nonlethal 66-millimeter munition.

M315 LVOSS

Figure L-21. LVOSS and M315

L-36. The M315 turret-mounted, 66-millimeter, multipurpose, adjustable grenade launcher installation kit was developed to provide a vehicle-mounted nonlethal platform. The system is capable of delivering nonlethal payloads to support a variety of mission requirements and can be mounted on various vehicles (M1025, M1026, M966, M114, or ASV M1117) equipped with a machine gun or a TOW II missile system mount. The system electronically fires 66-millimeter cartridges from four adjustable firing tubes.

L-37. The LVOSS, 66-millimeter, nonlethal grenade (L96/97 XM98/99) is an area target munition that can be fired from the standard LVOSS/M315 launcher *(Figure L-22)*. It can be mounted on selected armament carrier HMMWVs and can be fired from any 66-millimeter, smoke-launching system found on most armored vehicles.

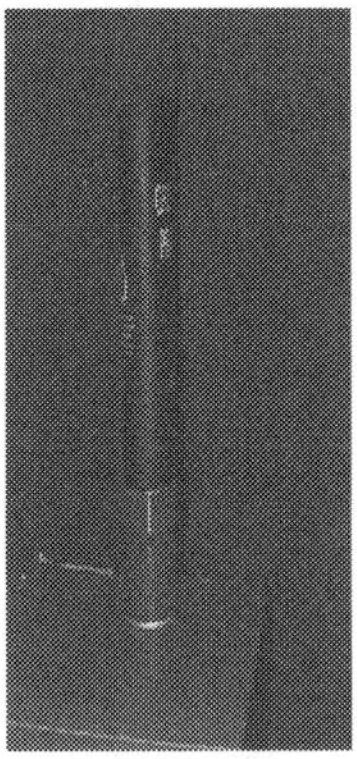

Figure L-22. LVOSS/M315 66-Millimeter Nonlethal Grenade

L-38. These rounds are designed to be fired from 80 to 100 meters. They deliver a flash bang diversionary warning (M98), a payload of rubber nonpenetrating projectiles (M99) (affecting a 10- to 20-meter circular area), or CS/cinnamic acid (CA) practice below the point of burst (L96A1/L97A1). These munitions lack the ability to be precision-delivered and are meant to affect a large number of people at long standoff ranges. This provides convoys and crowd control formations a long range support weapon to affect crowds beyond the range of shoulder-fired nonlethal weapons. Use the rounds to provide supporting nonlethal fires to crowd control formations.

M5 Modular Crowd Control Munitions

L-39. The MCCM *(Figure L-23)* munition is similar in operation to a claymore mine, but it delivers nonlethal effects to the threat by delivering a strong, nonpenetrating blow to the body with multiple submunitions (600 rubber balls). This round has a wide range of capabilities for tactical, law enforcement, I/R, and US military detention operations. In crowd control, it provides a nonlethal counterpersonnel capability that can be used to break contact, enforce a buffer zone (standoff distance), or demonstrate a show of force.

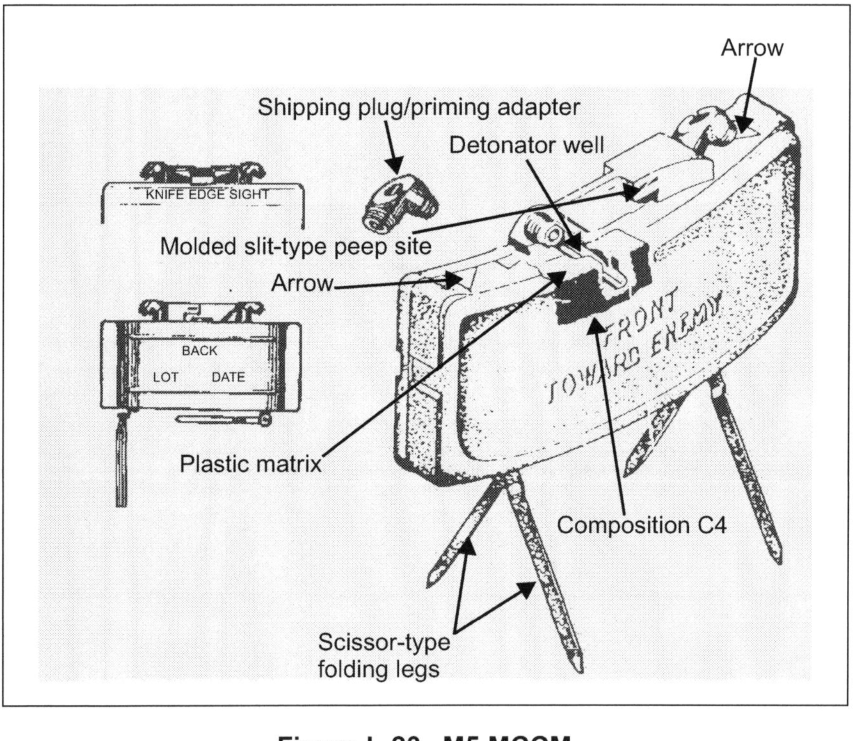

Figure L-23. M5 MCCM

L-40. This round is designed to be fired at the center mass of adult threats at ranges of 5 to 15 meters. It has a shot arc covering between 60 to 80 degrees (laterally).

WARNING
Shots fired at subjects closer than 5 meters may cause injury.

FRIENDLY COMMUNICATION, SINGLE-CHANNEL, GROUND-TO-AIR RADIO SYSTEM (SINCGARS)

L-41. SINCGARS is the primary communication system for MP. It is a series of 2-way FM radio sets that use an 18-element keypad for tuning 2,320 channels. This includes 8 preset channels in the single-channel mode and 6 preset channels in the jam-resistant, frequency-hopping mode. *Table L-3* and *Table L-4* provide the planning ranges for SINCGARS.

Table L-3. Voice Transmission Maximum Planning Ranges

Type of Radio	RF Switch Position	Planning Ranges
Manpack or vehicular	LO M HI	200 to 400 meters 400 meters to 5 kilometers 5 to 10 kilometers
Vehicular only	PA	10 to 40 kilometers

L-42. To increase the transmission range of SINCGARS, connect the OE-254 antenna group. It is an omnidirectional, biconical antenna designed for broadband operation, without field adjustment, from 30 to 88 megahertz, up to 350 watts. The following is the tabulated data transmission range of the OE-254:

- Between two OE-254 antenna groups–
  - Average terrain: 36 miles (57.9 kilometers).

  - Difficult terrain: 30 miles (48.3 kilometers).
- Between an OE-254 antenna group and a vehicular whip antenna–
  - Average terrain: 30 miles (48.3 kilometers).
  - Difficult terrain: 25 miles (40.3 kilometers).

Table L-4. Data Transmission Maximum Planning Ranges

Type of Radio	Baud Rate Used	RF Switch Position	Planning Ranges* (In Kilometers)
Manpack/ vehicular	600 to 4,800 baud per second (BPS)	HI	3 to 5
(short range)	16,000 BPS (16 kilobauds per second [KBPS])	HI	1 to 3
Vehicular	600 to 2,400 BPS	PA	5 to 25
(long range)	4,800 BPS	PA	5 to 22
	16,000 BPS (16 KBPS)	PA	3 to 10
***Ranges are based on the line of sight and are average for normal conditions. Ranges depend on the location, the sighting, the weather, and the surrounding noise level, among other factors. The use of the OE-254 antenna increases the range for both voice and data transmissions. Enemy jamming and mutual interference conditions degrade these ranges. In data transmission, the use of a lower baud rate increases the range.**			

THREAT WEAPONS AND EQUIPMENT

L-43. In any tactical situation, it is important that MP be able to properly identify threat weapons, vehicles, and aircraft. MP report threat location, activity, and direction of travel to higher HQ by the fastest means available.

L-44. *Tables L-5* through *L-12, pages L-27* through *L-52,* outline the most common features of threat weapons, equipment, and vehicles that MP may encounter in a hostile environment. Special operations forces and airborne assault, reconnaissance, and insurgent units often use these weapons.

Table L-5. Small Arms

Small Arms Characteristics	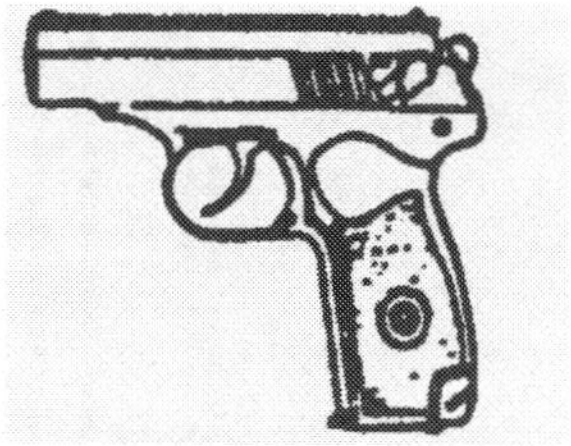 9-Millimeter PM Pistol	 5.45-Millimeter AK-74 Assault Rifle
Range, effective and maximum (in meters)	50	500/1,000
Rate of fire, practical and cyclic (rounds per minute)	30	100/600
Ammunition type	9- by 18-millimeter ball	5.45- by 39-millimeter ball, ball tracer, incendiary T
Fire mode	Semiautomatic	Selective semiautomatic or fully automatic

Table L-5. Small Arms (Continued)

Small Arms Characteristics	 5.45-Millimeter Squad Machine Gun (RPK)-74 Light MG	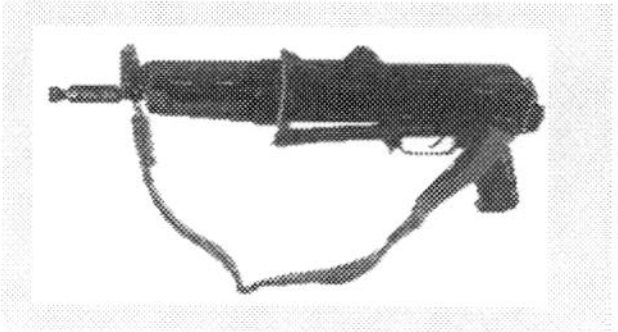 5.45-Millimeter AKSU-74 Submachine Gun (SMG)	 7.62-Millimeter Soviet Sniper Rifle (SVD)
Range, effective and maximum (in meters)	800/2,500	300/1,500	1,300 with scope, (800 without)/ 3,800
Rate of fire, practical and cyclic (rounds per minute)	150/600	150/600	30
Ammunition type	5.45- by 39-millimeter ball, ball tracer, incendiary T	5.45- by 39-millimeter ball, ball tracer, incendiary T	7.62- by 54R-millimeter light or heavy ball, steel core, tracer, AT, incendiary "rose" sniper bullet
Fire mode	Selective semiautomatic or fully automatic	Selective semiautomatic or fully automatic	Semiautomatic

Table L-5. Small Arms (Continued)

Machine Gun Characteristics	 RPK-74	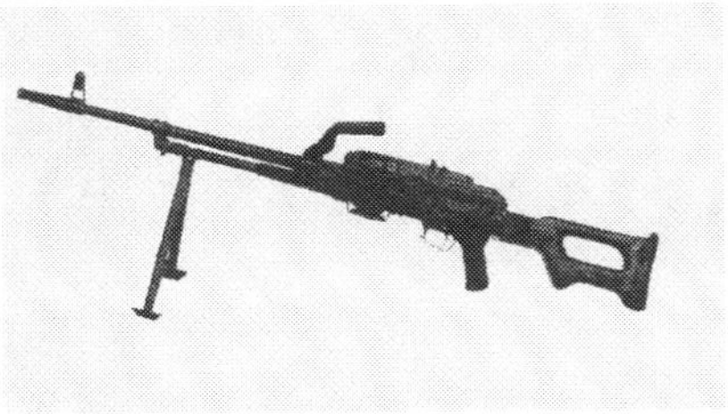 PKM
Range, effective and maximum (in meters)	800/2,500	1,000/3,800
Rate of fire, cyclic and practical (rounds per minute)	600/150	650/250
Ammunition type	5.45- by 39-millimeter, rimless	7.622- by 54-millimeter ball, ball tracer, API-T, incendiary
Fire mode	Selective	NA
Armor penetration (in millimeters)	NA	8 millimeters at 500

Table L-6. Grenade Launchers

Grenade Launcher Characteristics	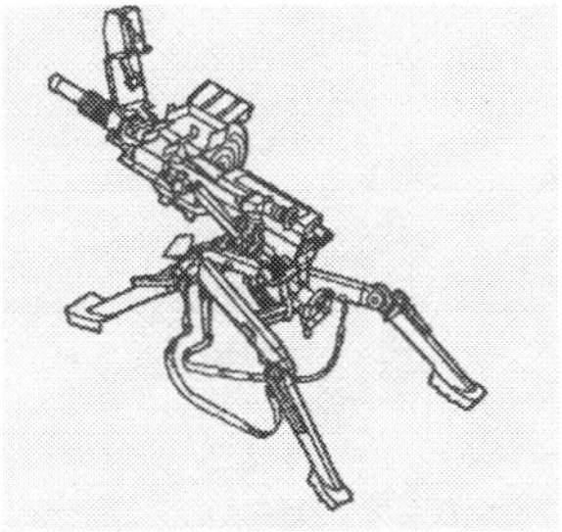 Armored Gun System (AGS)-17	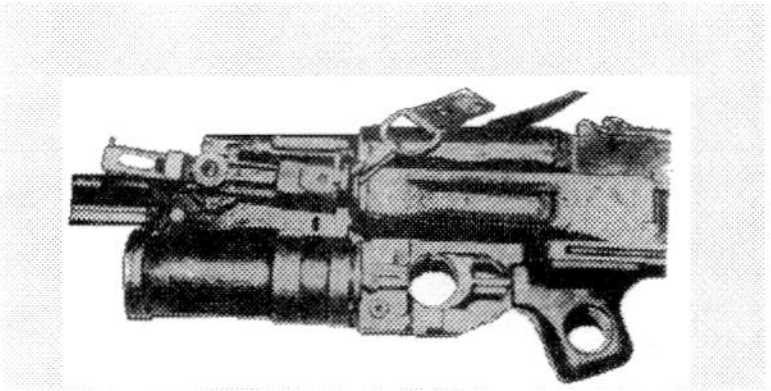 BG-15/GP-25
Range, effective and maximum (in meters)	1,200 indirect fire, 700 direct fire/1,730	4 to 400/400
Rate of fire, practical and cyclic (rounds per minute)	60 to 100/100 to 450	5
Elevation (in degrees)	7 to 87	NA
Traverse (in degrees)	30	NA
Platform	Tripod, vehicle, or helicopter	Under barrel grenade launcher
Fire mode	Selective, semiautomatic, or fully automatic	Single shot
Feed	29-round belt in drum magazine	Muzzle loaded
Ammunition type	30-millimeter, HE fragment	40-millimeter, HE fragment, bounding HE fragment
Caliber (in millimeters)	30	40
Crew	3	1
Remarks	15-meter lethal area of burst	Bounding HE fragment strikes the ground and "bounds" up 1.5 to 2 meters before exploding

Table L-7. Rocket-Propelled Grenades

AT Grenade Launcher Characteristics	 Rocket-Propelled Grenade (RPG)-7V					 RPG-16D
Projectile	PG-7	PG-7M	PG-VR	PG-VL	OG-7/7V	PG-16
Range, moving and stationary target (in meters)	300/500	NA	NA	NA	NA	500/800
Armor penetration (in millimeters)	330	330	>750	>PG-7M <PG-VR	NA	375
Ammunition type	High-explosive, antitank (HEAT)	HEAT	Tandem HEAT	HEAT	HE fragment	HEAT
Warhead caliber (in millimeters)	85	72	105	93	NA	58.3
Rate of fire (rounds per minute)	6	6	6	6	6	6
Crew	NA	NA	NA	NA	NA	2
Tube caliber (in millimeters)	40	40	40	40	40	58.3

Table L-7. Rocket-Propelled Grenades (Continued)

AT Grenade Launcher Characteristics	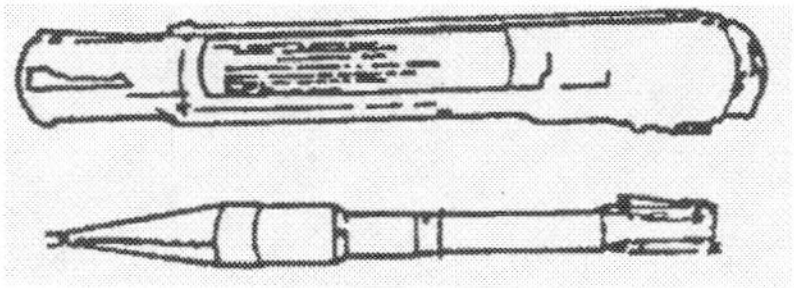 RPG-18 and 22		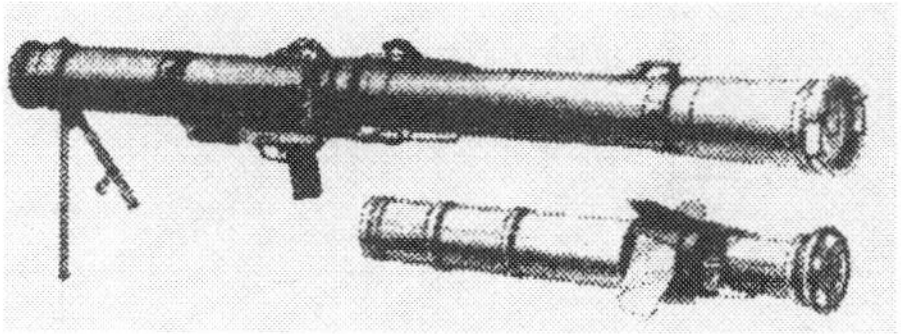 RPO-A
	RPG-18	RPG-22	RPO-A
Projectile			
Range, moving and stationary target (in meters)	200	250	200 direct fire 1,000 indirect fire
Armor penetration (in millimeters)	360	390	NA
Ammunition type	HEAT	HEAT	FAE
Rate of fire (rounds per minute)	NA	NA	2
Crew	1	NA	NA
Tube caliber (in millimeters)	64	73	93

Table L-8. Recoilless Rifles

AT Gun and Rifle Characteristics	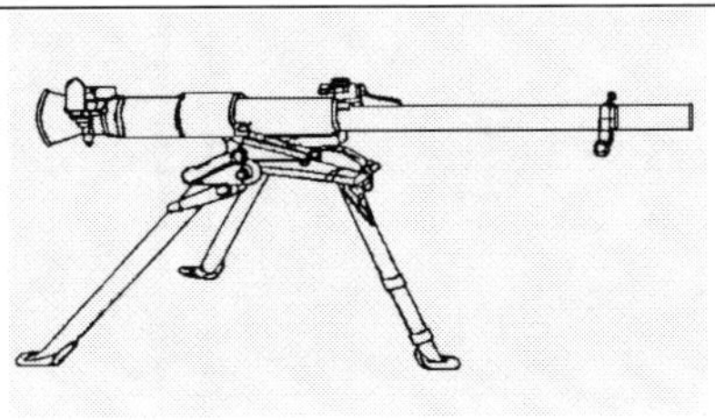 Self-Propelled Gun (SPG-9) Recoilless Rifle	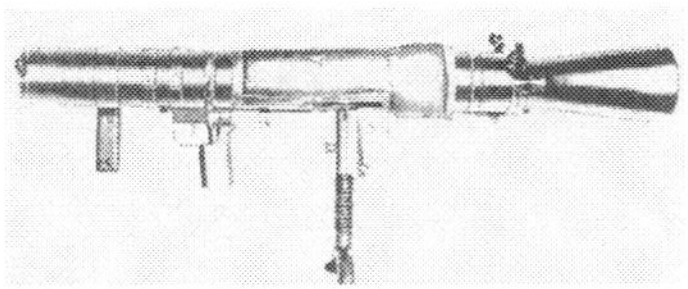 84-Millimeter Carl Gustaf Recoilless Rifle
Range, HEAT (in meters)	1,000	500
HEAT, rocket assist (RA)	NA	250
HE	NA	500
Indirect fire	NA	1,000 HE, 1,300-meter smoke, 2,300-meter illumination
Armor penetration (in millimeters)	400	500 HEAT, 900 HEAT, RA
Rate of fire, maximum and sustained (rounds per minute)	6	6
Fire control	Direct-view optics II (DVO)	3x DVO
Ammunition type	HEAT, HE, rocket-assisted projectile (RAP), armor piercing (AP) AT hollow charge	HEAT, HEAT RA, illumination, smoke, high explosive, dual purpose (HEDP), HE
Crew	3 or 4	

Table L-8. Recoilless Rifles (Continued)

Infantry Fighting Vehicle (IFV) Characteristics	 **Bronevaya Maschina Piekhota, Armored Vehicle, Infantry (BMP)-1P IFV**	 **BMP-2 IFV**	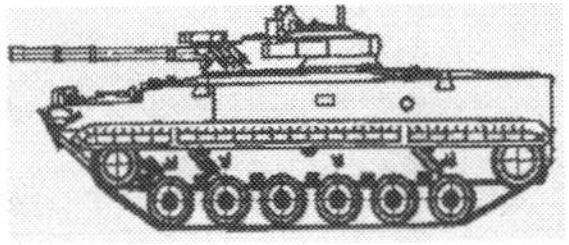 **BMP-3 IFV**
Main Armament (Caliber and Model)	73-millimeter 2A38 gun	30-millimeter 2A42 gun	100-millimeter 2A70
Ammunition type	HEAT FS, HE fragment	High explosive incendiary (HEI), high explosive tracer (HET), APT	HE fragment, HE
Range, effective (in meters)	800	1,500 light armor, 3,000 air, 2,500 ground	4,000 HE fragment
Rate of fire (rounds per minute) sustained/maximum	10	300/500	10/15

Table L-8. Recoilless Rifles (Continued)

IFV Characteristics	BMP-1P IFV (Continued)	BMP-2 IFV (Continued)	BMP-3 IFV (Continued)
Secondary Armament	AT-4a/5a ATGM	AT-4a/5a ATGM	AT-10 ATGM
Range, effective (in meters)	2,000/4,000	2,000/4,000	5,000
Rate of fire, cyclic and practical (rounds per minute)			
Penetration (millimeters at meter range)	600/650	600/650	660
Auxiliary Armament	7.62-millimeter PKT MG		30-millimeter 2A72 gun
Range, effective (in meters)			
Ammunition type			APT, HE, HET
Auxiliary Armament			
Model		7.62-millimeter PKT MG	3- by 7.62-millimeter PKT MG
Commander	IR	IR	IR
Speed, road and off road (kilometers per hour)	65/45	65/50	70/50

Table L-8. Recoilless Rifles (Continued)

IFV Characteristics	BMP-1P IFV (Continued)	BMP-2 IFV (Continued)	BMP-3 IFV (Continued)
Vehicle Characteristics (Continued)			
Range, road and off road (in kilometers)	600/570	600	600
Ground clearance (in millimeters)	390	420	190 to 510
Armor, hull/turret (in millimeters)	19/23	19/23	
Dimensions (length by width by height, in meters)	6.735 by 2.94 by 2.068	6.86 by 3.13 by 2.45	7.2 by 3.2 by 2.6
Crew and passengers	3/8	3/7	3/7

Table L-8. Recoilless Rifles (Continued)

IFV Characteristics	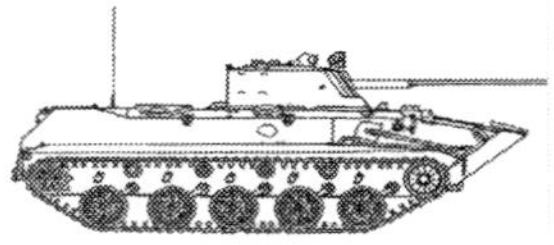 BMD-2	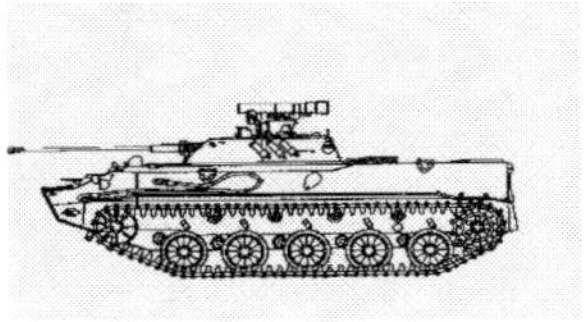 BMD-3	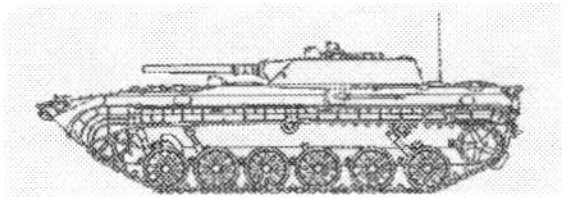 BRM
Main Armament (Caliber and Model)	30-millimeter 2A42 gun	30-millimeter 2A42 gun	73-millimeter 2A38 gun
Ammunition type	APT, fragment T, HEI	APT, fragment T, HEI	HEAT FS, FE fragment
Range, effective (in meters)	1,500 light armor, 2,000 air, 4,000 soft skin	2,000 AP, 4,000 HE	800
Rate of fire (rounds per minute) sustained and maximum	240/600	240/600	10
Secondary Armament	AT-4/5 ATGM	AT-4a/5a ATGM	7.62-millimeter PKT MG
Range, effective (in meters)	2,000/4,000	2,000/4,000	1,000
Rate of fire, cyclic and practical (rounds per minute)			650/250
Penetration (millimeters at meter range)	600 at 650	600 at 650	8 at 500 meters

Table L-8. Recoilles Rifles (Continued)

IFV Characteristics	BMD-2 (Continued)	BMD-3 (Continued)	BRM (Continued)
Auxiliary Armament		30-millimeter, AG-17 grenade launcher	
Model	2- by 7.62-millimeter PKT MG	7.62-millimeter PKT MG	
Vehicle Characteristics			
Night sights		Active IR	Second generation
Speed, road and off road (kilometers per hour)	60/35/10	70/45/10	70/10
Range, road and off road (in kilometers)	500/350	500	500
Armor, hull and turret (in millimeters)		16	10
Dimensions (length by width by height, in meters)	5.4 by 2.63 by 1.615	6.1 by 3.134 by 2.25	6.75 by 2.97 by 1.98
Crew and passengers	2/5	2/5	3/6

Table L-9. Armored Personnel Carriers (APCs)

APC Characteristics	 **BTR-60PB**	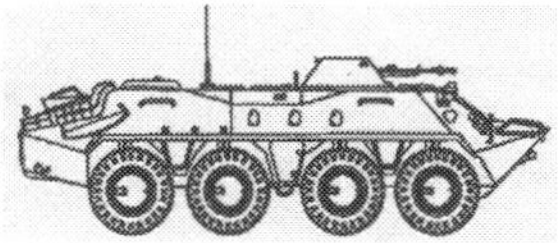 **BTR-70**	 **BTR-80**
Main Armament (Caliber and Model)	14.5-millimeter Soviet 14.5-millimeter, heavy, MG (KPVT)	14.5-millimeter MG (KPVT)	14.5-millimeter MG (KPVT)
Range, effective (in meters)	2,000 ground, 1,400 air	2,000 ground, 1,400 air	2,000 ground, 1,400 air
Rate of fire (rounds per minute) sustained and maximum	80/600	80/600	80/600
Secondary Armament	7.62-millimeter PKT MG	7.62-millimeter PKT MG	7.62-millimeter PKT MG
Range, effective (in meters)	1,000	1,000	1,000
Rate of fire, cyclic and practical (rounds per minute)	650/250	650/250	650/250
Penetration (millimeters at meter range)	8 at 500 meters	8 at 500 meters	8 at 500 meters

Table L-9. APCs (Continued)

APC Characteristics	BTR-60PB (Continued)	BTR-70 (Continued)	BTR-80 (Continued)
Vehicle Characteristics			
Commander	IR	IR	IR
Driver	IR	IR	IR
Range, road and off road (in kilometers)	500	450	800
Armor, hull and turret (in millimeters)		10/7	
Dimensions (length by width by height, in meters)	7.5 by 2.8 by 2.23	7.85 by 2.8 by 2.45	7.62 by 2.9 by 2.35
Crew and passengers	3/7	2/8	3/7

Table L-9. APCs (Continued)

APC Characteristics	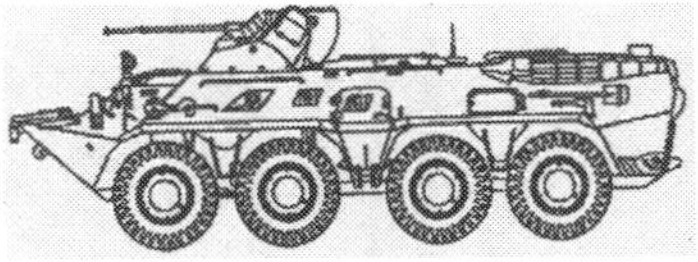 BTR-80A	 BTR-D	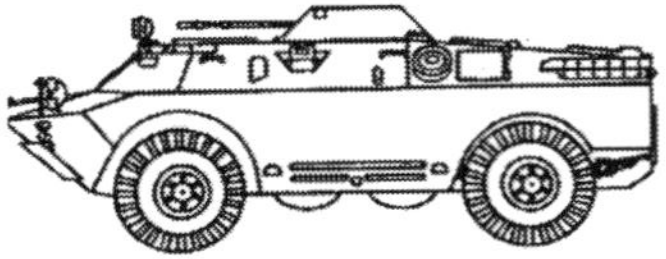 BRDM-2 Reconnaissance
Main Armament (Caliber and Model)	30-millimeter 2A72 gun	2 by 7.62 PKT MG	14.5-millimeter KPVT MG
Range, effective (in meters)	2,000 APT, 4,000 HEI, 800 night	1,000	2,000 ground, 1,400 air
Rate of fire (rounds per minute) sustained and maximum		250/650	80/600
Secondary Armament	7.62-millimeter PKT MG		7.62-millimeter PKT MG
Range, effective (in meters)	1,000		1,000
Rate of fire, cyclic and practical (rounds per minute)	650/250		650/250
Penetration (millimeters at meter range)	8 at 500 meters		8 at 500 meters

Table L-9. APCs (Continued)

APC Characteristics	BTR-80A (Continued)	BTR-D (Continued)	BRDM-2 Reconnaissance (Continued)
Vehicle Characteristics			
Night sights			
Range, road and off road (in kilometers)	800	500/350	750
Armor, hull and turret (in millimeters)			
Dimensions (length by width by height, in meters)		5.883 by 2.63 by 1.67	5.75 by 2.262 by 2.31
Crew and passengers	2/8	3/10	4
Remarks		Airborne assault. Can have AGS-17 grenade launcher. BTR-RD has 5 AT-4 antitank guided missiles (ATGMs), dismountable to 2 tripods.	

Table L-9. APCs (Continued)

APC Characteristics	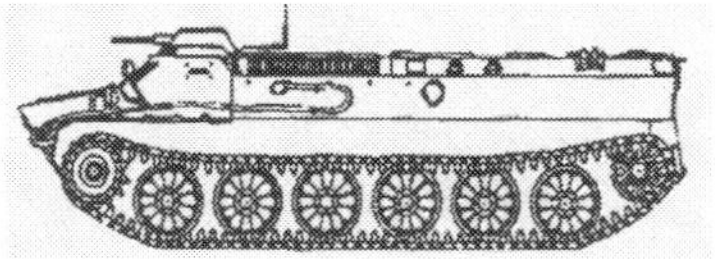 MTLB
Main Armament (Caliber and Model)	7.62 PKT MG
Range, effective (in meters)	1,000
Rate of fire (rounds per minute) sustained and maximum	250/650
Vehicle Characteristics	
Night sights	
Range, road and off road (in kilometers)	
Armor, hull and turret (in millimeters)	7/7
Dimensions (length by width by height, in meters)	6.45 by 2.85 by 1.87
Crew and passengers	2/11
Remarks	

Table L-10. ATGM

ATGM Characteristics	AT-3c SAGGER 9K11 Malyutka	AT-4a/b SPIGOT 9K111/9K111M Fagot/Faktoriya	AT-5 a/b SPANDREL 9K133/9K133m Konkurs/ Konkurs-M	AT-6 a/b/c SPIRAL 9K114 Shturm
Range (in meters)	500 to 3,000	70 to 2,000/75 to 3,500	70 to 4,000/75 to 4,000	400 to 5,000/400 to 5,000/400 to 6,000
Flight time to maximum range (in seconds)	23 to 26	11/19.5	19	13.3/16/18.6
Guidance and command link	Wire MCLOS, 9M14-2: wire SACLOS	Wire SACLOS	Wire SACLOS	Radio freguency (RF) SACLOS
Warheads	HEAT, 9M14-2: tandem HEAT	HEAT	HEAT/tandem HEAT	HEAT or FAE
Remarks	The AT-3 continues to be improved and proliferated worldwide. The M14-2 missile is effective against reactive armor.	Thermal sight detection range is 3,600 meters, with identification at 2,000 meters.	Thermal sight detection range is 3,600 meters, with identification at 2,000 meters.	FAE ranges 400 to 5,000 meters.

Table L-10. ATGM (Continued)

ATGM Characteristics	AT-7 SAXHORN 9K115 Metis	AT-8 SONGSTER 9K112 Kobra	AT-9 9K120 Ataka	AT-10 STABBER 9K116-1/9K116 Bastion/Kastet
Range (in meters)	40 to 1,000	100 to 4,000	400 to 6,000	100 to 5,000
Flight time to maximum range (in seconds)	6	10	14.5	15
Penetration (rolled hardened armor (RHA)/behind extended range artillery (ERA), in millimeters)	500	800	1,000/800	660
Guidance and command link	Wire SACLOS	RF (30 gigahertz) SACLOS		Laser beam rider SACLOS
Warheads	HEAT	HEAT	Tandem HEAT, blast, FAE, antihelicopter	HEAT
Remarks	Thermal sight detection range is 3,200 meters, with identification at 1,600 meters. Emplace and displace times are 12/20 seconds.	Fired through the main gun tube.		Fired through the main gun tube. Off BMP-3 range is 100 to 4,000 meters, but it can fire on the move.

Table L-10. ATGM (Continued)

ATGM Characteristics	AT-11a/b SNIPER 9K119/9K119M Svir/Invar	AT-12 STABBER 9K116-2 Sheksna	AT-13 9K115-2 Metis-M	AT-14 Kornet
Range (in meters)	100 to 5,000	100 to 5,000	80 to 1,500	100-5,500
Flight time to maximum range (in seconds)	15	12 to 4,000	8.4	22
Penetration (RHA/ behind ERA in millimeters)	770 AT-11a, 1,050 AT-11b	800	1,000/800	1,200/980
Guidance and command link	Laser beam rider SACLOS	Laser beam rider SACLOS	Wire SACLOS	Laser beam rider SACLOS
Warheads	HEAT/tandem HEAT	HEAT	Tandem HEAT, FAE	Tandem HEAT, FAE
Remarks	Fired through the main gun tube. T-72 variants have a 4,000-meter maximum range.	Fired from the halt through the main gun tube.	FAE has greater power than 152-millimeter artillery round.	Thermal sight range is 4,000-meter detection and 3,500-meter identification. Maximum altitude of engaged helicopters is 3,000 meters.

Table L-10. ATGM (Continued)

ATGM Characteristics	AT-16 9K121 VIKhR	Milan (France [FR]) 1/2/2T/3	Tow (US) 1/ITOW/ 2a/2b	HOT (FR/Germany [GE])
Range (in meters)	1,000 to 10,000	20 to 2,000	3,750/3,750/65-3,750/200 to 3,750	75 to 4,000
Flight time to maximum range (in seconds)	23	12.5	2a: 20, 2b: 21	16.5
Penetration (RHA/ behind ERA in millimeters)	1,050/900	600/1,060/1,200/ 1200	500, ITOW:800, 2:920, 2a 1,000	850, 2:950
Guidance and command link	Laser beam rider SACLOS (lock on before launch)	1 and 2: SACLOS wire SACLOS IR	SACLOS wire	SACLOS wire
Warheads	Tandem HEAT			
Remarks	Maximum airborne target speed is 800 kilometers per hour.	Milan 3 medium range ATGM has IR command link resistant to jamming and tandem warhead.	TOW 2b is fly over, shoot down, top attack, with self-forming penetrant 2nd generation forward looking infrared (FLIR).	HOT 3 long-range ATGM has a tandem warhead, bispectral day and night sights, and may be mounted on posts, vehicles, or helicopters.

Table L-11. Rotary Wing Aircraft

Rotary Wing Aircraft Characteristics	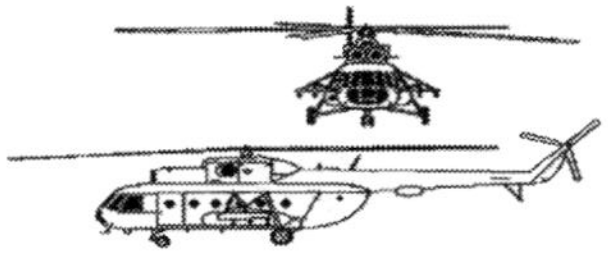 **Mi-8 HIP**	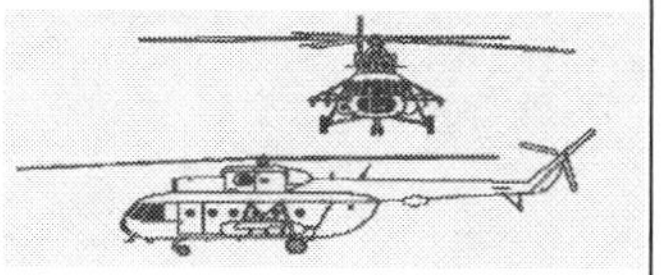 **Mi-17 HIP H**	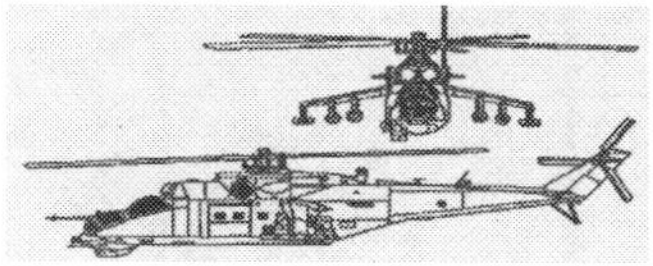 **Mi-24 HIND D/E/F**
Mission	Utility, transport	Utility, transport	Attack, close support
Gun	12.7 millimeter	12.7 millimeters	30-millimeter HIND F 12.7-millimeter HIND D/E
ATGM	AT-2c	AT-2c, AT-3c	4-16 AT-2c HIND D AT-6c HIND E/F
Rockets	57 millimeters	57 millimeters	57 millimeters or 80 millimeters
Bombs	250 kilograms, 500 kilograms	250 kilograms, 500 kilograms	250 kilograms, 500 kilograms

Table L-11. Rotary Wing Aircraft (Continued)

Rotary Wing Aircraft Characteristics	Mi-8 HIP (Continued)	Mi-17 HIP H (Continued)	Mi-24 HIND D/E/F (Continued)
Range, normal payload (in kilometers)	460	495	480
Crew	3	3	2
Passengers	24	24	8
Remarks	Troops can fire personal weapons through windows. It has four external hard points.	Improved version of Mi-8MT, have upgraded engines and six external hard points.	
	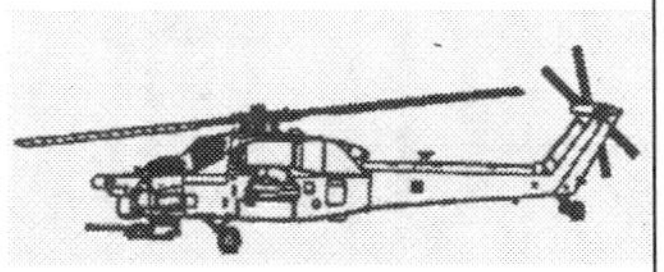 **Mi-28N HAVOC**	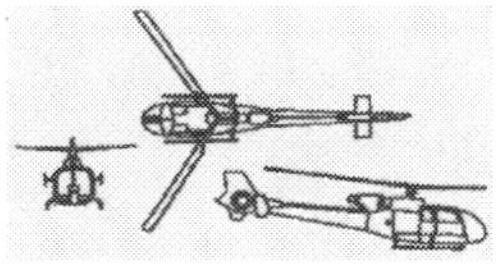 **Gazelle (FR)**	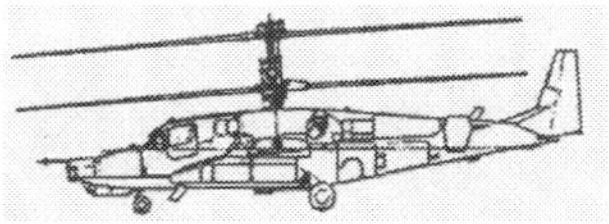 **Ka-50 HOKUM "Black Shark" or "Werewolf"**
Mission	Attack, close support	Light attack utility	Attack, close support
Gun	23 millimeters 30 millimeters	2 by 7.62 millimeters	2A42 30-millimeter cannon

Table L-11. Rotary Wing Aircraft (Continued)

Rotary Wing Aircraft Characteristics	Mi-28N HAVOC	Gazelle (FR)	Ka-50 HOKUM
ATGM	16 AT-6/9/16	AT-3, HOT	AT-16
Rockets	57 millimeters	5 millimeters 68 millimeters 2.75 inches	80 millimeters
Bombs	NA	NA	250 kilograms, 500 kilograms
Range, normal payload (in kilometers)	470 1,100 with drop tanks	710	455
Crew	2 or 3	1 or 2	1
Passengers	NA	2	NA
Remarks	This system is not operationally fielded in any armed force.	NA	The armored cockpit withstands 23-millimeter fire and the windscreen withstands 12.7-millimeter fire.

Table L-11. Rotary Wing Aircraft (Continued)

Rotary Wing Aircraft Characteristics	MD 500 (US)	UH-1H (US)	BO-105 (GE)	Lynx (UK)
Mission	General purpose, light attack	Utility	General purpose, light attack	General purpose, attack
Gun	7.62 millimeters	7.62 millimeters	20 millimeters	2 by 20 millimeters
ATGM	TOW	NA	HOT TOW	HOT Hellfire TOW
Rockets	70 millimeters	NA	NA	Sura, 80 millimeters
Range, normal payload (in kilometers)	540	465	318	540
Crew	2	2	2	2
Passengers	2	11	3	10

Table L-12. Combat Aircraft

Combat Aircraft Characteristics	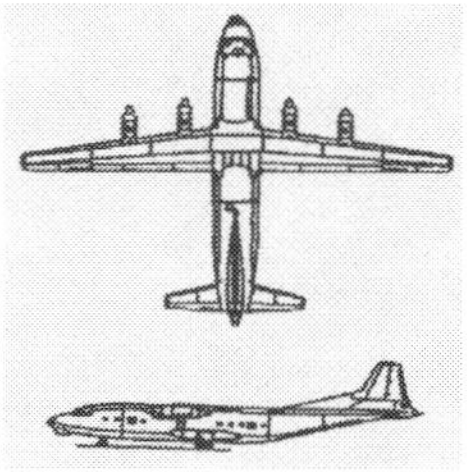 An-12 CUB	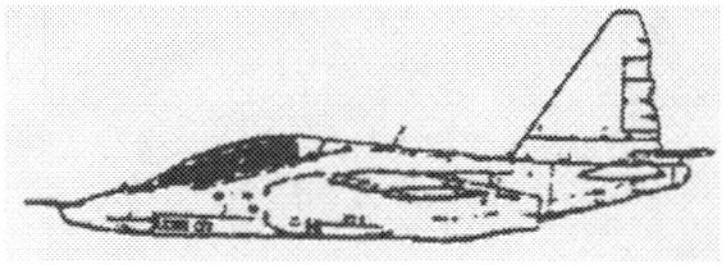 SU-25 FROGFOOT
Mission	Medium transport aircraft	CAS
Gun(s)	NR-23 23 millimeters in the tail turret	GSh-30-2, 30 millimeters (250 rounds), AO-17a, 30-millimeter pods
ATGMs		16x AT-16 or AT-9
Rockets		57-millimeter S-5 pod, 80-millimeter S-8 pod (up to 8 pods)
Air-to-surface missiles		AS-7, AS-10, AS-11, AS-14, AS-17
Bombs		100 kilograms, 350 kilograms, 500 kilo-grams (up to 4,000 kilograms)
Combat radius (in kilometers)	1,500 to 1,800	495
Payload, paratroops	90	
Remarks	It can operate from dirt strips.	SU-25M: AT, SU-25UB: trainer, ceiling 10,000 meters, 4,344 kilograms maximum payload, 6.5 gram limit.

Table L-12. Combat Aircraft (Continued)

Combat Aircraft Characteristics	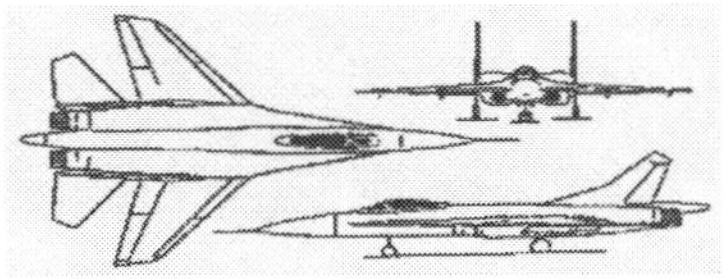 **SU-27 Flanker B**
Mission	Fighter/intercepter, fighter bomber variant
Gun(s)	GSh-301, 30-millimeter cannon
ATGMs	
Rockets	80-millimeter S-8 or 120-millimeter S-25
Air-to-air missiles	R-27, R-73, AA-10, AA-11
Bombs	100 kilograms, 250 kilograms, 500 kilograms
Combat radius (in kilometers)	1,125/1,950 with tanks

Glossary

∞	infinity symbol
"	inch(es)
'	foot; feet
<	more than
>	less than
°	degree
?	unknown information
AA	assembly area
AAR	after-action review
AB	A map's reference point.
ABD	air base defense
ACE	ammunition, casualty, and equipment
ADC	area damage control
AFJI	Air Force joint instruction
AFM	Air Force manual
AFSF	Air Force security force
AGS	armored gun system
AI	area of interest
alt	alternate
ammo	ammunition
ANCD	automated net control device

AO	area of operation
AOE	Army of Excellence
AOR	area of responsibility
AP	armor piercing
APC	armored personnel carrier
APL	antipersonnel land mines
APT	armor piercing tracer
AR	Army regulation; automatic rifle (SAW)
AS	area security
ASAP	all-source analysis product
ASCC	Army service component command
ASI	addition skill identifier
ASP	ammunition supply point
ASV	armor security vehicle
AT	antitank
AT/FP	antiterrorism/force protection
ATGM	antitank guided missiles
ATTN	attention
AWOL	absent without leave
BCC	battlefield circulation control
BCOC	base cluster operations center
BDOC	base defense operations center
BDU	battle dress uniform

BHL	battle handover line
BII	basic-issue item
bldg	building
BMP	Bronevaya Maschina Piekhota, armored vehicle, infantry
BPS	baud per second
BSA	brigade support area
BT	bomb trench
c	distance of the centerline of the road
C	clay; distance from the centerline
C^2	command and control
C4	composition C4
CA	cinnamic acid; civil affairs
CAC	crossing area commander
CAM	chemical agent monitor
CANA	convulsant antidote for nerve agents
CAS	close air support
CASEVAC	casualty evacuation
CCIR	commander's critical information requirements
cdr	commander
CERTEX	certification exercise
cGyph	centigray per hour
CI	civilian internee

CID	criminal investigation division
℄	centerline
cm	centimeter
co	company
COA	course of action
COMMEX	communication exercise
COMMZ	communications zone
CONEX	container express
CONUS	continental United States
CP	command post
CPOG	chemical protective overgarment
CPU	chemical protective undergarment
CR	Dibenz (B, f)-1, 4-oxazepine
CS	combat support; chlorobenzul-malononitrile
CSC	crossing site commander
CSS	combat-service support
CTOC	corps tactical operations center
CTUS	customs territory of the United States
DA	Department of the Army
DAP	decontamination apparatus
DC	dislocated civilian
DCG	degrees grid north

DD	Department of Defense
Dec	December
DED	detailed equipment decontamination
DFAS-IN	Defense Finance and Accounting Service-Indianapolis
DFC	defense force commander
DGG	degrees grid north
DGM	degrees magnetic north
DGT	degrees true north
DM	designated marksman
DNVT	digital nonsecure voice terminal
DOD	Department of Defense
DS	direct support; decontamination solution
DS2	decontamination solution number 2
DSA	division support area
DTD	detailed troop decontamination
DTG	date-time group
DTO	division transportation office
DVO	direct-view optics
DZ	drop zones
E	east
EA	engagement area
EAC	echelon above corps

EEFI	essential elements of friendly information
EEP	engineer equipment park
EIC	end item code
ENDEX	end of exercise
EO	executive order
EOC	emergency operations center
EPW	enemy prisoner of war
ERA	extended range artillery
ERP	engineer regulating points
FA	field artillery
FAE	fuel air explosive
FCF	field confinement facility
FDC	fire direction center
FEBA	forward edge of the battle area
FFIR	friendly force information requirements
FIST	fire support team
FLIR	forward looking infrared
FM	field manual
FMFM	fleet Marine force manual
FMJ	full metal jacket
FN	foreign nation
FO	forward observer

FOX	M93 NBC Reconnaissance System
FP	force protection
FPF	final protective fire
FPL	final protective line
FR	France
FRAG	fragment
FRAGO	fragmentary order
FSO	fire support officer
G	gravel; M203
G2	Assistant Chief of Staff, G2 (Intelligence)
G5	Assistant Chief of Staff, G5 (Civil Affairs)
gal	gallon
GB	guard bunker
GE	Germany
GM	grid to magnetic
GMG	grenade machine gun
GN	grid north
GP	general purpose
GPS	Global Positioning System
GS	general support
GTA	graphic training aid
GVW	gross vehicle weight

HAVOC	Soviet mi-28 aircraft
H_d	horizontal distance
HE	high explosive
HEAT	high explosive antitank
HEDP	high explosive, dual purpose
HEI	high-explosive incendiary
HET	high-explosive tracer
HHC	headquarters and headquarters company
HI	high
HIND	Soviet helicopter
HIP	Soviet medium lift helicopter
HMMWV	high mobility multipurpose wheeled vehicle
HN	host nation
HOKUM	Soviet Ka-? aircraft
HOT	air-to-ground antitank missile, also call Euromi
HTS	highway traffic section
hwy	highway
HQ	headquarters
IBCT	initial brigade combat team
ICITAP	International Criminal Investigative Training Assistance Program

ICM	improved conventional munitions
ID	identification
IEDK	individual equipment decontamination kit
IFV	infantry fighting vehicle
IPB	intelligence preparation of the battlefield
IPC	interpersonal communication
I/R	internment and resettlement
IR	infrared
IRIC	Internment Resettlement Information Center
JRA	joint rear area
Jul	July
KBPS	kilobaud per second
km	kilometers
KPVT	Soviet 14.5-millimeter heavy machine gun
L	location
L&O	law and order
LAW	light antiarmor weapon
lb	pound(s)
LC	line of contact
LCE	load-carrying equipment
LD	line of departure

ldr	leader
LMTV	light medium tactical vehicle
LO	low
LOC	lines of communication
LOGSEC	logistics security
LOTS	logistics over the shore
LP	listening post
LRA	local reproduction authorized
LVOSS	Light Vehicle Obscuration Smoke System
LZ	landing zones
m	meter; perpendicular distance from the center of the tape to the centerline of the road
M	medium; mud
MANSCEN	Maneuver Support Center
MBA	main battle area
MCCM	modular crowd control munitions
MCLOS	manual command line of sight
MCO	Marine Corps orders
MCRP	Marine Corps warfighting publication
MCT	movements control team
MCWP	Marine Corps warfighting publication
MDMP	military decision-making process

MEDEVAC	medical evacuation
MELIOS	mini eyesafe laser infrared observation set
MEVA	mission-essential or vulnerable area
METL	mission-essential task list
METT-TC	mission, enemy, troops, terrain, time available, and civilian considerations
MG	machine gun
MHZ	mega hertz
MI	military intelligence
MIJI	meaconing; intrusion; jamming; interference
mil	military
MILES	multiple-integrated laser engagement system
MILVAN	military-owned demountable container
min	minute
MLC	military load classification
MLG	mils grid north
MLM	mils magnetic north
MLT	mils true north
mm	millimeter
MMS	maneuver and mobility support
MOPP	mission-oriented protection posture

MOUT	military operations on urbanized terrain
MP	military police
mph	miles per hour
MPI	military police investigation
MRE	meal, ready-to-eat
MRO	medical regulating officer
MSR	main supply route
MTMC	military traffic management command
MTF	medical treatment facility
MTP	mission training plan
MWD	military working dogs
N	north
NA	not applicable
NAAK	nerve agent auto injector kit
NAI	named area of interest
NATO	North Atlantic Treaty Organization
NBC	nuclear, biological, chemical
NBCWRS	Nuclear, Biological, Chemical Warning and Reporting System
NCO	noncommissioned officer
NCOIC	noncommissioned officer in charge
NE	northeast
NEO	noncombatant evacuation operations

NGO	nongovernmental organization
NLP	nonlethal procedure
NLW	nonlethal weapons
no	number
NP	nonpersistent
NSN	national stock number
NTC	National Training Center
NVD	night vision device
NVG	night vision goggle
NW	northwest
NWP	Naval warfare publication
OB	obstruction
OC	observer/controller; oleoresin capsicum
OCOKA	observation and fields of fire, cover and concealment, obstacles, key terrain, and avenues of approach
OCONUS	outside the continental United States
Oct	October
OD	other detainees
OEG	operational exposure guidance
OF	observed fire
OIC	officer in charge
OP	observation post
OPCON	operational control

OPFOR	opposing forces
OPLAN	operation plan
OPNAVINST	Chief of Naval Operations Instruction
OPORD	operation order
OPSEC	operations security
ORP	objective rally point
OVM	on-vehicle material
P	pedestrian; artificial paving; persistent
PA	power amplifier
PAO	Public Affairs Office
PC	point of curvature
PCI	precombat inspection
PCIR	police and criminal information requirements
PDDE	power-driven decontamination equipment
PDF	principal direction of fire; portable document format
PERSCOM	personnel command
PEWS	Platoon Early-Warning System
PIAP	police information assessment process
PIO	police intelligence operations
PIR	priority intelligence requirements

PKT	The standard vehicle with a mounted light machine gun and is the basic Soviet coaxial AFV weapon.
PL	phase line
PLGR	precision lightweight global positioning system receiver
plt	platoon
PM	provost marshal
PMCS	preventive-maintenance checks and services
PMO	Provost Marshals Office
POE	port of embarkation
POL	petroleum, oils, and lubricants
PRC	populace and resource control
PSG	platoon sergeant
PSYOP	psychological operations
PT	point of tangency
PVAB	portable vehicle-arresting barrier
PVO	private volunteer organization
pyro	pyrotechnics
R	radio; rock; radius of curve; radar set
R&S	reconnaissance and surveillance
RA	rocket assist
RADAR	radio detection and ranging

RADIAC	radiation detection, indication, and computation
RAOC	rear-area operations center
RAP	rocket-assisted projectile
RCA	riot control agents
RES	radiation exposure status
RF	radio frequency
RHA	rolled hardened armor
RL	release line; railroad bridge
ROE	rules of engagement
ROI	rules of interaction
RORO	roll-on/roll-off
RP	release point; retained person
RPG	rocket-propelled grenade
RPK	squad machine gun
RTO	radio telephone operator
S	size; sand; south
S2	Intelligence Officer (US Army)
S3	Operations and Training Officer (US Army)
S4	Supply Officer (US Army)
S5	Civil Affairs Officer (US Army)
S&I	standardization and interoperability
SACLOS	semiautomatic command line of sight

SAGGER	An AT-3 ATGM.
SALUTE	size, activity, location, unit, time, and equipment
SAM	surface-to-air missile
SAW	squad automatic weapon
SAXHORN	A nickname for the Soviet AT-7-ATGM.
SE	southeast
SF	security forces
SINCGARS	Single-Channel, Ground-to-Air Radio System
SIR	serious incident report
SITREP	situation report
SJA	Staff Judge Advocate
SM	soldier's manual
SMCI	senior military customs inspector
SMG	submachine gun
SNAP	size, nature, activity, protection
SONGSTER	The nickname for the Soviet AT-8 ATGM.
SPANDREL	An AT-5 ATGM.
SPIGOT	An AT-4 ATGM.
SPIRAL	A Soviet AT-6 ATGM.
SOC	security operations center
SOFA	Status of Forces Agreement

SOI	signal operating instructions
SOP	standard operating procedure
SP	start point
SPG	self-propelled gun
SPOTREP	spot report
sqd	squad
SRT	special-reaction team
SSCO	small-scale contingencies operations
SSN	social security number
STABBER	A nickname for the Soviet AT-10 ATGM.
STANAG	standardization agreement
STX	situational training exercise
SVD	Soviet sniper rifle
SW	southwest
T	time; snow blockage
T&E	traversing and elevating
TA	training area
TACSOP	tactical standing operating procedure
TAI	traffic accident investigation
TC	training circular
TCF	tactical combat force
TCMD	transportation control and movement document

TCP	traffic control post
TEM	training execution model
TL	team leader
TLP	troop-leading procedures
TM	technical manual
tm	team
TO	table of organization
TOC	tactical operations center
TOE	table(s) of organization and equipment
TOW	tube-launched, optically tracked, wire-guided missile
TP	training practice
TRADOC	United States Army Training and Doctrine Command
TRL	traffic regulating line
TRP	target reference point
TRS	transportation railway service
TSC	theater support command
TTP	tactics, techniques, and procedures
U	unit
UCMJ	Uniform Code of Military Justice
UIN	unit identification number
UK	United Kingdom
UO	urban operation

US	United States
USACIDC	United States Army Criminal Investigation Division Command
USAF	United States Air Force
USAPA	United States Army Publishing Agency
USATSC	United States Army Training Support Center
USCG	United States Coast Guard
USDA	United States Department of Agriculture
USDOJ	United States Department of Justice
UTM	universal transverse mercator
UXO	unexploded ordnance
V	vehicular
V_d	vertical distance
VHS	video home system
VP	both vehicle and pedestrian
W	flooding; west
WO	warning order
Y	limited, all-weather route, significant, serious, regular, or recurrent flooding or snow blockage
X	all-weather route
Z	Zulu; fair-weather route

Bibliography

AR 190-8. *Enemy Prisoners of War, Retained Personnel, Civilian Internees and other Detainees {OPNAVINST 3461.6; AFJI 31-304; MCO 3461-1}*. 1 October 1997.

AR 190-11. *Physical Security of Arms, Ammunition and Explosives*. 30 September 1993.

AR 190-12. *Military Police Working Dogs*. 30 September 1993.

AR 190-14. *Carrying of Firearms and Use of Force for Law Enforcement and Security Duties*. 12 March 1983.

AR 190-29. *Misdemeanors and Uniform Violation Notices Referred to US Magistrates or District Courts*. 1 March 1984.

AR 190-30. *Military Police Investigations*. 1 June 1978.

AR 190-40. *Serious Incident Report*. 30 November 1993.

AR 190-47. *The Army Corrections System*. 15 August 1996.

AR 190-51. *Security of Unclassified Army Property (Sensitive and Nonsensitive)*. 30 September 1993.

AR 525-13. *Antiterrorism Force Protection (AT/FP): Security of Personnel, Information, and Critical Resources*. 10 September 1998.

AR 700-93. *Processing and Shipping DOD Sponsored Retrograde Materiel Destined for Shipment to the United States, Its Territories, Trusts, and Possessions*. 15 May 1980.

DA Form 1156. *Casualty Feeder Report.* 1 June 1966.

DA Form 1248. *Road Reconnaissance Report.* 1 July 1960.

DA Form 1249. *Bridge Reconnaissance Report.* 1 July 1960.

DA Form 1251. *Ford Reconnaissance Report.* 1 January 1955.

DA Form 1355-1-R. *Hasty Protective Row Minefield Record (LRA)* 00 October 1997.

DA Form 1594. *Daily Staff Journal or Duty Officer's Log.* 1 November 1962.

DA Form 1711-R. *Engineer Reconnaissance Report (LRA).* 00 May 1985.

DA Form 1911. *Altitude and Azimuth (SIN-COS)* 00 February 1957.

DA Form 1971-R. *Radiological Data Sheet-Monitoring or Point Technique (LRA).* 00 September 1994.

DA Form 2028. *Recommended Changes to Publications and Blank Forms.* 1 February 1974.

DA Form 2404. *Equipment Inspection and Maintenance Worksheet.* 1 April 1979.

DA Form 2823. *Sworn Statement.* 1 December 1998.

DA Form 3881. *Rights Warning Procedures / Waiver Certificate.* 00 November 1989.

DA Form 3975. *Military Police Report.* 00 December 1998.

DA Form 4137. *Evidence/Property Custody Document.* 01 July 1976.

DA Form 5367-R. *Personnel Status Report (LRA).* 00 December 1984.

DA Form 5517-R. *Standard Range Cards (LRA).* 00 February 1986.

DA Pamphlet 190-12. *Military Working Dog Program.* 30 September 1993.

DA Pamphlet 190-51. *Risk Analysis for Army Property.* 30 September 1993.

DD Form 1265. *Request for Convoy Clearance.* 1 January 1959.

DD Form 1408. *Armed Forces Traffic Ticket (Book, Consisting of 25 Three-Part Sets).* 00 December 1987.

DD Form 1911. *Material Courier Receipt.* 00 May 1982.

DD Form 2707. *Confinement Order (Available on DOD Website).* 00 November 1999.

DD Form 2708. *Receipt of Inmate or Detained Person* 00 November 1999.

DD Form 2745. *Enemy Prisoner of War (EPW) Capture Tag.* 00 May 1996.

DFAS-IN 37-1. *Finance and Accounting Policy Implementation.* January 2000.

Geneva Convention Relative to the Treatment of Prisoners of War. August 1949.

FM 100-14. *Risk Management.* 23 April 1998.

FM 101-5. *Staff Organization and Operations.* 31 May 1997.

FM 101-5-1. *Operational Terms and Graphics*

{MCRP 5-2A}. 30 September 1997.

FM 101-5-2. *US Army Report and Message Formats.* 29 June 1999.

FM 17-95. *Cavalry Operations.* 24 December 1996.

FM 17-98. *Scout Platoon.* 10 April 1999.

FM 19-10. *The Military Police Law and Order Operations.* 30 September 1987.

FM 19-15. *Civil Disturbances.* 25 November 1985.

FM 19-20. *Law Enforcement Investigations.* 25 November 1985.

FM 19-25. *Military Police Traffic Operations.* 30 September 1977.

FM 20-32. *Mine/Countermine Operations.* 29 May 1998.

FM 2-11. *First Aid for Soldiers.* 27 October 1988.

FM 21-26. *Map Reading and Land Navigation.* 7 May 1993.

FM 21-60. *Visual Signs.* 30 September 1987.

FM 21-75. *Combat Skills of the Soldier.* 3 August 1984.

FM 23-25. *Light Antiarmor Weapons.* 17 August 1994.

FM 25-100. *Training the Force.* 15 November 1988.

FM 25-101. *Battle Focused Training.* 30 September 1990.

FM 27-10. *The Law of Land Warfare.* 18 July 1956.

FM 3-19.1. *Military Police Operations.* 22 March 2001.

FM 3-19.30. *Physical Security.* 8 January 2001.

FM 3-19.40. *Military Police Interment/Resettlement Operations.* 1 August 2001.

FM 3-3. *Chemical and Biological Contamination Avoidance {FMFM 11-17}.* 16 November 1992.

FM 3-4. *NBC Protection {FMFM 11-9}.* 29 May 1992.

FM 3-5. *NBC Decontamination {MCWP 3-37.3}.* 28 July 2000.

FM 3-6. *Field Behavior of NBC Agents (Including Smoke and Incendiaries) {AFM 105-7; FMFM 7-11-11}.* 3 November 1986.

FM 3-34.2. *Combined-Arms Breaching Operations.* 31 August 00.

FM 34-2. *Collection Management and Synchronization Planning.* 8 March 1994.

FM 34-130. *Intelligence Preparation of the Battlefield.* 8 July 1994.

FM 41-10. *Civil Affairs Operations.* 14 February 2000.

FM 44-8. *Combined Arms for the Air Defense.* 1 June 1999.

FM 5-170. *Engineer Reconnaissance.* 5 May 1998.

FM 55-10. *Movement Control.* 9 February 1999.

FM 55-20. *Rail Transport in a Theater of Operations.* 1 June 2000.

FM 6-20. *Fire Support in the Airland Battle.* 17 May 1988.

FM 6-30. *Tactics, Techniques, and Procedures for Observed Fire.* 16 July 1991.

FM 7-8. *Infantry Rifle Platoon and Squad.* 22 April 1992.

FM 7-10. *The Infantry Rifle Company.* 14 December 1990.

FM 90-10-1. *An Infantryman's Guide to Combat in Built-Up Areas.* 12 May 1993.

FM 90-40. *Multiservice Procedures for the Tactical Employment of Nonlethal Weapons (NLP) (S&I, CDR, USATSC, ATTN: ATIC-TMSD-T (Bldg {MCRP 3-15.8; NWP 3-07-31; USCG Pub 3-07.31}.* 6 October 1998.

Geneva Convention Relative to the Treatment of Prisoners of War. August 1949.

GTA 19-7-1. *Enemy Prisoner of War (EPW) Basic Commands.* 1989.

Section 660 of the Foreign Assistance Act.

STANAG 1059. *Distinguishing Letters for Geographical Entities for Use in NATO.* 17 April 2000.

STANAG 2010. *Military Load Classification Markings.* 18 November 1980.

TC 25-10. *A Leader's Guide to Lane Training.* 26 August 1996.

TM 11-5860-202-10. *Operator's Manual for Mini Eyesafe Laser Infrared Observation Set (MELIOS) AN/PVS-6 (NSN 5860-01-35-8551) (EIC: N/A).* 1 September 1995.

TM 11-6665-251-10. *Operator's Manual for Radiac Set AN/VDR-2 (NSN 6665-01-222-1425).* 1 March 1988.

TM 3-6665-312-12&P. *Operator's and Organizational Maintenance Manual Including Repair Parts and Special Tools List for M8A1 Automatic Chemical Agent Alarm (NSN 6665-01-105-5623) and*

Auxiliary Equipment M10 Power Supply (6665-00-859-2225); M10A1 Power Supply (6665-00-093-2739); M228 High Profile Mounting Kit (6665-00-859-2212) and M182 Low Profile Mounting Kit (6665-00-110-9492) {TO 11H2-17-1}. 17 March 1985.

TM 5-6665-298-10. *Operator's Manual for AN/PSS-12 Mine Detecting Set {TM 5-6665-298-10}*. 28 February 1995.

Index

PREFACE

1. Scope

This publication describes multi-Service tactics, techniques, and procedures (MTTP) for consideration and use during the tactical employment of nonlethal weapons (NLW) in support of warfighting personnel conducting training and tactical operations. This publication—

a. Provides an overview of NLW and its relationship to deadly force.

b. Provides NLW system description.

c. Describes the capability requirements of NLW.

d. Discusses fundamental concepts and training requirements involved with NLW.

e. Discusses the tactical employment considerations of NLW.

f. Discusses lessons learned from previous NLW use.

2. Purpose

a. This publication provides a single-source, consolidated reference on the tactical employment of NLW and supporting systems.

b. This publication is not intended to restrict the authority of the commander from organizing the force and executing the mission in a manner deemed appropriate.

3. Application

a. This publication provides commanders and their staffs unclassified guidance for NLW employment and planning. Commanders and staffs can use this publication to aid in the tactical employment of NLW during exercises and contingencies.

b. The United States (U.S.) Army, Marine Corps, Navy, Air Force, and Coast Guard approved this multi-Service publication for use.

4. Implementation Plan

Participating service command offices of primary responsibility will review this publication, validate the information and reference, and incorporate it in service manuals, regulations, and curricula follows.

Army. The Army will incorporate the procedures in this publication in U.S. Army training and doctrinal publications as directed by the Commander, U.S. Army Training and Doctrine Command (TRADOC). Distribution of this publication is in accordance with (IAW) initial distribution number (IDN) 115757.

Marine Corps. The Marine Corps will incorporate the procedures in this publication in U.S. Marine Corps doctrinal and training publications as directed by the Commanding General, U.S. Marine Corps Combat Development Command (MCCDC). Distribution of this publication is IAW Marine Corps Doctrinal Publication System (MCPDS).

Navy. The Navy will incorporate the procedures in this publication in U.S. Navy doctrinal and training publications as directed by the Commander, Navy Warfare Development Command (NWDC). Distribution of this publication is IAW Military Standard Requisition-

MARINE CORPS PCN: 143 000057 00
Air Force Distribution: F

ing and Issue Procedures (MILSTRIP) Desk Guide and Navy Standing Operating Procedure (NAVSOP) Pub 409.

Air Force. The Air Force will validate and incorporate appropriate procedures IAW applicable governing procedures. Distribution is IAW Air Force Instruction (AFI) 33-360.

Coast Guard. The Coast Guard will incorporate the procedures in this publication in U.S. Coast Guard doctrinal and training publications as directed by the Assistant Commandant for Operations, Director of Operations Policy, U.S. Coast Guard (USCG). Distribution of this publication is IAW Coast Guard standing operating procedures.

5. User Information

a. TRADOC, MCCDC, NWDC, Air Force Doctrine Center (AFDC), USCG, and the Air Land Sea Application (ALSA) Center developed this publication with the joint participation of the approving service commands. ALSA will review and update this publication as necessary.

b. This publication reflects current joint and service doctrine, command and control organizations, facilities, personnel, responsibilities, and procedures regarding NLW employment. Changes in service protocol, appropriately reflected in joint and service publications, will likewise be incorporated in revisions of this document.

c. We encourage recommended changes for improving this publication. Key comments to the specific page and paragraph and provide rationale for each recommendation directly to—

FM 3-22.40 (FM 90-40) — U.S. Army Training and Doctrine Command, Fort Monroe, Virginia

MCWP 3-15.8 — U.S. Marine Corps Combat Development Command, Quantico, Virginia

NTTP 3-07.3.2 — U.S. Navy Warfare Development Command, Newport, Rhode Island

AFTTP(I) 3-2.45 — Headquarters, Air Force Doctrine Center, Maxwell Air Force Base, Alabama

USCG Pub 3-07.31 — U.S. Coast Guard, Washington, D.C.

15 January 2003

NLW

Tactical Employment of Nonlethal Weapons

TABLE OF CONTENTS

Page

*This publication supersedes FM 90-40, MCRP 3-15.8, NWP 3-07.31, USCG Pub 3-07.31, dated 6 October 1998.

FIGURES

Tables

EXECUTIVE SUMMARY

Tactical Employment of Nonlethal Weapons (NLW)

Today's world events mandate a need to project nonlethal force across all levels of war to enable our warfighters and leaders to effectively deal with a host of traditional as well as nontraditional threats. Now more than ever, the minimal level of tolerance for collateral damage and loss of human life, coupled with the tendency for the typical adversary to exploit the rules of engagement (ROE) to his benefit, necessitates an effective and flexible application of force through nonlethal weapons. Much like a rheostat switch in which power can be dialed up or down as desired, NLW provide tools to allow a commander to employ sufficient force to accomplish an objective without requiring the destruction of an enemy or the habitat. The intent of employing NLW is not to add another step in the progression of escalation with an adversary, but to add another tool to use anywhere along that continuum. This publication provides initial guidance for employing NLW in a tactical environment.

Overview of NLW

Chapter I describes the background, definition, and the concept description of NLW. This chapter discusses the development and employment of NLW, and explains the relationship of NLW to deadly force capabilities. The final portion of the chapter reviews policy per Office of Secretary of Defense instructions on employing NLW.

Purpose, System Description, and Organization

Chapter II describes the purpose of NLW and gives a general description of currently available commercial or government off-the-shelf equipment and munitions. The chapter also addresses organization structure, to include system location/distribution, personnel requirements, and mission requirements/tactical structure.

New Capability Requirements

Chapter III describes the applicability of NLW across the range of military operations. Additionally, this chapter discusses the core capabilities as stated in the Joint Mission Area Analysis (JMAA) as described in the Department of Defense Joint Concept for NLW.

Training With the NLW Capability Set

Chapter IV describes training requirements, topics, methods, locations, and a program of instruction from the Interservice Nonlethal Individual Weapons Instructor Course currently available for NLW options. This chapter also addresses additional training available from civilian law enforcement agencies.

Employment of NLW Capabilities

Chapter V establishes the fundamental understanding of the force continuum needed to employ NLW. It outlines employment prerequisites, mission-planning factors, and personnel requirements needed to employ NLW options. It further delineates NLW tasks according to the JMAA and provides considerations for preparation for unit tactical operations. This chapter also discusses employment objectives for NLW and ROE development. This chapter ends with multiple examples of service-specific crowd control formations used by units equipped with NLW.

Lessons Learned

Chapter VI focuses on NLW lessons learned from recent military operations. The chapter discusses how nonlethal technology has changed the way forces are looking at their adversaries in using NLW. It also covers the use of NLW along the force continuum in conjunction with ROE and the media.

PROGRAM PARTICIPANTS

The following commands and agencies participated in developing this publication:

Joint

U.S. Joint Forces Command, Suffolk, VA
U.S. European Command, J3/J5, APO AE
U.S. Pacific Command, J3832, Camp H. M. Smith, HI
U.S. Central Command, J3, MacDill AFB, FL
U.S. Southern Command, SCJ5-PS, Miami, FL
U.S. Space Command, SPJ5X, Peterson AFB, CO
U.S. Special Operations Command, HQ USSOCOM, SOOP-PTE, MacDill AFB, FL
Joint Nonlethal Weapons Directorate, Quantico, VA
Interservice Nonlethal Individual Weapons Instructor Course, Fort Leonard Wood, MO

Army

Headquarters, Department of the Army, ODCSOPS (DAMO-FDT), Pentagon, Washington, D.C.
U.S. Army Training and Doctrine Command, DCSDCS (ATDO-A), Fort Monroe, VA
U.S. Army Training and Doctrine Command, DCSDEV (ATCD-SN), Fort Monroe, VA
U.S. Army Infantry Center, DCD, Fort Benning, GA
U.S. Army Military Police School, ATTN: DOTD, MANSCEN (ATZT-DT-WM-M), Fort Leonard Wood, MO
USAJFKSWCS, DOTD, Fort Bragg, NC
National Ground Intelligence Center, Charlottesville, VA

Marine Corps

Marine Corps Combat Development Command, Quantico, VA
MARCORSYSCOM, Quantico, VA
USMC Warfighting Lab, Quantico, VA
MCSFT Company, MCSF Battalion, Chesapeake, VA
2d FAST MCSF Battalion, Yorktown, VA
1st MARDIV Military Police Company, HQ Battalion, Camp Pendleton, CA
Marine Corps Detachment Military Police and Chemical School, Fort Leonard Wood, MO
Special Operations Training Group, I Marine Expeditionary Force, Camp Pendleton, CA
Special Operations Training Group, II Marine Expeditionary Force, Camp Lejeune, NC

Navy

Commander, Navy Warfare Development Command (NWDC), Newport, RI
USN Surface Warfare Center, Crane, IN
USN Surface Warfare Center, Dahlgren, VA

Air Force

AFDC/DJ, Langley AFB, VA
Headquarters, ACC/SFOT, Langley AFB, VA
Air Force Security Forces Center/SFOR, Lackland, AFB, TX
Headquarters, USAF/XOF, Washington, D.C.

Coast Guard

Commandant, U.S. Coast Guard, (G-OPL-2), Washington, D.C.

Chapter I

Overview of Nonlethal Weapons

1. Background

a. In today's operational environment, United States (U.S.) forces regularly perform missions that were almost inconceivable a few decades ago. In this new environment, lethal firepower or the threat of its use may no longer be the default solution to all crises or problems. Senior leaders face a new level of public and media sensitivity and scrutiny concerning the proper role of the military as an instrument of national power. Field commanders must understand these sensitivities and attempt to achieve an appropriate military force. Junior leaders must apply the resulting decisions wisely, often in changing situations filled with uncertainty and danger.

b. Throughout history, changes in culture and technology have influenced the character of military force and the manner in which it is employed. These changes are an attempt to maximize military force in the force continuum. In the complex and changing political and social domain of the modern world, nonlethal capabilities may offer the opportunity to increase or maximize the military as an instrument of national power.

c. Nonlethal weapons (NLW) must now be considered an integral part of joint operational capabilities. Joint forces anticipate that NLW technology will advance and provide additional options for force commanders.

2. Definition

a. Department of Defense Directive (DODD) 3000.3, Policy for Nonlethal Weapons, 9 July 1996, defines NLW as "weapons that are explicitly designed and primarily employed so as to incapacitate personnel or materiel while minimizing fatalities, permanent injury to personnel, and undesired damage to property and the environment." Furthermore, "unlike conventional lethal weapons that destroy their targets principally through blast, penetration, and fragmentation, nonlethal weapons employ means other than gross physical destruction to prevent the target from functioning. Nonlethal weapons are intended to have relatively reversible effects on personnel and materiel."

b. The term "nonlethal" does not mean zero mortality or nonpermanent damage; these are goals and not guarantees of these weapons. NLW can add flexibility to combat operations and enhance force protection by providing an environment in which friendly troops can engage threatening targets with reduced risk of noncombatant casualties and collateral damage.

3. Concept Description

Nonlethal capabilities within the context of the service capability sets (see Appendix E) may provide armed forces additional offensive and defensive tools to assist in the incapacitation and destruction of an enemy, capture of important intelligence sources, crowd control, civil disturbances, enforcing mandates, and force protection in accordance with the specified rules of engagement (ROE) for a specific operation. These operations include small-scale contingencies (SSC), military operations other than war (MOOTW), stability and support operations (SASO), and civil disturbances. The U.S. Army's Task Force (TF) FALCON in Kosovo and U.S. armed forces located at Camp X-RAY in Guantanamo Bay, Cuba are superb examples of the use of NLW to prevent escalation within an operation.

4. Background on Development and Employment

a. Military forces have long used nonlethal force to influence behavior of people and nations, defeat adversaries with minimum use of lethal capabilities, and weaken adversaries to defeat them with conventional arms. Examples of nonlethal means include a show of force, physical obstacles, riot control agents (RCA), noise to create or enhance psychological effects, smoke and obscurants to mask operations or defeat homing and guidance mechanisms, and using light to disorient combatants. These nonlethal means will remain relevant in future operations and are a part of an evolutionary process of weapons development. Joint operations in Somalia provided increased awareness of the need to incorporate nonlethal options into the fighting capabilities. Therefore, the Department of Defense (DOD) established the Joint Nonlethal Weapons Program in 1996 to research, develop, and acquire these capabilities.

b. Increased interaction between U.S. forces and civilian populations has become a feature of the contemporary operational landscape that is likely to remain the case in the future. Three factors account for this forecast:

(1) Worldwide patterns of population growth and migration have resulted in an expansion of urban culture within established industrialized nations, as well as many pre-industrial and emerging third world societies (60 percent of the world's population live in built-up areas). The prevalence of urbanization in many crisis-prone regions of the world creates the potential for large, vulnerable groups of noncombatants to be entrapped in volatile confrontations involving the deployment of military forces in military operations in urban terrain (MOUT).

(2) U.S. forces increasingly operate in the challenging environment known as MOOTW; a category that includes such missions as humanitarian assistance, disaster relief, noncombatant evacuations, and the various types of peace operations. These operations commonly involve close and continual interaction between U.S. forces and noncombatant civilians. Some MOOTW scenarios include the presence of paramilitary forces, armed factions, or rogue elements that present a constant, but uncertain threat to joint forces. In these situations, the mission of military forces includes preventive tasks. In other words, U.S. forces enhance mission accomplishment by preventing individuals or factions from carrying out specified undesirable activities, such as rioting, looting, attacking, harassing, or threatening. Often, the adversary blends in with the local population of innocent citizens. At other times, sectors of the local population may rise against U.S. forces and become active participants in acts of violence. Factional alignments, the level of violence, and the threat to mission accomplishment, may change frequently with little or no warning. Under such circumstances, the identity of adversaries is uncertain and the use of deadly force for purposes other than self-defense may be constrained by ROE or the judgment of the on-scene commander.

(3) From a defense-in-depth or situational understanding perspective, the force protection, homeland security, and the antiterrorism missions, require that forces assess the use of NLW as a response to potential asymmetric attacks. Asymmetric threats run the spectrum of adversarial use of innovative or nontraditional strategies, tactics, or technologies, to exploit vulnerabilities and strike a decisive blow while avoiding the strengths of an opponent. The objectives of these strikes are intended to undermine national will. Terrorist attacks such as those directed against the USS Cole, World Trade Center, and Pentagon, highlight the severity of the asymmetric threat and its resultant effects.

5. Relationship of NLW to Deadly Force

a. NLW provide "the lesser means" that could be considered prior to employing deadly force. The services definition of deadly force is as follows:

(1) U.S. Army – "Force that a person uses causing, or that a person knows, or should know, will create a substantial risk of causing death or serious bodily harm." (Army Regulation (AR) 190–14, Military Police Carrying of Firearms and Use of Force for Law Enforcement and Security Duties, 12 March 1993).

(2) USMC – "That force which a person uses with the purpose of causing death or serious bodily harm or which a reasonable and prudent person would consider likely to create substantial risk of causing death or serious bodily harm. It will be used as a last resort when all lesser means have failed." Marine Corps Order (MCO) 5500.6F, Arming of Security and Law Enforcement Personnel and the Use of Force, 20 July 1995.

(3) USN – "Force that a person uses causing, or that a person knows or should know would create a substantial risk of causing death or serious bodily harm." Secretary of The Navy Instruction (SECNAVINST) 5500.29B, Use of Deadly Force and the Carrying of Firearms by Personnel of the Department of the Navy in Conjunction with Law Enforcement, Security Duties and Personal Protection, 29 September 1992.

(4) USAF – "Force that causes, or that you know has a substantial risk of causing death or serious bodily harm." Air Force Instruction (AFI) 31-207, Arming and Use of Force by Air Force Personnel, 1 September 1999.

(5) USCG – "Deadly force is any force that is likely to cause death or other serious physical injury." (COMDTINST M16247.1B, Maritime Law Enforcement Manual, 9 August 2001.)

b. The commitment of military power to resolve crises has traditionally involved the use of deadly force or the implicit or explicit threat of the use of deadly force. Military units are trained, organized, and equipped for this purpose. A force armed only with traditional military weapons normally has two options for enforcing compliance: threating deadly force and applying deadly force. This limitation creates a critical vulnerability that belligerents may quickly discern and use to their advantage.

c. Nonlethal capabilities provide a wider range of options that augment, but do not replace traditional means of deadly force. The option to resort to deadly force must always remain available when the commander believes it is appropriate to the mission. DODD 3000.3, states "the availability of NLW shall not limit a commander's inherent authority and obligation to use all necessary means available and to take all appropriate action in self-defense." The existence of nonlethal capabilities does not represent the potential for "nonlethal war," and unrealistic expectations to that effect must be vigorously avoided. Noncombatant casualties, to include serious injuries and fatalities, are often an unavoidable outcome when employing military power, regardless of NLW availability. Augmenting deadly force with NLW should be fundamental to planning and executing any operation where the commander is contemplating deadly force.

d. The commander must clearly articulate the roles of NLW within the ROE. NLW is an additional means of employing force throughout the force continuum for the particular purpose of reducing the probability of death or serious injury to noncombatants or belligerents. Using deadly force must always remain an inherent right of individuals in instances when they, their fellow servicemen, or personnel in their charge are threatened with death or serious bodily harm. NLW add flexibility to combat operations and enhance force protection by providing the field commander with options to engage threatening targets with limited risk of noncombatant casualties and collateral damage.

e. Commanders and public affairs officers (PAO) must prepare personnel to address media questions and concerns regarding the role of NLW. Operational experience indicates that new capabilities give rise to significant media interest. Personnel participating in interviews or briefings must be prepared to address the role of NLW. It must be clear that the presence of NLW in no way indicates abandoning the option to employ deadly force in appropriate circumstances.

6. Advantages of Employing NLW

NLW provide commanders the flexibility to influence the situation favorably with increased safety to friendly forces and reduced risk of both noncombatant fatalities and collateral damage. Some advantages follow:

a. NLW can be more humane and consistent with the political and social implications implicit in humanitarian and peacekeeping missions.

b. The force that properly employs nonlethal options can gain advantages over those who rely on lethal options alone. Because the degree of provocation required to employ NLW is substantially less than lethal options, the force can take a more proactive position that may diminish the likelihood of having a situation escalate to a point where deadly force is required to resolve a conflict.

c. NLW options can be less likely to provoke others. However, commanders should be cautioned using NLW does not preclude an undesired result.

d. Demonstrated restraint can greatly diminish feelings of anger and remorse when deadly force is required after nonlethal options fail.

e. Using NLW can facilitate post incident stabilization by reducing populace alienation and collateral damage.

f. NLW can reduce the possibility of injury to friendly forces compared to forces without NLW.

7. Policy Per Office of the Secretary of Defense Instructions on Employment of NLW

a. DODD 3000.3 states, "nonlethal weapons, doctrine, and concepts of operation shall be designed to reinforce deterrence and expand the range of options available to commanders. Nonlethal weapons should enhance the capability of U.S. Forces to accomplish the following objectives:

(1) Discourage, delay, or prevent hostile actions.

(2) Limit escalation.

(3) Take military action in situations where use of lethal force is not the preferred option.

(4) Better protect our forces.

(5) Temporarily disable equipment, facilities, and personnel.

(6) Nonlethal weapons should also be designed to help decrease the post-conflict costs of reconstruction."

b. DODD 3000.3 continues by stating, "The availability of nonlethal weapons shall not limit a commander's inherent authority and obligation to use all necessary means available and to take all appropriate action in self-defense. Neither the presence nor the potential effect of nonlethal weapons shall constitute an obligation for their employment or a higher standard for employment of force than provided for by applicable law. In all cases, the United States retains the option for immediate use of lethal weapons, when appropriate, consistent with international law."

c. DODD 3000.3 further states, "Nonlethal weapons shall not be required to have a zero probability of producing fatalities or permanent injuries. However, while complete avoidance of these effects is not guaranteed or expected, when properly employed, nonlethal weapons should significantly reduce them as compared with physically destroying the same target." Nonlethal weapons may be used in conjunction with lethal weapon systems to enhance the latter's effectiveness and efficiency in military operations. This shall apply across the range of military operations to include those situations where overwhelming force is employed."

d. According to paragraph 3.21 of the Interium Guidance issued by the Deputy Secretary of Defense, 30 October 2002, all developmental weapon systems must be submitted for legal review and must receive favorable findings before they can be incorporated into the force. This review includes examination for consistency with applicable international and domestic laws, including the laws of war, various arms control treaties, and protocols. U.S. forces can be assured that any NLW they are issued have passed this critical test. Medical and policy reviews should also be considered prior to fielding any new weapon systems. Commanders and operators should consult their specific ROE for guidance.

e. Some capabilities, although designed to minimize fatalities and serious injuries, may have effects that could actually discourage their use. Therefore, the basic decision to employ military force in defense of national interests is usually a matter of intense public concern. The manner in which that force is exercised is subjected to the same scrutiny.

f. The presence of NLW on the battlefield does not mean they must be used. It also does not mean they must be used prior to the use of lethal force. In all cases, the U.S. retains the option for immediate use of lethal weapons when the on-scene commander deems it appropriate. Additionally, the immediate use of lethal weapons must be consistent with U.S. and international law and the ROE.

Chapter II

PURPOSE, SYSTEM DESCRIPTION, AND ORGANIZATION

1. Purpose

The purpose of NLW is to enhance the ability of military commanders to conduct operations across the spectrum of conflict. Over the past several years, military operations have gravitated towards the lower end of this spectrum and have involved U.S. forces in various missions. These missions include SSC, SASO, MOOTW, and civil disturbances where the ROE are inherently restrictive. NLW are designed to address the special requirements of low-intensity environments. They provide tools that enable the individual, team, or unit, to anticipate and/or respond to provocation with more appropriate means than through the use of deadly force. Essentially, NLW provide U.S. forces with a near-term capability for effectively controlling the nontraditional battlefield, within the constraints levied by ROE, by reducing the risk to friendly forces, mitigating casualties, and minimizing collateral damage.

2. System Description

a. NLW Capability Set. The services have, and are continuing to procure, versatile packages of NLW comprised of commercial off-the-shelf (COTS) and government off-the-shelf (GOTS) equipment and munitions. The government is also exploring advanced technologies for future procurements. In addition to addressing contingency requirements, the services' NLW capability sets provide limited sustainment training ammunition and equipment.

b. NLW Capability Set—Component Categories. Each service determines the contents of its own capability set. Examples of the services current configurations are shown in Appendix E. The NLW capability set components can be divided into four distinct categories: personnel protectors, personnel effectors, mission enhancers, and training devices. The following examples are not all-inclusive.

(1) Personnel Protectors. Personnel protectors include items such as face shields, riot shields, and shin guards that protect the individual from blunt trauma injuries inflicted by thrown objects, clubs, etc.

(2) Personnel Effectors. Personnel effectors include items such as riot batons, malodorants, stingball grenades, pepper sprays, and other kinetic rounds (e.g. sponge grenades) designed to discourage, disorient, or incapacitate individuals or groups.

(3) Mission Enhancers. Mission enhancers include items such as bullhorns, spotlights, caltrops, isotropic radiators (stun/flash), and barriers/entanglers. These items are designed to facilitate target identification and crowd control, and to limit personnel and vehicular movement.

(4) Training Devices. Training devices include items such as training suits, training rounds, training batons, and inert pepper sprays. They are designed to facilitate realistic hands-on scenario training in preparation for operations.

c. System Relationships and Inter/Intraoperability. The NLW capability set is designed to provide an operating force (ranging from squad through battalion-sized units [see Appendix E]) with tactically deployable nonlethal equipment. Fielding of this capability set enhances the operation of systems and equipment already fielded and planned within the joint community. The NLW capability set components are compatible with military equipment that is already fielded or planned for future fielding. The components of the NLW capability set are supported within current military logistic capabilities.

d. Augmentation of Organic Equipment. The operational intent of the capability set is to augment an operating force with the required nonlethal equipment to enhance its force projection capability. The NLW capability set is additional equipment that must be incorporated into lift and load planning.

3. Organization

Organization/organizational structure is not anticipated to change as a result of NLW capability sets being added to the inventory.

a. System Location/Distribution. The United States Army (USA), United States Marine Corps (USMC), and United States Navy (USN) have procured and fielded capability sets to the operating forces, security forces, bases, and stations. All services have current plans to field additional capability sets as required. At the time of this printing, the United States Air Force (USAF) has identified its capability sets to field in the near future. The United States Coast Guard (USCG) also intends to field NLW.

b. NLW Capability Set Composition. Appendix E describes the current and proposed capability sets.

c. Personnel Requirements. To date, no change or increase in military specialty codes have been determined in order to field, operate, and maintain the NLW capability set. Each service determines its own training personnel requirements.

d. Mission Requirements/Tactical Structure.

(1) The mission requirements and tactical structure pertinent to the employment of the NLW will vary depending on the type of contingency and subsequent task organization of the employing unit. Mission requirements, derived from commander's guidance, should be satisfied through the appropriate combination of lethal and nonlethal capabilities.

(2) NLW provide commanders with a means of shaping their area of operations (AO) short of employing deadly force. However, it is imperative that U.S. forces retain the means and ability to employ lethal force in the event that nonlethal force fails to achieve desired outcomes. For example, NLW employment may provoke a lethal response from the adversary. Without the imminent threat of lethal force, nonlethal munitions will prove to be less effective and possibly even counterproductive. More importantly, denying U.S. forces a lethal capability would place them at unacceptable risk. If nonlethal force is employed without the threat of lethal force, the belligerent may opt to exploit a perceived vulnerability. In this case, the use of NLW becomes counterproductive, because it serves only to aggravate the threat and causes the situation to deteriorate.

Chapter III

NLW CAPABILITY REQUIREMENTS

1. Applicability Across the Range of Military Operations

a. U.S. forces conduct military operations across a large spectrum. This spectrum spans from humanitarian assistance and peacekeeping to strategic nuclear exchange. Since the end of the Cold War, the U.S. military has been increasingly involved with operations at the lower end of the spectrum. These operations focus on deterring war, resolving conflict, supporting civil authorities and promoting peace. NLW may be useful across many categories of these operations. The range of military operations includes various levels of combat "intensity," and NLW provide one more force option to deal with the varying intensity. MOOTW may be found across the range of military operations and often include situations where noncombatants are mingled with the adversaries or involved in acts of violence, such as rioting or looting. In such circumstances, NLW provide commanders the flexibility to influence the situation favorably with reduced risk of noncombatant fatalities and collateral damage. NLW also provide a means by which to ascertain the intent of individuals/belligerents. Some characteristics of operations that present the opportunity to use NLW are—

(1) Close contact between friendly forces and noncombatant civilians.

(2) The presence of paramilitary or irregular forces.

(3) Armed civilian gangs, unlawful combatants, or militia.

b. These characteristics present a real but ill-defined threat. Under such circumstances, the commander's judgment or ROE may restrict using deadly force for other than self-defense. NLW may provide U.S. forces flexibility to resolve situations where the use of deadly force is restricted. These weapons provide flexibility by—

(1) Allowing U.S. forces to apply measured military force.

(2) Reducing risk of serious noncombatant causalities.

(3) Accomplishing the mission.

(4) Providing force protection.

(5) Retaining the initiative.

(6) Allowing U.S. forces to remain relevant and capable in increasingly complex and chaotic environments.

c. The need to reduce the risks of serious injury to personnel is not limited to crowd control scenarios. In MOUT, some of the local civilian populace may remain in an urban area in the midst of battle. The traditional solution to such challenges has been restricting ROE. Nonlethal capabilities may provide commanders the flexibility to adapt a fluid approach to MOUT, which allows their subordinates the freedom of action to employ appropriate levels of measured military force to accomplish their mission.

d. Another example of NLW applications in MOOTW is in peace enforcement. In these operations, NLW are used in an area denial role to separate belligerents without resorting to deadly force. They are also used to support humanitarian relief organizations in protecting food and medical distribution sites. The presence of NLW will not limit a commander's inherent authority and obligation to use all necessary means available and to take appropriate action in self-defense.

2. Core Capabilities

Core capabilities are those fundamental competencies that enable U.S. forces to achieve desired operational objectives. NLW provide a flexible means to protect friendly forces, high-value assets, and influence the actions of potential adversaries and/or noncombatants. These goals can be achieved without resorting to lethal force and in a manner that may minimize collateral damage. The core capabilities associated with nonlethal effects fall into three major categories as stated within the Joint Mission Area Analysis (JMAA) for Nonlethal Weapons (approved December 2000 by the Joint Requirements Oversight Council): counterpersonnel, countermateriel, and countercapability.

a. Counterpersonnel Capabilities. The DOD Joint Concept for Nonlethal Weapons concept paper, 5 January 1998, describes required nonlethal counterpersonnel capabilities for U.S. forces as follows:

(1) "The application of military force to accomplish a mission with reduced risk of fatalities or serious casualties among noncombatants—or even, in some instances, among enemy forces." There are several specific nonlethal counterpersonnel capabilities to be considered. These include, but are not limited to, the means to influence the behavior and activities of a potentially hostile crowd and the capability to bring a mob engaged in a riot under control. While there are many similarities in these two capabilities, each involves unique challenges, which may require radically different solutions.

(2) "Capability to incapacitate personnel." For the purposes of this document, "incapacitation" is achieved when the weapon's effects render personnel unable to physically or mentally (real or perceived) act in a hostile or threatening manner. In keeping with the guiding principles for NLW, this incapacitation should be reversible through the passage of time. The focus will be on capabilities that affect both individuals and groups of people.

(3) "Capability to deny personnel access to an area (land, sea, or air)." NLW can provide new possibilities for barrier planning in any type of military operation, including systems that produce physical or mental discomfort to those who enter the denied area.

(4) "Capability to clear facilities and structures of personnel." The Joint Concept for NLW states this capability "will facilitate MOUT by reducing the risks of noncombatant casualties and collateral damage while simultaneously minimizing the advantages accruing to an enemy defending a built up area." This capability is applicable to all clearing missions that may be conducted (e.g., Maritime Interception Operations [MIO]) and could assist in providing enhanced force protection.

(5) "Capability to seize personnel, which could include some combination of technologies inherent in other core capabilities, for example, incapacitation and entangling devices (which may aid in seizing personnel)." This capability is intended to augment lethal means used to capture specified individuals, such as enemy combatants or persons who are inciting a mob or crowd to violence.

b. Countermateriel Capabilities. The Joint Concept for Nonlethal Weapons describes countermateriel capabilities as follows:

(1) Rendering equipment and facilities unusable without complete destruction by attacking only weapons of war and supporting infrastructure while minimizing risk to noncombatants. This could enhance operations by defusing potentially volatile situations where conventional military means might otherwise prove counterproductive.

(2) Disabling or denying the use of vehicles, vessels, and aircraft entry into targeted areas or access within an AO. This capability may include physical barriers that impede or stop movement. Alternative systems may alter combustion properties of fuels

and viscosity of lubricants, embrittle or decay materials, attack tires, gaskets, or hoses, or act as adhesives to fuse metal parts. The requirement also includes the capability to facilitate boarding and clearing operations.

c. Counter Capability Capabilities. For U.S. forces counter capability capabilities are described within the JMAA for NLW as follows:

(1) "Disable or neutralize facilities and systems." This includes a nonlethal capability for disabling/neutralizing electrical generating facilities, command, control, communications, computers, intelligence, surveillance, and reconnaissance (C4ISR) systems, Integrated Air Defense Systems (IADS), weapons systems, optical sensors, electrical sensors, and navigation capabilities with such controlled effects as to allow for selective, precise engagement. Other needs include deceiving reconnaissance with illusions of barriers, obstacles, and/or forces, and concealing/covering friendly activities or movements, particularly within complex urban terrain.

(2) "Deny the use of weapons of mass destruction (WMD)." Mission needs include rendering a WMD inoperative; containing the potential release of deadly agents/contaminants; and preventing or neutralizing the production, storage, deployment (transport), employment, and delivery of WMD. Nonlethal capabilities in combating WMD should be designed to permit engagement within a populated or sensitive terrain, prior to the initiation of armed conflict.

3. Limitations

The limitations of NLW are effectiveness, perception, and legality.

a. The effectiveness of NLW is dependent on factors such as motivation, age, environmental and human factors, range of engagement, and numerous others. NLW can be lethal if improperly applied. Furthermore, countermeasures for thwarting virtually all nonlethal options are usually apparent, quickly learned, and readily available. Because they are not intended to kill, nonlethal options teach adversaries what to avoid in the future. Small unit commanders must prepare to keep one step ahead of belligerents.

b. The perceptual limitation is that personnel may misunderstand the appropriate applications of NLW across the range of military operations. The incorrect perception that NLW will allow wars and MOOTW to be prosecuted without casualties may lead to conflicting expectations between political and military leaders. These conflicts can result in U.S. forces being vulnerable to misplaced or inconsistent ROE. All leaders, political and military, involved in planning and executing military missions must understand there are no "nonlethal operations."

c. Some proposed NLW might be forbidden by law or policy. Accordingly, it is essential that all NLW developments be evaluated by appropriate authorities to ensure they comply with the law of war, U.S. law, and U.S. treaty obligations. Use of NLW containing chemical agents, for instance, must be evaluated in the context of the Chemical Weapons Convention. Also, using RCA in an armed conflict requires Presidential approval.

Chapter IV

TRAINING

1. Fundamental Concepts and Training Requirements

The successful accomplishment of any operation in which nonlethal measures are employed requires extensive preparation, of which individual, unit, medical support, and unit training are vital parts. Training should be designed to give individuals an understanding of the entire subject area and enable them to function efficiently as members of a unit. This training must be intensive and realistic. Training with the NLW capability set, and other NLW, is critical to employ these weapons with confidence. The training flow should be—train leaders, train instructors, train unit, exercise. Qualified instructors should conduct initial unit training. Unit sustainment training should be conducted per individual service and unit requirements.

2. Training of Instructors

a. Verbal communication skills, open-hand control techniques, RCA, effects of impact weapons, and civil disturbance tactics, techniques, and procedures (TTP) are areas trained to standard at the Interservice Nonlethal Individual Weapons Instructor Course (INIWIC) held at the Marine Corps Detachment, Fort Leonard Wood, Missouri. INIWIC is an Interservice Training Review Organization (ITRO) course, and quotas are available to any service through normal training reservation channels. A course outline is provided in Appendix C. This is the only formal DOD NLW instructor-training course. The training requirements outlined in the INIWIC represent the minimum training standards to include in developing NLW and tactics instructors. Additionally, the United States Army Military Police School (USAMPS) has developed a training support package that includes Army-specific TTP for small units (company level and below) in using NLW capabilities. The reference for this is the U.S. Army's website at: www.wood.army.mil.

b. The INIWIC is a "train-the-trainer" program to provide commanders NLW instructors within their command. Furthermore, the military services are not precluded from developing their own training based upon the INIWIC curriculum. The commander can then tailor the NLW instruction to fit mission needs. Establishing the service-level INIWIC serves as an assurance measure, giving the commander a quantifiable means of certifying his nonlethal instructors.

3. Nonlethal Training Topics for Deploying Units

Each service and unit will have unique training requirements based on mission type, unit size, deployment area, etc. As a minimum, the training instructor should cover the following topics in any training for units deploying with nonlethal capabilities. They are part of the INIWIC curriculum.

NOTE: U.S. Army training standards are contained in Army Regulation (AR) 350-38, Training Devices Policies and Management, 15 October 1993; USMC training standards for NLW are contained in MCO 1510.112, Individual Training Standards for Nonlethal Weapons, 27 May 1998.

a. Force Continuum. This is an introduction to the federal force continuum model as outlined by MCO 5500.6F, Arming of Law Enforcement and Security Personnel and the Use of Deadly Force; AR 190-14; AFI 31-207, Arming and Use of Force by Air Force Personnel (1 September 1999) (or other service orders). Identifying the proper levels of force, understanding levels of resistance, and how nonlethal technologies affect the force continuum are the minimum topics the instructor should cover.

b. Crowd Dynamics/Crowd Control. This is an introduction to the differences between crowds, mobs, and riots. This course teaches the student basic crowd control techniques that will be easily applied to various situations. The student should be familiarized with tactics and techniques found in field manual (FM) 19-15, MCWP 3-11.1, Rifle Company, TC 90-1, and Air Force manual (AFMAN) 31-201V6, Civil Disturbance (1 April 2002), but will also consider nontraditional and small unit application.

c. Communication Skills. This is an introduction to the stages of conflict management, verbal aggression, nonverbal communication, physical aggression, physiological diversions, and proper mental conditioning. The first NLW skill an individual must have is the ability to communicate. General A. M. Gray, former Commandant of the Marine Corps, once said, "A warrior's most formidable weapon is his mind." The results of an individual's mental processes are evident through actions and words, and the words and communication skills used in dealing with people are truly the first line of nonlethal options.

d. Oleoresin Capsicum (OC) Aerosol and other RCA Training. This is an introduction to the uses of OC/pepper spray and other RCA. The student should have an appreciation for decontamination requirements, storage, and shelf life of OC/RCA. The instructor will also teach legal and tactical considerations. INIWIC certified instructors should teach existing training programs that include live agent training.

e. Open-Hand Control. This is an introduction to pressure point control techniques, unarmed self-defense measures, weapon retention techniques, and other submission/restraint/search techniques. The instructor should cover handcuffing/flex-cuffing for contact team members.

NOTE: USMC plans to replace the current open-hand defensive tactics training with the USMC Martial Arts Program.

f. Impact Weapons. This is an introduction to using the rigid straight baton, collapsible straight batons, side handle batons, and/or riot control batons.

g. ROE/Law of Armed Conflict (LOAC). This is an introduction to the ROE/LOAC and their relationship. (Refer to DODD 5100.77, AR 27-1, and NWP 1-14M).

h. Nonlethal Munitions—Employment and Tactics. This subcourse is an introduction to the capabilities and employment of available NLW. NLW provide additional unit force protection capabilities. Introduction to tactics is outlined in Chapter 5. Appendix D is a sample unit training schedule covering the topics listed above. U.S. Army Training and Doctrine Command (TRADOC) has developed training support packages that will include company level and below TTP.

4. Leader Training

a. An understanding of the topics covered in the INIWIC curriculum is essential. However, commanders/leaders should also possess an in-depth understanding of the ROE, their implementation, as well as the politics, culture, and demographics of the area to which they are deploying.

b. Commanders should involve their medical personnel, public affairs (PA) representatives, civil affairs (CA), psychological operations (PSYOP), staff judge advocate

(SJA), engineers, explosive ordnance disposal (EOD), and military police (MP) in training exercises when possible. Involvement by these special staff sections will greatly enhance the quality and realism of the training. Medical personnel will better understand OC decontamination and the possible injuries they may encounter as a result of NLW employment. Because of the international attention that NLW receive, all personnel will benefit from media training by PA personnel. CA and PSYOP personnel can provide area-specific cultural, ethnic, and political briefs to enhance awareness of the challenges within the AO. SJA personnel can clarify complex ROE questions and assist in training on ROE-related issues. Additionally, supply and logistic staff members also require training in NLW maintenance and environmental issues unique to the equipment and munitions.

5. Unit Training

The INIWIC is primarily designed to produce instructors who will train individuals in using and employing NLW. Collective unit tactical training is the unit's responsibility; however, a NLW sample schedule based on INIWIC training is provided in Appendix D. Units should conduct collective training before deploying with NLW. Unit standard operating procedures (SOP) and tactics may require adjustment to enable the employment of NLW and new equipment for the success of specific missions. All collective training should include nonlethal options and decision making. Whenever NLW are employed during training, the instructor must also conduct lethal capability training. Troops must understand the integration of nonlethal and lethal capabilities and always be prepared to protect themselves, as required, with lethal force. It is impossible to accurately predict the response of an individual or crowd.

6. Training Disclaimer

Initial unit training should not be conducted without qualified instructors. Non-type classified and COTS nonlethal munitions cannot be fired without Army Materiel Command (AMC), Army Training Support Center (ATSC), Crane Naval Weapons Center, or Air Force Munitions Center type classifying the ammunition. The instructor should conduct unit sustainment training in accordance with (IAW) service guidelines.

7. Additional Training

a. Additional training is left to the discretion of the unit but should be encouraged in order to add credibility to the training package. Some recommended training certifications for instructors include, but are not limited to—

(1) Management of Aggressive Behavior (MOAB)
R.E.B. Training International, Inc.
PO Box 845, Stoddard, NH 03464

(2) Oleoresin Capsicum Aerosol Training (OCAT)
R.E.B. Training International, Inc.
PO Box 845, Stoddard, NH 03464

(3) Monadnock Defensive Tactics System (MDTS)
Training calendar available at the Monadnock website:
http://www.batons.com/traincal.html

b. These courses add viable credentials to the instructors as perceived by civilian and military alike. In addition, the instructor can issue certification cards for the area which an individual has been certified. Certifications in various areas offers additional protection in the event of a lawsuit or claim against the government by ensuring the individuals that have been trained within your unit are completely capable.

8. Summary

NLW and tactics training are not meant to be stand-alone training—units must anticipate the requirement for NLW in their missions and train accordingly. NLW provide commanders with a graduated response in the force continuum.

Chapter V

EMPLOYMENT

1. Background

a. A military unit deployed to an area will normally conduct operations IAW current doctrine. The environment may contain both noncombatants and combatants. Nonlethal capabilities provide the military force with tools to engage adversaries and minimize collateral damage. Fundamental to employing NLW is a thorough understanding of the force continuum. As stated before, nonlethal capabilities provide a wider range of options that augment traditional means of deadly force but do not replace them. Nonlethal capabilities allow a commander the flexibility to increase and decrease the amount of force applied to accomplish a mission. The force continuum is generally continuous and seamless, yet a careful examination reveals five broad categories:

(1) Threats. The force continuum begins with threats. These can be either implied or expressed. An implied threat may be manifested through presence by forming into riot-control formations, fixing bayonets, or donning protective masks. An expressed threat occurs when a commander makes known the consequences of defiance.

(2) Denial options. The equipment used in denial options are usually placed in highly visible areas and often marked with placards. Examples include concertina, caltrops, barbed wire, or other obstacles.

(3) Munitions that cause physical discomfort/incapacitate. Munitions that cause physical discomfort, but fall short of inflicting trauma, include flash bangs, RCA—tear gas (O-Chlorobenzyl-Malononitrile [CS]), and OC. Although the discomfort or injury may be substantially less than impact munitions, employing these munitions requires an appropriate use of force assessment. Factors such as training, discipline, prejudices, emotions, and judgment all play a part in their application.

(4) Munitions that inflict blunt trauma. Munitions that inflict blunt trauma will inflict painful or debilitating injuries. These munitions constitute the upper end of nonlethal options within the force continuum. Examples include foam batons, stingball grenades, beanbags, sponge projectiles, and rubber ball rounds.

(5) Lethal munitions. Lethal weapons are at the highest level of force continuum. Lethal options should always be regarded as part of the force continuum and not as a separate option altogether. ROE, along with judgment and experience, helps on-scene commanders decide when using lethal force is authorized and appropriate.

b. Nonlethal capabilities include using existing systems that were not designed specifically as NLW. Nonlethal capabilities can minimize fatalities, permanent injury to personnel, and undesired damage to property, facilities, and the environment. Examples include PSYOP, CA, military working dogs, and PA. However, this chapter limits discussion primarily to using weapons and components explicitly designed as part of a NLW capability set. This focus is not intended to limit the commander's option to use whatever means are available to accomplish the mission without unnecessary lethality.

2. Mission Planning Factors

Incorporating NLW into a plan does not change the fundamental principles of mission planning. Commanders should use NLW as an operational enhancement. The first concern of a commander is successfully completing the mission. ROE are a critical consideration in the commander's planning and guide him in determining the appropriate use of nonlethal and lethal force. The availability of NLW does not relieve a force from its obligation to defend itself. Therefore, it must be equipped and trained with both lethal and nonlethal means.

3. NLW Tasks

a. NLW tasks were derived from the combatant commanders and service mission needs statements and are codified in the JMAA for Nonlethal Weapons. These tasks are grouped under three core capabilities: counterpersonnel, countermateriel, and countercapability.

(1) Counterpersonnel:

(a) Control crowds.

(b) Incapacitate personnel.

(c) Deny an area to personnel.

(d) Clear facilities/structures of personnel.

(2) Countermateriel:

(a) Deny an area to vehicles, vessels, and aircraft.

(b) Disable/neutralize vehicles, vessels, aircraft, and equipment.

(3) Countercapability:

(a) Disable or neutralize facilities and systems.

(b) Deny use of WMD.

b. Although not all encompassing, the following list of employment considerations is designed to assist commanders and staffs in formulating plans for use of NLW:

(1) General considerations:

(a) Understand and define ROE.

(b) Commander develops identification of friend or foe (IFF) procedures.

(c) Distribute NLW munitions to troops in advance.

(d) NLW options require lethal overwatch fire.

(e) Ensure area is well lighted to include perimeter.

(f) Employ night vision devices outside the perimeter to identify personnel.

(g) Utilize appropriate media to inform civilians to stay away from denied areas.

(h) Enforce positive access control into perimeter (personnel/vehicles).

(i) Ensure posted signs are easily understood (symbols/pictures). If interpreter support available, post warning signs in native language.

(j) Use entanglements such as caltrops, tanglefoot, barbed wire, and concertina/razor tape as these disrupt/slow intrusion attempts.

(k) Employ sensors, if available.

(l) Employ aerial or naval pickets.

(2) Counterpersonnel considerations:

(a) Employing kinetic rounds at less than 15 feet/5 meters may result in fatal outcomes.

(b) Target area at 15 to 40 feet should be center mass for blunt munitions.

(c) Head shots may be lethal.

(d) Kinetic rounds will not be skip fired.

(e) Kinetic rounds are effective against selected individual targets and can be used to disperse individuals.

(f) Considering differing rates of fires; the M203 and shotgun are complementary in a nonlethal role, depending upon the munitions used and their effective ranges.

(g) Employ riot batons only after proper training.

(h) Use RCA to disrupt/disperse crowds.

(i) Use RCA markers (RCA with a nontoxic, water-soluble marking dye) to mark agitators for future identification and apprehension.

(j) RCA greatly reduce visibility of control force.

(k) Closely control the employment of RCA.

(l) Disperse RCA to face and eyes for full effect.

(m) Employ optics with designated marksmen to identify mob leaders.

(n) Searchlights dazzle/disorient individuals (night only).

(o) Use bullhorn for communicating with crowd and to control formations.

(p) Employ recovery/apprehension/snatch teams to apprehend agitators.

(q) Use flex cuffs for agitators.

(r) Prepare disposition plan for cleared personnel.

(s) Isolate area to prevent reinforcements.

(t) Consider future friendly use of cleared facility.

(u) Prepare security plan for cleared facility.

(v) Channel passive traffic away from formations.

(w) Maintain rapid reaction force/quick reaction force (RRF/QRF).

(3) Countermateriel/countercapability considerations:

(a) Use marking agents to mark a vessel for later interception.

(b) Consider materiel handling equipment required to evacuate vessel/airframe.

c. Selected NLW tasks are further addressed in Tables V-1 through V-7. The tables are not intended to be all-inclusive but merely a stimulus to planning. The equipment listed is taken from the currently available systems designed for nonlethal purposes. The user should not limit his nonlethal options to this specific equipment and should be aware that there are many existing and emerging systems and potential uses of current technology and common objects to achieve the desired mission end state.

Table V-1. Crowd Control

Tasks	*NLW Planning Factors*	*Equipment*	*Employment Considerations*	*Notes*
Crowd control	Administration/ Personnel • SJA / claims planning • Chaplain affairs • Casualty notification • Detainee disposition plan • PAO/media planning • Information dissemination	Support Equipment • Riot face shield • Full length riot shield • Expandable riot baton • Optics • Portable bullhorn • Vest	• Distribute to troops in advance • Employ riot baton only after proper training • Employ optics with designated marksman (DM) to identify (ID) mob leaders • Use bullhorn for communicating with crowd and to control formations	• Strictly define ROE • NLW options require lethal cover fire
Crowd control	Intelligence • Intelligence preparation of the battlespace (IPB) • Background on country and culture • Route/map recon • Sensor emplacement for advance notification • Information on Mob leaders • Patrols and listening post (LP)/observation post (OP) • Determine reason for the crowd	Kinetic Rounds • 12-gauge (ga) bean bag round • 12-gauge rubber bullet • 12-gauge launching cartridge • 40-mm rubber baton • 40-mm wooden baton **NOTE:** USMC no longer uses wooden baton. • 40-mm stinger grenade • Stun grenade • Flash bang	• Do not employ kinetic rounds at less than 15 feet due to possible fatal outcome • Target area at 15 to 40 feet should be center mass • Head shots are not acceptable • Kinetic rounds will not be skip fired • Kinetic rounds effective against selected targets and to disperse individuals • M203 and shotgun are nonlethal complementary; M203 has low rate of fire; area target. Shotgun has high rate of fire; point target	

Table V-1. Crowd Control

Tasks	*NLW Planning Factors*	*Equipment*	*Employment Considerations*	*Notes*
Crowd control	Operations • Tactical Integration of NLW and lethal fires • Fire Support Element (FSE) plans to include RCA • Allocation of NLW munitions • Integration of NLW fires and maneuver • Engineer barrier planning • Understand mission (deny access/disperse/ monitor)	RCA • OC Dispenser • Team OC dispenser • High volume dispenser	• Disrupt/disperse crowds • Disperse RCA to face and eyes for full effect • Employ snatch teams to apprehend agitators	
Crowd control	Logistics • Distribution of NLW supplies • Security of rear area facilities • Resupply of NLW munitions • Procurement of medical supplies • Hospital planning • Medic planning for tactical units	Riot Control • Xenon Searchlight • Flex cuffs • RCA markers	• Searchlights dazzle/disorient individuals (night only) • Flex cuffs used for agitators • RCA markers are used to mark agitators for future ID	

Table V-2. Incapacitate Personnel

Task	*NLW Planning Factors*	*Equipment*	*Employment*	*Notes*
Incapacitate personnel	Administration/Personnel • SJA/claims planning • Media plan	Support Equipment • Riot face shield • Full length riot shield • Expandable riot baton • Rifleman's combat optic • Vest	• Distribute to troops in advance • Do not employ riot baton to the head • Employ optic with DM to ID mob leaders and cover with lethal fire	• Commander needs to define incapacitation mission • Take a balanced approach to public dissemination of NLW so counter measures cannot be rapidly developed • Cover NLW options by lethal force
Incapacitate personnel	Intelligence • ID types of insurgents and demonstrators • ID causes and factions • ID mob ringleaders	Kinetic Rounds • 12-gauge bean bag round • 12-gauge rubber bullet • 12-gauge launching cartridge • 40-mm rubber baton • 40-mm wooden baton **NOTE:** USMC no longer uses wooden baton. • 40-mm stinger grenade • Stun grenade • Flash bang • M203 and shotguns are nonlethal complementary. M203 has a low rate of fire, area target. Shotgun has a high rate of fire, point target	• Do not employ kinetic rounds at less than 15 feet due to possible fatal outcome • Target area at 15-40 feet should be center mass • Head shots are not acceptable • Kinetic rounds will not be skip fired • Kinetic rounds effective against selected targets and to disperse individuals • M203 and shotguns are nonlethal complementary. M203 has a low rate of fire, area target. Shotgun has a high rate of fire, point target	

Table V-2. Incapacitate Personnel

Task	*NLW Planning Factors*	*Equipment*	*Employment*	*Notes*
Incapacitate personnel	Operations • Prepare and confirm ROE for NLW • Integration of NLW and deadly force • Allocate NLW and designate unit use • Integrate NLW fires and maneuver	RCA • OC dispenser • Team OC dispenser • High-volume • OC dispenser	• Disrupt/disperse crowds • Disperse RCA to face and eyes for full effect	
Incapacitate personnel	Logistics • Distribute NLW munitions • ID special storage requirements • Resupply NLW munitions	Riot Control • Xenon • searchlight • Flex cuffs • Chemical markers	• Searchlights dazzle/disorient individuals (night only) • Use flex cuffs for agitators • Use RCA markers to mark agitators for incapacitation	
Incapacitate personnel	Civil Affairs • Collect civilian intelligence • Care for injured personnel • Detain civilian personnel			

Table V-3. Deny Area to Personnel

Task	*NLW Planning Factors*	*Equipment*	*Employment*	*Notes*
Deny area to personnel	Administration/ Personnel • SJA/claims planning • Media plan	Support Equipment • Area lights • Area sensors • Night vision • Access control • Warning signs written in host nation language	• Ensure area is well lighted to include perimeter • Employ night vision devices outside the perimeter • Positively ID accessed personnel • Employ warning signs in native language	• Utilize appropriate media to warn civilians away from denied area

Table V-3. Deny Area to Personnel

Task	*NLW Planning Factors*	*Equipment*	*Employment*	*Notes*
Deny area to personnel	Intelligence • Terrain and map study • Threat ID • Sensor emplacement around area • IPB			
Deny area to personnel	Operations • Prepare and confirm ROE • Integrate NLW and deadly force • Allocate NLW and designate unit use • Determine scope of mission • Determine access procedures • Use of area (friendly or total denial)			
Deny area to personnel	Logistics • Distribute NLW munitions • ID special storage requirements • Resupply NLW munitions • Security of rear area facilities.			
Deny area to personnel	Civil Affairs • Collect civilian intelligence	Entanglements • Caltrops • Tanglefoot • Barbed wire • Foams • Concertina/ razor tape • Stinger spike system	• Disrupt/slow intrusion attempts • Maintain suitable reaction force • Channel passive traffic away • Use foam to cover caltrops	

Table V-4. Deny Area to Vehicles

Task	*NLW Planning Factors*	*Equipment*	*Employment*	*Notes*
Deny area to vehicles	Administration/ Personnel • SJA/claims planning • Media plan	Support Equipment • Area lights • Area sensors • Night vision • Access control • Warning signs written in host nation language	• Ensure area is well lighted to include perimeter • Employ night vision devices outside the perimeter • Positively ID accessed vehicles	• Commander needs to define denial mission • Utilize appropriate media to warn civilians away from denied area
Deny area to vehicles	Intelligence • Terrain and map study • Threat ID • Sensor emplacement around area • IPB • Traffic analysis			
Deny area to vehicles	Operations • Prepare and confirm ROE • Allocate NLW and designate unit use • Determine scope of mission • Determine access procedures • Determine MP/ security forces requirements • Use of area (friendly or total denial)			• Commanders must consider flying operations when employing RCA. Area(s) should be reusable in short order.
Deny area to vehicles	Logistics • Distribute NLW supplies • Be prepared to evacuate stalled vehicles			
Deny area to vehicles	Civil Affairs • Collection of civilian intelligence	Entanglements • Caltrops • Tanglefoot • Barriers • Organic equipment • Barbed wire • Foams	• Disrupt/slow intrusion attempts • Maintain suitable reaction force • Channel passive traffic away • Use foam to cover caltrops	• Commanders must consider flying operations when employing NLW. NLW should not hinder flying operations.

Table V-5. Deny/Neutralize Vehicles, Aircraft, Vessels, and Facilities

Task	***NLW Planning Factors***	***Equipment***	***Employment***	***Notes***
Deny/ neutralize vehicles, aircraft, vessels, facilities	Administration/Personnel • SJA/claims planning • Media plan	Support Equipment • Sensors to track approaching vehicles, aircraft, and vessels • Night vision equipment • Materiel handling equipment • Portable bullhorn	• Employ ground and water sensors, if required • Employ aerial or naval pickets • Use materiel handling equipment to evacuate vessel	• Commander needs to define neutralization mission • Element of surprise remains a critical factor in mission accomplishment • Commander needs to develop IFF procedures
Deny/ neutralize vehicles, aircraft, vessels, facilities	Intelligence • Sensor emplacement if required • Threat ID • IPB • Analyze target area			
Deny/neutralize vehicles, aircraft, vessels, facilities	Operations • Prepare and confirm ROE • Integrate NLW and deadly force • Allocate NLW and designate unit use • Determine scope of mission • Determine access procedures • Determine boarding procedures (vessels and aircraft) • Determine apprehension procedures • Determine decontamination (DECON) procedures	RCA • Marking agent	• Use marking agents to mark a vessel for later interception	

Table V-5. Deny/Neutralize Vehicles, Aircraft, Vessels, and Facilities

Task	***NLW Planning Factors***	***Equipment***	***Employment***	***Notes***
Deny / neutralize vehicles, aircraft, vessels, facilities	Logistics • Distribute NLW munitions • ID special storage requirements • Resupply NLW munitions • Determine vessel/aircraft disposition	Riot Control • Xenon searchlight	• Use RCA to incapacitate crew members • Searchlights can disorient personnel or illuminate target	• Commanders must consider flying operations when employing RCA • Area(s) should be reusable in short order.
Deny / neutralize vehicles, aircraft, vessels, facilities	Civil Affairs • Collect civilian intelligence • Handle displaced civilian personnel	Entanglements • Caltrops • Foams • Speed bumps • Nets • PVAB	• Use caltrops to disable vehicle • Use combination of foam and caltrops to preserve the element of surprise • Channel passive traffic away • Use nets to foul props	
Deny/neutralize vehicles, aircraft, vessels, facilities	Medical • Treating injured civilian personnel			

Table V-6. Seize Personnel

Task	*NLW Planning Factors*	*Equipment*	*Employment*	*Notes*
Seize personnel	Administration/ Personnel • SJA/claims planning • PAO media plan	Support Equipment • Optics	• Employ optics with DM to ID personnel to be seized • Cover personnel employing nonlethal force	• Commander needs to define seizure mission and strictly enforce ROE • Element of surprise is critical for mission accomplishment • Speed is critical • Seizing personnel can be surgical with special operations personnel or conducted within a crowd to seize an agitator
Seize personnel	Operations • Prepare and confirm ROE • Integrate NLW and deadly force • Allocate NLW and designate unit use • Integrate NLW fires and maneuver • Breaching and assault planning • Determine holding area/processing center	RCA • OC dispenser • CS	• Use RCA to incapacitate personnel • Spray into eyes and face for full effect	
Seize personnel	Logistics • Distribute NLW munitions • ID special storage requirements • Resupply NLW munitions • Plan transportation	Riot Control • Xenon searchlight • Flex cuffs • RCA markers	• Searchlights dazzle/disorient individuals • Use flex cuffs for apprehension • Use RCA markers to mark agitators for seizure	

Table V-6. Seize Personnel

Task	*NLW Planning Factors*	*Equipment*	*Employment*	*Notes*
Seize personnel	Civil Affairs • Collect civilian intelligence • Care for injured personnel • Detain civilian personnel			
Seize personnel	Medical • Medical evacuation (MEDEVAC) planning			

Table V-7. Clear Facilities of Personnel

Tasks	*NLW Planning Factors*	*Equipment*	*Employment Considerations*	*Notes*
Clear facilities of personnel	Administration/ Personnel • SJA/claims planning • Chaplain affairs • Casualty notification • PAO media plan	Support Equipment • Riot face shield • Full-length riot shield • Expandable riot baton • Optics • Portable bullhorn	• Distribute to troops in advance • Do not employ riot baton to the head • Employ optic with DM to ID occupant leaders and provide lethal cover fire • Use bullhorn used for communicating with occupants	• Surprise is a critical element for mission accomplishment • Commander must strictly define ROE • Due to close quarter tactics, the force must be instantly ready to utilize deadly force if required

Table V-7. Clear Facilities of Personnel

Tasks	***NLW Planning Factors***	***Equipment***	***Employment Considerations***	***Notes***
Clear facilities of personnel	Intelligence • IPB • Background on facility, access, power plant • Route/map recon • Background info on building occupants • Determine any outside support for building occupants	Kinetic Rounds • 12-gauge bean bag round • 12-gauge rubber bullet • 12-gauge launching cartridge • 40-mm rubber baton • 40-mm wooden baton **NOTE:** USMC no longer uses wooden baton. • 40-mm stinger Stun grenade • Flash bang	• Do not employ kinetic rounds at less than 15 feet due to possible fatal outcome • Targeted personnel at 15 to 40 feet should be center mass • Head shots are not acceptable • Kinetic rounds will not be skip fired • Kinetic rounds effective against selected targets and to disperse individuals • Stun grenade effects can be amplified in buildings • Visibility will be reduced once kinetic rounds are employed • M203 and shotguns are nonlethal complementary. M203 has a low rate of fire, area target. Shotgun has a high rate of fire, point target	• Using RCA greatly reduces visibility • Use of RCA is restricted to certain circumstances; verify that proper authority has been given before use

Table V-7. Clear Facilities of Personnel

Tasks	*NLW Planning Factors*	*Equipment*	*Employment Considerations*	*Notes*
Clear facilities of personnel	Operations • Tactically integrate NLW and lethal fires • FSE plans to include RCA • Allocate NLW munitions • Integrate NLW fires and maneuver • Determine assault and breaching plan • Plan for isolation of facility	RCA • OC dispenser • Team OC dispenser • High-volume dispenser	• Disrupt/disperse crowds • Apply RCA to face and eyes for full effect	• Isolate area to prevent reinforcements • Prepare disposition plan for cleared personnel • Consider use of facility • Prepare security plan for cleared facility
Clear facilities of personnel	Logistics • Distribute NLW supplies • Resupply NLW munitions • Plan for transportation of displaced civilians • Plan facility maintenance or repair	Riot Control • Xenon searchlight • Flex cuffs • RCA markers • Flash bangs	• Searchlights dazzle/disorient individuals • Use flex cuffs for agitators • Use RCA markers to mark agitators for future ID • Flash bangs to dazzle/disorient individuals	
Clear facilities of personnel	Civil Affairs • Collect civilian intelligence • Care for injured civilians • Coordinate plans for controlling civilians			
Clear facilities of personnel	Medical • Hospital planning • Medic planning for tactical units • Procurement of medical supplies			

4. Staff Planning/Responsibilities

a. Administration. A staff composed of the G-1/S-1, PAO, SJA, and others as needed, is responsible for advising the commander on NLW employment and ROE. Additionally, administrative personnel are responsible for processing all civilian detainees and claims that arise from NLW employment. As a minimum, the staff must consider the following:

(1) SJA/claims planning.

(2) Chaplain.

(3) Casualty notification.

(4) Detainee disposition plan.

(5) PAO/media planning.

(6) Information dissemination.

b. Intelligence. The G-2/S-2 office is responsible for conducting the IPB process and providing information on the adversary's culture, political stability, and level of dedication to the cause, motivation, leadership, capabilities, equipment, and tactics. The G-2/S-2 should locate native language speakers to assist the interrogator-translator team.

(1) Background on country and culture.

(2) Route/map reconnaissance.

(3) Place sensors for advance notification.

(4) Information on leaders.

(5) LP/OP.

(6) Determine reason for the crowd.

(7) Identify threat.

(8) Analyze traffic.

c. Operations. The G-3/S-3 office is responsible for unit training, nonlethal certification of units, mission analysis, maneuver plans, and CA planning when required.

(1) Tactically integrate NLW and lethal fires.

(2) Fire support plans to include using RCA.

(3) Allocate NLW munitions.

(4) Plan engineer barrier.

d. Logistics. The G-4/S-4 office is responsible for logistics considerations, ship-to-shore movement of weapons and ammunition, storage and maintenance of equipment, resupply, transportation ashore, and other support requirements inherent in using NLW.

(1) Distribute NLW supplies.

(2) Security of rear area facilities.

(3) Resupply NLW munitions.

(4) Special maintenance requirements for NLW equipment and munitions.

e. Communications Officer. The G-6/S-6 staff is responsible for installing, operating, and maintaining communications from units confronting belligerents to the tactical operation center (TOC) or combat operations center (COC). The staff must plan for the effects NLW will have on communications to include proper frequencies.

f. PAO. The PAO provides liaison with outside media and participates in the staff planning process. PAO is also responsible for ensuring the NLW media plan and news releases consider the perceptions of the native media and belligerents or adversaries. PA personnel must receive training on NLW technologies, employment and capabilities, acute and long-term effects, impact on the environment, and legal approval procedures (at a minimum) so they can respond to media and public inquiries.

g. Medical. Medical personnel must be aware of the effects of NLW. Medical personnel should also be familiar with and able to provide decontamination of RCA as well as administer antidotes in cases of adverse reactions to riot control and other agents.

(1) MEDEVAC planning.

(2) Procuring medical supplies.

(3) Hospital planning.

(4) Medical planning for tactical units.

h. Chaplain. The commander should consider chaplain support for counseling and/or religious services for military and civilian personnel. Additionally, chaplains should provide information regarding the religious and cultural background of the local populace.

i. CA. The CA office is responsible for planning and coordinating all aspects of CA activities and providing assistance in gathering intelligence from civilian sources.

(1) Handle displaced civilian personnel.

(2) Coordinate plans for control of civilians.

j. PSYOP. Tactical PSYOP forces serve as an ideal tool for the commander to resolve issues of armed resistance and local interference with military operations while encouraging dialogue and cooperation with noncombatants. This force provides dissemination of PSYOP products (loudspeaker messages, leaflets, posters, etc.) and conducts face-to-face communications with target audience. PSYOP forces provide the supported commanders with nonlethal leverage by influencing foreign populations and expressing information subjectively to influence attitudes and behavior, thereby obtaining compliance or noninterference. They provide public information to foreign populations to support humanitarian activities, ease suffering, and restore or maintain civil order. They serve as the supported commander's voice to foreign populations to convey intent, countering enemy propaganda, misinformation, and opposing information to correctly portray friendly intent and actions. Finally, PSYOP forces deny others the ability to polarize public opinion and political will against the United States and its allies within an AO.

5. Tactics

The commander should include the following considerations in preparing for unit tactical operations involving NLW:

a. Avoid reconfiguring small-unit organization, except for special configurations such as—

(1) Special teams.

(2) Attachments.

(3) Crowd control formations.

(4) Security organizations.

b. Standardize postures in unit tactical SOP. Take the following steps to add a higher volume of specialty impact munitions and equipment within squads, platoons, or units:

(1) Avoid arming troops involved in apprehending designated personnel with long weapons (i.e., shotguns, rifles) because they can be easily grabbed/pulled away by individuals.

(2) Add weapons systems (shotguns) that provide a greater nonlethal capability. When loaded with nonlethal munitions, the longer reloading time of the M203 grenade launcher is protected by coverage from the high rate of fire from shotgun.

(3) Add nonlethal munitions to existing organic weapons systems (M203 grenade launcher).

(4) Avoid arming troops in the front lines of crowd control formations with long weapons.

(5) Focus riot-control capabilities in specialized reaction forces.

(6) Establish base line NLW postures.

(7) Establish command relations for special units.

(8) Maximize distance and barriers.

(9) Create NLW range cards for static positions.

(10) Provide the force with available OC/RCA consistent with ROE.

(11) Always mutually support forces utilizing NLW with lethal means.

(12) Consider environmental conditions (heat/cold/wind) and their effect on the performance of nonlethal munitions. Proper storage, periodic inspections, and rotation of NLW ammunition are critical to maintaining the effectiveness and viability of nonlethal munitions.

c. Unit Weapons, Unit Configurations, and Ammunition Considerations.

(1) Individual. Designate individuals as nonlethal shooters. Nonlethal shooters should also carry lethal munitions for use as required. Lethal rounds must be marked and carried in a manner to prevent confusing nonlethal rounds with lethal rounds.

(2) Squad. Squad leaders should dictate who carries nonlethal grenades and maintain their weapons loaded with lethal ammunition. Ideally, the squad should not change its task organization to accommodate nonlethal equipment, but should designate nonlethal shooters.

(3) Patrols. Commanders should not plan a nonlethal patrol, but should plan a combat/security patrol with a nonlethal capability when mission dictates.

(4) Static Positions. Individuals in static positions should have their weapons loaded with lethal munitions (mission, enemy, terrain and weather, troops and support available–time available and civil considerations dependent) (METT-T/METT-TC). Additionally, they should carry nonlethal munitions. Both shotgun and M203 work well from static positions and should complement each other in a nonlethal mode.

(5) Recovery Teams. Recovery teams have the primary responsibility of retrieving and controlling designated personnel from the crowd; the commander should establish a recovery team before deploying. The force should train team members in unarmed self-defense, open-hand control, and hand cuffing/flex cuffing. It should equip recovery teams with personal protection gear, 9-mm weapon, and flex cuffs. Due to the physical nature of recovery teams, it is important not to arm team members with long rifles. If security for the recovery team is an issue, the commander should attach a security element.

(6) Crowd Control Formations. The unit should establish crowd control formations within a minimum response time. Because of the physical nature of crowd control, individuals in these formations, if possible, should not carry long rifles. These formations should have inherent nonlethal capability. The unit must provide lethal coverage for the entire formation.

d. Individual Elements of a Formation. These elements are: base element, support element, and command element. The formation needs to provide versatility. Restrictions must never be set on numbers, equipment, or changes to the standard formations.

(1) Base Element. Role of the base element (front line):

(a) Base element serves as the front line or the base/foundation of the formation (can be divided into two or three sections or ranks).

(b) Front line consists of the shield holders.

(c) Angle the bottom of the shield to channel flammable liquids away from the body.

(d) Distance or interval between the shields differs based upon the range to the crowd. The closer the crowd, the tighter the interval between the shields.

(e) Provide frontal security.

(f) Hold a solid line.

(g) Provide intimidation factor.

(h) Fire team leaders are in the line holding shields.

(i) Secondary (tertiary) line(s) can consist of the baton holders and/or nonlethal gunners.

(j) Baton holders will—

- Strike with batons.
- Spray OC or disperse smoke.
- Serve as spotters for the NLW gunners.
- Control and detain subjects pulled from the crowd.

(k) Nonlethal gunners will—

- Consist of team leader with M203/M16 (provides nonlethal cover with M203 40-mm munitions and lethal cover with M16).
- Position themselves between shields working with baton holders.
- Serve as immediate control for the team.
- Maintain continuous muzzle awareness of the weapon.
- Provide immediate support to the shields.

(2) Support Element. Role and employment of the support element:

(a) Provide the means to employ lethal and nonlethal capabilities.

(b) Can consist of two ranks: M16/M203 (2nd rank) Mossberg 12-gauge shotguns (3rd rank).

(c) Provide security for search and recovery team.

(d) Consists of munitions teams and/or recovery teams.

(e) Munitions teams provide the means to employ lethal and nonlethal capabilities consisting of M16/M203 and Mossberg 12-gauge shotguns.

(f) Recovery teams provide ability to detain and search rioters downed by nonlethal munitions. Recovery teams will—

- Extract vehicles or personnel from the crowd who are in immediate danger.
- Detain and escort downed rioters to rear of the formation.
- Restrain and search detained individuals.
- Communicate with the base element.
- Spray or employ RCA as needed.
- Fill in the gaps in the front line as needed.
- Insert into a confined or congested area too large for the formation.
- Use a soft base line that capitalizes on the munitions fired to deploy forward of the shields in order to drive the crowd back, retrieve personnel/gear, and apprehend key riot personnel. Employed in pairs (1 x shield man/1 x shotgun).

NOTE: Refer to Marine Corps Security Force (MCSF) formation examples.

---

(3) Command Element. Role of the command element is to provide command and signal to the formation and consists of the platoon leader/flight commander/ platoon sergeant/flight sergeant, and squad leaders.

(a) Platoon Leader/Flight Commander, Platoon Sergeant/Flight Sergeant—

- Pass verbal commands.
- Maintain order within formations.
- Gauge and evaluate use of munitions and resupply considerations.

(b) Squad Leaders—

- Effectively control the squad.
- Pass information up the chain.
- Employ OC as directed.

(c) Designated Marksman (DM). During a nonlethal engagement, using DM provides confidence and safety to those facing a riot. The DM in an overwatch position scans the crowd, identifies threats and designated personnel for recovery, and fires lethal rounds if a lethal threat is presented. Additionally, DM are ideally suited for flank security and counter sniper operations.

6. Employment Objectives

a. Innovative employment methods of NLW and ingenuity take on critical significance. NLW are employed in concert with other nonlethal and lethal systems to create a synergistic effect.

b. The measures for countering many nonlethal options are usually apparent, quickly learned, and readily available. Because they are not intended to kill, nonlethal options "teach" an adversary what to avoid in the future. Small unit commanders must prepare to keep one step ahead of motivated belligerents.

c. Many nonlethal munitions have both maximum effective and minimum safe ranges. Individuals short of the minimum safe range may suffer severe injuries or death while the effects of most nonlethal munitions are greatly decreased at longer ranges. To be effective, an adversary must be engaged within the optimum range of the munition. Figure V-1 illustrates the ranges for a variety of nonlethal munitions and systems.

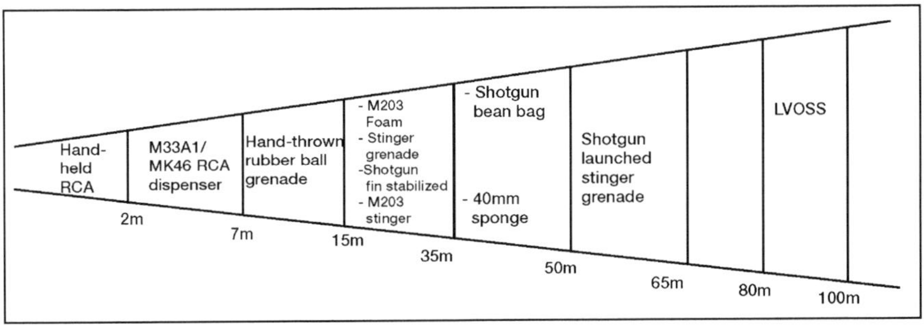

Figure V-1. Nonlethal Weapons Ranges

7. Rules of Engagement

a. U.S. forces follow directives issued by competent military authority that delineate the circumstances and limitations under which deadly force will be initiated against other forces. Intrinsic to the effective use of nonlethal options is the authorization to employ them. ROE are the primary means by which the chain of command can provide guidance to deployed forces in peacetime for handling crises and, in wartime, to help control the level of hostilities.

b. The commander should distinguish ROE from tactical control measures, threat conditions, and arming orders/weapons conditions. Although arming orders are sometimes listed on the same quick reference card as ROE for easy reference, the commander should distinguish ROE from other policies and directives developed, such as weapons confiscation rules.

c. The commander should delegate the decision to use NLW against an adversary to the lowest possible level, preferably to the platoon, flight, or squad level. However, this requires that all personnel, not just leaders, have a clear understanding of ROE and his intent.

d. Using lethal force, employed under the standing ROE, will never be denied. At no time will forces be deployed without the ability to defend themselves against a lethal threat nor will they forego normal training, arming, and equipping for combat. Nonlethal options are a complement to, not a replacement for, lethal force and seek to expand a proactive response across the range of military operations.

e. The following figures are examples of service-specific formations. These examples are not all-inclusive, and the tactical situation will always dictate actual unit employment.

FM 19-15, *Civil Disturbance*; MCWP 3-11.1, *Rifle Company;* and MCWP 3-11.2, *Marine Rifle Squad* provide extensive detail about small-unit formations, their use, and other considerations. Figures V-2 to V-7 are from I and II Marine Expeditionary Force (MEF) Special Operations Training Groups (SOTG) and represent a notional infantry platoon. Figures V-8 to V-13 represent a notional U.S. Army infantry platoon. Figures V-14 to V-21 are additional platoon formations submitted by Marine Corps Security Force Training Company.

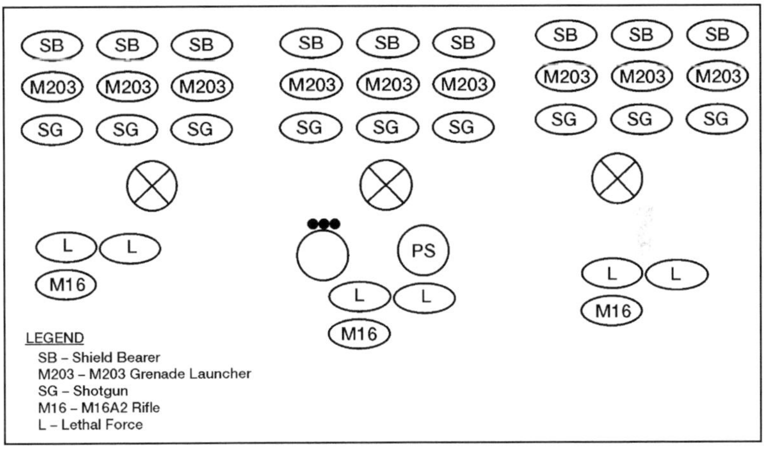

Figure V-2. Platoon in Depth

NOTE: The platoon is the smallest unit capable of independent maneuver. The platoon in depth formation shown above provides the following benefits:

- In-depth support with three ranks.
- Complementary support with the shotguns covering the slower reload time for the M203.
- Additional personnel from each squad to form the recovery teams.
- Each squad is capable of moving as an independent formation forward of the platoon line, if situation dictates.
- Other tactical formations are applicable—echelon, wedge, column.

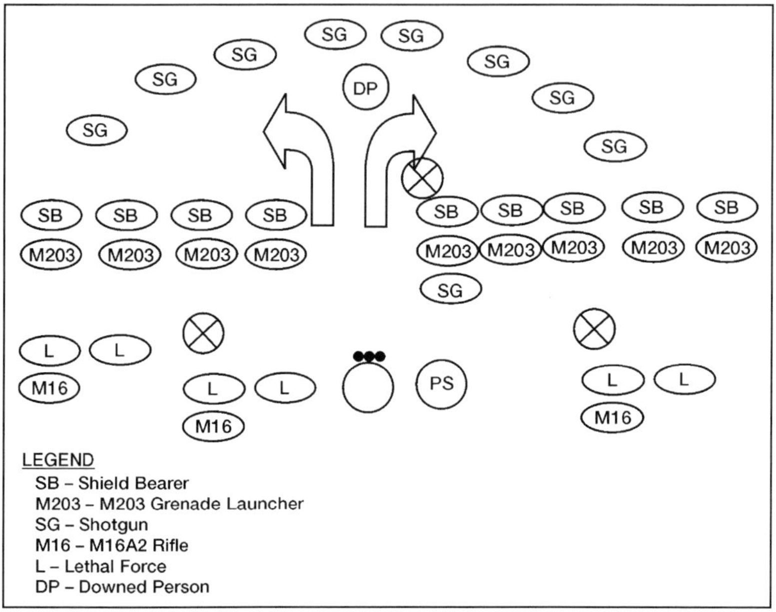

Figure V-3. Platoon Recovery Team Deployment—Security Element

NOTE: Security element deploys around downed person.

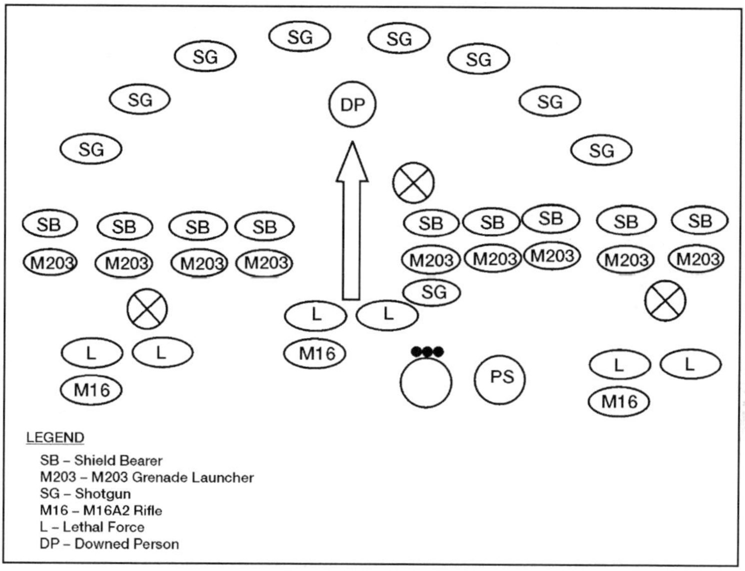

Figure V-4. Platoon Recovery Team Deployment—Recovery Team

NOTE: Recovery team deploys to detain, search, and recover downed person.

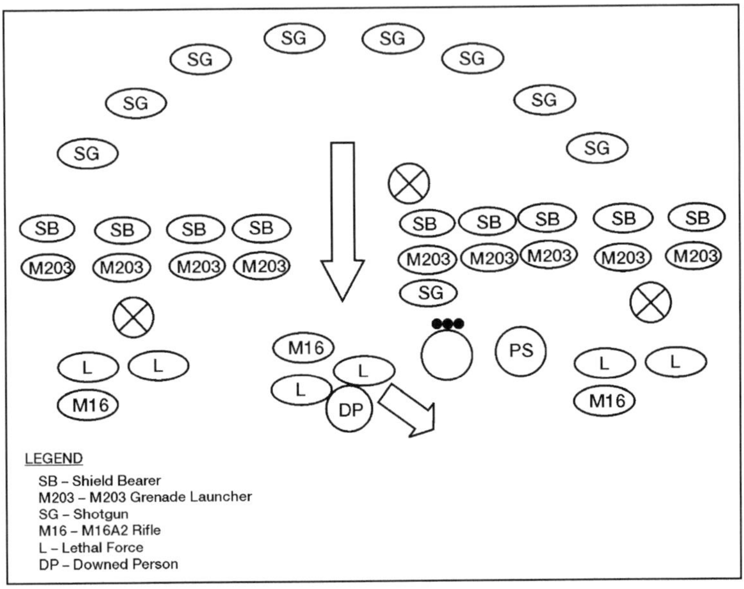

Figure V-5. Platoon Recovery Team Deployment—Recovery Team Redeploy

---

NOTE: Recovery team redeploys with detained person.

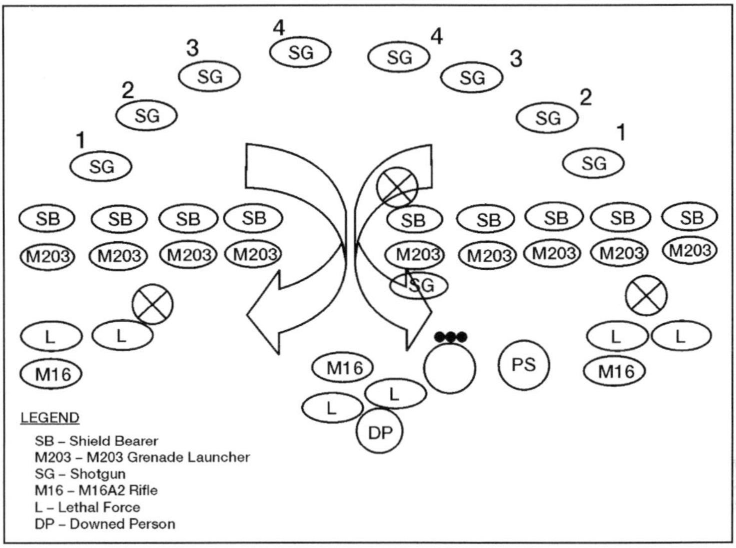

Figure V-6. Platoon Recovery Team Deployment—Security Element Redeploy

NOTE: Security element redeploys back to 3rd rank "by the numbers." Team leader calls "4s in," then "3s in," etc. The last person in calls "last man in" and formation closes the gap.

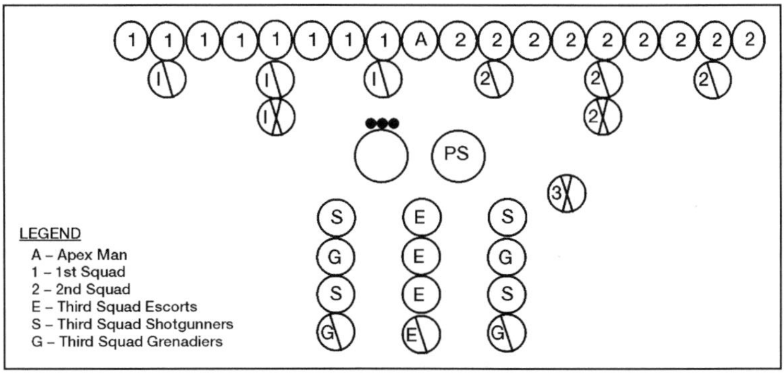

Figure V-7. Platoon On Line—Munitions Squad in Close Support to Cover Flanks

NOTE: Platoon in close support. NLW and lethal weapons are sited to cover the flanks of the formation, as well as support the front line.

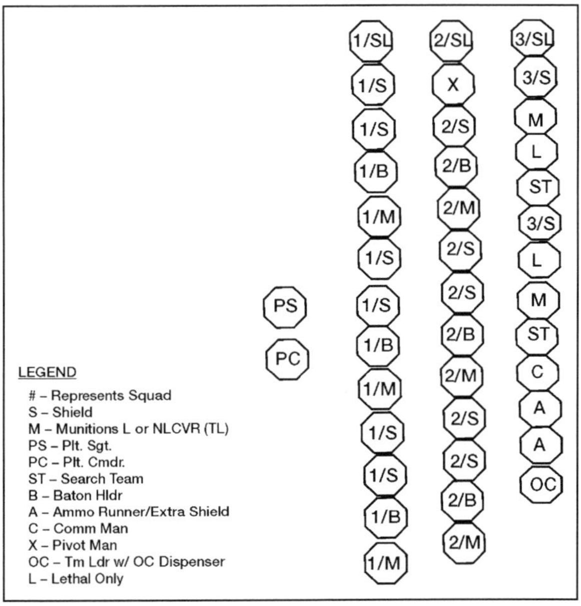

Figure V-8. Platoon Column

NOTE: Platoon in column formation. For administrative movement only.

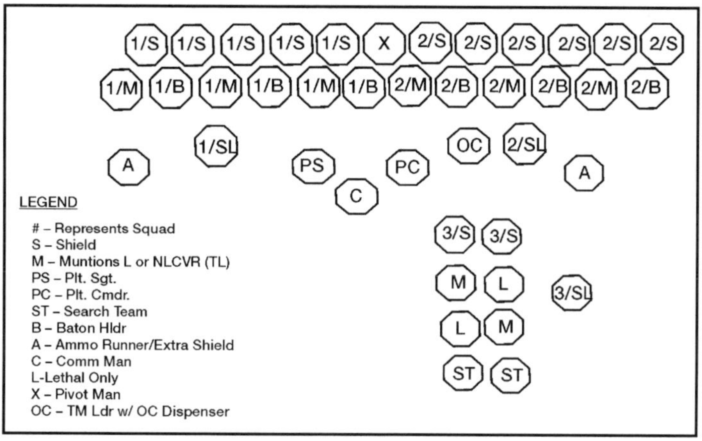

Figure V-9. Platoon On Line

NOTE: Platoon on line—basic formation for the employment of NLW.

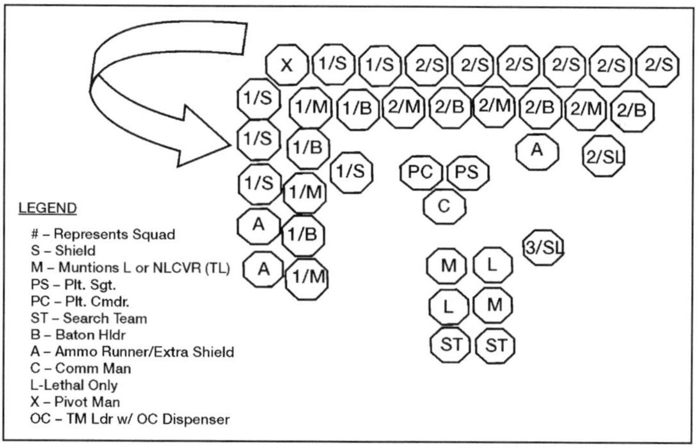

Figure V-10. Platoon Lateral Support, Left (Right)

NOTE: Platoon in lateral support to protect flanks.

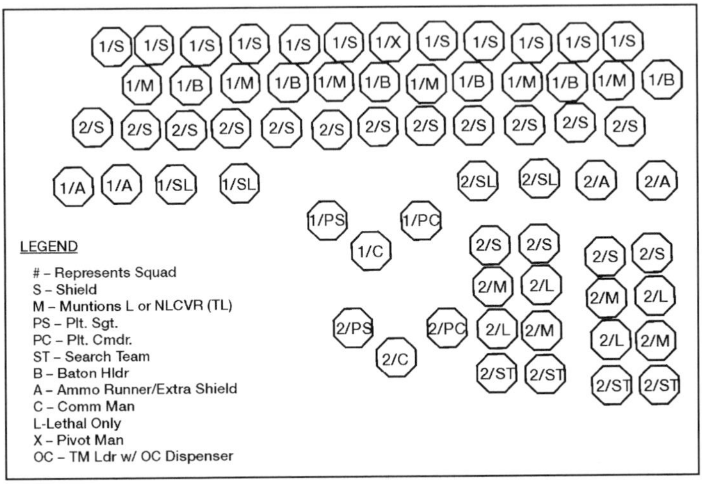

Figure V-11. Two Platoons in Close Support

NOTE: Two platoons in close support—tied in for mutual advantage.

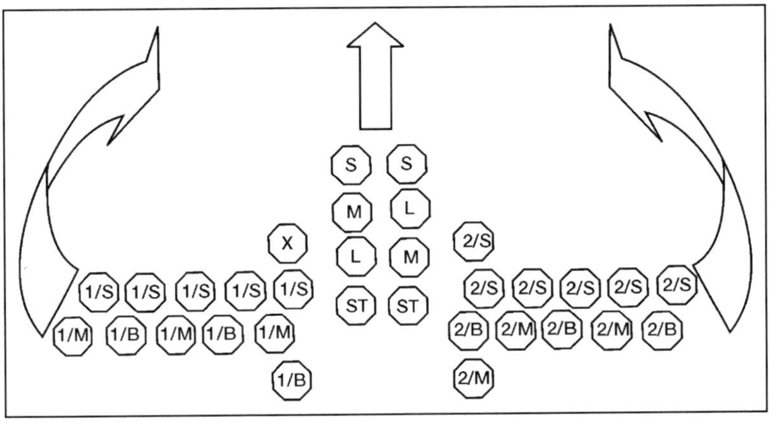

Figure V-12. Recovery Team Deployment

NOTE: Recovery team deploys from formation—remainder of platoon can deploy around the recovery team from either flank.

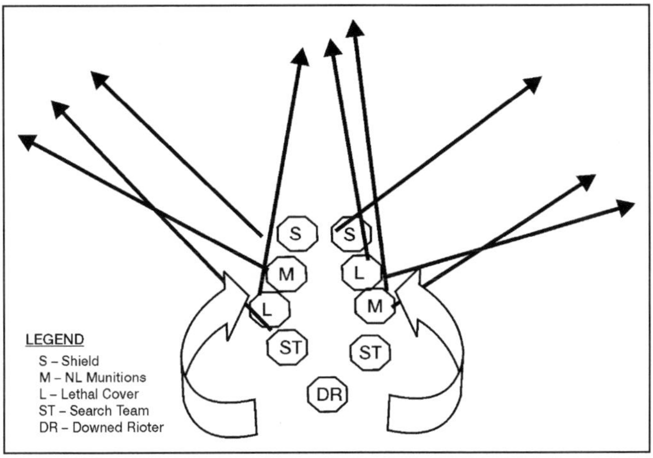

Figure V-13. Recovery Team Actions and Security Posture

NOTE: Recovery team forms security perimeter around downed rioter, conducts search and detain actions. Once actions are complete, recovery team redeploys.

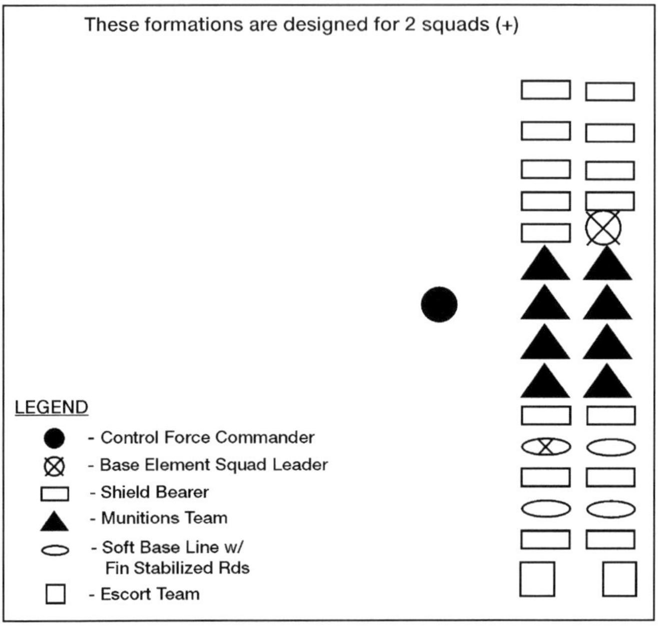

Figure V-14. Platoon in Column (USMC)

NOTE: These formations are designed for 2 squads (+). This allows the platoon to move rapidly when there is no chance of being engaged with the crowd.

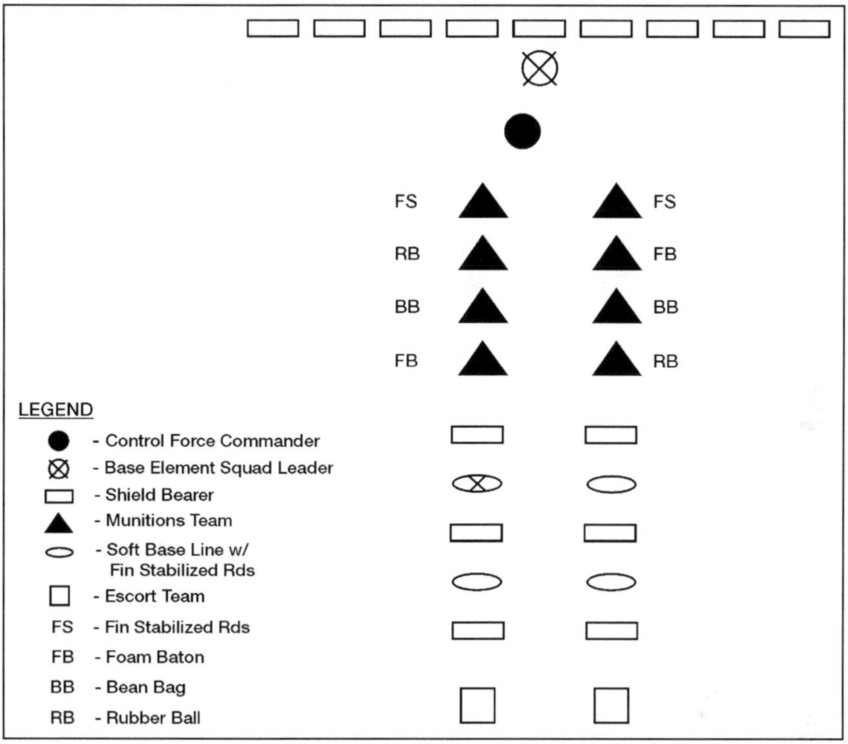

Figure V-15. Platoon in General Support (USMC)

NOTE: The control force commander moves into this formation approximately 60 to 70 meters from the crowd, as the crowd could start to engage the control force with thrown objects.

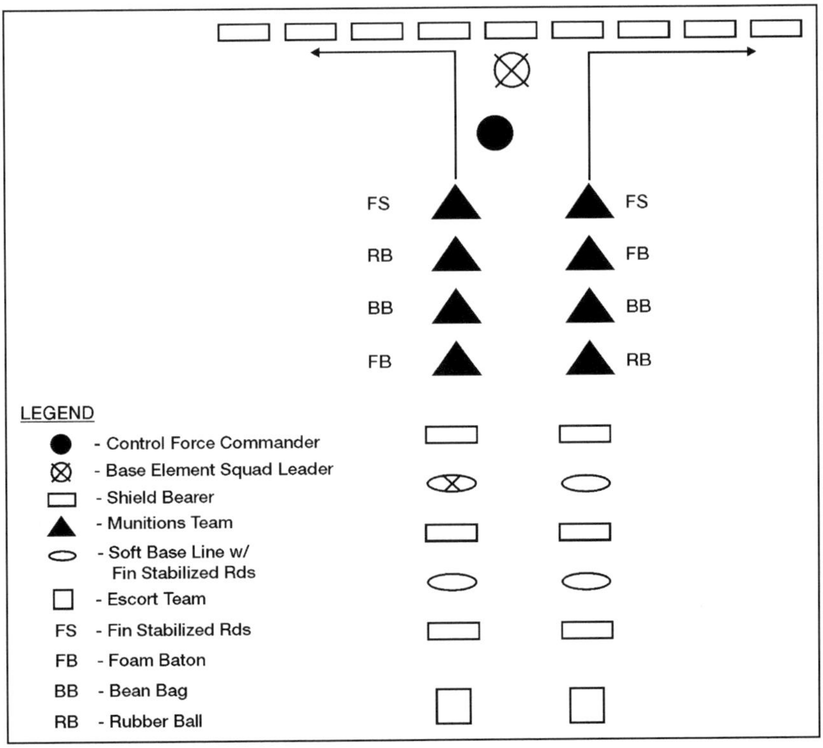

Figure V-16. Platoon in Close Support (Start) (USMC)

NOTE: The control force commander moves into this formation when he feels ready to engage the crowd with nonlethal munitions.

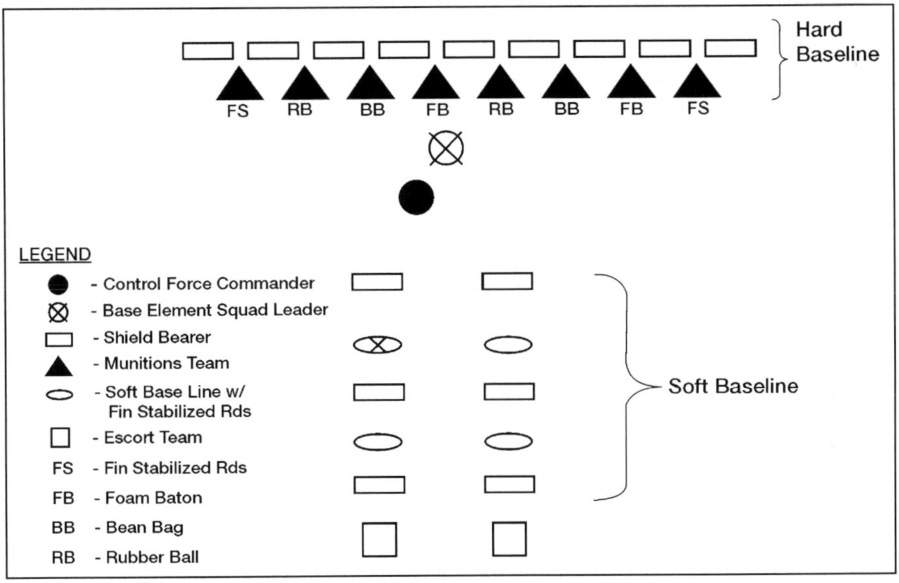

Figure V-17. Platoon in Close Support (Finish)

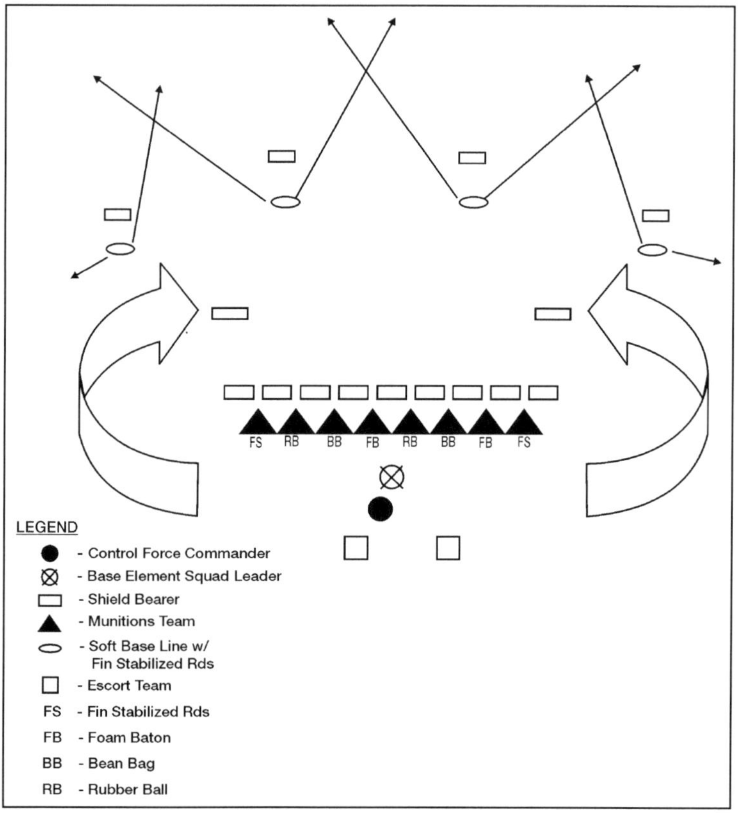

Figure V-18. Deploying Soft Baseline

NOTE: The soft baseline affords the control force commander much flexibility. The primary mission of this formation is to disperse the crowd back using speed and extreme violence of action. By deploying the soft baseline, the commander is able to maneuver the hard baseline where needed (around the barrier plan or obstacles.) The soft baseline is deployed by a predetermined signal (bullhorn siren, whistle blasts, vehicle horn) and will be retrieved using the same signal. This signal conditions the crowd to expect violence of action and causes them to disperse. The soft baseline should never be deployed into a fresh crowd; the crowd needs to be distracted by nonlethal munitions. The formula found to work is distract with stingball grenades, followed by a volley of shotguns and M203 munitions, then send the soft baseline out to disperse or move the crowd back. (Sting them, shoot them, soften them up.)

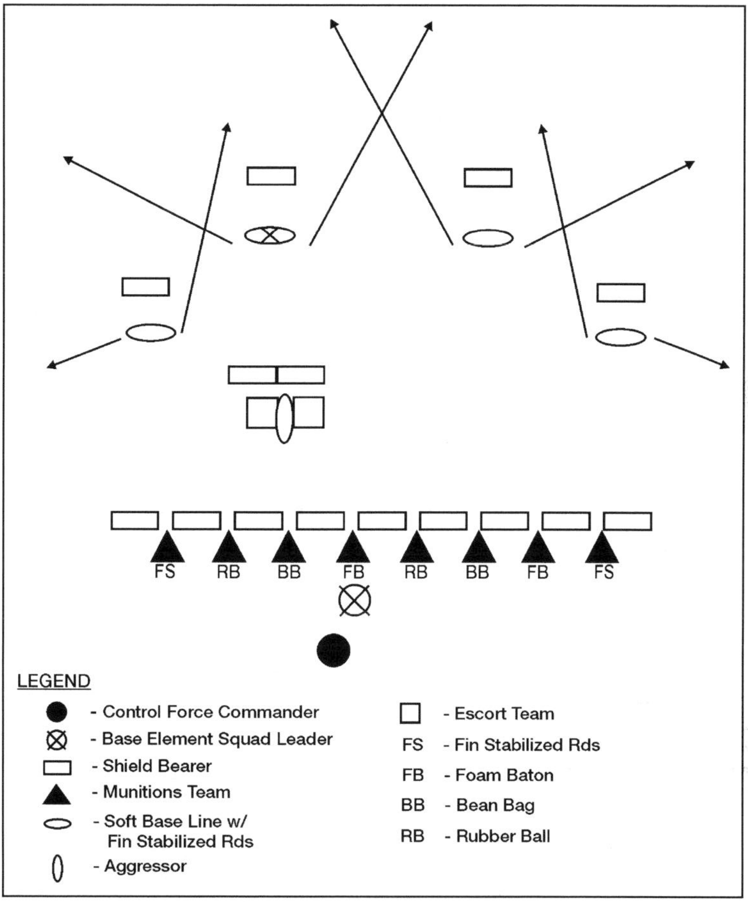

Figure V-19. Deploying Escort Team (Start)

NOTE: When the control force commander identifies the aggressor to retrieve, the command is, "Escort team left/right," then the signal. The signal is the same to deploy the soft baseline. As the soft baseline deploys, the two shields without gunners provide protection for the escort team. The shields remain with the escort team until the soft baseline is called back in, at which time the shields go back to the soft baseline.

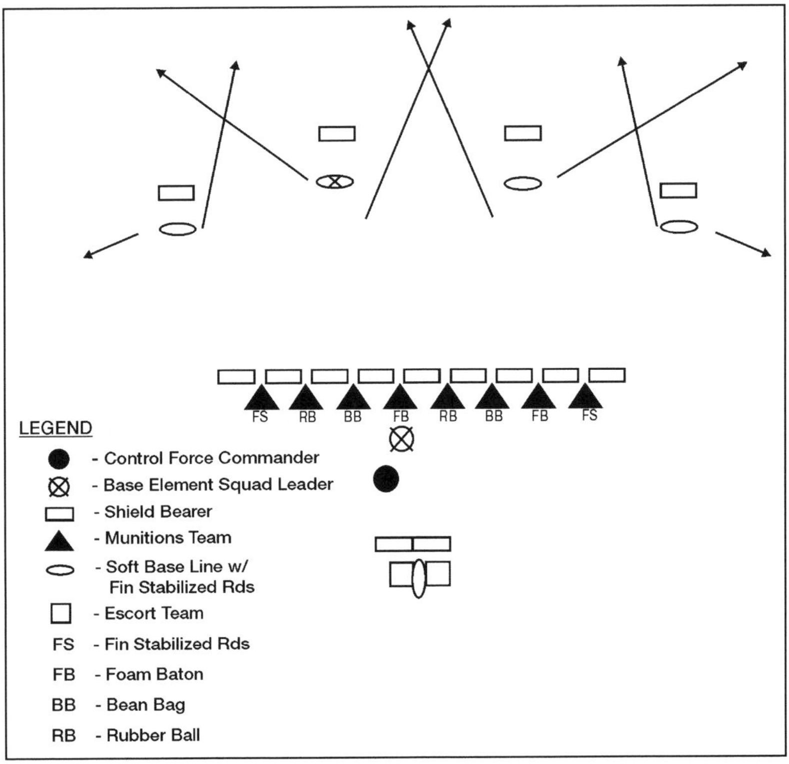

Figure V-20. Deploying Escort Team (Finish)

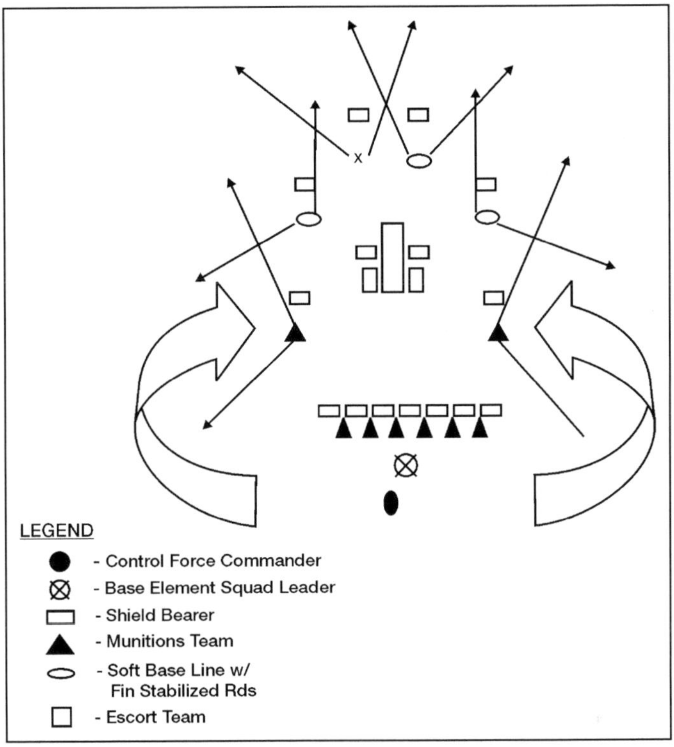

Figure V-21. Recovery of Downed Personnel or Stranded Vehicle (Big Johnson)

NOTE: In the event the crowd attacks personnel or a vehicle, the control force commander can call for a "Big Johnson." In a Big Johnson, the soft base-line deploys as usual except now the shield and shotgunners on the flank of the hard baseline also deploy creating a 190-degree perimeter around the personnel or vehicle. This maneuver must be executed quickly and with extreme violence of action because the crowd already has its hands on individuals in the force. The escort team deploys with the Big Johnson in order to escort or carry injured individuals. The shields without gunners provide protection for the escort team.

Chapter VI

LESSONS LEARNED

1. Background

Although lessons learned from employing NLW are still being studied and impacts analyzed, it is beneficial to look at the preliminary lessons gained from the NLW experience. These lessons are far from conclusive, because U.S. forces have yet to see the first large-scale employment of NLW. However, U.S. forces can gain tremendous insight from the experience of others and use what has been learned to guide future NLW employment.

2. Lessons Learned

a. Force Continuum. Using nonlethal options needs to be based on the principle of a graduated and proportional response, using the decisive force necessary to respond to the threat.

b. Attempts to legislate and regulate every aspect of NLW employment can prove disadvantageous to the on-scene commander. During the evacuation of U.N. forces in Somalia (OPERATION UNITED SHIELD, 1995), the ROE made an arbitrary and unnecessary distinction between situations justifying deadly force and those that justified only lesser means. Crowd control items were restricted to situations requiring deadly force. Many soldiers, airmen, and Marines asked, "why resort to nonlethal means at all if I have to wait until deadly force is authorized before I can shoot them with a beanbag?" Additionally, the use of sticky foam was limited to deadly force situations because of the perception that adversaries might suffocate from its application. This stipulation effectively eliminated sticky foam as a nonlethal option to the forces in Somalia.

c. With ROE that take into account the force continuum and encourage graduated response to nonlethal threats, the commander can employ NLW to enhance his flexibility, deterrence ability, and reaction to uncertain situations. To accomplish the mission with minimum loss of life and property, commanders need to have ROE that do not restrict them from properly employing NLW.

Lessons Learned from OPERATION JOINT GUARDIAN 1999 (Battalion Landing Team (BLT) 3/8 26 Marine Expeditionary Unit (MEU) Peace Operations in Kosovo:

ROE significantly impaired the employment of NLW. For example, OC was not authorized because it was considered a chemical agent. Nonlethal direct fire munitions were not authorized during the initial occupation of the province. This significantly hampered the individual Marine's ability to neutralize a threat where deadly force would not have been justified but action was required. Overly restrictive ROE thus negated the flexibility of response that NLW are supposed to provide. The intent of deploying NLW is to provide the operator the capability and flexibility to neutralize a threat without having to apply deadly force. Thus, the ROE for NLW should ideally be less restrictive, not the same or more restrictive, than those for lethal force.

It is often enticing for authors of ROE involving the use of NLW to base those ROE on the principle of minimum force necessary. This may be appropriate for base and station or homeland security ROE; however, for expeditionary operations the decision to mandate minimum force deserves critical consideration. Commanders in military operations have traditionally gained advantage over their enemy not only by the actual combat power that they bring to the fight, but also from that enemy's perceptions and beliefs about the intensity and resolve with which that combat power will be applied. When ROE are written stating that the confrontations will be addressed using the minimum force necessary, this advantage is lost. ROE developed for expeditionary nonlethal operations should be based on the concept of decisive and compelling force, rather than the minimum force. This will maintain an adversary's hesitancy to harass and endanger U.S. forces even after that adversary has tested or probed the response of those U.S. forces. Further, this will allow synergy to be gained through the complimentary employment of lethal and NLW by providing more response options to the small unit leader.

3. Employment

a. Through the use of modeling, simulation, and field exercises, new concepts for employing NLW are continually being developed. The force should train with NLW to the same degree as for lethal weapons. In this way, units become confident when using nonlethal options in chaotic and uncertain situations. The commander should push the decision to employ NLW to the lowest possible tactical level. Commanders who intend to use NLW in achieving military objectives must recognize the fluidity of the situations in which they are likely to be employed. The decisionmaking cycle is short, stressful, and the level of violence can move in both directions along the force continuum. In quick succession, a small unit leader might employ nonlethal munitions, decide to employ deadly force to an immediate specific threat that arises, then return to using nonlethal means in response to unarmed hostile elements.

b. A mixture of nonlethal "tools" has been determined to be advantageous in an operational scenario. As the situation develops, commanders can apply different nonlethal responses with respect to range/distance capability and severity of effect. For example, to disperse a crowd, area fire weapons would precede use of close-in point target nonlethal munitions.

c. An important point to remember is that NLW are just one element of a graduated response available to the force commander. Other factors make up a large contribution to the success of U.S. forces, including barrier plans and PSYOP. An example of this operation would be using barriers such as roadblocks or concertina wire, warning commands to inform a mob of potential responses based on their actions, and only then taking actions through escalated employment of NLW. These continued responses would involve smoke, RCA, and blunt munitions. A specific verbal warning should precede every escalation of nonlethal means.

d. Whenever NLW are employed, a separate lethal capability must always be in place. Troops must always be prepared to protect themselves as required by lethal means, because it is impossible to accurately predict the response of an individual or mob to NLW employment.

Lessons Learned from OPERATION ESSENTIAL HARVEST 2001 (U.S. Embassy Reinforcement operations conducted at the U.S. Embassy Skopje Macedonia by 5th Platoon, 2nd Fleet Antiterrorism Security Team [FAST] USMC):

The anticipated threats were violent demonstrations directed against U.S. facilities in the AO. In this case, USMC forces tasked to reinforce the embassy were equipped with a compliment of NLW. ROE took into account the capabilities and limitations of the NLW. The ROE and the NLW systems on the ground were coordinated and thus compatible. In addition, it was found that integrated employment of different nonlethal munitions greatly enhanced that force's ability to suppress the threats encountered (large rioting mobs). Random security patrols reinforced with military working dogs proved very effective. Reinforcement with multiple M203s provided USMC units with both a lethal and nonlethal capability in the same weapons system. The M203 was particularly useful as it gave the individual Marine the means to quickly transition between a lethal (5.56-mm) and nonlethal (40-mm nonlethal) capability. A basic understanding of the capabilities and limitations of nonlethal munitions enabled a relatively small force (40-50 Marines) to counter a sizable threat many times its size (500-1000 rioters). Although NLW were not used, the capability and preparation of the units provided the adequate deterrence that prevented the situation form escalating.

4. Publicity and Media

a. The commander can use the media to his advantage, as they will inform the local population of the military's capabilities and responses available. During OPERATION UNITED SHIELD in Somalia, the local media was informed of the Marines' nonlethal options and "new technologies" in order to discourage hostile actions. The media was encouraged to report this to the population. Intelligence had indicated the Somalis were planning to mix civilians with armed clansmen, creating hostile mobs. In fact, feuding warlords had previously successfully demonstrated this strategy on several occasions. Lack of deadly mob incidents implies Somalis were not anxious to test this new technology. Evidently, publicity had directly contributed to mission accomplishment.

b. The public dissemination of the presence of NLW capabilities must strike a balance between information sufficient to discourage hostile actions and information that enables the local population to learn too much of the capabilities of NLW. Too much information allows adversaries to develop countermeasures or avoid the NLW effects. Commanders who employ NLW systems must involve the PAO from the beginning of planning. The PAO must be familiar with the objectives of the mission, commander's intent, ROE, NLW and lethal weapons employed, and be prepared to advise the commander on the best methods of dealing with outside media. The commander should fully integrate the PA plan with PSYOP objectives and the concept of operations. Additionally, he must work the media to ensure the "sensationalism" of such a new technology does not create unrealistic expectations where he can be unfairly judged or constrained.

5. Capabilities and Impact on Mission Accomplishment

a. The commander should understand the unique capabilities of NLW and use them to accomplish the mission. He should use NLW in conjunction with the planned employment of the force and within established ROE to minimize loss of life and property while accomplishing the mission. The force requires early hands-on training to recognize the capabilities and limitations of NLW. Appendix D has sample training plans developed to meet these objectives.

b. The force commander must fully integrate planning for using nonlethal capabilities as quickly as possible in the staff planning cycle. In order to have a full range of alternatives for use against unarmed but hostile elements, the force needs to deliver nonlethal systems, complete necessary ROE and legal review, and make sufficient quantities of each NLW readily accessible to the force. The choice of nonlethal capabilities should be based on several factors and the choice can be made after addressing these questions:

(1) First, are the NLW systems available and can they be delivered on time?

(2) Second, are they compatible with, and do they complement existing weapons and training?

(3) Third, are there clear and unambiguous guidelines and ROE for employment?

(4) Finally, can the NLW be effectively employed to save lives and contribute to mission accomplishment?

c. Leaders need to anticipate local civilians developing effective countermeasures to their NLW and tactics. Therefore, prior to beginning their mission, leaders must determine and rehearse actions to take when nonlethal munitions are effectively countered.

d. Using nonlethal and lethal force are not mutually exclusive. In many instances, the commander can gain synergistic effects by using a combination of both. The following are some examples of this:

(1) Utilizing NLW to separate agitators from noncombatants with simultaneous employment of lethal force against hostiles or combatants where deadly force is justified.

(2) Utilizing NLW when making entry into structures or rooms to gain time in order to discern temper and intent of occupants prior to applying deadly force.

6. Knowing Your Adversary

a. Although commanders desire thorough intelligence of any potential adversary, such a goal may be difficult to reach in a NLW employment environment. During OPERATION UNITED SHIELD in Somalia and OPERATION UPHOLD DEMOCRACY (1995) in Haiti, a study of the local population's culture, history, and customs enabled commanders to anticipate and control many situations before they got out of hand. On the other hand, during OPERATION UPHOLD DEMOCRACY, poor intelligence and a lack of understanding between the two cultures led to the Haitians and Cubans being lodged together in Guantanamo Bay. This situation led to riots between factions who had a deep dislike for each other. By understanding a particular crowd's motivation, friendly forces may be better able to take necessary steps to prevent escalation resulting in the use of deadly force. Early training in crowd dynamics and riot control will enhance the effective use of NLW.

Lessons Learned from U.S. Peace Operations, 739th MP Battalion (Bn), January 2001

An Army MP platoon received a mission to clear a rioting crowd of 300 civilian locals from a main supply route. The platoon placed vehicle-mounted machine guns in overwatch positions to allow quick transition from NLW to conventional rounds. 12-gauge shotgun teams armed with fin stabilized rounds carried M9 pistols for lethal backup. Two volleys of shotgun and M203 nonlethal rounds successfully dispersed the crowd. The platoon had utilized INIWIC graduates to conduct preparatory training on using NLW. The NLW were extremely effective in pushing back the crowd and causing it to disperse. The 40-mm rubber ball round was moderately effective at close range, but very inaccurate at longer ranges. Shotguns were more effective than the M203 due to their high rate of fire and quick-loading capabilities.

b. Specific lessons learned during this operation are as follows:

(1) Placing interpreters is critical. The solution in this incident was to place interpreters approximately 25 meters behind formation so they could deploy forward as needed. Interpreters were often needed to communicate with crowds and with adjacent forces and coalition units that were organic to Kosovo forces (KFOR) that also responded to the incident.

(2) Leaders as well as soldiers should prepare to speak to the media. The command should have a PAO representative on hand to deal with the media.

(3) The commander should thoroughly brief the transition to lethal force from NLW.

(4) U.S. commanders in Kosovo engaged in the decisionmaking process had disparate levels of knowledge of nonlethal capabilities, constraints, and operational considerations. Therefore, leader training and orientation requires more emphasis.

(5) ROE were vague regarding the use of NLW. In many operations, NLW use is delegated to the on-site commander. Delegating to the lowest possible level is the preferred method.

(6) A unit needs a mixture of nonlethal rounds based on the size of crowd, composition of the crowd, available weapons systems, and synergistic effect of selected nonlethal combinations.

(7) When a RRF/QRF arrives on station with NLW to support a force equipped only with lethal means, the commander should address the following command and control concerns:

(a) Who makes the decision to employ NLW?

(b) Is the reinforced commander familiar with NLW?

(c) Who controls the reinforced unit?

(d) How does this relationship change as the situation develops?

7. Summary

a. The wide range of military operations in which U.S. forces are involved has demanded the introduction of NLW into the arsenal of weaponry available to commanders. NLW provide a graduated and proportional response to threats posed by a variety of adversaries. The ROE must be consistent with the rapidly changing situation of today's chaotic and uncertain military operations. This reality demands a high level of training in properly employing NLW.

b. Effective use of NLW involves a synergy between PSYOP activities, intelligence gathering, and professional media/PA plan. NLW effectiveness is enhanced through integration of these activities to reduce the chance adversaries will be able to develop effective countermeasures. NLW, when combined with lethal weapons, provide the commander with a force response across a wide range of military operations.

Appendix A

DOD Nonlethal Weapons and Equipment Item Description

The following is a cumulative list of NLW and equipment currently fielded within the DOD. Each service has various components of the following list in its inventory. A list of the service's items is located in Appendix E.

 **Figure A-1. Riot Face Shield**	1. Riot Face Shield: The riot face shield is a non-ballistic face shield designed to protect the service member's face from objects thrown or attacks from non-ballistic weapons. (Figure A-1)
No picture available.	2. Ballistic Face Shield: The ballistic face shield provides the service member with face protection from a wide range of threats (i.e. debris, liquids, handthrown objects). The shield is manufactured from acrylic and bullet resistant materials to provide Level IIIA protection (9-mm and .44 magnum pistol rounds), weighs 3.4 pounds and is compatible with the PASGT helmet and M-17/M-40 protective masks.
 **Figure A-2. Riot Body Shield**	3. Riot Body Shield: The riot body shield is a non-ballistic body shield designed to protect the service member's body from objects thrown or attacks from non-ballistic weapons. (Figure A-2)
No picture available.	4. Ballistic Body Shield with Light Kit: The ballistic shield provides the service member with torso protection from a wide range of threats (i.e. debris, liquids, hand thrown objects). The shield is manufactured from Spectra Shield resistant material painted anti-reflective flat black, and is capable of providing Level IIIA protection (9-mm and .44 magnum pistol rounds). The shield weighs 18 pounds and is 20 x 36 inches in size with a view port of 4 x 16 inches.

 Figure A-3. Riot Shin Guard	5. Riot Shin Guard: The riot shin guard is a non-ballistic shin guard designed to protect the service member's shins from objects thrown or attacks from non-ballistic weapons. (Figure A-3)
No picture available.	6. Ballistic Shin Guards: Ballistic shin guards provide the service member with leg protection from a wide range of threats (i.e. debris, liquids, hand thrown objects). The guards are manufactured from Kevlar KM2 and provide Level IIIA protection (9-mm and .44 magnum pistol rounds). They are available in three sizes and weigh between 7 to 10 pounds.
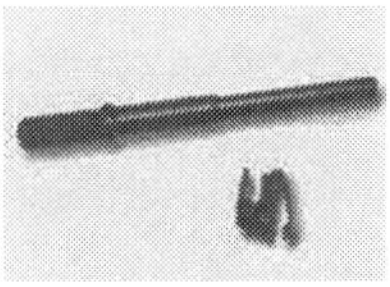 Figure A-4. Expandable Riot Baton (with holster)	7. Expandable Riot Baton (with holster): The expandable riot baton provides the service member a nonlethal means of crowd control and self-defense. The baton comes with a mounting device that attaches the baton to a belt. The baton is 24 to 36 inches in length. (Figure A-4)
No picture available.	8. Wooden Baton: The wooden baton provides self-defense for the service member and the option between physical force and lethal force. The baton comes with a mounting device that attaches the baton to a belt. The baton is 36 inches in length.
No picture available.	9. Practice Riot Batons: The practice riot batons provide the service member with a safe dynamic training environment for properly using a riot baton. The batons are available in a variety of lengths.
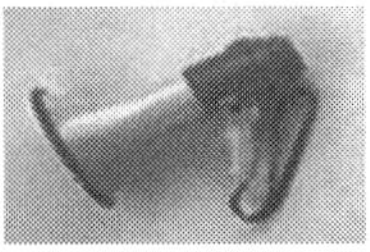 Figure A-5. Portable Bullhorn	10. Portable Bullhorn: The portable bullhorn provides the service member the capability to control his forces by projecting his voice above the noise and commotion created by crowds and mobs. (Figure A-5)
No picture available	11. Ground-Mounted Public Address System: The ground-mounted public address system is a communication enhancement device for conducting crowd control tactics. The system facilitates communication with the crowd in the absence of PSYOP support and assists with communication of commands to troops engaged in crowd-over-crowd noise.
No picture available.	12. M7 Individual Voice Amplification System: The M7 individual voice amplification system is a communication enhancement device for conducting crowd control tactics while wearing the M40 protective mask and using riot control agents. The M7 is fitted to the M40 protective mask to facilitate oral communication and increase the user's ability to communicate using radios and other devices.

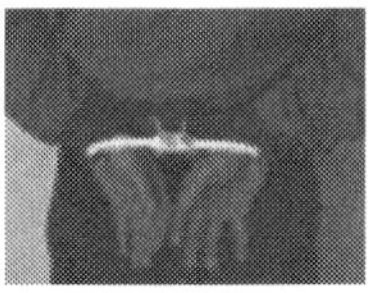 **Figure A-6. Disposable Handcuff**	13. Disposable Handcuff: The disposable handcuff provides service members with a means of restraining and controlling suspects, prisoners, and other individuals encountered during crowd control. The disposable handcuff is designed for one-time use. (Figure A-6)
 **Figure A-7. Training Disposable Handcuff**	14. Training Disposable Handcuff: The training disposable handcuff provides service members with a means for training in handcuffing procedures prior to actual use. The training disposable handcuff is red and reusable. (Figure A-7)
 **Figure A-8. MK-4 Pepper Spray Pouch**	15. MK-4 Pepper Spray Pouch: The MK-4 pepper spray pouch is designed to carry the MK-4 pepper spray canister. (Figure A-8)
 **Figure A-9. MK-4 Individual Live OC Pepper Spray**	16. MK-4 Individual Live OC Pepper Spray: The MK-4 individual pepper spray has a range of up to 12 feet in a target-specific stream. It contains 30 to 35 half-second bursts. (Figure A-9)
No picture available.	17. M-36 Individual Riot Control Disperser: M-36 Individual Riot Control Disperser has a range of up to 12 feet in a target-specific stream. It contains 30 to 35 half-second bursts. It is loaded with either CN or CS agent.
 **Figure A-10. MK-9 Pepper Spray Pouch**	18. MK-9 Pepper Spray Pouch: The MK-9 pepper spray pouch is designed to carry the MK-9 pepper spray canister. (Figure A-10)

Figure A-11. MK-9 Squad Live OC Pepper Spray

19. MK-9 Squad Live OC Pepper Spray: The MK-9 squad pepper spray has a range of up to 15 feet in a target-specific fog. It contains 8 to10 half-second high volume bursts. (Figure A-11)

Figure A-12. MK-46 Platoon Live (Horizontal) OC Pepper Spray

20. MK-46 Platoon Live (Horizontal) OC Pepper Spray: The MK-46 platoon pepper spray has a range of up to 25 to 30 feet in a target-specific dispersed spray pattern. It contains up to 12 one-second high volume bursts. (Figure A-12)

Figure A-13. MK-46 Sling

21. MK-46 Sling: The MK-46 sling gives a service member the ability to sling the MK-46 canister on the shoulder. (Figure A-13)

Figure A-14. MK-46 Hardware Kit

22. MK-46 Hardware Kit: The MK-46 hardware kit consists of a transfer tube, funnel, and transfer bottle. A compressed nitrogen source is also required to utilize the kit. (Figure A-14)

Figure A-15. MK-46 Refill Live OC Pepper Spray

23. MK-46 Refill Live OC Pepper Spray: The MK-46 refill solution comes in a one-gallon container. (Figure A-15)

Figure A-16. MK-4 Individual Inert OC Pepper Spray

24. MK-4 Individual Inert OC Pepper Spray: The MK-4 inert individual pepper spray has a range of up to 12 feet in a target-specific stream. It contains 30 to 35 half-second bursts. It provides the service member with an inert canister for familiarization and training purposes. The inert device is loaded with a non-irritant, non-toxic formulation payload. The device may also contain a non-toxic, water-soluble marker to allow immediate feedback on content delivery to target. The MK-4 device is marked "Inert." (Figure A-16)

Figure A-17. MK-9 Squad Inert OC Pepper Spray

25. MK-9 Squad Inert OC Pepper Spray: The MK-9 squad inert pepper spray has a range of up to 15 feet in a target-specific fog. It contains 8 to 10 half-second high volume bursts. It provides the service member with an inert canister for familiarization and training purposes. The inert device is loaded with a non-irritant, non-toxic formulation payload. The device may also contain a non-toxic, water-soluble marker to allow immediate feedback on content delivery to target. The inert MK-9 device is prominently marked "Inert" with a yellow sticker attached. (Figure A-17)

Figure A-18. MK-46 Platoon Inert OC Pepper Spray

26. MK-46 Platoon Inert OC Pepper Spray: The MK-46 platoon inert pepper spray has a range of up to 25 to 30 feet in a target-specific dispersed spray pattern. It contains up to 12 one-second high volume bursts. It provides the service member with an inert canister for familiarization and training purposes. The inert device is loaded win a non-irritant, non-toxic formulation payload. The device may also contain a non-toxic, water-soluble marker to allow immediate feedback on content delivery to target. The inert MK-46 device is prominently marked "Inert" with a yellow sticker attached. (Figure A-18)

Figure A-19. MK-46 Refill Inert OC Pepper Spray

27. MK-46 Refill Inert OC Pepper Spray: The MK-46 refill solution comes in a one-gallon container. (Figure A-19)

Figure A-20. FIST Training Suit

28. FIST Training Suit: The FIST training suit is made of closed-cell shock-absorbent foam that allows service members to hone their riot baton skills through realistic hand-to-hand engagement training. The FIST suit can absorb blows inflicted by the current expandable and wood riot baton. It provides protection for the head, face, hands, and legs, without significant degradation of the wearer's mobility. The FIST suit consists of a helmet, chest and back protector, bicep and forearm protectors, groin and buttock protectors, thigh and shin protectors, gloves, and carry bag. (Figure A-20)

NOTE: FIST is the current manufacturer of this training suit).

Figure A-21. Riot Training Strike Bag

29. Riot Training Strike Bag: The riot training strike bag gives the service member a strike bag for training and proficiency in open-hand control techniques. (Figure A-21)

Figure A-22. Caltrop

30. Caltrop: The caltrop is a field expedient area denial system. Securing approximately 50 caltrops to 550 military cord best employs it. This allows a service member to quickly toss/recover the system, using the caltrops to deny vehicle/pedestrian movement in a designated area. The four prongs are approximately two inches tall and will puncture tires or boots. (Figure A-22)

Figure A-23. Road Side Spike Strip

31. Road Side Spike Strip: The spike strip is designed to puncture a pneumatic tire as the vehicle makes contact with the spike strip. The spikes pivot into the tire and the tip-guards are pushed down. The spikes are imbedded deeply into the tire. The spikes are pulled out of the unit and remain in the tire. On the second revolution they will be pushed directly into the tire, allowing air to escape slowly without causing a blowout or sudden loss of control. The average tire will pick up 4 to 6 spikes with the deflation time of about 12 to 20 seconds. It allows a service member to quickly toss/recover the system, impeding vehicle movement in a designated area. It allows a force to provide additional physical security in areas where the caltrop cannot be tactically employed. (Figure A-23)

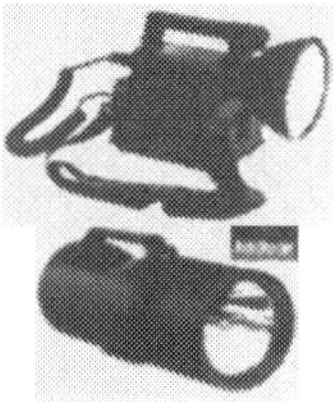

Figure A-24. High-Intensity Search Light

32. The high-intensity searchlight is a rugged, lightweight, hand-held, high-intensity spotlight for use during low-light conditions. Two different high-intensity lights were fielded. The Maxabeam light (shown on top) was fielded in 1997. It was replaced by the Nighthunter light (shown underneath) in 2000. (Figure A-24)

Figure A-25. Individual Light Holster (6P)

33. Individual Light Holster (6P): The 6P individual light holster is designed to hold the 6P individual light. (Figure A-25)

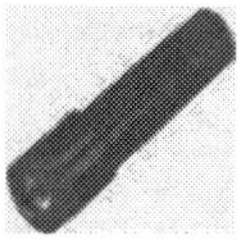

Figure A-26. 6P Individual Light

34. 6P Individual Light: The 6P individual light is issued to service members within the platoon. (Figure A-26)

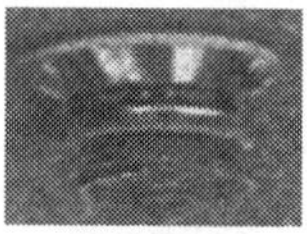

Figure A-27. 6P Individual Light Accessory Kit

35. 6P Individual Light Accessory Kit: The 6P individual light accessory kit includes a spare light bulb installation kit. (Figure A-27)

Figure A-28. 3V Lithium Batteries

36. 3V Lithium Batteries: Two 3V Lithium Batteries are required to operate the 6P individual light. (Figure A-28)

Figure A-29. 12-Gauge Buttstock Cuff

37. 12-Gauge Buttstock Cuff: The buttstock cuff is designed to fasten to the buttstock of most shotguns to hold seven 12-gauge cartridges for quick loading of the shotgun. (Figure A-29)

Figure A-30. 12-Gauge Utility Pouch

38. 12-Gauge Utility Pouch: The 12-gauge utility pouch is designed to hold twenty-five 12-gauge munitions and attaches to the service member's combat gear. (Figure A-30)

Figure A-31. 40-mm Carrying Pouch

39. 40-mm Carrying Pouch: The 40-mm carrying pouch is partitioned in a manner to provide the service member with a means of carrying not less than six 40-mm nonlethal rounds in separate compartments. It attaches to the service member's combat gear. (Figure A-31)

Figure A-32. Diversionary/Rubber Ball Grenade Pouch

40. Diversionary/Rubber Ball Grenade Pouch: The diversionary/rubber ball grenade pouch is an ammunition carrying pouch partitioned in a manner to provide the service member with a means of holding up to six rubber ball grenades or six flash bang grenades in separate compartments. (Figure A-32)

Figure A-33. 12-Gauge Mossberg 500A2/590 Shotgun

41. 12-Gauge Mossberg 500A2/590 Shotgun: The 12-gauge Mossberg 500A2 or 590 shotgun are the only 12-gauge shotguns currently authorized for firing 12-gauge nonlethal projectile munitions. The launching cup for the rubber ball grenade will attach to the Mossberg 500A2 Shotgun. The launch cup will not attach to the Mossberg 590 shotgun. Until the Joint Service Combat shotgun can effectively fire nonlethal 12-gauge rounds, the 12-gauge Mossberg 500A2 or 590 shotgun configuration meets the requirement of the NLW force. (Figure A-33)

Figure A-34. 12-Gauge M870 (modified) Version 2 (USCG)

42. 12-Gauge M870 (modified) Version 2 (USCG): The modified Remington 870 shotgun has a fixed synthetic stock and pistol grip. The pistol grip gives the shooter more control of the weapon and helps dissipate recoil from the shoulder. The synthetic materials used are more durable and lighter weight than the wood stocks used on the Standard M870. The barrel is fitted with a rifled choke tube for firing slugs (solid projectiles) to increase accuracy. It is also equipped with an Advanced Combat Optical Gunsight that was developed for use in "both-eyes-open" close quarters battle. The advantage is rapid target acquisition without the need to align front and rear sights. Batteries do not power the illumination sight; it contains tritium gas.

In the Coast Guard, a yellow "stock sock" will be used whenever the shotgun is loaded with nonlethal munitions. This removable nylon sleeve is used to identify the weapon as a nonlethal delivery system. (Figure A-34)

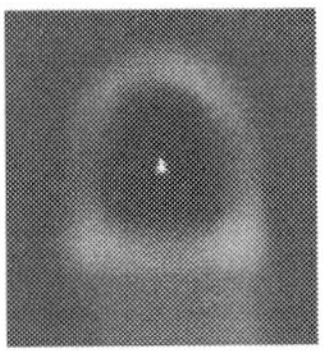

Figure A-35. Advanced Combat Optical Gunsight (ACOG) Reflex Sight (USCG)

43. Advanced Combat Optical Gunsight (ACOG) Reflex Sight (USCG): Modified shotguns are equipped with this site. (Figure A-34)

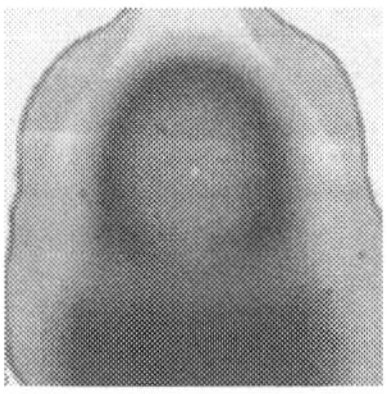

Figure A-36. ACOG Reflex Sight (USCG)

44. ACOG Reflex Sight (USCG): The dual-illuminated, Trijicon® Reflex sight was developed for the military for use in both-eyes-open close quarters battle (CQB). The advantage is quick target acquisition thus eliminating the need to align front and rear sights. Batteries do not power the illumination sight; it contains tritium gas. (Figure A-35)

Equipment Currently Being Fielded

1. 40-mm Crowd Dispersal Cartridge (CDC):

This direct fire, low-hazard, non-shrapnel-producing, multiple ball projectile provides friendly forces the capability to conduct crowd control. The CDC uses nonlethal, blunt trauma to stop, confuse, disorient, and/or temporarily incapacitate personnel and increase standoff distance between friendly forces and the hostile crowd. Primarily designed as an area effect munition, the CDC can be used on individuals. The M203 40-mm grenade launcher fires the CDC and contains 48, .48 caliber rubber balls. Effective range is 10 to 30 meters. (Figures A-37 and A-38)

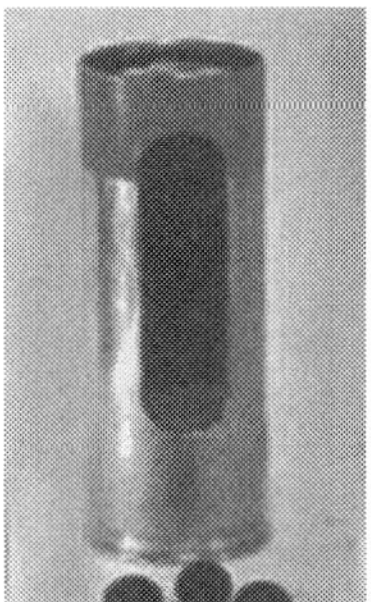

Figure A-37. Crowd Dispersal Cartridge

Figure A-38. Soldier Employing CDC with M203

2. Portable Vehicle Arresting Barrier (PVAB):

PVAB is a lightweight, portable, manually emplaced and recoverable barrier entrapment device to create, on command, a barrier that stops vehicles driven by known or suspected hostile forces with minimal damage to the vehicle or its occupants. It is used at dismount points, checkpoints, and roadblocks to prevent unauthorized access by wheeled vehicles into or out of areas (e.g. airfields, ports, command posts [CP]). Two individuals can easily emplace it, in approximately four hours. In stand-by mode, PVAB folds down into something similar to a speed bump (vehicles pass freely). In capture mode, the net springs up on two hydraulic pistons, to maximum height in less than a second. PVAB can be set up to span 12 to 24 feet and can stop a light truck (up to 7,500 pounds) at speeds up to 45 miles per hour, with a stopping distance of less than 200 feet PVAB can be containerized: 53.75 cubic feet (7 boxes, 4 large-3 small)—1,044 pounds on approximately one pallet. (Figures A-39 and A-40)

Figure A-39. PVAB in Action (Far View)

Figure A-40. PVAB in Action Near View

3. Modified Crowd Control Munition—Ground Emplaced (MCCM-GE):

The MCCM-GE (Figure A-41) is a nonlethal munition, housed in a casing similar to the current claymore mine, which will stop, confuse, disorient and/or temporarily incapacitate area targets/personnel at close range. The force can use it to provide crowd control by dispersing hostile crowds and force protection by integrating it into a layered perimeter defense. It delivers 600, .32 caliber rubber balls as well as a flash bang effect. The MCCM-GE gives the commander or small unit leader greater flexibility to apply nonlethal force as a first line of defense against hostile noncombatants. Effective range is 5 to 15 meters. (Figure A-42)

Figure A-41. Modified Crowd Control Munition-Ground Emplaced (MCCM-GE)

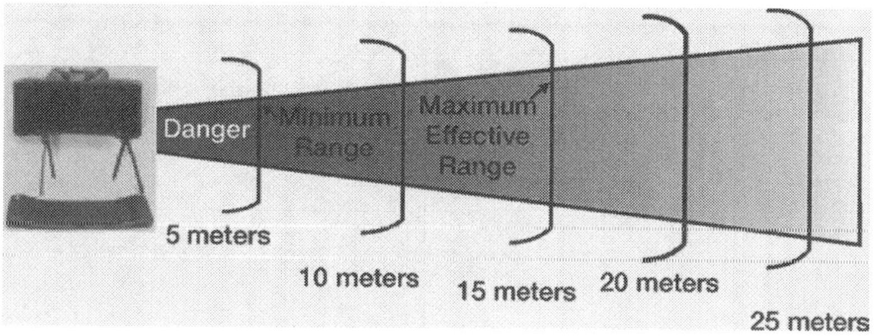

Figure A-42. MCCM-GE Range Chart

4. 66-mm Vehicle Mounted Nonlethal Grenade Launcher (VMNLGL):

The VMNLGL is a nonlethal grenade-launching device mounted on high mobility multipurpose wheeled vehicle (HMMWV). It is an indirect fire-support system that can deliver a 66-mm distraction device (flash-bang), CS, or blunt trauma (sting ball) payload. The distraction grenade uses a pyrotechnic charge for crowd control through audio and visual stimuli. The blunt trauma grenade uses a pyrotechnic charge to discharge 140, .32 caliber (cal.), rubber balls to achieve crowd control through audio, visual, and physical stimuli. The force uses VMNLGL to enhance the ability of friendly forces to conduct force protection without direct contact with hostile or threatening crowds. It enforces standoff distances and deters potential threats. It can disperse or subdue crowds, keep personnel from surrounding or climbing on patrol vehicles, slow or stop threatening personnel or vehicles by confusing, disorienting, and/or temporarily incapacitating individuals for capture. The grenade is fired from the grenade launchers traditionally found on tracked vehicles. Vehicle-launched grenades have three submunitions designed to ground burst. Effective range is 50 to 100 meters. An adjustable bracket allows engagement at ranges of 50 meters, 75 meters, and 100 meters. (Figure A-44)

Figure A-43. 66-mm Vehicle Mounted Nonlethal Grenade Launcher (VMNLGL)

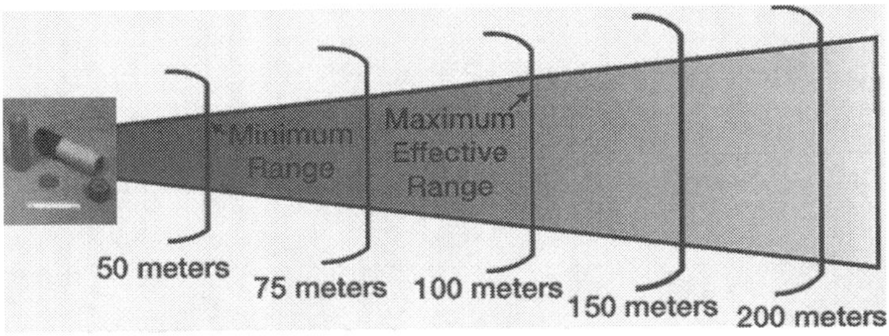

Figure A-44. VMNLGL Range Chart

5. Running Gear Entanglement System (RGES)

The Running Gear Entanglement System (RGES) is series of short, separate, looped polyethylene lines that are attached to a single, long polyethylene base line. This entanglement net fouls the propellers of planing hull small boats. The capability to manually deploy the net from a soft-sided bag has been fielded for USCG use in limited numbers. The USCG is in the process of incorporating three other variants of RGES into its inventory of nonlethal capabilities. These include aerial-delivered RGES, surface launched RGES to deploy from vessels, and static barrier RGES. The static barrier RGES has broad application into support homeland security, physical security and force protection missions. The surface launched RGES has direct application in the USCG's law enforcement mission wherever the requirement exists to stop a non-compliant vessel. Both of these systems are expected to be fielded in the fall/winter of CY02. The air launched version is currently still in the concept development phase and is not yet fully developed. (Figure A-45)

Figure A-45. Running Gear Entanglement System (RGES)

a. RGES Surface Delivery System: The USCG can use stern-mounted, manually deployed RGES from a cutter's small boat to stop slow-moving migrant-type craft. (Figure A-46)

(1) Uses COTS compressed air cylinder launcher.

(2) Removable, fix-mounted launcher on side of craft.

(3) Launch projects device ahead of target.

Figure A-46. RGES Surface Delivery System

b. RGES Barrier

(1) Provides static barrier around ship or other waterside asset.

(2) Provides a visible demarcation line, establishes zones, deters intrusions. (Figure A-47)

Figure A-47. RGES Barrier Employment

c. RGES Aerial Delivery System (ADS): RGES is aerial deployed from the ADS ahead of target aboard a H60 helicopter (Figure A-48). Device should float in the water ahead of target ensuring target runs over device. Device fouls propeller or rudder until target stops and awaits boarding.

(1) Uses COTS compressed air cylinder launcher

(2) Removable, fix-mounted launcher on helicopter

(3) Launch projects device ahead of target. (Figures A-49)

Figure A-48. RGES Aerial Delivery System (ADS)

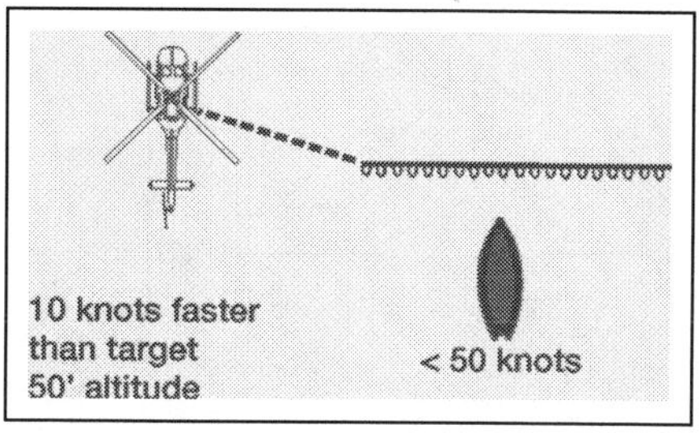

Figure A-49. RGES Aerial Delivery System (ADS)

6. Warp Cutter

a. Warp cutting is a nonlethal use of force, which has been safely and effectively employed by several countries to enforce fisheries laws in their respective exclusive economic zones (EEZ). The warp cutter is a device that closely resembles a grappling hook or small anchor. It has hardened steel blades attached to the crux of the grapnels. It is towed behind an enforcement vessel and across the warps of a non-compliant fishing vessel. As the cutting device engages the warps, it slices through them, separating the net from the fishing vessel.

b. The USCG uses the warp cutter to compel foreign vessels suspected of illegally fishing in the U.S. EEZ to comply with lawful orders to stop and be boarded. The USCG may cut the trawl warps of non-compliant vessels as a means of compelling them to stop for a boarding.

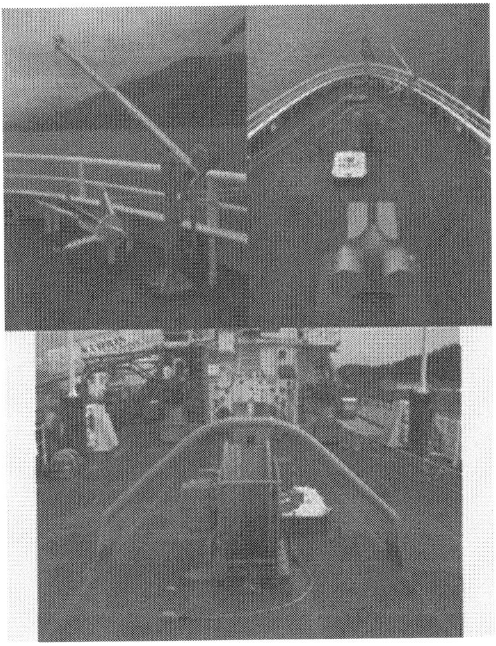

Figure A-50. Warp Cutter

Appendix B

Summary: Specialty Impact Munitions

12-Gauge Launch Cartridge

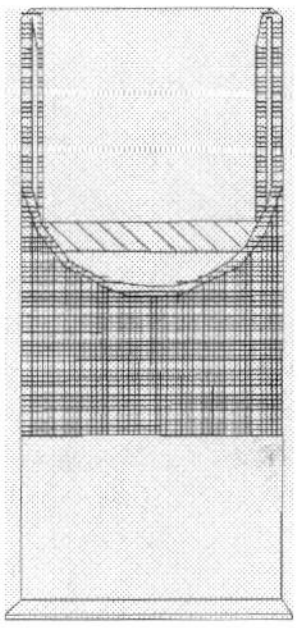

Using Services: USN, USMC
Status: Fielded 1998
DODIC: AA30
NSN: 1305-01-464-8389
Acquisition Activity: MARCORSYCOM—NSWC Crane, IN
Further Information:
Commercial: MARCORSYSCOM (703) 784-9395
NSWC Crane (812) 854-5801

Color .... Clear case
Diameter 12-gauge
Projectile .... N/A
Length .... 2.00 inches (51 mm)
Total Weight .... 13.6 grams
Propellant Weight .... 46.3 grains (black powder)
Muzzle Velocity .... 100 feet/second (30.5 m/sec)—rubber ball grenade
Acceptance Accuracy .... 200 feet (61 m) + 25 feet (7.6 m) at 30-degree launch angle using rubber ball grenade and 12-gauge launch cup

Projectile Specification: N/A

- Use 12-gauge launch round in combination with rubber ball grenade and 12-gauge launch cup to hurl a grenade a distance of 200 feet (61 m) + 25 feet (7.6 m) when launch angle is approximately 30 degrees.
- Chamber launch cartridge only after installing grenade in launch cup and pulling safety pin from grenade.

Figure B-1. 12-Gauge Launch Cartridge

12-Gauge Bean Bag Cartridge

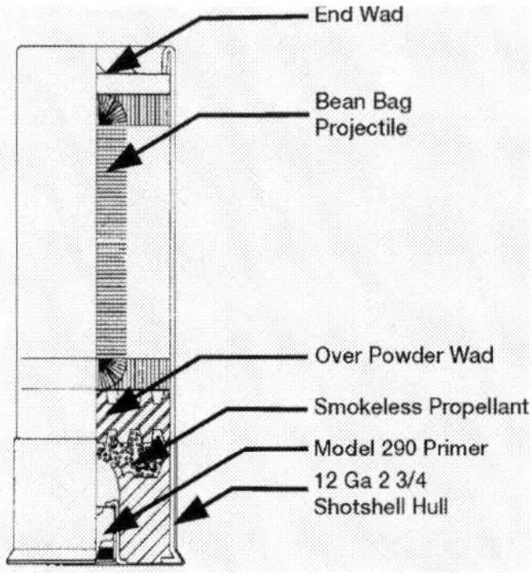

Using Services: USN, USMC

DODIC: AA29

NSN: 1305-01-454-0191

Status: Fielded 1998

Acquisition Activity: MARCORSYCOM—NSWC Crane, IN

Further Information:

Commercial: MARCORSYSCOM (703) 784-9395
NSWC Crane (812) 854-5801

Color .... Clear or red case
Diameter .... 12 gauge
Projectile .... 1 bean bag
Length .... 2.45 inches (62 mm)
Total Weight .... 49.9 grams
Propellant Weight .... 3.4 grains
Muzzle Velocity .... 300 + 25 feet/second (91.4 + 7.62 m/sec)
Acceptance Accuracy .... 2 feet (.6 m) X 6 feet (1.8 m) at 100 feet (30.3 m) with Mossberg 500A2

Shotgun Projectile Specification:

Shape .... Square bag or tube sock
Material .... Nylon or cotton-polyester
Filler .... #9 lead shot
Filler weight .... 44.4 grams (maximum)

- Fatal injuries are possible at employment distances of less than 20 feet (6 m)
- At distances of 20 feet (6 m) to 40 feet (12 m), target area should be center mass
- Training needs to reinforce that head shots are NOT ACCEPTABLE
- Do not skip fire this round! Round becomes unpredictable upon striking the ground
- Round is most effective against individually selected targets (point round)
- End wad has low light marker

Figure B-2. 12-Gauge Bean Bag Cartridge

12-Gauge Rubber Fin Stabilized Cartridge

Using Services: USN, USMC, USCG

Status: Fielded 1998

DODIC: AA31

NSN: 1305-01-454-0189

Acquisition Activity: MARCORSYCOM—NSWC Crane, IN

Further Information:

Commercial: MARCORSYSCOM (703) 784-9395
NSWC Crane (812) 854-5801

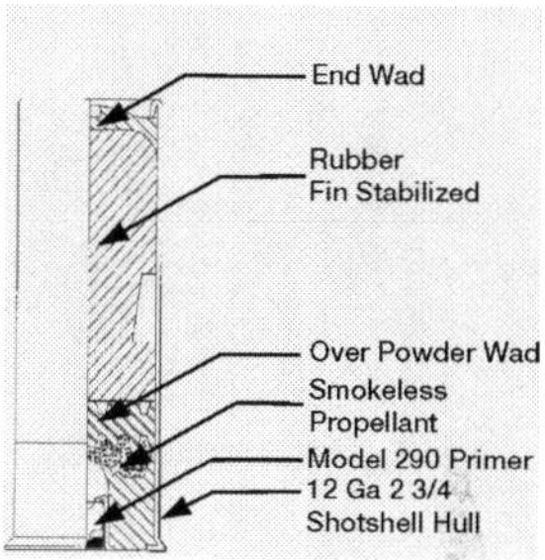

Color ....Clear or red case
Diameter ....12 gauge
Projectile ....1 rubber projectile
Length ....2.45 inches (62 mm)
Total Weigh ....9.1 grams
Propellant Weight.................4.8 grains
Muzzle Velocity.....................530 + 65 feet/second
(161.5 + 19.8 m/sec)
Acceptance Accuracy..........2 feet (.6 m) X 6 feet (1.8 m) at 100 feet (30.3 m) with Mossberg 500 A2 shotgun
Projectile Specification: ......Caliber 12 gauge
Material.................................Rubber
Hardness ....Durometer 70-80 Shore "A" scale
Weight ....7.8 grams
Temperature Limits (demonstrated):
Upper limit...125° F
Lower limit...0° F

Shipping Data: Hazard classification/division and storage compatibility group...1.4S
DOT Class ....ORM-D
DOT Markings Cartridge......Small Arms
UNO serial number...............0012

- Fatal injuries are possible at employment distances of less than 20 feet (6 m); USCG: 15 yards
- At distances of 20 feet (6 m) to 40 feet (12 m), target area should be center mass; USCG: 15 to 30 yards
- Training needs to reinforce that head shots are NOT ACCEPTABLE
- Do not skip fire this round! Round becomes unpredictable upon striking the ground
- Round is most effective against individually selected targets (point round)
- End wad has low light marking

Figure B-3. 12-Gauge Rubber Fin Stabilized Cartridge

40-mm Foam Rubber Baton Cartridge

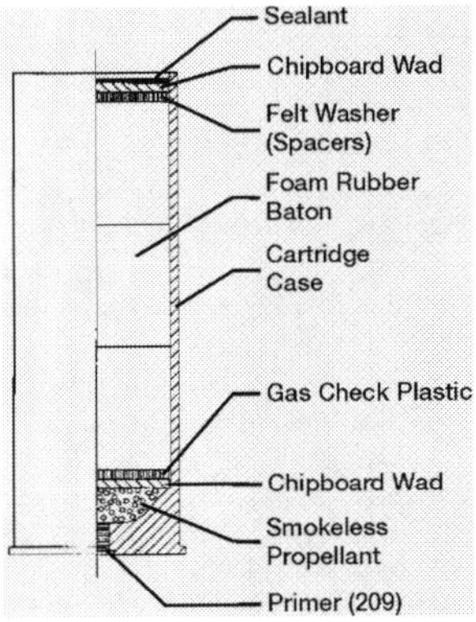

Using Services: USN, USMC

Status: Fielded 1998

DODIC: BA07

NSN: 1310-01-453-9168

Acquisition Activity: MARCORSYSCOM—NSWC Crane, IN

Further Information:

Commercial: MARCORSYSCOM (703) 784-9395
NSWC Crane (812) 854-5801

Color.......................................Aluminum (case)
Diameter................................40 millimeter
Projectile...............................3 foam batons
Length ....4.78 inches (121.4 mm)
Total Weight..........................201 grams
Propellant Weight.................13.0 grains
Muzzle Velocity.....................325 + 35 feet/second (99 + 10.6 m/sec)
Acceptance Accuracy..........6 feet (1.8 m) X 3 feet (.9 m) at 50 feet (15.2 m) with M203

Projectile Specifications:

Number.............................3
Material............................Foam rubber
Hardness..........................Durometer 25 + 10 on Shore "A" scale
Weight...............................250 to 280 grains

- Fatal injuries are possible at employment distances of less than 20 feet (6 m)
- At distances of 20 feet (6 m) to 40 feet (12 m), target area should be center mass
- Training needs to reinforce that head shots are NOT ACCEPTABLE
- Round is most effective against point targets (individual personnel)
- End wad has low light marker

Figure B-4. 40-mm Foam Rubber Baton Cartridge

40-mm Rubber Ball Cartridge

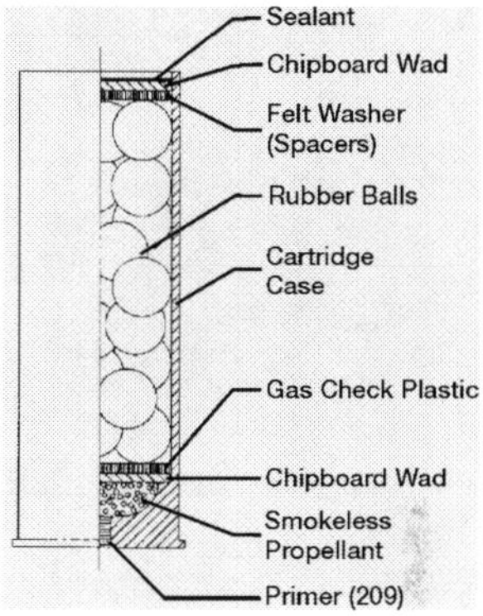

Using Services: USN, USMC

Status: Fielded 1998

DODIC: BA08

NSN: 1310-01-453-9154

Acquisition Activity: MARCORSYSCOM—NSWC Crane, IN

Further Information:

Commercial: MARCORSYSCOM (703) 784-9395

Color .... Aluminum (case)
Diameter .... 40 millimeter
Projectile .... 24 rubber balls
Length .... 4.78 inches (121.4 mm)
Total Weight .... 201 grams
Propellant Weight .... 7.3 grains
Muzzle Velocity .... 325 + 35 feet/second (99 + 10.6 m/sec)
Acceptance Accuracy .... 6 feet (1.8 m) x 9 feet (2.7 m) at 50 feet (15.2 m) with M203NSWC Crane (812) 854-5801

Projectile Specifications:

Caliber 60
Number .... 22 (minimum)
Material .... Rubber
Hardness .... Durometer 60 + 10 on Shore "A" scale
Weight .... 2.2 grams each

- Fatal injuries are possible at employment distances of less than 20 feet (6 m)
- At distances of 20 feet (6 m) to 40 feet (12 m), target area should be center mass
- Training needs to reinforce that head shots are NOT ACCEPTABLE
- Round is most effective against area targets (multiple personnel)
- End wad has low light marker

Figure B-5. 40-mm Rubber Ball Cartridge

Rubber Ball Grenade

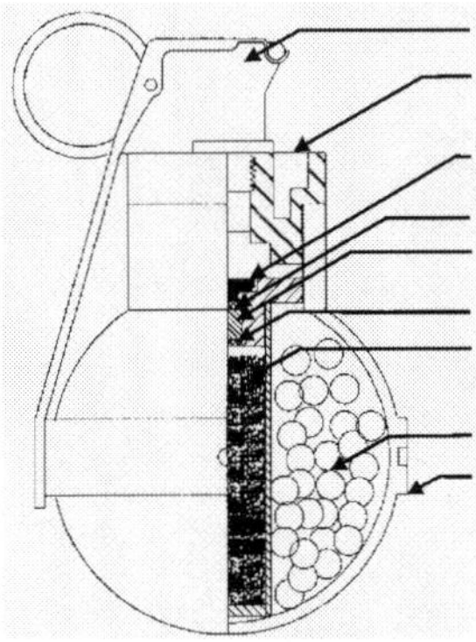

Using Services: USN, USMC

Status: Fielded 1998

DODIC: GG04

NSN: 1310-01-454-0132

Acquisition Activity: MARCORSYSCOM—NSWC Crane, IN

Further Information:

Commercial: MARCORSYSCOM (703) 784-9395
NSWC Crane (812) 854-5801

Color .... Black with white lettering
Diameter .... 3.12 inches (79 mm)
Projectiles .... 100 rubber balls (minimum)
Height .... 5.25 inches (133.4 mm)
Total Weight .... 290 grams
Explosive Weight .... 59 grams
Total Time Delay .... 3.0 + 0.5 seconds

Employment Distance:
Hand Thrown .... 50 feet (15.2 m)
Shotgun Launched .... 200 feet (61 m)

Projectile Specifications:
Number .... 100 (minimum)
Material .... Rubber
Hardness .... Durometer 70 to 80 on Shore "A" scale
Diameter .... 0.25 inches (6.4 mm) (minimum)

- Fuze separates from the body 1.3 to 1.7 seconds after primer ignition.
- Second delay causes burster charge detonation 3.0 + 0.5 seconds average after primer ignition. Rubber balls are then dispersed in a 50-feet (15-m) circular pattern.
- 12-gauge shotgun with launch cup and launch cartridge can launch grenade 200 feet (61 m) + 25 feet (7.6 m)

Figure B-6. Rubber Ball Grenade

Practice Grenade Body

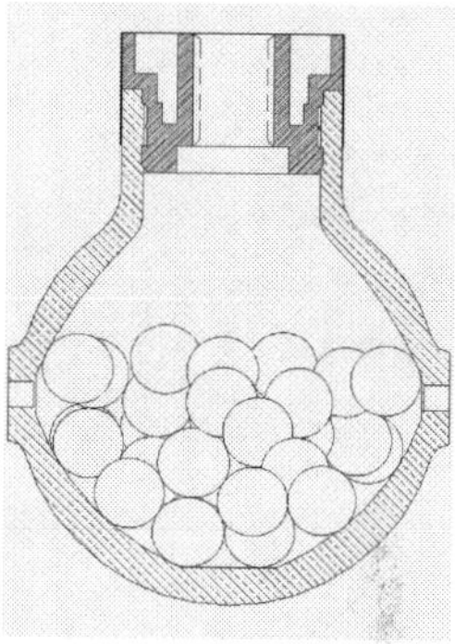

Using Services: USMC
Status: Fielded 2002
DODIC: GG05
NSN: Pending
Acquisition Activity: MARCORSYSCOM – NSWC Crane, IN
Further Information:

Commercial: MARCORSYSCOM (703) 784-9395
NSWC Crane (812) 854-5801

Diameter ....3.12 inches (79 mm)
Projectiles ....N/A
Height ....4.2 inches (106 mm)
Total Weight...........................208 grams
Explosive Weight..................N/A (inert)
Total Time Delay...................3.0 + 0.5 seconds when used with practice grenade fuze

Employment Distance:
Hand Thrown...................50 feet (15.2 m)
Shotgun Launched ....200 feet (61 m) using Mossberg launch cup and 12-gauge launch cartridge (AA30).

- Practice grenade body is reusable up to 25 times and is a component of a practice grenade assembly, consisting of the body and a practice grenade fuze.
- Practice grenade assembly accurately replicates the GG04 rubber ball grenade and is used for training purposes.

Figure B-7. Practice Grenade Body

Practice Grenade Fuze

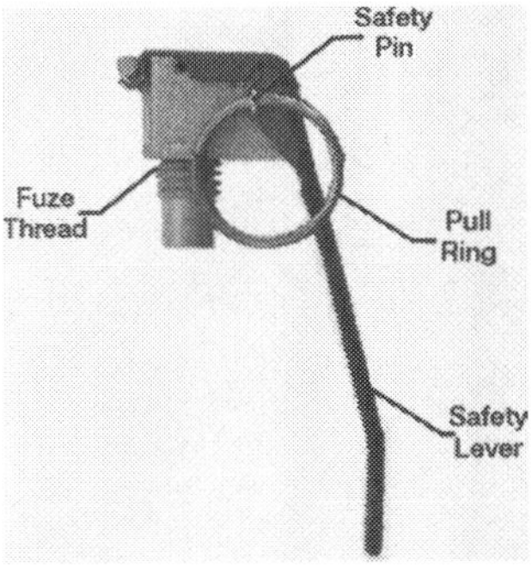

Using Services: USMC

Status: Fielded 2002

DODIC: Pending

NSN: Pending

Acquisition Activity: MARCORSYSCOM—NSWC Crane, IN

Further Information:

Commercial: MARCORSYSCOM
(703) 784-9395
NSWC Crane (812) 854-5801

Color .... Blue safety lever with brown stripe
Projectiles .... N/A
Height .... 4.25 inches (107 mm)
Total Weight.............................. 82 grams
Explosive Weight....................... Not available
Total Time Delay........................ 3.0 + 0.5 seconds

Employment Distance when used with GG05 Practice Grenade Body:

Hand Thrown .... 50 feet (15.2 m)
Shotgun Launched........... 200 feet (61 m)—using Mossberg launch cup and 12-gauge launch cartridge (AA30).

- Employ practice grenade fuze with the practice grenade body to create a practice grenade assembly consisting of the body and a practice grenade fuze.
- Practice grenade assembly accurately replicates the GG04 rubber ball grenade and is used for training purposes.

Figure B-8. Practice Grenade Fuze

Nonlethal Grenade Launch Cup

Using Service: USMC

Status: Fielded 2002

DODIC: N/A

NSN: Pending

Acquisition Activity: MARCORSYSCOM—NSWC Crane, IN

Further Information: Commercial: MARCORSYSCOM
(703) 784-2006 Ext. 2733
NSWC Crane (812) 854-5801

NOTE: Cup body shown is representative of cup bodies used on all shotguns. Adapter shown is subject to change depending on the shotgun used and design improvements.

Color .... Black
Diameter (internal) .... 3.12 inches (79 mm)
Length (without adapter) ..... 4 inches (102 mm)
Weight .... 397 grams (maximum)

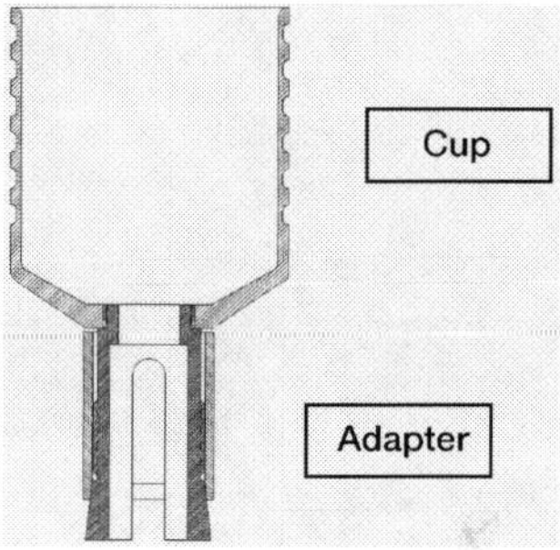

Cross Section View

- Use launch cup in concert with AA30 launch cartridge and either the GG04 rubber ball grenade or GG05 practice grenade assembly
- Attach launch cup and remove from the shotgun without using tools
- Launch cup can withstand accidental detonation of rubber ball grenade without fragmenting

Figure B-9. Nonlethal Grenade Launch Cup

Diversionary Charge, MK 141, MOD 0

Using Services: USN, USMC

Status: Fielded 1997

DODIC: DWBS

NSN: 1375-01-387-5748

Acquisition Activity: MARCORSYSCOM—NSWC Crane, IN

Further Information:

Commercial: MARCORSYSCOM (703) 784-9395
NSWC Crane (812) 854-3934

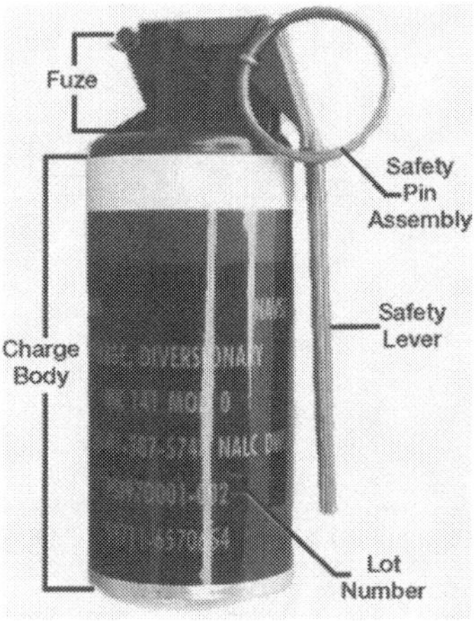

Color .... Brown with white stripe
Diameter .... 1.77 inches (45 mm)
Projectile .... N/A
Length .... 5.214 inches (132.4 mm)
Total Weight........................... 99.8 grams
Net Explosive Weight........... 17.8 grains
Light Output........................... 2 million candlepower for 4 milliseconds
Noise Output.......................... 176 to 185 decibels
Material.................................. Molded and machined polyurethane

Time Delay:
Initial .... 1.26 seconds (minimum)
Total .... 1.34 to 1.84 seconds

- Diversionary charge can cause serious injury if held at time of detonation
- Light and noise most effective in confined space to stun individuals
- Most often employed for hostage rescue

Figure B-10. Diversionary Charge, MK 141, MOD 0

Cartridge, 12-Gauge Rubber Sting Ball™

Using Services: USCG

Status: Fielded

DODIC: N/A Model No. 2552

NSN: Commercial Procurement

Acquisitions Activity: Commandant (G-OCU)

Further Information: Commandant (G-OCU) (202) 267-1522

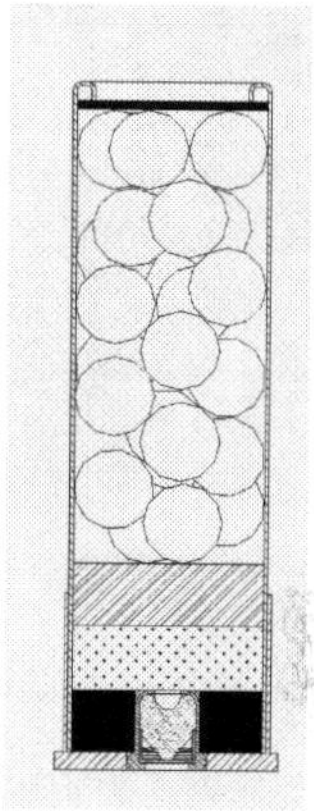

Model Number .... 31-caliber sting-ball
Type .... Multiple projectile rubber pellets
Range .... 10 yards (9 m)
Projectiles .... 18 (average)
Cartridge Material................. Plastic
Projectile Weight .... .35 grams each
Projectile Material.................. .31-caliber rubber pellets
Overall Height........................ 2.75 inches (70 mm)
Overall Weight .... 18 grams
Compatibility.......................... 12-gauge shotguns
Velocity.................................. 400 feet/second

USCG

- Fatal injuries are possible at employment distances of less than 5 yards.
- At distances of 20 feet (6 m) to 40 feet (12 m), target area should be center mass
- This round is most effective against area targets with a maximum effective range of 15 yards
- Training needs to reinforce that head shots are NOT ACCEPTABLE
- Round is most effective against area targets (multiple personnel)
- Stock Sock: a yellow "stock sock" will be used whenever the modified 12-gauge Remington 870 shotgun is loaded with nonlethal munitions. This removable nylon sleeve is used to identify the weapon as a nonlethal delivery system.

- All specifications are average and are subject to change without notice
- 2552SPEC

Figure B-11. Cartridge, 12-Gauge Rubber Sting Ball™

Cartridge, 12-Gauge Sting Ball™ High Velocity

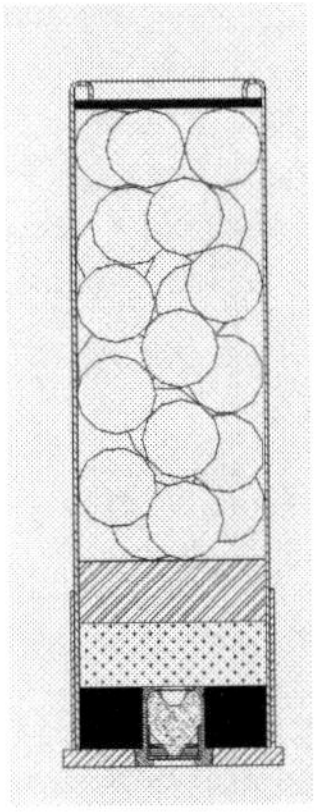

Using Services: USCG
Status: Fielded
DODIC: N/A Model No. 2553
NSN: Commercial Procurement
Acquisitions Activity: Commandant (G-OCU)
Further Information: Commandant (G-OCU) (202) 267-1522

Model Number	31-caliber sting-ball
Type	Multiple projectile rubber pellets
Range	10 yards (9 m)
Projectiles	18 (average)
Cartridge Material	Plastic
Projectile Weight	.35 grams each
Projectile Material	.31-caliber rubber pellets
Overall Height	2.4 to 2.5 inches (61 – 63 mm)
Overall Weight	18 grams
Compatibility	12-gauge shotguns
Velocity	900 feet/second

USCG

- Fatal injuries are possible at employment distances of less than 5 yards.
- At distances of 20 feet (6 m) to 40 feet (12 m), target area should be center mass
- Maximum effective range is 15 yards
- Training needs to reinforce that head shots are NOT ACCEPTABLE
- Round is most effective against area targets (multiple personnel)
- Stock Sock: a yellow "stock sock" will be used whenever the modified 12-gauge Remington 870 shotgun is loaded with nonlethal munitions. This removable nylon sleeve is used to identify the weapon as a nonlethal delivery system.

- All specifications are average and are subject to change without notice.
- 2553SPEC

Figure B-12. Cartridge, 12-Gauge Sting Ball™ High Velocity

M1006, 40-mm Nonlethal Cartridge (Sponge Grenade)

Using Services: U.S. Army

Status: Materiel Released April 2000

DODIC: BA06

NSN: 1310-01-452-1190

Acquisition Activity: Acquisition Center, Picatinny Arsenal, NJ

Further Information:

Commercial: Picatinny Arsenal NL Systems Integrator: (973) 724-6948
Product Director: (973) 724-6283

Color:

Projectile ....Black, olive green (foam rubber)
Cartridge Case ....Opaque
Diameter ....40 millimeter
Length ....5.272 inches (13.4 cm) (maximum)
Total Weight..........................68 grams (0.15 lbs)
Propellant Weight.................2 grains
Muzzle Velocity.....................265 + 20 feet/second (81 + 6 m/sec)
Acceptance Accuracy..........>85 % accuracy at 30 meters (E-silhouette) and >60 % accuracy at 50 meters (E-silhouette) when fired from an M16A2/M203 or a M4/M203
Engagement Range..............10 meters to 50 meters
Projectile Specification:
Shape ....40-mm bullet shaped foam rubber
Material................................Foam rubber
Weight ....28.5 grams (.06 lbs)

- Fatal injuries are possible at employment distances of less than 32.8 feet (10 m)
- At distances of 10 meters to 50 meters, target area should be center mass
- Training needs to reinforce that head shots are NOT ACCEPTABLE
- Do not skip fire this round! Round becomes unpredictable upon striking the ground
- Round is most effective against individually selected targets (point round)

Figure B-13. M1006, 40-mm Nonlethal Cartridge (Sponge Grenade)

M1012, 12-Gauge Shotgun: Nonlethal, Point Control Cartridge

Using Services: U.S. Army

Status: Type Classified, June 2000

DODAC: AA51

NSN: 1305-01-470-2405

Acquisition Activity: Acquisition Center, Picatinny Arsenal, NJ

Further Information:

Commercial: Picatinny Arsenal NL Systems Integrator: (973) 724-6948
Product Director: (973) 724-6086

Color:

Cartridge Case: .... Clear

Diameter .... 12 gauge

Length .... 2.45 inches (62.2 mm)

Total Weight .... 500 grains

Propellant Weight .... 6.6 grains smokeless (Red Dot)

Muzzle Velocity .... 500 feet/second (152.4 m/sec)

Acceptance Accuracy .... 80 % accuracy at 10 and 20 meters (E-silhouette) fired out of Mossberg 500, Mossberg 590, and Winchester 1200

Engagement range .... 10 to 20 meters

Projectile Specification:

Shape .... Fin stabilized 'bomblet' shape

Material .... Rubber

Weight .... 0.2 ounces (5.8 gm)

Hardness .... 75 Durometer "A" Scale

- Fatal injuries are possible at employment distances of less than 33 feet (10 m)
- At distances of 10 to 20 meters, target area should be center mass
- Training needs to reinforce that head shots are NOT ACCEPTABLE
- Round is most effective against individually selected targets (point round)

Figure B-14. M1012, 12-Gauge Shotgun: Nonlethal, Point Control Cartridge

M1013, 12-Gauge Shotgun: Nonlethal, Crowd Dispersal Cartridge

Using Services: U.S. Army
Status: Type Classified June 2000
DODAC: AA52
NSN: 1305-01-470-2139
Acquisition Activity: Acquisition Center, Picatinny Arsenal, NJ
Further Information:
Commercial: Picatinny Arsenal NL Systems Integrator: (973) 724-6948
Product Director :(973) 724-6086

Color:
Projectile ....Black rubber
Cartridge Case ....Clear
Diameter ....12 gauge
Projectile ....18 rubber balls
Length ....2.37 inches (60.2 mm)
Total Weight...........................500 grains (32.4 gm)
Propellant Weight.................14 grains smokeless (Red Dot)
Muzzle Velocity.....................900 feet/second (274 m/sec)
Acceptance Accuracy..........80 % accuracy for a 2-by-3-meter target at 10 and 20 meters fired out of Mossberg 500, Mossberg 590, and Winchester 1200
Engagement range ....10 to 20 meters
Projectile Specification:
Shape................................32-gauge (0.32 in) ball
Material.............................PVC rubber compound
Weight...............................0.01 ounces (0.25 gm)
Hardness ....75 Durometer "A" Scale

- Fatal injuries are possible at employment distances of less than 32.8 feet (10 m)
- At distances of 10 to 20 meters, target area should be center mass
- Training needs to reinforce that head shots are NOT ACCEPTABLE

Figure B-15. M1013, 12-Gauge Shotgun: Nonlethal, Crowd Dispersal Cartridge

M1029, 40-mm Nonlethal Crowd Dispersal Grenade

Using Services: U.S. Army

Status: Type Classified May 2001

DODIC: BA13

NSN: 1310-01-475-0628

Acquisition Activity: Acquisition Center, Picatinny Arsenal, NJ

Further Information:

Commercial: Picatinny Arsenal NL Systems Integrator: (973) 724-6948
Product Director: (973) 724-6283

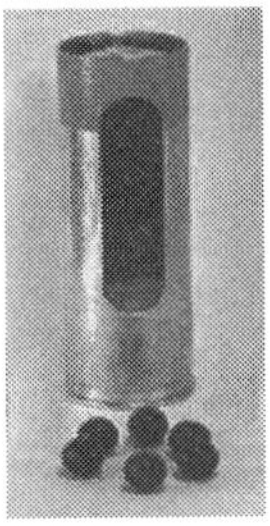

Color:

- Projectile .... Black rubber
- Cartridge Case .... Aluminum

Diameter .... 40 millimeter

Projectile .... 48 rubber balls

Length .... 4.8 inches (12.2 cm)

Total Weight .... 202.22 grams (7.0 oz)

Propellant Weight .... 8.5 grains smokeless

Muzzle Velocity .... 450 feet/second

Acceptance Accuracy .... > 90 % hit accuracy 4 out of 5 (shoulder to shoulder) E-silhouettes target at 30 meters and when fired from an M16A2/M203 or a M4/M203

Projectile Specification:

- Shape .... 48-gauge ball (.48 in diameter)
- Material .... Rubber
- Weight .... 1.3 grams (.046 oz)
- Hardness .... Durometer 60, scale "A"

- Fatal injuries are possible at employment distances of less than 32.8 feet (10 m)
- At distances of 10 to 30 meters, target area should be center mass of group of individuals
- Round is most effective when used to disperse groups of individuals

Figure B-16. M1029, 40-mm Nonlethal Crowd Dispersal Grenade

L96A1*, Grenade, Discharger: Antiriot, Irritant, CS

Using Services: U.S. Army, USMC

Status: Materiel Release, 4 January 2002

DODIC: FZ14

NSN: 1330-01-459-4018

Acquisition Activity: Acquisition Center, Picatinny Arsenal, NJ

Further Information:

Commercial: Picatinny Arsenal NL Systems Integrator: (973) 724-6948
Project Officer: (410) 436-2139

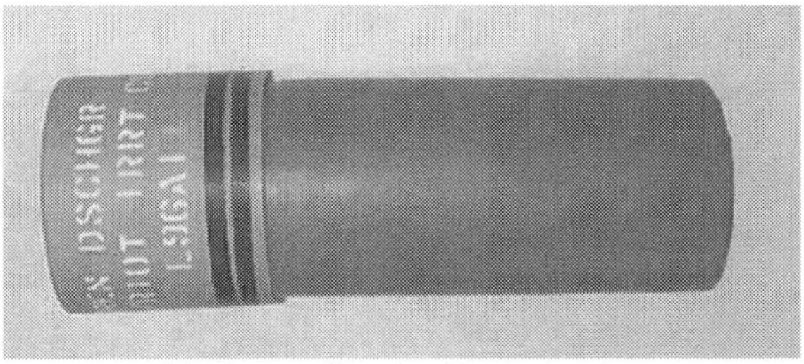

Color ....Light gray with red and brown bands
Diameter ....66 millimeters
Projectile ....23 canisters filled with CS compound
Length ....185 millimeters (7.28 in)
Total Weight..........................568 grams (1.25 lbs)
Propellant Weight................ 3.8 grams (.008 lbs)
Muzzle Velocity..................... 35.8 meters/second (117.5 ft/s)
Acceptance Accuracy..........65 to 95 meters
Deployment Range...............65 to 95 meters
Projectile Specification:

NOTE: It is not accurate to refer to the submunitions of the L96A1 grenade as projectiles. The antiriot material (CS) is released from the submunitions as they burn on the ground.

Shape...............................Cylindrical
Material...........................Aluminum
Weight..............................11.3 grams (.025 lbs)

* The L97A1 Grenade is the training grenade for the L96A1 Grenade.

Figure B-17. L96A1, Grenade, Discharger: Antiriot, Irritant, CS

L97A1, Grenade, Discharger: Antiriot, Practice

Using Services: U.S. Army, USMC

Status: Material Released, 4 January 02

DODIC: FZ15

NSN: 1330-01-459-4032

Acquisition Activity: Acquisition Center, Picatinny Arsenal, NJ

Further Information:

Commercial: Picatinny Arsenal NL Systems Integrator: (973) 724-6948 Project Officer: (410) 436-2139)

Color .... Blue with brown and green bands
Diameter .... 66 millimeters
Projectile .... 23 canisters filled with Cinnamic Acid (CA)
Length .... 185 millimeters (7.28 in)
Total Weight.......................... 568 grams (1.25 lbs)
Propellant Weight................. 3.8 grams (.008 lbs)
Muzzle Velocity..................... 35.8 meters/second (117.5 ft/s)
Acceptance Accuracy.......... 65 to 95 meters
Deployment Range............... 65 to 95 meters
Projectile Specification:

NOTE: It is not accurate to refer to the submunitions of the L97A1 grenade as projectiles. The antiriot, practice material (CA) is released from the submunitions as they burn on the ground.

Shape............................... Cylindrical
Material .... Aluminum
Weight............................. 11.3 gram (.025 lbs)

- CA smoke can cause irritation to the eyes, mucous membranes, and skin. Personnel entering a CA cloud must wear appropriate face protection such as an M17- or M40-series protective mask.
- The burning grenade canisters could start a fire in dry grasses. Keep fire extinguisher available.
- During firing training, personnel within 16 meters must wear single hearing protection. This includes personnel within vehicle.
- Do not fire grenades when personnel or equipment are within 150 meters covering a 90-degree arc around a firing discharger.
- Do not fire grenades at the 50- or 75-meter position on the M315 adjustable aiming bracket. Fire only at the 100-meter setting.

Figure B-18. L97A1, Grenade, Discharger: Antiriot, Practice

Grenade, Launcher: Nonlethal, Distraction, M98

Using Services: U.S. Army, USMC

Status: Type Classified, 17 October 2001

DODIC: FZ16

NSN: 1330-01-484-7773

Acquisition Activity: Acquisition Center, Picatinny Arsenal, NJ

Further Information: Commercial: Picatinny Arsenal
NL Systems Integrator:
(973) 724-6948
Project Officer: (410) 436-2139

Color .... Green and black with brown band
Diameter .... 66 millimeters
Projectile .... 3 bursting submunitions, with pyrotechnic charges for audio and visual stimuli
Length .... 25.25 centimeters (9.94 in)
Total Weight .... 725.7 grams (1.6 lbs)
Propellant Weight .... 5.0 grams (.011 lbs)
Muzzle Velocity .... 48.9 meters/seconds (160.4 ft/s)
Acceptance Accuracy .... 100 to 120 meters
Deployment Radius .... 10 to 15 meters
Projectile Specification:

NOTE: It is not accurate to refer to the submunitons of the M98 grenade as projectiles. These submunitions are ground bursting and the payload of inert ballast is just for flight stabilization and to deliver the audio/visual stimuli to the required range.

Filler .... Inert Ballast

- There is a remote chance of starting a fire. Keep fire extinguisher available.
- Personnel within 17 meters of a firing discharger must wear single hearing protection. This includes personnel within vehicle.
- Avoid looking directly at a bursting grenade; it could result in temporary loss of vision, lasting 2 to 5 minutes.
- Personnel within 35 meters of an operating launcher or 0.5 meters of a bursting canister should wear safety or ballistic-type eye protection, military clothing with long sleeves and elevated shirt collar, and helmet headgear.
- Do not handle or load M98s within 50 meters of a frequency transmitter emitting at 175 MHz, 448 MHz, 503 MHz and 1430 MHz.
- Do Not fire grenades when personnel or equipment are within 160 meters covering a 90-degree arc around a firing discharger.

Figure B-19. Grenade, Launcher: Nonlethal, Distraction, M98

M99, Grenade, Launcher: Nonlethal, Blunt Trauma

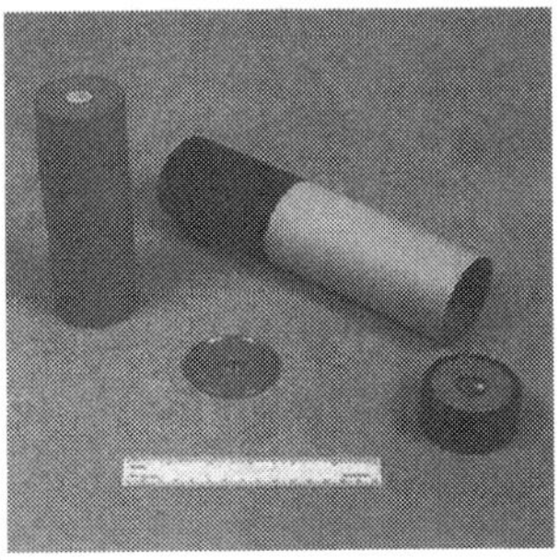

Using Services: U.S. Army, USMC

Status: Type Classified—17 October 2001

DODIC: FZ17

NSN: 1330-01-484-7775

Acquisition Activity: Acquisition Center, Picatinny Arsenal, NJ

Further Information:

Commercial: Picatinny Arsenal NL Systems Integrator: (973) 724-6948
Project Officer: (410-436-2139)

Color....................................Green and black with brown band
Diameter.................................66 millimeters
Projectile...............................3 bursting submunitions, each with 140 balls
Length...................................25.25 centimeters (9.94 in)
Total Weight.............................725.7 grams (1.6 lbs)
Propellant Weight........................5.0 grams (.011 lbs)
Muzzle Velocity..........................48.9 meters/second (160.4 ft/s)
Acceptance Accuracy......................100 to 120 meters
Deployment Radius........................5 to 10 meters
Projectile Specification:

NOTE: It is not accurate to refer to the submunitions of the M99 grenade as projectiles. These submunitions are ground bursting and the payload of inert ballast is just for flight stabilization and to deliver the audio/visual stimuli to the required range.

Shape.....................................32-caliber (5/16 in) ball
Material..................................PVC
Weight....................................0.40 grams (.014 oz)
Hardness..................................90 Durometer

- There is a remote chance of starting a fire. Keep fire extinguisher available.
- Personnel within 17 meters of a firing discharger must wear single hearing protection. This includes personnel within vehicle.
- Avoid looking directly at a bursting grenade; it could result in temporary loss of vision, lasting 2 to 5 minutes
- Personnel within 35 meters of an operating launcher or 0.5 meters of a bursting canister should wear safety or ballistic-type eye protection, military clothing with long sleeves and elevated shirt collar, and helmet headgear.
- Do not handle or load M98s within 50 meters of a frequency transmitter emitting at 175 MHz, 448 MHz, 503 MHz and 1430 MHz.
- DO NOT fire grenades when personnel or equipment are within 160 meters covering a 90-degree arc around a firing discharger.
- To minimize risk of eye injury, use the M99 primarily against crowds of adults.

Figure B-20. M99, Grenade, Launcher: Nonlethal, Blunt Trauma

Appendix C

Interserviee Nonlethal Individual Weapons Instructor Course

1. Introduction

This appendix provides an overview of the Interservice Nonlethal Individual Weapons Instructor Course (INIWIC). The INIWIC is a "train-the-trainer" course taught in 15 subcourses and divided into four sections. Section I provides INIWIC descriptive data that includes the purpose, scope, peacetime and mobilization training requirements, student prerequisites, and equipment/ammunition requirements necessary to successfully complete the course. Section II lists the total training days and hours required per academic subcourse. Section III gives a detailed description of each subcourse and terminal learning objectives that each student must meet to complete the course. Section IV is a listing of 15 annexes that comprise the INIWIC Program of Instruction (POI). Access the entire INIWIC, with history, updates, and lesson plans at the following Internet address:
http://www.wood.army.mil/mcdet/niwic/niwic.htm.

2. Course Descriptive Data

a. Course Title: Interservice Nonlethal Individual Weapons Instructor Course

b. Location: Marine Corps Detachment, Fort Leonard Wood, Missouri

c. Marine Corps service school code: A16H5A3

d. Other service course number: USAF—L5AZA3P071019

e. Military assistance program articles and service list number: P166810

f. Purpose: To certify selected DOD/DOT service persons as nonlethal individual weapons instructors.

g. Scope: This is a train-the-trainer course designed for E-4s and above. The objective of this course is to prepare students to train individuals and groups in the utilization and employment of NLW and to serve as the unit NLW subject matter expert. This course is designed for all occupational specialties.

h. Length: 10 training days

i. Curriculum Breakdown: 100 total hours.

(1) 2.00 Demonstration.

(2) 24.00 Demonstration/practical exercise.

(3) 22.50 Lecture.

(4) 5.00 Lecture/demonstration.

(5) 4.00 Live-fire exercise.

(6) 26.50 Practical application.

(7) 6.00 Performance exam.

(8) 4.00 Written exam.

(9) 6.00 Administrative time.

j. Maximum Class Capacity: 30.

k. Optimum Class Capacity: 25.

l. Minimum Class Capacity: 15.

m. Class Frequency: 12.

n. Prerequisites: Students must be selected by commanders to be nonlethal weapons instructors and be an E-4 or above.

o. MOS Received: None.

p. Quota Control: Formal Schools Branch, Training and Education Command, Quantico, Virginia.

q. Funding: Service or command funded.

r. Reporting Instructions: All students report with orders to the Fort Leonard Wood, Missouri, Billeting Office, Bldg # 470 no later than 2400 on the day prior to class convening. Government quarters are available; however, messing is not.

s. Instructor Staffing: Two USMC instructors and two U.S. Army instructors.

t. School Overhead: Course OIC and chief instructor

Training Support: Training support is provided by the USMC for all common training support items. Service-specific training items are provided by the respective service.

Table C-1. Required Equipment For INIWIC

Item	*Quantity*
Personnel Protectors:	
Non-Ballistic Riot Face Shield	40
Non-Ballistic Full Length Riot Shield	40
Non-Ballistic Shin Guards	40
Ballistic Riot Face Shield	20
Ballistic Full Length Riot Shield	20
Ballistic Shin Guards	20
Kevlar Helmets	40
Flak Jacket/Body Armor	40
Personnel Effectors:	
Disposable Handcuffs (Flex Cuffs)	300
12-Gauge Area (rubber ball)	1400
12-Gauge Beanbag	1400
12-Gauge Fin Stabilized	1400
12-Gauge Flash-bang	1400
12-Gauge OO Buck	xxxx
12-Gauge 7 1/2 shot	xxxx
12-Gauge Launching Cup Cartridge	1400
40-mm Area (rubber ball)	1400
40-mm Foam Baton (3 foam sub-munitions)	1400
40-mm Sponge Grenade	1400
MK141 Diversionary Grenade	600

Table C-1. Required Equipment For INIWIC

Item	*Quantity*
M84 Reloadable Diversionary Grenade	600
Sting Ball Hand Thrown or Shot launched Grenade	600
Mossberg 500A2 Shotgun Launch Cup	15
M203 Grenade Launcher	15
Mossberg 500 3" Shotgun	15
36" Riot Oak Baton w/Belt Ring	40
Monadnock® 24-36" Expandable Baton w/holster	40
M36 RCA Dispenser	40
Mk-4 Personal OC Dispenser	40
Mk-9 Team OC Dispenser	40
Mk-46 High Capacity High Output OC Dispenser	20
MK-46 Hardware Kit (Refill Kit)	10
M33A1 Squad Riot Control Agent Dispenser	10
Modular Crowd Control Munitions Ground Emplaced (MCCM-GE)	20
66-mm Vehicle Launched Nonlethal Grenade (VLNLG)	3
66-mm NL munitions	30
Mission Enhancers:	
Portable Bullhorn	5
40-mm Carrying Pouch	40
Diversionary Rubber Ball Grenade Pouch	40
12-Gauge Butt Cuff carriers	40
12-Gauge Utility 25-Round Pouch	40
Mk-4 Individual Carrying Pouch	40
Portable Public Address System w/Microphone	3
12-Volt Hand-Held Spotlight	5
Night Hunter High Intensity Light System w/Kit	5
12-Gauge High Intensity Light Set	5
Caltrops	5
Road Side Spike Strip	5
Portable Vehicle Arrest Barrier (PVAB)	3
Training Devices:	
Impact Training Suit	6
Striking Bags	15
Training Batons (24 in)	40
Training Batons (36 in)	40
Practice Grenade Body	40
Practice Grenade Fuse	540
• OPFOR Training Munitions (foreign objects)	xxxx
• Target Requirements (full or partial torso)	40
MK-4 Personal OC Dispenser (training)	300
MK-9 Team OC Dispenser (training)	300
Training Disposable Handcuffs (flex cuffs)	100
12-Gauge Dummy Rounds	100

Table C-2. INIWIC Class V Requirements

Type	*#Per Student*	*# Demo Rounds*	*Total Per Class*	*Total Per Year*
12-Gauge Area (rubber ball)	5	5	155	1400
12-Gauge Beanbag	5	5	155	1400
12-Gauge Fin Stabilized	5	5	155	1400
40-mm Area (rubber ball)	5	5	155	1400
40-mm Foam Baton	5	5	155	1400
Sting Ball Hand Thrown or Shotgun Launched Grenades	2	2	62	600
MK141 Diversion Devices	2	2	62	600

3. Training Days Summary

This course consists of 10 training days. The curriculum is broken down into three categories.

Table C-3. INIWIC Curriculum

NLW Tactics	***Hours***
Force Continuum/ROE	2.00
Crowd Dynamics/Control/Formations	14.00
Communication Skills	2.00
MOOTW	2.00
Tactical Considerations	2.00
	22.00
NLW Systems	***Hours***
Oleoresin Capsicum	4.00
Empty Hand Control Techniques	19.00
Riot Baton Techniques	8.00
Nonlethal Munitions and Equipment	8.00
Portable Vehicle Barrier	5.00
66-mm Grenade Launcher	2.00
Modular Crowd Control Munition	2.00
	48.00
Nonlethal Weapons Instructor	***Hours***
Techniques of Military Instruction	10.00
NLW Scenario Exercise	10.00
	20.00
Subtotal =	**90.00**
Examinations =	***4.00***
Administration =	***6.00***
Total Hours =	***100.00***

4. Course Body

The course body consists of 15 subcourses. A quick synopsis of each subcourse is included below.

a. **Instructor Development**. This subcourse teaches and reinforces basic military instructor methodology. It identifies characteristics of good instructors and four ways they can improve their teaching ability. Additionally, it illustrates rules of conducting military instruction as well as the steps in planning a lesson and teaching classes.

b. **Force Continuum**. This subcourse introduces the five levels of resistance as well as identifying the levels of the force continuum. It illustrates influencing the use of force decisions and the definitions relating to deadly force. Lastly, it identifies the three preconditions for deadly force and seven authorized occasions for use of deadly force.

c. **Crowd Dynamics/Crowd Control**. This subcourse outlines the differences between crowds, mobs, and riots, and teaches the student basic crowd control techniques to apply in various situations. Upon completion, the student will be able to instruct others in both the behavioral aspects of a crowd during a civil disturbance and also various techniques of controlling a civil disturbance. It will identify factors that affect both individual and crowd behavior and it identifies all crowd control equipment needed, or essential, for mission accomplishment.

d. **Empty Hand Control Techniques**. This subcourse identifies safety factors to consider when instructing empty hand control techniques by illustrating the four Ms (mind, mass, movement, muscle) of self-defense, psychomotor skills, and phases of psychomotor skills. It illustrates levels of training and identifies common pressure points of the human body. It identifies hinges, appropriate angles of approach, and recommended patterns of movement. Instructors will learn blocking techniques, striking skills, control holds, escort positions, and restraint devices.

e. **Baton Techniques**. This subcourse identifies riot baton nomenclature and safety standards when conducting riot baton training. It identifies methods of teaching psychomotor skills and the phases of teaching psychomotor skills. It highlights baton-striking techniques, blocking techniques, and retention techniques.

f. **Military Operations Other Than War (MOOTW)**. This subcourse defines operations other than war and the difference between MOOTW and declarations of war. It helps identify the primacy of political objectives and the strategic aspects of MOOTW. It illustrates the range of operations and the principles of operations, as well as the types of military operations other than war.

g. **Rules of Engagement (ROE)**. This subcourse defines ROE and specific terms that apply to these rules. It also defines the purpose of ROE and provides the student an understanding of policies underlying the need for ROE. It identifies three sources for ROE and the two parts of ROE that will not change. It explains how ROE relate to using and applying NLW.

h. **Oleoresin Capsicum (OC) Training**. This subcourse explains how the three levels of OC are created, as well as the nomenclatures for all aerosol canisters. It highlights the different spray patterns, methods of delivery, and cautions concerning the traumatic hydraulic needle effect. It explains employment considerations and the three levels of OC contamination. Additionally, it explains the physical and mental effects of OC and trains students on different grips, draw methods, and proper stances when employing OC. It explains procedures for decontamination, recovery, and first aid.

i. **Types of Formations**. This subcourse explains the elements and individual responsibilities within a riot control formation. Additionally, it trains proper grip, balance, holding, and retention of the riot control shield. It highlights leadership requirements by teaching different commands and cadence for conducting riot control formations. It illustrates the different types of formations and the support elements of a riot control operation

j. **Tactical Considerations**. This subcourse addresses the preparations necessary for crowd control operations and the graduated response of a crowd control force. It identifies support and information required from host nation police and civil authorities. It describes the employment measures of military forces in crowd control and MOOTW environments. It highlights the actions, optional measures, and guidance for using NLW in crowd control and MOOTW environments.

k. **Nonlethal Munitions and Employment**. This subcourse identifies the types of nonlethal munitions and their intended purpose. It illustrates how nonlethal munitions fall within established "use of force" continuums. It identifies the types, characteristics, and ranges of all currently employed nonlethal munitions.

l. **Communications Skills**. This subcourse identifies the areas of nonverbal communication employed to determine an individual's state of mind. This, along with an understanding of the importance of personal space, gives the student the tools to resolve conflict with the amount of force as applied in the force continuum.

m. **Portable Vehicle Arrest Barrier**. This subcourse teaches the employment and maintenance of a force protection asset assigned to the nonlethal capability set. The PVAB is designed to stop a 7,500-pound vehicle going 45 mph. The PVAB protects gates and check points.

n. **Modular Crowd Control Munition**. This subcourse teaches the employment of the MCCM; the nonlethal version of the claymore mine. The MCCM is a force protection asset designed to help protect static defensive positions or check points.

o. **M-315 66-mm Grenade Launcher**. This subcourse teaches the installation, employment, and maintenance of the M-315. The M-315 increases the current maximum range of nonlethal sting ball munitions from approximately 50 meters to 100 meters. Additionally, this weapons platform can deploy flash bangs, smoke, and CS munitions.

Appendix D

Sample: Unit Training Schedules/Firing Tables

Nonlethal Individual Weapons Training Course Training Schedule: Unit Training Schedule Number 1.

Training Day #1					
Time	***Subject***	***Location***	***Uniform***	***Instr***	***Ref/Notes***
0800	Administration Time	Classroom	U1		
0830	Force Continuum	Classroom	U1		
1000	ROE	Classroom	U1		
1200	Chow	Mess Hall	U1		
1300	MOOTW	Classroom	U1		
1500	Communication Skills	Classroom	U1		
1700	Secure/COB				

Administration:

Uniforms:
U1: BDU/utility uniform
U2: BDU/utility uniform with LBE/harness and canteens
U3: BDU/utility uniform with LBE/harness, canteens, flak jacket, and helmet
U4: BDU/utility uniform with LBE/harness, canteens, flak jacket, helmet, and assigned weapon

Transportation:
Weapons:
Ammunition:
Required Equipment:
Screen
Video projector with central processing unit
Televisions
VCR
Miscellaneous:
All personnel are required to have a notebook, pencil, and pen.
Handouts prepared and passed out.

Training Day #2					
Time	*Subject*	*Location*	*Uniform*	*Instr*	*Ref/Notes*
0800	Nonlethal Equipment Brief	Classroom	U1		
0900	Pressure Point Control Techniques/ Open Hand Control Techniques	Classroom	U1		
1000	Pressure Point Control Techniques/ Open Hand Control Techniques (Practical Application)	PT Field	U2		
1200	Chow				
1300	Pressure Point Control Techniques/ Open Hand Control Techniques (Practical Application)	PT Field	U2		
1700	Debrief/Secure/COB				

Administration:

Uniforms:
U1: BDU/utility uniform
U2: BDU/utility uniform with LBE/harness and canteens
U3: BDU/utility uniform with LBE/harness, canteens, flak jacket, and helmet
U4: BDU/utility uniform with LBE/harness, canteens, flak jacket, helmet, and assigned weapon

Transportation:

Weapons:

Ammunition:

Required Equipment:
Screen
Video projector with central processing unit
Flexicuffs (one set of cuffs per two students)
Riot Training Bags (one bag per two students)

Miscellaneous:
All personnel are required to have a notebook, pencil, and pen.
Handouts prepared and passed out.
Medical personnel on station.

Training Day #3					
Time	*Subject*	*Location*	*Uniform*	*Instr*	*Ref/Notes*
0800	Pressure Point Control	Classroom	U1		
	Techniques/Open Hand Control Techniques	Classroom	U1		
0900	Pressure Point Control Techniques/Open Hand Control Techniques (Practical Application)	PT Field	U1		
1200	Chow				
1300	Pressure Point Control Techniques/Open Hand Control Techniques (Practical Application)	PT Field	U2		
1700	Debrief/Secure/COB				

Administration:

Uniforms:
U1: BDU/utility uniform
U2: BDU/utility uniform with LBE/harness and canteens
U3: BDU/utility uniform with LBE/harness, canteens, flak jacket, and helmet
U4: BDU/utility uniform with LBE/harness, canteens, flak jacket, helmet, and assigned weapon

Transportation:

Weapons:

Ammunition:

Required Equipment:
Screen
Video projector with central processing unit
Flex cuffs (one set of cuffs per 2 students)
Riot Training Bags (one bag per 2 students)

Miscellaneous:
All personnel are required to have a notebook, pencil, and pen.
Handouts prepared and passed out.
Medical personnel on station.

Training Day #4					
Time	*Subject*	*Location*	*Uniform*	*Instr*	*Ref/Notes*
0800	Impact Weapons	Classroom	U1		
0900	Impact Weapons (Practical Application)	PT Field	U2		
1200	Chow				
1300	Impact Weapons (Practical Application)	PT Field	U2		
1700	Debrief/Secure/COB				

Administration:

Uniforms:
U1: BDU/utility uniform
U2: BDU/utility uniform with LBE/harness and canteens
U3: BDU/utility uniform with LBE/harness, canteens, flak jacket, and helmet
U4: BDU/utility uniform with LBE/harness, canteens, flak jacket, helmet, and assigned weapon

Transportation:

Weapons:
Baton (one baton per 2 students)
Training baton (one baton per 2 students)

Ammunition:

Required Equipment:
Screen
Video projector with central processing unit
Flex cuffs (one set of cuffs per 2 students)
Riot training bags (one bag per 2 students)
Three (3) riot training suits

Miscellaneous:
All personnel are required to have a notebook, pencil, and pen.
Handouts prepared and passed out.
Medical personnel on station.

Training Day #5					
Time	***Subject***	***Location***	***Uniform***	***Instr***	***Ref/Notes***
0800	Impact Weapons (Practical Application)	PT Field	U2		
1000	Crowd Dynamics/Crowd Control	Classroom	U1		
1130	Chow				
1230	Crowd Dynamics/Crowd Control	Classroom	U1		
1330	Riot Control Formations	Classroom	U1		
1430	Oleoresin Capsicum (OC) Aerosol Training	Classroom	U1		
1630	Debrief/Secure/COB				

Administration:

Uniforms:
U1: BDU/utility uniform
U2: BDU/utility uniform with LBE/harness and canteens
U3: BDU/utility uniform with LBE/harness, canteens, flak jacket, and helmet
U4: BDU/utility uniform with LBE/harness, canteens, flak jacket, helmet, and assigned weapon

Transportation:

Weapons:
Training baton (one baton per 2 students)
Baton (one baton per 2 students)

Ammunition:
Inert OC for classroom instruction

Required Equipment:
Riot training bags (one bag per 2 students)
Screen
Video projector with central processing unit
Three (3) riot training suits

Miscellaneous:
All personnel are required to have a notebook, pencil, and pen.
Handouts prepared and passed out.

Training Day #6					
Time	***Subject***	***Location***	***Uniform***	***Instr***	***Ref/Notes***
0800	OC Aerosol Training (Practical Application)	PT Field	U2		
1200	Chow				
1300	OC Aerosol Training (Practical Application)	PT Field	U2		
TBD	Debrief/Secure/COB				

Administration:

Uniforms:
U1: BDU/utility uniform
U2: BDU/utility uniform with LBE/harness and canteens
U3: BDU/utility uniform with LBE/harness, canteens, flak jacket, and helmet
U4: BDU/utility uniform with LBE/harness, canteens, flak jacket, helmet, and assigned weapon

Transportation:

Weapons:

Ammunition:
MK 4.inert OC (one per student)
MK 4 individual can OC (2 cans)

Required Equipment:
Five (5) large trash cans with liners
Two (2) water hoses with running water

Miscellaneous:
Medical personnel on station.

Training Day #7					
Time	*Subject*	*Location*	*Uniform*	*Instr*	*Ref/Notes*
0730	Movement to Range		U4		
0800	Nonlethal Munitions and Employment	Range Classroom	U1		
1030	Chow	Range			
1100	Safety Brief/Range Brief	Range	U4		
1130	Nonlethal Munitions (Live Fire)	Range	U4		
TBD	Evening Chow				
TBD	Safety Brief/Range Brief	Range	U4		
1800	Night Exercise Sun Set +30)	Range	U4		
TBD	Debrief/Clean Up	Range	U1		
TBD	Movement to Rear				
TBD	Secure/ COB				

Administration:

Uniforms:
U1: BDU/utility uniform
U2: BDU/utility uniform with LBE/harness and canteens
U3: BDU/utility uniform with LBE/harness, canteens, flak jacket, and helmet
U4: BDU/utility uniform with LBE/harness, canteens, flak jacket, helmet, and assigned weapon

Transportation: Yes

Weapons:
Assigned Weapons

Ammunition:
12-gauge rubber ball
12-gauge fin stabilized
12-gauge bean bag
12-gauge launching cup cartridge
40-mm foam baton
40-mm rubber ball
40-mm sponge grenade
Stingball grenades

Required Equipment:
Eight (8) 12-Gauge stinger launcher cup
Fifteen (15) targets
Eye and hearing protection
Fifteen (15) full-length riot shields
All students require munitions pouches, shin guards, face shields
Two (2) MRE per student
Video projector with central processing unit
Three (3) high-intensity light
Thirty (30) individual tactical lights

Miscellaneous:
Range safety officer.
Medical personnel on station.
Secondary means of communication (radios).
Conduct operational risk assessment.

Training Day #8					
Time	***Subject***	***Location***	***Uniform***	***Instr***	***Ref/Notes***
0700	Movement to Training Area #1		U2		
0730	Portable Vehicle Arresting Barrier (PVAB)	TA#1	U1		
0900	PVAB (Practical Application)	TA#1	U2		
1200	Chow				
1300	PVAB (Practical Application)	TA#1	U2		
1500	Debrief/Movement to Rear				
1530	Preventive Maintenance of PVAB				
1600	Secure/COB				

Administration:

Uniforms:
U1: BDU/utility uniform
U2: BDU/utility uniform with LBE/harness and canteens
U3: BDU/utility uniform with LBE/harness, canteens, flak jacket, and helmet
U4: BDU/utility uniform with LBE/harness, canteens, flak jacket, helmet, and assigned weapon

Transportation: Yes

Weapons:

Ammunition:

Required Equipment:
One (1) portable vehicle-arresting barrier (PVAB)
Video projector with central processing unit

Miscellaneous:
(1) MRE per student.

Training Day #9					
Time	***Subject***	***Location***	***Uniform***	***Instr***	***Ref/Notes***
0730	Movement to Range		U4		
0800	Modular Crowd Control Munition-Ground Emplaced (MCCM-GE)	Range Classroom	U1		
0900	Safety Brief/Range Brief	Range	U4		
0930	MCCM-GE (Practical Application	Range	U4		
1030	Chow				
1130	Vehicle Mounted Nonlethal Grenade Launcher (VMNLGL)	Range Classroom	U4		
1300	VMNGL	Range	U4		
1730	Debrief/Movement to Rear				
1800	Secure/COB				

Administration:

Uniforms:
U1: BDU/utility uniform
U2: BDU/utility uniform with LBE/harness and canteens
U3: BDU/utility uniform with LBE/harness, canteens, flak jacket, and helmet
U4: BDU/utility uniform with LBE/harness, canteens, flak jacket, helmet, and assigned weapon

Transportation: Yes

Weapons:

Vehicle Mounted Nonlethal Grenade Launcher (VMNLGL)

Ammunition:

(2) MCCM-GE, (2) 66-mm Riot Control Training Grenade (L97E1)

Required Equipment:
Eye and ear protection
Video projector with central processing unit

Miscellaneous:
Range safety officer.
Medical personnel on station.
Secondary means of communication.
Two (2) MRE per student.

Training Day #10					
Time	***Subject***	***Location***	***Uniform***	***Instr***	***Ref/Notes***
0700	Movement Down Range				U4
0730	Tactical Considerations	MOUT Classroom		U1	
0930	Safety Brief/Range Brief	MOUT Facility		U4	
1000	Tactical Exercise* (Practical Application)	MOUT Facility		U4	
1200	Chow				
1300	Tactical Exercise (Practical Application)	MOUT Facility		U4	
1700	Debrief/Movement to Rear				
1730	Weapons Maintenance				
1800	Secure/COB				

Administration:

Uniforms:
U1: BDU/utility uniform
U2: BDU/utility uniform with LBE/harness and canteens
U3: BDU/utility uniform with LBE/harness, canteens, flak jacket, and helmet
U4: BDU/utility uniform with LBE/harness, canteens, flak jacket, helmet, and assigned weapon

Transportation: Yes

Weapons:
Assigned weapon
Inert OC individual (one per student)
Training batons (one per student)
Fifty (50) practice grenades

Ammunition:

Required Equipment:
Three (3) riot training suit
All students require munitions pouches, flex cuffs, shin guards, face shields
Video projector with central processing unit
Fifteen (15) full length riot shields

Miscellaneous:
Range safety officer.
Medical personnel present.
Secondary means of communication.
(2) MRE's per student.

UNITED STATES MARINE CORPS
Marine Corps Detachment
Fort Leonard Wood, MO 65473

1.	LESSON DESIGNATOR:	Interservice Nonlethal Individual Weapons Instructor Course (INWIC). 13A (INIWIC.13A)
2.	LESSON TITLE:	Scenario Practical Exercise
3.	SCHOOL/COURSE:	Nonlethal Individual Weapons Instructor Course
4.	STUDY ASSIGNMENT:	Review course curriculum
5.	ESTIMATED TIME:	8 hours
6.	INSTRUCTORS REQUIRED:	(1) primary, (1) alternate per ratio: 1:10
7.	INSTRUCTOR REFERENCES:	INIWIC POI
8.	DATE PREPARED:	September 2001
9.	PRESENTATION METHOD:	Practical Exercise
10.	TRAINING AIDS:	Warning Order and five paragraph operations order for exercise, rubber rifles/shotguns, batons, shields, Inert MK4 OC canister, bullhorn, radio
11.	STUDENT EQUIPMENT:	(1) 782 gear
		(1) Kevlar with face shield
		(1) Shin guards

5. Unit Training Schedule Number 2

Table D-1. NLW Training Course 5-Day Training Schedule

Day/date	*Time*	*Event*	*Location*	*Instructor*	*Notes*
Mon/07 Jan 02	0730-0745	Welcome Aboard	Bldg RR202	MSgt Dixon	
T-1	0745-0845	Force Continuum	SOTG	TBD	L
	0900-1015	MOAB	SOTG	TBD	L
	1030-1130	OC	SOTG	TBD	L
	1130-1300	Noon Meal	Mess Hall	Company GySgt	
	1300-1400	OC Practical Application	SOTG	TBD	PA
	1400-1700	OC Certification	SOTG	TBD	PA
Tues/08 Jan 02	0730-0830	ROE	SOTG	TBD	L
T-2	0845-0945	Crowd Control	SOTG	TBD	L
	1000-1130	MDTS	SOTG	TBD	L, PA
	1130-1300	Noon Meal	Mess Hall	Company GySgt	
	1300-1700	MDTS	SOTG	TBD	L, PA
Wed/09 Jan 02	0730-1130	Batons	SOTG	TBD	L, PA
T-3	1130-1300	Noon Meal	Mess Hall	Company GySgt	
	1300-1500	Batons	SOTG	TBD	L, PA
	1500-1700	Formations	SOTG	TBD	L
Thurs/10 Jan 02	0730-1130	Formations	SOTG	TBD	PA
T-4	1130-1300	Noon Meal	Mess Hall	Company GySgt	
	1300-1700	Nonlethal Munitions/shotguns/ M2O3/sustainment training	Square bay/M203	TBD	L, PA
Fri/11 Jan 02	0730-1130	Nonlethal Munitions/stinger ball grenades/shotgun/M2 03	Square Bay/M203	TBD	PA
T-5	1130-1300	Noon Meal	Mess Hall	Company GySgt	
	1300-1700	Nonlethal Munitions/stinger ball grenades/shotgun/M2 03	Nonlethal Range 1&2	TBD	PA

Notes:

L = Lecture
PA = Practical Application

MDTS = Manadnock Defensive Tactics System
MOAB = Management of Aggressive Behavior
SOTG = Special Operations Training Group (USMC training unit)

6. Sample Unit Training Schedule Number 3

Table D-2. NLW Training Course 10-Day Training Schedule

DAY/DATE	*TIME*	*EVENT*	*LOCATION*	*INSTRUCTOR*	*NOTES*
Mon/29 Oct 01 T-1	0730-0745	Welcome Aboard	Bldg RR-202	MSgt Dixon	
	0745-0845	Force Continuum	PT Field	Sgt Whitworth	L
	0900-1015	MOAB	PT Field	GySgt German	L
	1030-1130	OC	PT Field	Sgt Hoskins	L
	1130-1300	Noon Meal	Mess Hall	Company Gunny	
	1300-1400	OC Practical Application	PT Field	Sgt Hoskins	PA
	1400-1700	OC Certification	PT Field	Sgt Hoskins	PA
Tues/30 Oct 01 T-2	0730-0830	ROE	PT Field	Cpl Fewox	L
	0845-0945	Crowd Control	PT Field	SSgt Parrish	L
	1000-1130	MDTS	PT Field	Sgt Johnson	L, PA
	1130-1300	Noon Meal	Mess Hall	Company Gunny	
	1300-1700	MDTS	PT Field	Sgt Johnson	L, PA
Wed/31 Oct 01 T-3	0730-1130	Batons	PT Field	Sgt Johnson	L, PA
	1130-1300	Noon Meal	Mess Hall	Company Gunny	
	1300-1500	Batons	PT Field	Sgt Hoskins	L, PA
	1500-1700	Formations	PT Field	GySgt German	L
Thur/1 Nov 01 T-4	0730-1130	Formations	PT Field	Sgt Whitworth	PA
	1130-1300	Noon Meal	Mess Hall	Company Gunny	
	1300-1700	Nonlethal Munitions	Nonlethal Range 1&2	GySgt German	L, PA
Fri/2 Nov 01 T-5	0730-1130	Nonlethal Munitions	Nonlethal Range 1&2	SSgt Parrish	PA
	1130-1300	Noon Meal	Mess Hall	Company Gunny	
	1300-1700	Nonlethal Munitions	Nonlethal Range 1&2	SSgt Parrish	PA
Mon/5 Nov 01 T-6	0730-1130	Nonlethal Grenades Formation Fire	Nonlethal Range 1&2	GySgt German	L, PA
	1130-1300	Noon Meal	Mess Hall	Company Gunny	
	1130-1700	Formation Fire/Sustainment Training	Nonlethal Range 1&2	SSgt Blackler	PA
Tues/6 Nov 01 T-7	0730-0830	Introduction to MOOTW	PT Field	Maj Desy	L
	0830-1130	Vehicle Checkpoints/Vehicle Searches	PT Field	Sgt Whitworth	L, PA
	1130-1300	Noon Meal	Mess Hall	Company Gunny	
	1300-1500	Vehicle	PT Field	SSgt Parrish	PA

Table D-2. NLW Training Course 10-Day Training Schedule					
DAY/DATE	***TIME***	***EVENT***	***LOCATION***	***INSTRUCTOR***	***NOTES***
	1500-1700	Checkpoints/Vehicle Searches Roadblocks/Personnel Searches	PT Field	SSgt Blackler	L, PA
Wed/7 Nov 01	0730-1130	Roadblock/Personnel Searches	PT Field	SSgt Blackler	PA
T-8	1130-1300	Noon Meal	Mess Hall	Company Gunny	
	1300	Operations Order	Nonlethal Classroom	MSgt Dixon	Company Staff
Thurs/8 Nov 01	0730-1000	Confirmation Brief	PT Field	Maj Desy	
T-9	1000-1200	Insert	Combat Town	SSgt Parrish	
	1200-0000	STX	Combat Town	SSgt Parrish	
Fri/9 Nov 01	0001-1200	STX	Combat Town	SSgt Parrish	
T-10	1200-1300	Cleanup/Debrief	Combat Town	Company Gunny	
	1300	Graduation	Combat Town	SOTG OIC	

Notes:
L = Lecture
PA = Practical Application
MOAB = Management of Aggressive Behavior
MDTS = Manadnock Defensive Tactics System
STX = Situational Training Exercise

Appendix E

NLW Capability Sets

This appendix provides the current NLW capability sets for the services. These sets continue to be upgraded and changed according to service needs and the availability of new technologies. The U.S. Army and USMC capability sets are standard for the service, while the USAF, USN, and USCG are still developing and fielding their sets.

1. U.S. Army NLW Capability Set

Table E-1 U.S. Army Nonlethal Capability Set (Version FY 2001)

	Name	*Quantity*	*Unit Cost*	*Item Total*
GSAPMDK5X250AFHC	Non-Ballistic Riot Face Shields	200	$42.50	$8,500.00
GSAA TIPMBS9	Non-Ballistic Riot Body Shields Paulson	150	$88.50	$13,275.00
GSAATIUSMCSG	Non-Ballistic Riot Shin Guards	150	$60.00	$9,000.00
PT702MTUSA	Ballistic Face Shields	10	$246.00	$2,460.00
PT2237	Ballistic Body Shields	10	$1,052.00	$10,520.00
GSASF6SL	Sure Fire Field Light	10	$185.82	$1,858.20
MON2800 w/BCS-1	36" Hickory Riot Baton with Grommet	200	$15.50	$3,100.00
EADBH	Eagle Duty Baton Holder OD	200	$5.50	$1,100.00
GSAATIA2M	Portable Bullhorn	10	$82.00	$820.00
AAPB35W1	Ground Mount Bullhorn	8	$815.00	$6,520.00
AAPBWH10001	Megavox Wireless MIC	8	$0.00	$0.00
AASS450	Megavox Stand	8	$0.00	$0.00
GSASF6P	Individual High Intensity Light	15	$42.74	$641.10
GSAATIEATAVLFP6POD	Eagle Sure Fire 6P Holster	15	$8.75	$131.25
GSASFP60	Individual High Intensity Light Accessory Kit	15	$12.53	$187.95
XENNCSUSA	High Intensity Light	8	$2,795.00	$22,360.00
NIKFLEXCUFF	Disposable Forearm/Ankle Cuffs (Single)	1000	$0.70	$700.00
NIKCUTTER	Flex Cuff Cutters	2	$9.00	$18.00
GSAATIEAMK4	MK-4 Pouch OD	200	$6.75	$1,350.00
GSADEF5149	Inert Individual Training (OC) Dispenser	200	$6.07	$1,214.00
GSASF622	12-Gauge Shotgun High Intensity Light	50	$139.65	$6,982.50
GSAATIEAUPMS	12-Gauge Utility (25-Round) Pouch	50	$23.65	$1,182.50
GSAATISSSCQTR	12-Gauge Buttcuff Carrier (6-Round)	50	$10.75	$537.50
GSAATISASDD60D	Diversionary/Sting Ball Grenade Pouch	25	$47.25	$1,181.25
GSAATIEA40MM6	40-mm Carrying Pouch	100	$13.65	$$1,365.00
GSAATIRSC3	Caltrop	250	$10.50	$2,625.00
GSAATIFS333	Riot Training Suit with Accessories	4	$638.50	$2,554.00
GSAMONUTB	Riot Training Bag	10	$80.00	$800.00
BNCCELLS	C-Cell Batteries	0	$1.00	$0.00
BNAAACELLS	AA Batteries	0	$0.75	$0.00
GSASFDL123A	3V Lithium Batteries	0	$2.85	$0.00
BN9V	9V Batteries	0	$2.00	$0.00
ATIPMI	Packaging Marking and Integration Fee			$20,300.00
	Hardware Tools		**Item Total**	**$125,683.25**
	U.S. Army NLW Capability Set	5	**$125,683.25**	**$628,416.25**

2. USMC Nonlethal Weapons Capability Set

a. USMC NLW Company-Size Set

Table E-2 U.S. Marine Corps FY02 Company-Size Nonlethal Capability Set

TAMCN C41202E	***NSN 9999-01-464-2285***	***SET A (Company Size)***		
NSN/Product ID	***Name***	***Quantity***	***Unit cost***	***Item Total***
GSAATINCSPMDK5	Non-Ballistic USMC Riot Face Shields (S/R)	200	$42.50	$8,500.00
GSAATIPMBS9	Non-Ballistic Riot Body Shields Paulson	60	$88.50	$5,310.00
GSAATIUSMCSG	Non-Ballistic Riot Shin Guards	200	$60.00	$12,000.00
GSAATI23/36	USMC 23"/36" Riot Baton (Expandable) (w/Holster)	200	$73.25	$14,650.00
GSAATITRAINING BATON	USMC Riot Baton (Practice) (Price TBD)	200	$20.00	$4,000.00
GSAATIA2M	Portable Bullhorn	12	$82.00	$984.00
GSAATIDC/W	USMC Disposable Handcuffs	400	$0.94	$376.00
GSAATIDC/T	USMC Disposable Handcuff Training Device	100	$0.94	$94.00
GSAATIEAMK4	MK-4 Pouch OD	200	$6.75	$1,350.00
GSADEF5049	MK-4 Individual Live (OC) Dispenser (Pepper Spray)	200	$7.29	$1,458.00
GSADEF5099	MK-9 Live Squad (OC) Dispenser (Pepper Spray)	40	$26.20	$1,048.00
GSATIEAMK9	MK-9 Pouch OD	40	$11.00	$440.00
GSAATIDEF5846L	MK-46 Live (OC) (Horizontal) (Pepper Spray)	12	$300.00	$3,600.00
GSAATIDEF1515	MK-46 Sling	12	$30.25	$363.00
GSAATIDEF5520	MK-46 Hardware Kit	4	$68.50	$274.00
GSAATIDEF5546	MK-46 Refill Unit (OC) Live (Pepper Spray)	4	$46.25	$185.00
GSADEF5149	MK-4 Inert Individual Training (OC) Dispenser	200	$6.07	$1,214.00
GSADEF5199	MK-9 Inert Squad Training (OC) Dispenser	40	$20.10	$804.00
GSAATIDEF5946L	MK-46 Inert Training (OC) (Horizontal) (w/Dye)	12	$240.00	$2,880.00
GSADEF5546I	MK-46 Refill Unit (OC) Inert Training (w/Dye)	4	$27.50	$110.00
GSAATIFS333	Riot Training Suit w/Accessories (F.I.S.T.)	12	$638.50	$7,662.00
GSAMONUTB	Riot Training Strike Bag	24	$80.00	$1,920.00
GSAATIRSC3	Caltrop	400	$10.50	$4,200.00
TBD	Stinger Road Side Spike Kit (15.5') (Price TBD)	12		
XENNLCSUSMC	High Intensity Light (Xenonics)	6	$2,975.00	$17,850.00
GSAATIEATAVLFP6POD	Eagle Sure Fire 6P Holster	60	$8.75	$525.00
GSASF6P	Individual High Intensity Light	60	$42.74	$2,564.40
GSASFP60	Individual High Intensity Light Accessory Kit	30	$12.53	$375.90
GSASFDL123A	3V Lithium Batteries	120	$2.85	$342.00
GSAATISSSCQTR	12-Gauge Buttstock Cuff	36	$10.75	$387.00
GSAATIEAUPMS	12-Gauge Utility (25-Round) Pouch	72	$23.65	$1,702.80
GSAATIEA40MM6	40-mm Carrying Pouch	36	$13.65	$491.40
GSAATISASDD60D	Diversionary/Sting Ball Grenade Pouch	60	$47.25	$2,835.00
CSI9512	12-Gauge Launching Cup (Availability and Price TBD)	36		
1005-01-371-5639/ 1005-01-251-8578	12-Gauge Mossberg 500A2 / 590 Shotgun	36	$250.00	$9,000.00
ATIPMI	Packaging Marking and Integration Fee			$11,578.00
	Hardware Totals		**Item Total**	**$121,073.50**

b. USMC NLW Platoon-Size Set

Table E-3 USMC FY02 NLW Capability Set—Platoon Size

TAMCN C41202E	***NSN 9999-01-464-2285***	***SET B (Platoon Size)***		
NSN/Product ID	***NAME***	***Quantity***	***Unit Cost***	***Item Total***
GSAATINCSPMDK5	**Non-Ballistic USMC Riot Face Shields (S/R)	50	$42.50	$2,125.00
GSAATIPMBS9	Non-Ballistic Riot Body Shields Paulson	15	$88.50	$1,327.50
GSAATIUSMCSG	Non-Ballistic Riot Shin Guards	50	$60.00	$3,000.00
GSAATI23/36	USMC 23"/36" Riot Baton (Expandable) (w/Holster)	50	$73.25	$3,662.50
GSAATITRAINING BATON	USMC Riot Baton (Practice) (Price TBD)	50	$20.00	$1,000.00
GSAATIA2M	Portable Bullhorn	3	$82.00	$246.00
GSAATIDC/W	USMC Disposable Handcuffs	100	$0.94	$94.00
GSAATIDC/T	USMC Disposable Handcuff Training Device	25	$0.94	$23.50
GSAATIEAMK4	MK-4 Pouch OD	50	$6.75	$337.50
GSADEF5049	MK-4 Individual Live (OC) Dispenser (Pepper Spray)	50	$7.29	$364.50
GSADEF5099	MK-9 Live Squad (OC) Dispenser (Pepper Spray)	10	$26.20	$262.00
GSATIEAMK9	MK-9 Pouch OD	10	$11.00	$110.00
GSAATIDEF5846L	MK-46 Live (OC) (Horizontal) (Pepper Spray)	3	$300.00	$900.00
GSAATIDEF1515	MK-46 Sling	3	$30.25	$90.75
GSAATIDEF5520	MK-46 Hardware Kit	2	$68.50	$137.00
GSAATIDEF5546	MK-46 Refill Unit (OC) Live (Pepper Spray)	2	$46.25	$92.50
GSADEF5149	MK-4 Inert Individual Training (OC) Dispenser	50	$6.07	$303.50
GSADEF5199	MK-9 Inert Squad Training (OC) Dispenser	10	$20.10	$201.00
GSAATIDEF5946L	MK-46 Inert Training (OC) (Horizontal) (w/Dye)	3	$240.00	$720.00
GSADEF5546I	MK-46 Refill Unit (OC) Inert Training (W/Dye)	2	$27.50	$55.00
GSAATIFS333	Riot Training Suit W/Accessories (F.I.S.T.)	3	$638.50	$1,915.50
GSAMONUTB	Riot Training Strike Bag	3	$80.00	$240.00
GSAATIRSC3	Caltrop	50	$10.50	$525.00
TBD	Stinger Road Side Spike Kit (15.5') (Price TBD)	3		
XENNLCSUSMC	High Intensity Light (Xenonics)	4	$2,975.00	$11,900.00
GSAATIEATAVLFP6POD	Eagle Sure Fire 6P Holster	50	$8.75	$437.50
GSASF6P	Individual High Intensity Light	50	$42.74	$2,137.00
GSASFP60	Individual High Intensity Light Accessory Kit	25	$12.53	$313.25
GSASFDL123A	3V Lithium Batteries	100	$2.85	$285.00
GSAATISSSCQTR	12-Gauge Buttstock Cuff	15	$10.75	$161.25
GSAATIEAUPMS	12-Gauge Utility (25 Round) Pouch	30	$23.65	$709.50
GSAATIEA40MM6	40-mm Carrying Pouch	50	$13.65	$682.50
GSAATISASDD60D	Diversionary/Sting Ball Grenade Pouch	50	$47.25	$2,362.50
CSI9512	12-Gauge Launching Cup (Availability and Price TBD)	15		
1005-01-371-5639/ 1005-01-251-8578	12-Gauge Mossberg 500A2 / 590 Shotgun	15	$250.00	$3,750.00
ATIPMI	Packaging Marking and Integration Fee			$3,000.00
	Hardware Totals		**Item Total**	**$43,471.25**

** Riot shield. Marine Corps security forces are currently evaluating to use in conjunction with each other as listed below. They are not currently deployed but are undergoing field evaluation.
(1) The six feet Armadillo Interlocking Shield for the hard baseline.
(2) The four feet.. Armadillo Interlocking Shield for the soft baseline.

3. USN NLW Capability Set

a. The USN is in the process of developing NLW capabilities and capability sets.* The sets will support the specific areas of operation, mission, and/or user.

b. The first version will address the force protection mission and will be based on other service-approved items and some longer-range commercial-off-the-shelf items currently being upgraded for military use. The mobile security forces, master at arms, and shipboard force protection units are the targeted users.

c. A notional capability set for the Mobile Security Forces is listed below.

Table E-4. Notional USN Capability Set

Name	*Nomenclature (Mfr. Representative)*	*Quantity*	*Unit Cost*	*Case Size*	*Cost Extended*
Fin Stabilize Round		1 Case	$5	100	$400
Stingshot		1 Case	$5	100	$400
Shotbag		1 Case	$5	100	$450
Flamethrower Round		1 Case	$8	100	$750
Tactical Blast Bursting Round		1 Case	$7	100	$650
Ultraflash Bursting Round		1 Case	$8	100	$750
Launchable Stun Munition	POP, Inc. T-444	1 Case	$60	24	$1,440
40-mm Sparkshot		1 Case	$35	24	$840
40-mm Stingshot		1 Case	$25	24	$600
40-mm Shotbag		1 Case	$25	24	$600
40-mm OC		1 Case	$28	24	$660
Stinger	POP, Inc. I-452	1 Case	$60	24	$1,440
Stun	POP, Inc. T-470	1 Case	$55	24	$1,320
Flash/Bang Multi	POP, Inc. D-451	1 Case	$65	24	$1,560
OC	POP, Inc. I-452	1 Case	$65	24	$1,560
12-Gauge Launching Blanks		200	$3		$600
Grenade Training Rounds		12	$65		$780
12-Gauge Training Rounds		24	$3		$72
40-mm Training Rounds		12	$20		$240
12-Gauge Shotguns	Standard Issue/Modified	12	$1,000		$12,000
12-Gauge Launching Adapter		6	$325		$1,950
Face Shields		15	$100		$1,500
Helmets		15	$200		$3,000

* In FY 01, the USN fielded an USMC capability set to Naval Air Station, Roosevelt Roads, Puerto Rico.

Table E-4. Notional USN Capability Set

Name	*Nomenclature (Mfr. Representative)*	*Quantity*	*Unit Cost*	*Case Size*	*Cost Extended*
Tactical Vest		15	$300		$4,500
Bullhorn		2	$100		$200
Medical Kit		2	$50		$100
Searchlights		2	$100		$200
Field Pack		15	$100		$1,500
Holsters/Belts		15	$50		$750
Ordnance Packaging, Handling, Storage and Transportation (PHS&T)					$1,500
Misc. (Certification, Training and Documentation)					$3,000
			Total Cost Basic Kit		**$45,312**

Table E-5. Notional USN Capability Set

Name	*Nomenclature (Mfr. Representative)*	*Quantity*	*Unit Cost*	*Case Size*	*Cost Extended*
Stingmore Mine (Std Body) with Rubber Ball Projectiles	POP, Inc. I-454	24	$350		$8,400
Stingmore Mine Stand		6	$100		$600
Electric Firing Boxes		8	$35		$280
Area Lighting					$1,000
Electric Generator					$500
Signs and Placards					$500
Personal NVGs and Opticals		6	$2,000		$12,000
Additional PHS&T					$2,000
Misc. (Certification, Training and Documentation)					$2,000
Total Cost Augmented Basic Kit					**$27,280**
Total Cost (Basic plus Augment)					$72,592
Portable Vehicle Arresting Barrier	U.S. Army	1	$40,000		$40,000
Running Gear Entanglement System (RGES)	NSWCDD	1000 ft system	$25,000		$25,000
Antimaterial Round (USCG)		24	$12		$288
Spike Strips		2	TBD		
Improved Sensor (s)					
RF			TBD		

Table E-5. Notional USN Capability Set

Name	*Nomenclature (Mfr. Representative)*	*Quantity*	*Unit Cost*	*Case Size*	*Cost Extended*
IR			TBD		
Acoustic			TBD		
Chemical Detection	SABRE 2000	1	$20,000		$20,000
Ordnance Packaging, Handling, Storage and Transportation (PHS&T)					$6,000
Misc. (Certification, Training and Documentation)					$2,000
Total Cost Vehicle Supplement					**$93,288**
Total Cost Basic, Augmented, and Vehicle Supplement					**$165,880**

4. USAF NLW Capability Set

Table E-6. USAF NLW Capability Set

NSN	*Item Description*	*EA*	*Item $*	*Item Total*
GSADEF5149	Inert Individual (OC) Training Dispenser	1	$6.07	$6.07
GSADEF5846H	Mid-Size Riot Control Dispenser (Chemical)	3	$318.75	$956.25
GSADEF5546	Riot Control Dispenser (OC) Agent Quarts	1	$46.25	$46.25
	12-Gauge Shotgun (Area CDC/Rubber Pellet)	36	$4.46	$160.56
	12-Gauge Shotgun (Point/Fin Stabilized)	36	$2.85	$102.60
CSI (USMC) GG04	Stun Hand (Sting Ball) Grenade	50	$35.00	$1,750.00
M84	Stun Hand Grenade/Flash-Bang Device	50	$78.00	$3,900.00
GSAPMDK5X/AF	Non-Ballistic Face Shields	13	$29.50	$383.50
GSAPMBS9	Non-Ballistic Full Length Riot Shield	13	$88.50	$1,150.50
GSAUSMCSG	Non-Ballistic Shin guards	13	$6.000	$780.00
GSAATI23/36	36" Riot Baton w/Holster	13	$73.25	$952.25
GSAA2	Portable Megaphone	1	$82.00	$82.00
EAAFCUSTOM	Tactical Vest Rough Estimate	3	$150.00	$450.00
HARG700	Riot Control Gloves	13	$52.20	$678.60
Champro	Protective Cups	13	$7.25	$94.25
	40-mm (Point) Sponge Round	54	$23.00	$1,242.00
	40-mm (Area) Crowd Dispersal Cartridge Grenade	54	$22.00	$1,188.00
GSAF333	Riot Baton Training Suit	1	$638.50	$638.50

Table E-6. USAF NLW Capability Set

NSN	*Item Description*	*EA*	*Item $*	*Item Total*
MONDC/W	Disposable Forearm/Ankle Cuffs (FLEX)	40	$0.94	$37.60
GSADEF5049/ EA	Individual (OC) Dispenser with Carrier	13	$7.29	$94.77
	12-Gauge Utility Pouch	2	$24.00	$48.00
	12-Gauge Buttstock Cuff	3	$11.00	$33.00
	40-mm Carrying Pouch	2	$14.00	$28.00
	Training Strike Bag	2	$80.00	$160.00
	Cost of Clip-Lok Boxes			$480.00
	Approximate freight shipped via commercial freight into CONUS			$1,500.00
			Total Estimate	**$16,942.70**
Proposal to add TASERS to USAF Capability Set after testing.				

5. Notional USCG NLW Capability Set

Table E-7. Notional USCG NLW Capability Set

Reference: (a) COMDTINST M16247.1B Maritime Law Enforcement Manual (MLEM) (b) Project Ericson Website	
Expandable Baton	
Restraint Devices	Hand Cuffs, Flex Cuffs, Leg Irons, Migrant Restraint Harness
Individual OC Spray	
Non-Ballistic Riot Face Shield	
Non-Ballistic Riot Body Shield	
Non-Ballistic Riot Shin Guard	
Running Gear Entanglement Systems	

The following information describes each of the components of the NLW capability set currently deployed by the USCG. (The USCG has not designated a standard set of nonlethal gear. The using unit determines the gear needs.)

6. Restraint Devices

a. Handcuffs are the preferred restraint devices when dealing with a small number of subjects. Handcuffs should be steel, chain-linked construction capable of being double locked, with nickel, blued, or stainless steel finish.

b. Using flexible handcuffs for the temporary restraint of subjects is also authorized. Since they do not require a key and are relatively easy to apply, flexible handcuffs are particularly useful and appropriate when dealing with large numbers of subjects (e.g. migrants).

c. Leg irons are authorized restraint devices and may be a preferred option for restraining subjects for extended periods of time.

d. Migrant restraint harnesses are authorized when dealing with large numbers of subjects.

7. Expandable Baton

The expandable baton is the only intermediate weapon authorized for USCG personnel to carry. It should have a steel shaft that is silver in color and corrosion resistant (e.g. chrome, nickel, stainless steel) and have no knurling or sharp edges. It should have a foam or similar, non-slip grip. It should be rigid when deployed and must be no less than 15 inches and no greater than 22 inches when expanded. Spring loaded or other trigger style batons are not authorized.

8. OC Pepper Spray

Only chemical irritants containing the active ingredient OC are authorized for USCG personnel to carry. The aerosol must contain no greater than 10 percent and no less than 5 percent OC concentration. It shall have a water-based carrier, be nonflammable, nontoxic and produce a straight stream delivery (not fog).

9. Helmet Riot Face Shield

Comparable to DOD type. Not currently deployed. Undergoing field evaluation.

10. Riot Shield

Two shields are currently being evaluated. A 24 x 48 inches clear lexan weighing 14 pounds and a smaller version measuring 20 x 36 inches. They are not currently deployed. Undergoing field evaluation.

REFERENCES

Joint

JP 1-02, *DOD Dictionary of Military and Associated Terms*, 12 Apr. 2001

JP 3-07, *Joint Doctrine for Military Operations Other Than War*, 16 Jun. 1995

JP 3-07.3, *JTTP for Peace Operations*, 12 Feb.1999

JP 3-57, *Doctrine for Joint Civil-Military Operations*, 8 Feb. 2001

JP 3-58, *Joint Doctrine for Military Deception*, 31 May 1996

JP 3-61, *Doctrine for Public Affairs in Joint Operations*, 14 May 1997

JP 4-0, *Joint Doctrine for Logistic Support of Joint Operations*, 6 Apr. 2000

Multi-Service

FM 3-100.12, MCRP 5-12.1C, NTTP 5-03.5, AFTTP(I) 3-2.34, *Risk Management* for *Multiservice Tactics, Techniques, and Procedures*, 15 Feb. 2001

FM 3-31.1, MCWP 3-36, *Army and Marine Corps Integration in Joint Operations*, 21 Nov. 2001

Army

AR 190-14, *Military Police Carrying of Firearms and Use of Force for Law Enforcement and Security Duties*, 12 Mar. 1993

AR 350-38, *Training Devices Policies and Management*, 15 Oct. 1993

AR 27-1, *Judge Advocate Legal Services*, 30 Sep. 1996

FM 3-0, *Operations*, 14 Jun. 2001

FM 3-05.30 (33-1), *Psychological Operations*, 19 Jun. 2000

FM 19-15, *Civil Disturbances*, 25 Nov. 1985

FM 27-10, *The Law of Land Warfare*, 18 Jul. 1956

FM 100-19, *Domestic Support Operations*, 1 Jul. 1993

TC 90-1, *Training for Urban Operations*, 1 Apr. 2002

TRADOC Pamphlet 525-5, *Force XXl Operations, A Concept for the Evolution of Full Dimensional Operations for the Strategic Army of the Twenty-First Century,* 1 Aug. 1994

TRADOC Pam 350-9, *TRADOC Training Devices for Army wide Use*, 30 Sep. 1993

TRADOC Regulation 350-70, *Systems Approach to Training Management, Processes, and Products*, 9 Mar. 1999.

TRADOC Regulation 385-2, chg.1, TRADOC Safety Program, 10 Oct. 2000.

Marine Corps

MCWP 3-33.4, *Domestic Support Operations*, Jul. 1993

I MEF FMF, After Action Report—UNITED SHIELD, 7 Jun. 1995

MCO 1510.112, *Individual Training Standards for Nonlethal Weapons*, 27 May 1998

MCO 5500.6F, *Arming of Security and Law Enforcement Personnel and the Use of Force*, 20 Jul. 1995

MCRP 5-12.1A, *The Law of Land Warfare.*

MCO 1630.4A, *Law Enforcement and Physical Security Activities*, 19 Jun. 1991.

MCO 1510.114, *Individual Training Standards (ITS) for Antiterrorism Force Protection (AT/FT) System*, 30 Jan. 1998.

MCWP 3-11.1, *Marine Rifle Company*, Feb. 1978

MCRP 3-11.1A, *Commander's Tactical Handbook*, 30 Jul. 1988. .

MCWP 3-11.2 *Marine Rifle Squad*

Marine Corps Gazette, *"Less-Lethal" Force in Operation UNITED SHIELD*, Sep. 1995.

Navy

Naval Doctrine Publication 1, Naval Warfare (U), 28 Mar. 1994

NWP 1-14M, *The Commander's Handbook on the Law of Naval Operations*, Oct. 1995

SECNAVINST 5500.29B, *Use of Deadly Force and the Carrying of Firearms by Personnel of the Department of the Navy in Conjunction with Law Enforcement Security Duties and Personal Protection*, 29 Sep. 1992

Air Force

AFI 31-201, *Security Police Standards and Procedures*, 4 Dec. 2001

AFI 31-207, *Arming and Use of Force by Air Force Personnel*, 1 Sep. 1999

AFMAN 31-201V6, *Civil Disturbance*, 17 May 2002

AFMAN 31-222, *Security Forces Use of Force Manual*, 1 Jun. 2001

Coast Guard

COMDTINST M16247.1B, Maritime Law Enforcement Manual, 9 Aug. 2001

Other

DODD 3000.3-R, *Policy for Nonlethal Weapons*, 9 Jul. 1996

DODD 3025.1, *Military Support to Civil Authorities (MSCA)*, 15 Jan. 1993

DODD 5100.77, *DOD Law of War Program*, 9 Dec. 1998

DODD 3025, *I-M Manual for Civil Emergencies*, 2 Jun. 1994

DODD 3025.12, *Military Assistance for Civil Disturbances*, 4 Feb. 1994

DODD 3025.13, *DOD Resources in Support of US Secret Service*, 13 Sep. 1985

DODI 5000.2, *Operation of the Defense Acquisition System*, 5 Apr. 2002

DODR 5000.1, *Defense Acquisition*, 23 Oct. 2000, Ch 1, 4 Jan. 2001

DOD Joint Concept for Nonlethal Weapons, 5 Jan. 1998

DODD 3025.12, *Military Assistance for Civil Authorities*, 18 Feb. 1997

DODD 3025.16, *Military Assistance for Civil Authorities Liaison Office (EPLO) Program,* 18 Dec. 2000

DODI 5000.2, *Operation of the Defense Acquisition System*, 5 Apr. 2002

Mission Need Statement (MNS), *Range of Nonlethal Force Capabilities (NLFC)*, 10 Oct. 1996

A Guide for Effective Peacetime Employment of Reserve Components Units and Individuals, OASD RA (M&P), Dec. 2000

Defense and Intelligence Abbreviations and Acronyms, Nov. 1997

Emergency Management Assistance Compact—Guide Book and Operating Procedures, National Emergency Management Association, Aug. 2000

Inside the Navy, Marine Corps Nonlethal Weapons Experiment Underway at Camp Pendleton, CA, Vol. 9, No. 29, Jul. 22, 1996

Military Support To Civil Authorities—Interagency Disaster Resource Guide, National Interagency Civil-Military Institute, Jan. 2001

National Guard Bureau Military Support to Civil Authorities Lessons Learned, 6 Sep. 2000

The Robert T. Stafford Disaster Relief and Emergency Assistance Act, as amended 42 USC 1521, Jul.2001

Joint Universal Lessons Learned System, Operation UNITED SHIELD

Nonlethal Weapons Joint Mission Area Analysis/Joint Mission Needs Analysis, Dec. 2000

Interservice Nonlethal Individual Weapons Instructor Course (INIWIC) Program of Instruction, Oct.2000

Presidential Decision Directive PDD-39 U.S. Policy on Counterterrorism 21 Jun.1995

PDD-62 Combating Terrorism—Fact Sheet, 22 May 1998

The Federal Response Plan Notice of Change 7, Feb.1997; FEMA, Number 229 Change 11

Federal Response Plan, Apr 1999

GLOSSARY

PART I – ABBREVIATIONS AND ACRONYMS

A	
ACOG	advanced combat optical gunsight
ADS	aerial delivery system
AFDC	Air Force Doctrine Center
AFI	Air Force instruction
AFMAN	Air Force manual
ALSA	Air Land Sea Application
AMC	Army Materiel Command
AO	area of operation
AR	Army regulation
ATSC	Army Training Support Center
B	
bldg	building
BLT	battalion landing team
bn	battalion
C	
C4ISR	command, control, communications, computers, intelligence, surveillance, and reconnaissance
CA	civil affairs
cal	caliber
CANS	compressed air net system
CDC	crowd disposal cartridge
CMC	Commandant of the Marine Corps
COB	close of business
COC	combat operations center
COMDTPUB	Commandant, Coast Guard Publication
COMDTINST	Commandant, Coast Guard Instruction
CONUS	continental United States
COTS	commercial off-the-shelf
CP	command post
CQB	close quarters battle
CS	O-Chlorobenzyl-Malononitrile (riot control agent (tear gas))
CSSOC	combat service support operations center
cu ft	cubic feet

D

DA	Department of the Army
dB	decibel
DECON	decontamination
demo	demonstration
DM	designated marksman
DOD	Department of Defense
DODD	Department of Defense Directive
DODI	Department of Defense Instruction
DODIC	Department of Defense Identification Code
DODAC	Department of Defense Ammunition Code
DOT	Department of Transportation
DP	downed person
DPB	deployable pursuit boat
DTD	dated

E

EADBH	Eagle Duty Baton Holder
EEZ	exclusive economic zone
EOD	explosive ordnance disposal

F

FAST	Fleet Antiterrorism Security Team
FDP	food distribution point
FLEX	forearm/ankle cuff
FM	field manual
four M's	mind, mass, movement, muscle
FSE	fire support element
ft	feet

G

G1	Army or Marine Corps component manpower or personnel staff officer (Army division or higher staff, Marine Corps brigade or higher staff)
G2	Army or Marine Corps component intelligence staff officer (Army division or higher staff, Marine Corps brigade or higher staff)
G3	Army or Marine Corps component operations staff officer (Army division or higher staff, Marine Corps brigade or higher staff)

G4	Army or Marine Corps component logistics staff officer (Army division or higher staff, Marine Corps brigade or higher staff)
ga	gauge
GOTS	government off-the-shelf
H	
HMMWV	high mobility multipurpose wheeled vehicle
I	
IADS	Integrated Air Defense System
IAW	in accordance with
ID	identify/identification
IFF	identification, friend or foe
in.	inch
INIWIC	Interservice Nonlethal Individual Weapons Instructor Course
IPB	intelligence preparation of the battlespace
ITRO	Interservice Training Organization
J	
JMAA	Joint Mission Area Analysis
JTTP	joint tactics, techniques, and procedures
K	
KFOR	Kosovo forces
L	
LBE	load-bearing equipment
LOAC	Law of Armed Conflict
LP	listening post
LVOSS	light visual optical screening system
M	
M16	M16A2 rifle
M203	M203 Grenade Launcher
MARCORSYCOM	Marine Corps System Command
MCCDC	Marine Corps Combat Development Command
MCCM-GE	modified crowd control munition—ground emplaced
MCO	Marine Corps Order
MCPDS	Marine Corps Doctrinal Publication System
MCRP	Marine Corps reference publication

MCSF	Marine Corps security force
MCWP	Marine Corps warfighting publication
MDST	Monadnock Defensive Tactics System
MEDEVAC	medical evacuation
MEF	Marine expeditionary force
METT-T	mission, enemy, terrain and weather, troops and support available – time available
METT-TC	mission, enemy, terrain and weather, troops and support available – time available and civil considerations
MEU	Marine expeditionary unit
MIO	Maritime interception operation
MLEM	Maritime Law Enforcement Manual
MILSTRIP	military
mm	millimeter
MNS	mission need statement
MOAB	management of aggressive behavior
MOOTW	military operations other than war
MOS	military occupational specialty
MOUT	military operations in urban terrain
MP	military police
mph	miles per hour
MRE	meal, ready to eat
MTTP	multi-Service tactics, techniques, and procedures

N

NATO	North Atlantic Treaty Organization
NAVSOP	Navy standing operating procedures
NDC	Naval Doctrine Command
NLFC	nonlethal force capabilities
NLW	nonlethal weapon
NSN	national stock number
NSWC	Naval Surface Warfare Command
NWDC	Naval Warfare Development Command
NWP	Naval Warfare Publication

O

OC	Oleoresin Capsicum
OCAT	Oleoresin Capsicum aerosol training
OCONUS	outside the continental United States
OD	olive drab
OP	observation post

OPFOR	opposing force

P

PA	public affairs
PAO	public affairs officer
PASGT	personnel armor system, ground troops
POI	program of instruction
PSYOP	psychological operations
PT	physical training
PVAB	portable vehicle arresting barrier
PVC	polyvinyl chloride

Q

QRF	quick reaction force

R

R&D	research and development
RCA	riot control agent
RGES	running gear entanglement system
ROE	rules of engagement
RRF	rapid reaction force

S

S1	battalion or brigade manpower or personnel staff officer (Army; Marine Corps battalion or regiment)
S2	battalion or brigade intelligence staff officer (Army; Marine Corps battalion or regiment)
S3	battalion or brigade operations staff officer (Army; Marine Corps battalion or regiment)
S4	battalion or brigade logistics staff officer (Army; Marine Corps battalion or regiment)
SASO	stability and support operations
SB	shield bearer
sec	second
SECNAVINST	Secretary of the Navy instruction
SG	shotgun
SJA	staff judge advocate
SNCO	staff noncommissioned officer
SOP	standard operating procedure
SOTG	Special Operations Training Group
SROE	standing rules of engagement
SSC	small-scale contingency

T	
TC	training circular
TF	task force
T/O	table of organization
TOC	tactical operations center
TRADOC	United States Army Training and Doctrine Command
TTP	tactics, techniques, and procedures
U	
UN	United Nations
US	United States
USA	United States Army
USAF	United States Air Force
USAMPS	United States Army Military Police School
USCG	United States Coast Guard
USMC	United States Marine Corps
USN	United States Navy
USS	United States Ship
V	
VCR	videocassette recorder
VLNL	vehicle launched nonlethal grenade
VMNLGL	vehicle mounted nonlethal grenade launcher
W	
w/	with
WMD	weapons of mass destruction

PART II – TERMS AND DEFINITIONS

Air Force specialty code (AFSC). Primary job of a U.S. Air Force officer or enlisted person. Analogous to the Army and Marine Corps military occupational skill (MOS).

area of operation (AO). An operational area defined by the joint force commander for land and naval forces. Areas of operation do not typically encompass the entire operational area of the joint force commander, but should be large enough for component commanders to accomplish their missions and protect their forces. (JP 1-02)

battalion landing team (BLT). In an amphibious operation, an infantry battalion normally reinforced by necessary combat and service elements; the basic unit for planning an assault landing. (JP 1-02)

civil affairs. Designate active and Reserve Component forces and units organized, trained, and equipped specifically to conduct civil affairs activities and to support civil-military operations. (JP 1-02)

civil-military operations. The activities of a commander that establish, maintain, influence, or exploit relations between military forces, governmental and nongovernmental civilian organizations and authorities, and the civilian populace in a friendly, neutral, or hostile operational area in order to facilitate military operations, to consolidate and achieve operational U.S. objectives. Civil-military operations may include performance by military forces of activities and functions normally the responsibility of the local, regional, or national government. These activities may occur prior to, during, or subsequent to other military actions. They may also occur, if directed, in the absence of other military operations. Civil-military operations may be performed by designated civil affairs, by other military forces, or by a combination of civil affairs and other forces. (JP 1-02)

combat service support. The essential capabilities, functions, activities, and tasks necessary to sustain all elements of operating forces in theater at all levels of war. Within the national and theater logistics systems, it includes, but is not limited to, that support rendered by service forces in ensuring the aspects of supply, maintenance, transportation, health services, and other services required by aviation and ground combat troops to permit those units to accomplish their missions in combat. Combat service support encompasses those activities at all levels of war that produce sustainment to all operating forces on the battlefield. (JP 1-02)

command and control. The exercise of authority and direction by a properly designated commander over assigned and attached forces in the accomplishment of the mission. Command and control functions are performed through an arrangement of personnel, equipment, communications, facilities, and procedures employed by a commander in planning, directing, coordinating, and controlling forces and operations in the accomplishment of the mission. (JP 1-02)

commercial off the shelf (COTS). Describes items procured for military use through civilian contractors.

core capability. The set of functions that define a significant, stand-alone, operationally effective and suitable military capability such that, should no further development occur, the user will have received a significant capability.

designated marksman (DM). A rifleman stationed in an overwatch position above friendly forces, armed with a standard infantry weapon appropriate to his service that is mounted with a high-powered scope and loaded with lethal ammunition. The designated marksman uses his vantage point to identify agitators to forces below, eliminate lethal threats, and monitor the battlefield.

dispersion. A scattered pastern of hits around the mean point of impact of bombs and projectiles dropped or fired under identical conditions. In chemical and biological operations, the dissemination of agents in liquid or aerosol form. (JP 1-02)

fleet Marine force (FMF). A balanced force of combined arms comprising land, air, and service elements of the U.S. Marine Corps. A fleet Marine force is an integral part of a U.S. Fleet and has the status of a type command. (JP 1-02)

force protection. Security program designed to protect service members, civilian employees, family members, facilities, and equipment, in all locations and situations, accomplished through planned and integrated application of combating terrorism, physical security, operations security, and personal protective services, and supported by intelligence, counterintelligence, and other security programs. (JP 1-02)

government off-the-shelf (GOTS). Analogous to COTS, but describing items already procured by the U.S. government and ready for issue to military units.

host nation. A nation that receives the forces and/or supplies of allied nations, coalition partners, and/or NATO organizations to be located on, to operate in, or to transit through its territory. (JP 1-02)

intelligence preparation of the battlespace (IPB). An analytical methodology employed to reduce uncertainties concerning the enemy, environment, and terrain for all types of operations. Intelligence preparation of the battlespace builds an extensive database for each potential area in which a unit may be required to operate. The database is then analyzed in detail to determine the impact of the enemy, environment, and terrain on operations, and present it in graphic form. Intelligence preparation of the battlespace is a continuing process. Also called IPB. (JP 1-02)

interrogator-translator team (ITT). Marine Corps unit, usually working as a small detachment to a larger infantry unit, highly trained in interrogation techniques and fluent in adversary native language.

joint force commander. A general term applied to a combatant commander, subunified commander, or joint task force commander authorized to exercise combatant command (command authority) or operational control over a joint force. (JP 1-02)

logistics. The science of planning and carrying out the movement and maintenance of forces. In its most comprehensive sense, those aspects of military operations which deal with:

a. Design and development, acquisition, storage, movement, distribution, maintenance, evacuation, and disposition of materiel.

b. Movement, evacuation, and hospitalization of personnel.

c. Acquisition of construction, maintenance, operation, and disposition of facilities.

d. Acquisition of furnishing of services. (JP 1-02)

low-intensity conflict. Political-military confrontation between contending states or groups below conventional war and above the routine, peaceful competition among states. It frequently involves protracted struggles of competing principles and ideologies. Low intensity conflict ranges from subversion to the use of armed force. It is waged by a combination of means employing political, economic, informational, and military instruments. Low-intensity conflicts are often localized, generally in the Third World, but contain regional and global security implications.

NOTE: Also known as small wars or "brushfire wars," low-intensity conflicts are rebellions, insurrections, and the like characterized by adversaries dedicated to a cause but lacking command and control structure and modern equipment.

Marine expeditionary force (MEF). The largest of the Marine air-ground task force (MAGTF) and the Marine Corps principal warfighting organization, particularly for larger crisis or contingencies. It is task-organized around a permanent command element and normally contains one or more Marine divisions, Marine aircraft wings, and Marine force service support groups. The Marine expeditionary force is capable of missions across the range of military operations, including amphibious assault and sustained operations ashore in any environment. It can operate from a sea base, a land base, or both. Also called MEF. See also aviation combat element; combat service support element; command element; Marine air-ground task force; Marine expeditionary force (forward); Marine expeditionary unit; special purpose Marine air-ground task force; task force. (JP 1-02)

maritime interception operation (MIO). Means by which an exclusion zone at sea is enforced. Involves the visit of merchant ships, by armed boarding parties if necessary, bound to, through, or out of a defined area, to examine each ship's papers and cargo and to search for evidence of contraband. The provisions must be established by a sanctioning body and must be applied to ships of all nationalities.

military operations other than war (MOOTW). Operations that encompass the use of military capabilities across the range of military operations short of war. These military actions can be applied to complement any combination of the other instruments of national power and occur before, during, and after war. (JP 1-02)

military occupational specialty (MOS). Primary job of military officer or enlisted personnel.

military operations in urban terrain (MOUT). All military actions planned and conducted on a topographical complex and its adjacent natural terrain where man-made construction is the dominant feature. It includes combat-in-cities, which is that portion of MOUT involving house-to-house and street-by-street fighting in towns and cities.

military police (MP). Marine Corps and Army police forces designated by specific MOS.

maritime pre-positioning force operation. A rapid deployment and assembly of a Marine expeditionary force in a secure area using a combination of strategic airlift and forward-deployed maritime pre-positioning ships. See also Marine expeditionary force; maritime pre-positioning ships (JP 1-02 and JP 4-01.6)

nonlethal weapons (NLW). Weapons that are explicitly designed and primarily employed so as to incapacitate personnel or materiel, while minimizing fatalities, permanent injury to personnel, and undesired damage to property and the environment.

a. Unlike conventional weapons that destroy their targets through blast, penetration, and fragmentation, nonlethal weapons employ means other than gross physical destruction to prevent the target from functioning.

b. Nonlethal weapons are intended to have one, or both, of the following characteristics:

(1) They have relatively reversible effects on personnel or materiel.

(2) They affect objects differently within their area of influence. (JP 1-02)

nonlethal weapon capability set. A set of systems containing weapon systems, munitions, and protective equipment. The components are explicitly designed and primarily employed to provide additional protection to friendly personnel and to incapacitate hostile personnel or material, while minimizing fatalities or permanent injury to intended targets and minimal damage to property and the environment. They are not required to have zero probability of producing fatalities or permanent injuries but are designed and employed in a manner that significantly reduces those probabilities over the traditional employment of currently fielded military weapon systems, munitions, and equipment.

nonlethal technology (N-LT). Synonymous with NLW, but also encompassing more advanced technology that is not yet available for use by military forces.

Oleoresin Capsicum (OC). Also known as "pepper spray", OC is made completely from organic materials and is FDA-approved for sale over-the-counter in the United States. Causes severe and immediate burning sensation to mucous membranes when sprayed into face, nose, and eyes.

psychological operations (PSYOP). Planned operations to convey selected information and indicators to foreign audiences to influence their emotions, motives, objective reasoning, and ultimately the behavior of foreign governments, organizations, groups, and individuals. The purpose of psychological operations is to induce or reinforce foreign attitudes and behavior favorable to the originator's objectives. (JP 1-02)

rules of engagement (ROE). Directives issued by competent military authority that delineate the circumstances and limitations under which U.S. forces will initiate and/or continue combat engagement with other forces encountered. See also law of war. (JP 1-02)

table of organization (T/O). Total manpower authorized for specific unit, organized by billet and rank. TO& E is a U.S. Army term which encompasses T/O and T/E.

training aids and devices. Any item developed or procured with the primary intent to assist in training and learning.

weapons of mass destruction (WMD). Weapons capable of a high order of destruction and/or of being used in such a manner as to destroy large numbers of people. Weapons of mass destruction can be high explosives or nuclear, biological, chemical, and radiological weapons, but exclude the means of transporting or propelling the weapon where such means is a separable and deliverable part of the weapon. (JP 1-02)

Index

Q

R

Made in United States
Orlando, FL
04 November 2026

90207702R00435